VOLUME SIXTEEN

Radioactivity in the Environment

ENVIRONMENTAL RADIONUCLIDES

RADIOACTIVITY IN THE ENVIRONMENT

A companion series to the Journal of Environmental Radioactivity

Series Editor
M.S. Baxter
Ampfield House
Clachan Seil
Argyll, Scotland, UK

Volume 1: Plutonium in the Environment
(A. Kudo, Editor)

Volume 2: Interactions of Microorganisms with Radionuclides
(F.R. Livens and M. Keith-Roach, Editors)

Volume 3: Radioactive Fallout after Nuclear Explosions and Accidents
(Yu.A. Izrael, Author)

Volume 4: Modelling Radioactivity in the Environment
(E.M. Scott, Editor)

Volume 5: Sedimentary Processes: Quantification Using Radionuclides
(J. Carroll and I. Lerche, Authors)

Volume 6: Marine Radioactivity
(H.D. Livingston, Editor)

Volume 7: The Natural Radiation Environment VII
(J.P. McLaughlin, S.E. Simopoulos and F. Steinhäusler, Editors)

Volume 8: Radionuclides in the Environment
(P.P. Povinec and J.A. Sanchez-Cabeza, Editors)

Volume 9: Deep Geological Disposal of Radioactive Waste
(W.R. Alexander and L.E. McKinley, Editors)

Volume 10: Radioactivity in the Terrestrial Environment
(G. Shaw, Editor)

Volume 11: Analysis of Environmental Radionuclides
(P.P. Povinec, Editor)

Volume 12: Radioactive Aerosols
(C. Papastefanou, Author)

Volume 13: U-Th Series Nuclides in Aquatic Systems
(S. Krishnaswami and J. Kirk Cochran, Editors)

Volume 14: Remediation of Contaminated Environments
(G. Voigt and S. Fesenko, Editors)

Volume 15: Airborne Radioactive Contamination in Inhabited Areas
(K.G. Andersson, Editor)

Volume 16: Environmental Radionuclides: Tracers and Timers of Terrestrial Processes
(K. Froehlich, Editor)

VOLUME SIXTEEN

Radioactivity in the Environment

ENVIRONMENTAL RADIONUCLIDES: TRACERS AND TIMERS OF TERRESTRIAL PROCESSES

Editor

Klaus Froehlich

Viktor-Wittner-Gasse 36/7, 1220 Vienna, Austria

ELSEVIER

Amsterdam • Boston • Heidelberg • London • New York • Oxford
Paris • San Diego • San Francisco • Singapore • Sydney • Tokyo

Elsevier
Radarweg 29, PO Box 211, 1000 AE Amsterdam, The Netherlands
Linacre House, Jordan Hill, Oxford OX2 8DP, UK

First edition 2010

British Library Cataloguing in Publication Data
A catalogue record for this book is available from the British Library

Library of Congress Cataloging-in-Publication Data
A catalog record for this book is available from the Library of Congress

ISBN: 978-0-08-043873-3
ISSN: 1569-4860

Printed and bound in Great Britain

10 11 12 13 14   10 9 8 7 6 5 4 3 2 1

Contents

Contributors

S.K. Bhattacharya
Physical Research Laboratory, Navarangpura, Ahmedabad, 380 009, India

Gordon T. Cook
Scottish Universities Environmental Research Centre (SUERC), Radiocarbon Dating Laboratory, Scottish Enterprise Technology Park, East Kilbride G75 0QF, UK

L. DeWayne Cecil
US Geological Survey, INL, Idaho Falls, ID USA

Klaus Froehlich
Viktor-Wittner-Gasse 36/7, 1220 Vienna, Austria

Prosenjit Ghosh
Centre for Earth Sciences, Indian Institute of Science, Bangalore, 560 012, India

Douglas D. Harkness
NERC Radiocarbon Laboratory, SUERC, UK

Susan Ivy-Ochs
Institute of Particle Physics, ETH Zurich, CH-8093 Zurich, Switzerland; Department of Geography, University of Zurich, CH-8057 Zurich, Switzerland

Thomas Kraemer
US Geological Survey, 430 National Center, Reston, VA, USA

Jozef Masarik
Department of Nuclear Physics and Biophysics, Faculty of Mathematics, Physics and Informatics, Comenius University, Mlynská dolina, 842 48 Bratislava, Slovakia

Robert L. Michel
US Geological Survey, MS 434, Menlo Park, CA, USA

Minh-Long Nguyen
Soil and Water Management and Crop Nutrition Section, Joint FAO/IAEA Division of Nuclear Techniques in Food and Agriculture, Department of Nuclear Sciences and Applications, International Atomic Energy Agency – IAEA, Wagramer Strasse 5, A-1400 Vienna, Austria

Mirjam Schaller
Department of Geological Sciences, University of Michigan, Ann Arbor, MI 48109.1005, USA

E. Marian Scott
Statistics Department, University of Glasgow, Glasgow G12 8QW, UK

Axel Suckow
Leibniz-Institut für Angewandte Geophysik (LIAG), Stilleweg 2, 30655, Hannover, Germany; Isotope Hydrology Laboratory, Department of Nuclear Sciences and Applications, International Atomic Energy Agency (IAEA), Wagramerstrasse 5, A-1440 Vienna, Austria

Ivan Sykora
Department of Nuclear Physics and Biophysics, Faculty of Mathematics, Physics and Informatics, Comenius University, Mlynská dolina, 842 48 Bratislava, Slovakia

Felipe Zapata
Soil and Water Management and Crop Nutrition Section, Joint FAO/IAEA Division of Nuclear Techniques in Food and Agriculture, Department of Nuclear Sciences and Applications, International Atomic Energy Agency – IAEA, Wagramer Strasse 5, A-1400 Vienna, Austria

Foreword

The functioning of the Earth's terrestrial systems is seriously affected by man-induced global changes. This is manifested by changes in the ecological functions of terrestrial systems such as surface water bodies (flood prevention), soils (fertility for food production) and groundwater (drinking water supply). The complex interrelations of these systems under the impact of global changes are still inadequately understood. Therefore, further efforts in studying these systems are required to adequately manage the human resources provided by them. Among the various approaches to be used in these studies, environmental radionuclides have shown great potential as tracers and timers of terrestrial processes. Radionuclides embedded in environmental substances serve as internal timers that count down from a specified time interval like an hourglass. The decay rates of environmental radionuclides cover timescales appropriate to the determination of the turnover times of lakes, recharge rates and flow velocities of groundwaters, atmospheric residence times of aerosols, exposure times of surface rocks and to date past climate changes. Environmental radionuclides are also excellent tracers because they are ubiquitous in terrestrial systems, can be detected extremely sensitively, yet occur in very low concentrations and thus do not affect the behaviour of the substance to be traced.

The book reviews the state of the art in using radionuclides as tracers and timers of processes and systems of the continental part of the Earth's environment. It is conceived as a companion to other volumes of this series dealing with radioactivity in the marine, atmospheric and terrestrial environments. However, in reviewing climatic studies, marine systems such as sediments and biota are also addressed. Similarly, it appears unavoidable to discuss the carbon cycle in a global context. Although the book focuses on environmental radionuclides, it also refers to stable environmental isotopes, which in many applications in hydrology and climatology are indispensable and complementary to radionuclides. In palaeoclimatology, for example stable isotopes provide proxy data on temperature and other climate parameters, while radionuclides present the timescales of climate records in archives such as corals, sediments, speleothems, ice cores and others. Similarly, in groundwater studies, stable isotopes indicate origins and mixing processes, while radionuclides address the ages of groundwaters, their recharge and flow regimes.

The book is intended as a monograph for specialists but should also serve (post)graduate students and researchers involved in studies of terrestrial systems and climate change in order to become familiar with the potential of environmental radionuclides in such fields. Therefore, the first two chapters of the book provide the background on environmental radionuclides required to understand their manifold applications in terrestrial systems studies. Chapter 1 describes the origins, sources and production of the various types of radionuclides including primordial, radiogenic, cosmogenic and anthropogenic radionuclides and their global distributions in the environment. Chapter 2 deals with the basic principles underlying the various applications of environmental radionuclides in studies of terrestrial systems.

Chapter 3 describes the use of natural and anthropogenic radionuclides to trace atmospheric gases and aerosols in transport, transformation and removal processes relevant to air quality and climate studies. Applications to determine atmospheric aerosol transit times and relevant parameter such as exchange rates between stratosphere and troposphere are also addressed.

Within the Earth's systems, the carbon cycle holds high rank, predominantly because of its life-supporting and climate-forcing functions. Considering tracers and timers of the carbon cycle, the isotopes of carbon (^{13}C and ^{14}C) are naturally and inevitably the most suitable tracers. Many processes within the carbon cycle and its interaction with other Earth's systems run on timescales in the order of the half-life of radiocarbon (^{14}C). In addition to naturally produced ^{14}C, bomb ^{14}C is used as a transient tracer in the carbon system providing invaluable insight into processes over the last 50–100 years when the largest man-made CO_2 perturbations took place. For all these reasons, a special chapter of the book (Chapter 4) has been devoted to radiocarbon as a tracer and timer of the global carbon cycle.

Environmental radionuclides, in combination with stable isotopes, geochemistry and other hydrological techniques, provide a powerful tool, often indispensable, for studying the cycling of water in continental hydrological systems (Chapter 5). The use of environmental radionuclides in surface water studies is reviewed in the first two sub-chapters; the third sub-chapter is devoted to groundwater and geothermal water. Taking into consideration the fact that most applications in groundwater and geothermal water studies require the combined use of radioactive and stable isotopes, a brief summary on stable isotopes is added to the third sub-chapter.

Radionuclides such as ^{10}Be, ^{14}C, ^{26}Al and ^{36}Cl are produced by cosmic rays in mineral grains in the surface layers of rocks and sediments. From their measurements, erosion rates, exposure ages and burial ages can be determined and these provide unique information on geomorphic and climatic changes. For example exposure dating of both boulders and bedrock surfaces can be used to constrain the timing of past ice margin fluctuations. A comprehensive review of the state of the art in this field that revolutionised geomorphology and changed it from a largely descriptive to a quantitative science is given in Chapter 6.

Landscape-changing processes such as soil erosion and associated sediment deposition are also addressed in Chapter 7 which focuses on the use of radiometric dating techniques, mainly based on measurements of ^{210}Pb and fallout radionuclides such as ^{137}Cs and ^{7}Be. These studies are of high socio-economic relevance, because land degradation by water and wind erosion is associated with irretrievable losses of soil and thus threatens food security. Chapter 8 shows that information on past climate changes is mainly based on water-borne stable isotope records of climate archives such as marine and lacustrine sediments, corals, speleothems and ice cores, to mention a few. Radionuclides, also enclosed in these archives, provide the timescales necessary for establishing the chronology of the Earth's climate. Recognising that a successful application of environmental radionuclides in studies of terrestrial systems requires adequate design of the experimental study as well as selection of appropriate sampling and analysis techniques, the final Chapter 9 of the book provides the background knowledge necessary to meet these requirements. It is also shown that advances in studies of terrestrial processes to a large extent depend on progress in developing new analytical techniques or in improving existing ones.

The editor is grateful to all authors for their collaboration and efforts in contributing to this volume and expresses his thanks to Murdoch Baxter, the Series Editor, for his continued encouragement and his patience during preparation of this book.

Klaus Froehlich
Viktor-Wittner-Gasse 36/7
1220 Vienna, Austria

CHAPTER 1

Origin and Distribution of Radionuclides in the Continental Environment

Jozef Masarik*

Contents

1. Introduction

Radioactivity and associated radiation exposure are sometimes conceived as environmental problems that have been created by modern

*Corresponding author. Tel.: +421-2-602-95-456; Fax: +421-2-654-25-882
E-mail address: masarik@fmph.uniba.sk

Department of Nuclear Physics and Biophysics, Faculty of Mathematics, Physics and Informatics, Comenius University, Mlynská dolina, 842 48 Bratislava, Slovakia

Radioactivity in the Environment, Volume 16
ISSN 1569-4860, DOI 10.1016/S1569-4860(09)01601-5

science and technology. However, our world is radioactive and has been since it was created. Natural radioactivity originates from extraterrestrial sources as well as from radioactive elements in the Earth's crust. Radionuclides are found naturally in air, water and soil. They are even found in us, being that we are products of our environment. Every day, we inhale and ingest radionuclides with air, food and water. Natural radioactivity is common in the rocks and soil that make up our planet, in water and oceans and in building materials and homes. There is nowhere on Earth that you cannot find natural radioactivity. Everything on the Earth evolved, for better or worse, in this radioactive environment. Evaluating the radioactivity at a given location, its variability in space and time has to be taken into account; this includes variation due to the geology of the region, chemical and physical mobility of natural radionuclides and man's impact on the environment.

Over 340 nuclides have been found in nature, of which about 80 are radioactive (radioactive elements or radionuclides). All nuclides with atomic numbers greater than 80 have radioactive isotopes, and all isotopes of elements with atomic numbers heavier than 83 are radioactive.

The radioactive nuclides (radionuclides) can be subdivided into three general groups:

- Primordial – from the creation of the Earth and their radioactive decay products
- Cosmogenic – produced by cosmic-ray interactions
- Manmade – produced by anthropogenic practices (mostly minor amounts compared to natural radioactivity).

Human awareness of the existence of ionising radiation dates only to the turn of the 19th to the 20th century when X-rays and unexpected ionising radiation from uranium ore were discovered by Wilhelm Roentgen (1895) and Henri Becquerel (1896) and Maria and Pierre Curie (1898), respectively.

2. Primordial and Natural Decay-Series Radionuclides

Matter in our solar system was created by several generations of nucleosynthesis processes including the explosion of a supernova near the protosolar nebula shortly before the formation of the Sun and the condensation of the planets of our solar system. A series of nuclear reactions, initiated primarily by protons and alpha particles, ignited the first phase of nucleosynthesis in which the elements up to the mass number $A = 60$ were created. Some of these nuclides were stable, and the others were radioactive. Almost all radioactive nuclides were beta particle emitters with half-lives much shorter than the age of the Earth. An important exception is ^{40}K.

Elements with atomic numbers $A \sim 60$–90 were mainly produced by neutron-capture reactions followed by beta decays. By emission of beta particles, the neutron-rich products of neutron-capture reactions decay into more stable isotopes. Sequences of neutron-capture reactions and beta decays led to the formation of heavy elements. Again, most of the radioactive nuclei that were formed in this way had short half-lives and decayed since they were formed. An important exception is ^{87}Rb.

Uranium and heavier nuclei with atomic number $Z \geq 92$ can also be built by neutron-capture reactions. In this case, the repulsive Coulomb force grows more rapidly than the corresponding attractive nuclear force, which makes alpha decay and nuclear fission possible. In the end, nuclei that are stable against beta decay are formed, but they can decay by alpha particle (alpha is ^{4}He nucleus) emission or nuclear fission. Some of these nuclei have half-lives comparable with the age of Earth and they are members of radioactive decay series discussed below. For very heavy nuclei, the sequence of neutron-capture reactions is terminated by nuclear fission and following series of decays that lead to the production of intermediate mass nuclei (with atomic mass number $A \sim 80$–160). These nuclides can become starting points of new sequences of neutron-capture reactions that can lead to the production of a further variety of atomic nuclei.

Most of the radionuclides produced in such nucleosynthesis reactions have decayed, only several dozen of them having half-lives of the order of the age of the Earth (4.5×10^9 years) or more having 'survived'. These naturally occurring terrestrial radionuclides, called primordial radionuclides (primordials), can be divided into two groups:

1. Primordials decaying directly into a stable nuclide (non-series radionuclides).
2. Primordials decaying through a chain (series) of radionuclides with shorter half-lives into a stable isotope of the element lead. These decay series include:
   a) the uranium series, which originates with ^{238}U
   b) the thorium series, which originates with ^{232}Th
   c) the actinium series, which originates with ^{235}U.

Considering the radiation exposure of man, only some of these natural radionuclides are important. The external radiation exposure is mainly caused by ^{226}Ra (or by uranium), ^{232}Th and ^{40}K in rocks and soil on the Earth's surface (the thickness of the layer is some few tens of centimetres). The global average dose rate is about $0.057\,\mathrm{mGy\,h^{-1}}$. Maximum values have been measured on monazite sand in Guarapari, Brazil (up to 50 mGy h^{-1}), and in Kerala, India (about 2 mGy h^{-1}), and on rocks with a high radium concentration in Ramsar, Iran (from 1 to 10 mGy h^{-1}) (Eisenbud and Gesell, 1997).

2.1. Non-series radionuclides

Two primordial non-series radionuclides (Table 1) that contribute substantially to the terrestrial radiation background dose are ^{40}K and ^{87}Rb.

Of three naturally occurring isotopes of potassium, ^{40}K is the only unstable one; it decays by beta emission to ^{40}Ca or (with a lower probability) by K capture to an excited state of ^{40}Ar followed by γ-ray emission to the ^{40}Ar ground state. It exists as a constant fraction (0.0117%) of stable potassium that is the seventh most abundant element in the crust of the Earth and the sixth most abundant element in solution in the oceans. Its contribution to external dose is variable, depending on its concentration in rocks and soil. The average concentration in the Earth's crust is 1.8 mg kg^{-1}, the corresponding specific activity being about 0.6 Bq g^{-1} in crustal rock. The potassium concentration varies over a wide range of values, from 0.1% in limestone to 3.5% in granite. The natural concentration of ^{40}K in seawater is about 11 Bq l^{-1} (Kohman and Saito, 1954).

Table 1 Non-series radionuclides.

Nuclide	Relative abundance (%)	Daughter nuclide	Half-life (years)	Major radiation
^{40}K	0.0118	^{40}Ar, ^{40}Ca	1.28×10^{9}	β^- (89%), EC (11%)
^{50}V	0.24	^{50}Ti, ^{50}Cr	6.00×10^{15}	β^- (30%), EC (70%)
^{87}Rb	27.85	^{87}Sr	4.70×10^{10}	β^-
^{115}In	96.67	^{115}Sn	5.00×10^{14}	
^{123}Te	0.87	^{123}Sb	1.20×10^{13}	
^{138}La	0.089	^{138}Ba, ^{138}Ce	1.10×10^{11}	
^{142}Ce	11.7	^{138}Ba	5.00×10^{15}	
^{144}Nd	23.8	^{140}Ce	2.40×10^{15}	
^{147}Sm	15.1	^{143}Nd	1.06×10^{11}	
^{148}Sm	11.35	^{144}Nd	1.20×10^{13}	
^{149}Sm	14.0	^{145}Nd	4.00×10^{14}	
^{152}Gd	0.205	^{148}Sm	1.10×10^{14}	
^{156}Dy	0.057	^{152}Gd	2.00×10^{14}	
^{174}Hf	0.163	^{170}Yb	4.30×10^{15}	
^{176}Lu	2.588	^{176}Hf	2.20×10^{10}	
^{187}Re	62.93	^{187}Os	4.00×10^{10}	
^{190}Pt	0.0127	^{186}Os	$7/00 \times 10^{11}$	
^{204}Pb	1.4	^{200}Hg	1.40×10^{17}	

Rubidium-87 is one of two rubidium isotopes found in nature; the other rubidium isotope, rubidium-85, is stable. Rubidium-87 is a pure beta emitter, and it makes up about 27.84% of natural rubidium; the latter is the 16th in the series of most abundant elements in crustal rock. Its average concentration in crustal rocks is $1\,mg\,kg^{-1}$ corresponding to a specific activity about $0.07\,Bq\,kg^{-1}$ (Pertsov, 1964). The ^{87}Rb content of ocean water was determined to be $0.1\,Bq\,l^{-1}$ (Mauchline and Templeton, 1964). Its contribution to the natural radiation exposure is negligible.

2.2. Decay-series radionuclides

From the four known decay series (Table 2), only three occur naturally: the uranium, thorium and actinium series. The fourth series is called the neptunium series because ^{237}Np is its longest-lived member. Some radionuclides (e.g. ^{241}Pu) which eventually decay into ^{237}Np are also considered members of the neptunium series (Table 2). Because of the relatively short half-life of ^{237}Np, this decay series is extinct. The only surviving member of the neptunium family is the nearly stable nuclide ^{209}Bi; its half-life has been estimated to be about 2.0×10^{18} years. Given the relatively low abundance of ^{235}U in natural uranium (less than 5% on an activity basis), the actinium series does not play a significant role in the terrestrial radioactivity inventory.

If not subjected to chemical or physical separation, each of these series attains a state of secular radioactive equilibrium (i.e. each nuclide has equal activity). Technological enrichment of parent nuclides as well as natural physical and chemical reactions often disturbs this equilibrium. Crustal concentrations of the first radionuclides of the three naturally occurring series are extremely small (parts per million, ppm); the short-lived decay progenies are present in such exceedingly minute concentrations that their behaviour does not always follow chemical (mass action) controls.

In nature, uranium isotopes and thorium isotopes undergo fission spontaneously or due to interactions with neutrons that originate from cosmic rays or other natural sources. Other heavy nuclides undergo spontaneous fission with half-lives $\sim 10^{14}$–10^{25} years. Some of the transuranic nuclides (e.g. ^{239}Pu, ^{237}Np) are produced in very small amounts as a result of interactions of naturally occurring neutrons (produced in fission of uranium, or by cosmic rays) with heavy nuclei.

Uranium is found in rocks and soil and in terrestrial and ocean water. Typical concentrations range from 0.03 ppm in ultrabasic igneous rocks to 120 ppm in phosphate rocks that have been considered for commercial exploitation. Uranium also occurs in many commercial products (fertilisers, steel products, etc.), food and human tissues. With some exceptions, uranium concentration in terrestrial water is relatively small. Natural uranium consists of four isotopes with mass numbers 230, 234, 235

Table 2 Nuclides of the decay series.

Nuclide	Half-life	Radiation	Nuclide	Half-life	Radiation
	Uranium series			Thorium series	
^{238}U	4.5×10^{9} years	α,γ	^{232}Th	1.4×10^{10} years	α,γ
^{234}Th	24 days	β,γ	^{228}Ra	6.7 years	β
^{234m}Pa	1.2 min	β,γ	^{228}Ac	6.13 h	β,γ
^{234}U	2.5×10^{5} years	α,γ	^{228}Th	1.9 years	α,γ
^{230}Th	8.0×10^{4} years	α,γ	^{224}Ra	3.64 days	α,γ
^{226}Ra	1,622 years	α,γ	^{220}Rn	55 s	α,γ
^{222}Rn	3.8 days	α,γ	^{216}Po	0.16 s	α
^{218}Po	3.05 min	α	^{212}Pb	10.6 h	β,γ
^{214}Pb	26.8 min	β,γ	^{212}Bi	60.5 min	β,α,γ
^{218}At	1.5–2.0 s	α	^{212}Po	3.04×10^{-7} s	α
^{214}Bi	19.7 min	β,α	^{208}Tl	3.1 min	β,γ
^{214}Po	1.64×10^{-4} s	α,γ	^{208}Pb	Stable	
^{210}Tl	1.3 min	β,γ			
^{210}Pb	22 years	β,γ			
^{210}Bi	5.0 days	β,α			
^{210}Po	138 days	α,γ			
^{206}Tl	4.2 min	β			
^{206}Pb	Stable				
	Actinium series			Neptunium series	
^{235}U	7.1×10^{8} years	α,γ	^{241}Pu	14.35 years	β,α,γ
^{231}Th	25.64 h	β,γ	^{237}Np	2.144×10^{6} years	α,γ
^{231}Pa	3.25×10^{4} years	α,γ	^{233}Pa	26.967 days	β,γ
^{227}Ac	21.6 years	β,α,γ	^{233}U	1.592×10^{5} years	α,γ
^{227}Th	18.2 days	α,γ	^{229}Th	7,300 years	α,γ
^{223}Fr	22 min	β,α,γ	^{225}Ra	14.9 days	β,γ
^{223}Ra	11.4 days	α,γ	^{225}Ac	10 days	α,γ
^{219}Rn	4.0 s	α,γ	^{221}Fr	4.9 min	α,γ
^{215}Po	1.77×10^{-3} s	α,β	^{217}Ac	6.9×10^{-8} s	α,γ
^{211}Pb	36.1 min	β,γ	^{213}Bi	2.5×10^{-2} s	β,α,γ
^{215}At	$\sim 10^{-4}$ s	α	^{213}Po	4.2×10^{-6} s	α,γ
^{211}Bi	2.16 min	α,β,γ	^{209}Tl	2.20 min	β,γ
^{211}Po	0.52 s	α,γ	^{209}Pb	3.253 h	β
^{207}Tl	4.79 min	β,γ	^{209}Bi	Almost stable	
^{207}Pb	Stable	–			

and 238. The most abundant isotope is ^{238}U at 99.28%; it is usually in equilibrium with ^{234}U (abundance 0.0058%). The abundance of ^{235}U is about 0.71%. The half-life of ^{230}U is 20.8 days and its natural abundance is negligible.

The typical thorium content of various rocks is 8.1–33 ppm for igneous rocks, with a mean value of 12 ppm (Faul, 1954), 6 ppm for sandstones (Taylor and McClennan, 1985) and ~1 ppm for limestones. The thorium content of igneous rocks is approximately four times higher than the uranium content, but since the specific activity of thorium is four times lower than that of uranium, the radioactivity due to these two nuclides is more or less the same.

Of the decay products (daughters) of the uranium isotopes, ^{226}Ra, ^{222}Rn and ^{210}Pb of the ^{238}U series are particularly important. ^{226}Ra is present in all types of rocks. The specific activities range from 16 Bq kg^{-1} in limestone to about 48 Bq kg^{-1} in igneous rocks (Rankama and Sahama, 1950). The radium content of surface waters is around 3–18 Bq m^{-3}, while in underground waters radium concentrations can be 10–50 times higher (Hess et al., 1985). The isotopic composition of radon and its concentration are strongly influenced by geological and geochemical factors; however, in most cases, the ratio between ^{226}Ra and ^{228}Ra (member of the thorium series) is around 1.

^{226}Ra decays by alpha emission to the inert gas ^{222}Rn, which decays further with a half-life of 3.8 days. ^{224}Ra, a descendant of ^{232}Th, decays by alpha emission to ^{220}Rn, which is also unstable and decays with a half-life of 54 s. Due to the difference in half-life, ^{222}Rn can diffuse from the soil to the atmosphere, while ^{220}Rn virtually decays within the soil. Many measurements at various locations on Earth show that the average concentration of radon in outdoor air is 3–18 Bq m^{-3}. The observed wide variations in radon concentration depend on many parameters, the most important including geographic position, water content and ice coverage of soil, and meteorological factors (which influence the rate of emanation and dilution in the atmosphere). The outdoor radon content is subject to diurnal and seasonal variations. Maximum concentrations are observed in the early morning hours; the late afternoon minima are about three times lower. Radon is also dissolved in water, where concentrations range from 1 to 1,000 kBq m^{-3} (Hess et al., 1980).

^{210}Pb is produced by the decay of ^{222}Rn through six short-lived alpha and beta emitters. It is a beta emitter with a half-life of 22 years and decays to ^{210}Po (Table 2). The ^{210}Pb produced in the atmosphere is washed out by precipitation and dry deposition, and eventually reaches soil, surface water and limnic and marine sediments. The ^{210}Pb content of the atmosphere varies between about 3×10^{-5} and 30×10^{-4} Bq m^{-3} (Magno et al., 1970; Preiss et al., 1996), and in rainwater its average concentration is 75 Bq m^{-3}.

3. Cosmogenic Radionuclides

Cosmic radiation penetrates all of the space, the source being primarily outside our solar system. The primary cosmic radiation consists of very-high-energy heavy particles, photons and muons. The interactions of cosmic-ray particles with the Earth's atmosphere and the Earth's surface produce a cascade of secondary particles and a variety of cosmogenic nuclides. This part of the review is devoted to cosmogenic nuclides produced in the atmosphere and surface layers. Technological development during recent decades has increased the detection sensitivity for long-lived cosmogenic radionuclides by several orders of magnitude and allows us now to analyse with high-resolution natural archives of different origin. The concentration of cosmogenic nuclides in these archives is the result of the interplay between four processes: production, decay, transport and deposition. In order to make full use of the information stored in these archives, a detailed knowledge of the source functions of the cosmogenic nuclides is necessary. We concentrate our attention in this part of the book only on the production processes.

Cosmogenic nuclides are produced by interaction of cosmic rays with atomic nuclei. Examples of cosmogenic nuclides are radioactive isotopes of the elements beryllium, carbon, aluminium, chlorine, calcium, and iodine. Their lifetimes range from seconds to thousands and even millions of years, and some of the isotopes are stable. Cosmogenic nuclides are produced in meteorites and other extraterrestrial materials, in the Earth's atmosphere, and on the surface (solid and liquid) of the Earth. The production rate of cosmogenic nuclides depends primarily on the cosmic-ray particle flux.

3.1. Cosmic-ray particle fluxes

Cosmic-ray particles refer both to the primary energetic particles of extraterrestrial origin that strike the Earth's atmosphere and to the secondary particles generated by their interaction with the atmosphere. Primary particles are attenuated in the upper atmosphere. Reactions take place and generate secondary particles. The cosmic radiation field at ground altitude (0–3 km) consists almost entirely of secondary particles.

The primary cosmic-ray flux at the Earth's orbit has two components: galactic cosmic ray (GCR) and solar cosmic ray (SCR). The GCR particles are a mixture of ~87% protons, ~12% alpha particles and ~1% of heavier nuclei with atomic numbers from 3 to ~90 (Simpson, 1983). The spectral distributions of all particles look quite similar if they are compared in units of energy per nucleon. The propagation of the GCR particles to the Earth is influenced by many interactions that lead to spatial and temporal variations. The dominant effect is the solar modulation, which is taken into account in the expression for the differential primary GCR proton flux (Lal, 1988).

From measurements of cosmogenic radionuclides with different half-lives and irradiation histories in meteorites, the average GCR flux was inferred to be constant within 10% during the last few million years

(Vogt et al., 1990). Using lunar experimental data (Masarik and Reedy, 1994), the effective flux of nucleons with energies above 10 MeV at Earth's orbit was determined to be approximately 4 nucleons $cm^{-2} s^{-1}$. Solar modulation is the dominant cause of the observed GCR variability. During a typical solar cycle, the low-energy part of the GCR particle flux ($E < 1$ GeV $nucleon^{-1}$) near the Earth varies by one order of magnitude. With increasing energy, the modulation effect becomes weaker. The modulation is caused by the interaction of incoming GCR particles with magnetic fields connected with the solar-wind plasma.

The SCRs consist of ~98% protons and ~2% heavier nuclei. The energies are typically in the range 1–100 MeV. The total SCR flux is subject to strong temporal variations, its average value being around 75–100 nucleons $cm^{-2} s^{-1}$. Because of their relatively low energies, they can cause nuclear reactions in the Earth's atmosphere only at high geomagnetic latitudes (above 60°N or S), and even there, the nuclide production is restricted to the very top of the atmosphere. The long-term average production of cosmogenic nuclides by SCR is not expected to be significant. Some huge solar-particle events produce proton fluxes much higher than the average (Shea and Smart, 1992), and they could contribute to the production of some cosmogenic nuclides (e.g. ^{7}Be and ^{36}Cl) that are observed in some polar ice layers from Greenland and Antarctica (Masarik and Reedy, 1995).

The atmosphere and the Earth's magnetic fields also act as shields against cosmic radiation, reducing the amount that reaches the Earth's surface. The deflection of incoming particles by the geomagnetic field depends on their electric charge, energy and angle of incidence. Below a critical energy, which is a function of geomagnetic latitude and angle of incidence, cosmic-ray particles cannot penetrate the Earth's atmosphere. This leads to a latitudinal dependence of the primary and secondary particle fluxes and consequently also of the production rate of cosmogenic nuclides, with higher values around the magnetic poles and lower values in the equatorial region.

Besides latitude, the cosmic-ray flux and consequently also the production rate of cosmogenic nuclides depend on the altitude. Primary cosmic rays incident on the top of the atmosphere consist mainly of protons with energies around 1 GeV. The characteristic feature of nuclear interactions at these energies is the production of secondary particles, many of which have enough energy to undergo further collisions in the atmosphere and contribute to the development of a particle cascade. The development of this cascade process leads to a decrease of the primary cosmic-ray particle flux and a strong increase of the number of secondary particles with increasing atmospheric depth. After production of several generations of particles, the cascade process ends when the energies of the particles become too low for further particle production. At sea level, secondary cosmic rays consist largely of neutrons and muons. Due to their high kinetic energies (up to many GeV), they reach the Earth's surface before they decay. Based on these features of particle-cascade development, one expects that the production rate begins to increase at the top of the atmosphere, reaches a maximum at a depth between 100 and 200 g cm^{-2} (depending on nuclide and latitude) and finally decreases gradually down to the Earth's surface.

After production, the nuclides are subject to different processes according to their geochemical properties. For example ^{14}C is oxidised to CO_2, and ^{10}Be becomes attached to aerosols. Then they are distributed from the site of their production following the trajectories of the air masses until finally a part of them becomes stored in natural archives. Because of the thermal structure of the stratosphere and its separation from the troposphere by the tropopause, the residence time of aerosols in the stratosphere is substantially longer (about 1–2 years) than in the troposphere (in the order of weeks). The concentration of the cosmogenic nuclides at a specific site is strongly dependent on the local precipitation rate.

Models have to be developed that describe the production of nuclides by the interaction of cosmic-ray particles with the main target elements of the atmosphere and their subsequent transport from the atmosphere into the various archives. Lacking reliable nuclear models and cross-sections, the first extensive and pioneering work in this field by Lal and Peters (1967) was based on data from direct observations limited to a few years. There have also been a number of model calculations devoted to particle and cosmogenic nuclide production in the atmosphere (Hess et al., 1961; Newkirk, 1963; Lingenfelter, 1963; Oeschger et al., 1969; Light et al., 1973; O'Brien, 1979; Blinov, 1988; Masarik and Reedy, 1995). The relatively good agreement between the calculated and measured ^{14}C production rates proved the reliability of the model approaches.

3.2. Cosmogenic nuclides in the Earth's atmosphere

The detection of naturally occurring radiocarbon in the Earth's atmosphere (Libby, 1946; Anderson et al., 1947) marked the onset of the search for cosmogenic nuclides in terrestrial and extraterrestrial samples. Table 3 lists the principal cosmogenic nuclides, their half-lives and global inventories. All these nuclides have relatively short half-lives, so that, if they were not being continually created by cosmic rays, they would not be present on the Earth today. Some other cosmogenic radionuclides are produced in the Earth's atmosphere: ^{80}Kr, ^{85}Kr, ^{32}Si, ^{39}Ar, ^{22}Na, ^{35}S, ^{37}Ar, ^{30}P, ^{33}P, ^{32}P, ^{28}Mg, ^{24}Na, ^{31}S, ^{37}S, ^{38}S, ^{31}Si, ^{18}F, ^{39}Cl, ^{38}Cl, ^{34m}Cl, ^{24}Ne and ^{28}Al. The production rates of ^{3}H, ^{7}Be, ^{10}Be, ^{14}C and ^{36}Cl have been determined, using the calculated or measured particle fluxes in the atmosphere (Tables 4 and 5).

Table 3 The principal and the most precisely studied cosmogenic nuclides in the Earth's atmosphere.

Nuclide	Half-life	Inventory (Bq)
^{3}H	12.3 years	1.3×10^{18}
^{7}Be	53.6 years	1.8×10^{17}
^{10}Be	1.5×10^{6} years	9.3×10^{16}
^{14}C	5,730 years	1.1×10^{19}
^{26}Al	7.30×10^{5} years	6.1×10^{13}
^{36}Cl	3.08×10^{5} years	9.9×10^{15}
^{81}Kr	2.13×10^{5} years	1.8×10^{12}

Table 4 Latitudinal and global average production rates (atoms $cm^{-2}\,s^{-1}$) in the Earth's atmosphere for long-term mean solar activity and the present magnetic field (Masarik and Beer 1999).

Nuclide	Latitude (degree)							
	0–10	10–20	20–30	30–40	40–50	50–60	60–90	Global average
^{3}H	0.118	0.129	0.159	0.224	0.380	0.565	0.602	0.281
^{7}Be	0.0148	0.0161	0.0197	0.0285	0.0478	0.071	0.0740	0.0354
^{10}Be	0.00775	0.00841	0.0104	0.0147	0.0248	0.0368	0.0386	0.0184
^{14}C	0.83	0.90	1.10	1.56	2.63	3.92	4.55	2.02
^{36}Cl[a]	0.791	0.86	1.05	1.51	2.52	3.75	3.96	1.88

[a]Units are 10^{-3} atoms $cm^{-2}\,s^{-1}$.

The most investigated cosmogenic nuclide in the Earth's atmosphere is ^{14}C (carbon-14). After production, it is quickly oxidised to CO_2. The equilibrium concentrations of ^{14}C in the atmosphere are controlled primarily by the exchange of CO_2 between the atmosphere and the ocean. The oceans are the major sink of atmospheric ^{14}C. The nuclear reaction ^{14}N (n,p) ^{14}C contributes more than 99% to the total production rate of ^{14}C, and thus it is by far the most important source of ^{14}C in the atmosphere. It produces 1.96 atoms $^{14}C\,kg^{-1}\,s^{-1}$ in the Earth's atmosphere. Integrating over the depth of the Earth's atmosphere, a global average production rate of 2.02 ^{14}C atoms $cm^{-2}\,s^{-1}$ is obtained. Its uncertainty of ~10% is mainly due to uncertainties and statistical errors in the neutron fluxes (the cross-sections are well measured). The above production rate is in good agreement with estimates of the radiocarbon production rate based on the analysis of the specific activity of ^{14}C. Damon et al. (1978) using this method obtained a value of 1.99 atoms $cm^{-2}\,s^{-1}$, while Lal and Peters (1967) obtained 1.8 atoms $cm^{-2}\,s^{-1}$, O'Brien (1979) obtained 1.75 atoms

Table 5 Principal cosmogenic nuclides produced in the Earth's atmosphere.

Nuclide	Production rate in atoms $cm^{-2} s^{-1}$	Reference
^{3}H	0.28[a]	Masarik and Beer (1999)
	0.26[a]	Masarik and Reedy (1995)
	0.19	Nir et al. (1966)
	0.255[a]	O'Brien (1979)
^{7}Be	0.035[a]	Masarik and Beer (1999)
	0.0185[a]	Oeschger et al. (1969)
	0.129[a]	Masarik and Reedy (1995)
	0.0810[a]	Lal and Peters (1967)
	0.0578	O'Brien (1979)
^{10}Be	0.016	Beer et al. (1994)
	0.026	Reyss et al. (1981)
	0.038	Monaghan et al. (1985/86)
	0.0184[a]	Masarik and Beer (1999)
	0.045[a]	Lal and Peters (1967)
	0.014[a]	Oeschger et al. (1969)
	0.026[a]	O'Brien (1979)
	0.0201[a]	Masarik and Reedy (1995)
	0.020[a]	Reyss et al. (1981)
^{14}C	2.02[a]	Masarik and Beer (1999)
	1.99	Damon et al. (1978)
	1.8[a]	Lal and Peters (1967)
	1.75[a]	O'Brien (1979)
	1.91	Masarik and Reedy (1995)
^{36}Cl	0.00188[a]	Masarik and Beer (1999)
	0.00220[a]	Oeschger et al. (1969)
	0.00118[a]	Masarik and Reedy (1995)
	0.0011[a]	Lal and Peters (1967)
	0.000901[a]	O'Brien (1979)

[a]Value calculated on the basis of a theoretical model.

$cm^{-2} s^{-1}$ and Masarik and Reedy (1995) obtained 1.91 atoms $cm^{-2} s^{-1}$ on the basis of other theoretical models.

The other radionuclides included in Table 4 are produced in nuclear spallation reactions. The uncertainties in calculated values for these nuclides are higher than for ^{14}C, mainly due to greater uncertainties in the corresponding excitation functions. The average errors are estimated to be within 30–50%. Comparing calculated production rates with those derived from measurements in various environmental systems, one has to keep in mind that the calculations represent global mean values, whereas the measured data mainly represent local conditions. The most recently calculated ^{10}Be production rate is 0.0184 atoms $cm^{-2} s^{-1}$ (Masarik and

Beer, 1999) and, thus, is close to the value measured in the Dye 3 ice core, 0.016 atoms $cm^{-2} s^{-1}$ (Beer et al., 1994), and also to the value derived from sediment cores, 0.026 atoms $cm^{-2} s^{-1}$ (Reyss et al., 1981). However, the calculated value is approximately half of that determined by ^{10}Be measurements in precipitation, 0.038 ± 0.008 atoms $cm^{-2} s^{-1}$ (Monaghan et al., 1985/86).[1] The newest model value for the ^{7}Be production rate comes to 0.035 atoms $cm^{-2} s^{-1}$ (Masarik and Beer, 1999).[2] The most recently calculated production rate of ^{36}Cl, by Masarik and Beer (1999), is 1.88×10^{-3} atoms $cm^{-2} s^{-1}$.[3]

For ^{3}H, a production rate of 0.28 atoms $cm^{-2} s^{-1}$ has been calculated by Masarik and Beer (1999).[4] The depth dependence of the production rates for ^{36}Cl and ^{10}Be is given in Figure 1. Using these dependencies and accounting for the variation of the stratospheric thickness with latitude, the fraction of the stratospheric production to the total atmospheric production is determined to be 56%, 54%, 51%, 54% and 55% for ^{10}Be, ^{36}Cl, ^{14}C, ^{7}Be and ^{3}H, respectively.

The latitudinal and global average production rates of cosmogenic nuclides depend on the interplay between the SCR modulation and the geomagnetic field intensity.[5] During a typical 11 years solar cycle (Schwabe cycle), the ratio between production at solar minimum and solar maximum is 1.34 for spallogenic radionuclides and 1.46 for ^{14}C.[6]

For many geophysical applications, it is useful to know the magnitude of variations of cosmogenic nuclide production rates due to variations of the geomagnetic field. From sedimentary paleomagnetic records, it is known that the intensity of the geomagnetic field varied in the past from almost zero to twice its present intensity (Guyodo and Valet, 1996). The ratios for the global average production rate for zero geomagnetic field and for a doubling of the present intensity are 2.7, 2.7 and 2.9 for ^{10}Be, ^{36}Cl and ^{14}C, respectively.[7]

[1] Older model estimates gave 0.045 (Lal and Peters, 1967), 0.014 (Oeschger et al., 1969), 0.026 (O'Brien, 1979), 0.0201 (Masarik and Reedy, 1995) and 0.020 atoms $cm^{-2} s^{-1}$ (Reyss et al., 1981).

[2] Other estimates were 0.0185 (Oeschger et al., 1969), 0.0129 (Masarik and Reedy, 1995), 0.0810 (Lal and Peters, 1967) and 0.0578 atoms $cm^{-2} s^{-1}$ (O'Brien, 1979).

[3] This value is in good agreement with an older estimate of 2.2×10^{-3} atoms $cm^{-2} s^{-1}$ by Oeschger et al. (1969), but substantially higher than 1.1×10^{-3} (Lal and Peters, 1967), 1.18×10^{-3} (Masarik and Reedy, 1995) and 9.01×10^{-4} atoms $cm^{-2} s^{-1}$ (O'Brien, 1979).

[4] The value agrees fairly well with 0.19 atoms $cm^{-2} s^{-1}$ calculated by Nir et al. (1966), 0.26 atoms $cm^{-2} s^{-1}$ obtained by Masarik and Reedy (1995) and by O'Brien (1979) whose calculated production rate was 0.255 atoms $cm^{-2} s^{-1}$.

[5] At low latitudes, the geomagnetic field reflects preferentially low-energy particles, and since the solar modulation effect on high-energy primary particles is small, the net effect is also small. In the case of high latitudes, on the other hand, all primary cosmic-ray particles can enter the Earth's atmosphere and interact with its atoms. Since the low-energy end of the spectrum is strongly modulated by the Sun and represents a considerable part of the total spectrum, the resulting effect is relatively large.

[6] The global production rates for cosmogenic nuclides produced in the Earth's atmosphere by spallation reactions during solar minimum are about 1.15 times higher than the average over a complete solar cycle, and the rates at solar maximum are 0.85 times the average rates. The equivalent quantities for ^{14}C are 1.19 and 0.82.

[7] In the extreme case (zero intensity geomagnetic field and solar activity minimum relative to a doubling of the geomagnetic field intensity and a solar activity maximum), the global average production ratios for ^{10}Be, ^{36}Cl and ^{14}C are 5.9, 6.0 and 8.6, respectively.

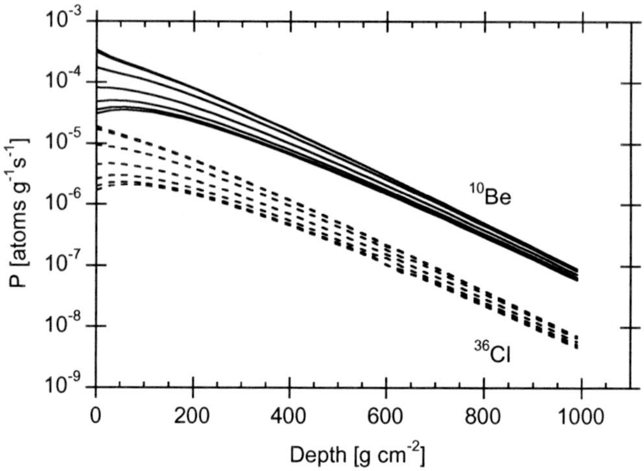

Figure 1 Change of the production rates of ^{10}Be and ^{36}Cl with depth in the atmosphere (from Masarik and Beer, 1999).

Most of the cosmogenic radionuclides are rapidly oxidised in the atmosphere and become attached to aerosol particles. These particles act as condensation nuclei for the formation of cloud droplets and eventually coagulate to form precipitation. About 10–20% of cosmic-ray-produced radionuclides are removed from the atmosphere by dry deposition on the Earth's surface.

The concentrations of cosmogenic radionuclides produced in the atmosphere vary in space (longitude) and time. Temporal variations follow diurnal, seasonal and sunspot cycles. Changes in the atmospheric concentrations of some cosmogenic radionuclide due to nuclear tests and releases from nuclear facilities will be discussed in Section 4.

3.3. Subsurface produced radionuclides

Accelerator mass spectrometry, developed more than two decades ago (Muller, 1977), has lowered the detection limit of some long-lived nuclides (e.g. ^{14}C, ^{10}Be, ^{36}Cl and ^{26}Al) by a factor of 10^6 compared to the conventional decay counting techniques, and thus has enabled measurement of the very low concentrations of cosmogenic nuclides produced in terrestrial rocks (see also Chapter 6). Due to the attenuation of cosmic-ray fluxes in the atmosphere, the *in situ* production rates in terrestrial rocks are about two orders of magnitude lower than in the upper atmosphere.

In general, cosmogenic radionuclide production depends on the chemical composition of the target. In the case of the atmosphere, there are only a few targets represented by its constituents N, O, Ar and Kr. In comparison to the atmosphere, the number of *in situ*-produced cosmogenic

Table 6 Routinely measured *in situ* terrestrial cosmogenic nuclides (Gosse and Phillips, 2001).

Nuclide	Half-life (years)	Main target elements
Radioactive		
^{10}Be	1.6×10^6	O, Mg, Si, Fe [N,O]
^{7}Be	53 days	O, Mg, Si, Fe [N,O]
^{14}C	5,730	O, Mg, Si, Fe [N,O]
^{26}Al	7.1×10^5	Si, Al, Fe [Ar]
^{36}Cl	3.0×10^5	Fe, Ca, K, Cl [Ar]
^{39}Ar	269	Fe, Ca, K [Ar]
^{41}Ca	1.0×10^5	Ca, Fe
^{53}Mn	3.7×10^6	Fe
^{60}Co	5.27	Co, Ni
^{129}I	1.6×10^7	Te, Ba. La, Ce [Xe]
Stable		
^{3}He	Stable	O, Mg, Si, Fe
^{21}Ne	Stable	Mg, Al. Si, Fe
^{38}Ar	Stable	Fe, Ca, K
^{126}Xe	Stable	Te, Ba, La, Ce, I

nuclides is much greater[8] because of the presence of higher mass number target elements in rocks. Table 6 compiles cosmogenic radioactive and stable nuclides that are produced in terrestrial rocks and have been used in earth sciences and related disciplines (see Chapter 6).

The production rate of *in situ* cosmogenic nuclides depends on the cosmic-ray particle flux, production mechanisms and temporal variability of solar and terrestrial influences. To estimate the source strengths of cosmic-ray-produced isotopes, Lal and Peters (1967) and Masarik and Beer (1999) discussed in detail the propagation of cosmic rays through the Earth's atmosphere. It has been shown that the production rates increase exponentially with decreasing pressure in the atmosphere, with a characteristic mean pressure change equivalent to 160–165 $g\,cm^{-2}$ (Lal and Peters, 1967; Masarik and Beer, 1999). At sea level, negative muon captures become significant and also the scattering of neutrons at the surface disturbs the exponential dependence. At larger depth below the surface ($\sim$1 m rock equivalent), negative muon captures, fast muon disintegrations and secondary neutrons produced in muon interactions are responsible for the production of cosmogenic nuclides. The production rates decrease according to a power law with increasing depth underground. With the present measurement techniques that enable counting down to about 10^5–10^6 atoms in a given

[8] About 60 *in situ*-produced cosmogenic radionuclides with half-lives exceeding 2 weeks have been measured.

sample, cosmogenic nuclides with nuclear disintegration rates larger than $10^{-11}\,g\,s^{-1}$ can be determined. This conclusion is based on the fact that rock samples integrate cosmic-ray exposures over long periods of time. For near-surface targets, effective cosmic-ray exposure ages of 10^4–10^6 years (Nishiizumi et al., 1993) can be determined, and for deeper samples, these ages can be longer than 10^7–10^8 years (Lal, 1991).

The production rate and the accumulation of cosmogenic nuclides in a given rock depend on its location (altitude and latitude), size and shape, and exposure history, especially the geometry of exposure of the upper surface layers. As noted above, the near-surface production rates decrease exponentially with a characteristic length of $160\,g\,cm^{-2}$ (about half a metre in a typical rock with density $2.7\,g\,cm^{-3}$). In a rock that is subject to erosion, the concentration of *in situ*-produced cosmogenic nuclides can be used to determine the erosion rate, or episodic surface losses. The faster the erosion rate, the smaller the net accumulation of a cosmogenic nuclide in the rock.

4. Anthropogenic Radionuclides

There are many radionuclides which have been produced by artificial nuclear reactions. Table 7 compiles some of these so-called anthropogenic radionuclides (and their sources), which have shown potential as tracers or time markers of terrestrial processes. The major sources of anthropogenic radionuclides and their dispersion in the environment are:

- production of nuclear fuels
- nuclear fuel reprocessing
- nuclear power reactors
- nuclear reactor accidents
- nuclear weapon explosions.

4.1. Production of nuclear fuels

The production of nuclear energy is based mainly on the fission of ^{235}U, which is present in natural uranium in a concentration of 0.7%. ^{238}U, which is the most abundant isotope of uranium, transmutes by neutron-capture reactions into ^{239}Pu that is also fissile. Thorium has thus far been a minor source of nuclear energy.

Nuclear fuel production starts with mining of uranium ores that takes place in underground workings and open pits. Radon and its daughter products are the only significant radionuclides released from the mines to the environment. The release of radon from a typical mine is comparable with the natural radon flux from about $1\,km^2$ of the Earth's surface (Holaday, 1959).

Table 7 Most important anthropogenic radionuclides.

Nuclide	Half-life	Main source
^{3}H	12.3 years	Nuclear weapon testing, fission reactors, reprocessing facilities, nuclear weapon manufacturing
^{14}C	5,730 years	Atmospheric nuclear weapon tests, nuclear industry, other
^{36}Cl	3.01×10^5 years	Atmospheric nuclear weapon tests
^{131}I	8.04 days	Fission product from nuclear weapon tests and fission reactors, medical applications
^{129}I	1.57×10^7 years	Fission product from nuclear weapon tests and fission reactors
^{137}Cs	30.17 years	Fission product from nuclear weapon tests and fission reactors
^{90}Sr	28.78 years	Fission product from nuclear weapon tests and fission reactors
^{99}Tc	2.11×10^5 years	Decay product of ^{99}Mo, used in medical applications
^{239}Pu	2.41×10^4 years	Produced by neutron bombardment of ^{238}U($^{238}U+n \rightarrow {}^{239}U \rightarrow {}^{239}Np+\beta \rightarrow {}^{239}Pu+\beta$)
^{85}Kr	10.73 years	Nuclear weapon testing, fission reactors, reprocessing facilities, nuclear weapon manufacturing

The second step in nuclear fuel production is milling, starting with uranium ore grinding, continuing with uranium separation by acid or alkaline leaching and ending with uranium concentration by ion exchange or solvent extraction. In this process, almost 95% of the uranium is extracted from the ore, almost all of the radioactive daughter products in the uranium series remaining in the tailings contained in slurries that are stocked in holding areas. Emanation of ^{222}Rn and dispersion of tailings by wind and water are the main sources of radionuclide input into the environment from milling processes. Radon production in tailings will continue for hundreds of thousands of years because they contain almost all the ^{230}Th (half-life 77,000 years) originally present in the uranium ore, and this is the parent of ^{226}Ra which in turn decays to ^{222}Rn. The ^{222}Rn emission rate from dry tailings is comparable with the rates of radon emission from normal soils (Schiger, 1986). The increased radon flux from the tailings piles is due to the few hundred times higher ^{226}Ra content of the tailings. Due to the small areas covered by tailings piles, they do not make a significant contribution to the global environment, their effects being strongly localised.

Milling of nuclear fuels is followed by refining. During refining, the uranium powders are converted by mechanical processes either to metal or some intermediate uranium compounds, such as UO_3 or UF_4. During this process, discharges of uranium dust to the environment can occur. This was a characteristic mainly of the early days of the nuclear fuel processing industry, and large amounts of uranium were discharged to the outside atmosphere.

The relatively high cost of uranium and the requirement for strict accountability make it unlikely that significant environmental contamination can occur during isotopic enrichment and fuel-element manufacture.

4.2. Fuel reprocessing

When the nuclear fuel in the reactor core is burned-out and taken off for reprocessing, only a small percentage of ^{235}U is consumed by nuclear fission, and the fraction of ^{238}U transmuted to ^{239}Pu and other transuranic elements is also small. The spent fuel is then chemically treated in reprocessing plant in order to:

- convert the fission products into a form suitable for long-term storage
- recover the remaining ^{235}U and the transuranic elements.

Radionuclides are released into the environment during the reprocessing of spent fuel in both normal plant operation and also in accidents.

The primary sources of gaseous radionuclides are fuel-element chopping and dissolution processes. Due to the relatively long storage time before fuel reprocessing, most of the ^{131}I has decayed (half-life 8.04 days); however, enough still remains and can escape to the ambient air. The isotope ^{129}I has a much longer half-life, of 1.7×10^7 years. It is produced in fission reactions with a high yield (1%) and is therefore present in reprocessed wastes in large amounts. Due to its long half-life, it eventually accumulates in the environment.

The radioactive gases ^{85}Kr and ^3H are released from spent fuel reprocessing plants. The amounts of ^{85}Kr are substantial, but have not been a source of significant exposure in the vicinity of reprocessing plants. A part of this ^{85}Kr is dissolved in atmospheric precipitation. With the filtrating precipitation, it reaches the groundwater and thus becomes a useful tool for identifying young groundwater (see also Chapter 5). The gaseous release of ^3H has been small compared to its release in the liquid form.

4.3. Power reactors

Since commissioning of the first nuclear reactor in December 1942, about 1,000 land-based reactors have been built throughout the world for nuclear research and applications as well as for production of plutonium and eventually for energy. Also almost 200 ships are powered by nuclear reactors.

There are several types of power reactors: light water reactors,[9] high-temperature gas-cooled reactors and breeder reactors. The release of radioactive substances into the environment depends on the type of the reactor.

During normal operation, light water reactors produce both gaseous and liquid wastes, some of which are released directly to the environment. These wastes are either direct fission products or radionuclides produced by activation of material in the primary loop and coolant. The major cause for release of fission products is leakage or diffusion through the fuel cladding. Minor contributions originate from traces of uranium present on the surface of fuel elements or other reactor components. In the coolant, radionuclides are present as dissolved or suspended solids or gases. The spectrum of radionuclides occurring in the coolant depends on the construction materials of the core, pressure vessel, pumps, piping and other components that are in contact with water. It also depends on impurities present in the water used for cooling. Cleaning systems are in operation that remove or concentrate these radionuclides for disposal. Neutron reactions with stainless steel, the important construction material of reactors, can produce radionuclides such as ^{51}Cr, ^{54}Mn, ^{55}Fe, ^{59}Fe and ^{60}Co, all of which have relatively long half-lives and thus ultimately have to be disposed of in nuclear waste repositories.

Concerning liquid wastes, tritium is produced in light water reactors in quantities that are relatively large compared to other radionuclides. Tritium is a pure beta emitter and is chemically bound in the water molecule. The sources of tritium are fission reactions (production rate of about 1 atom per 10,000 fissions) and spallation reactions following neutron irradiation of ^{10}B present in reactors that use ^{10}Be as a burnable poison. Under normal conditions, the radioisotopes of caesium and iodine are dominating in liquid wastes. In the event of fuel failure, the concentrations of radiostrontium can be increased.[10]

Gaseous wastes and their management depend on the reactor type. In light water reactors, after chemical treatment and about 3 months storage, the short-lived radionuclides are decayed and the remaining radioactivity is mainly due to ^{85}Kr, which can be released to the atmosphere or removed for off-site disposal. In the high-temperature gas-cooled reactors, the short-lived noble gases ^{89}Kr and ^{137}Xe, which decay to ^{89}Kr and ^{137}Cs, respectively, are the main products. After chemical treatment and some holding time in storage tanks, only insignificant amounts of ^{85}Kr and ^{131}I are released.

[9] Two types of light water reactors are used: pressurised (PWR) and boiling (BWR) water reactors. In the PWR, the steam which drives the turbo-generator does not pass through the reactor but receives its heat via the steam generator. In this way, the produced radionuclides are retained in the primary system. In the BWR, steam is provided directly from the reactor.

[10] The exact composition of liquid wastes in light water reactors depends on the construction and operation history of the reactor. Other sources of liquid waste include leaks in components of the primary loop and decontamination of various components and washing of various items used in power plants.

4.4. Reactor accidents

Reactor malfunctions can, in principle, cause releases of radioactivity into the environment. Although nearly 800 nuclides with mass numbers ranging from 72 to 160 are produced during reactor operation, not all of them are radioactive and others have very short half-lives that prevent their spread in the environment. Only about 50 of these nuclides have half-lives longer than 25 min. Significant amounts of radioactivity, especially of ^{131}I and ^{137}Cs, have been released by reactor accidents in which the core was damaged. In the Windscale accident (1957), the core of a plutonium production reactor was partly damaged by a fire and 700 TBq of radioactive material, mainly ^{131}I, was released to the environment via the cooling air. In the Idaho experimental reactor accident (1961), 0.4 TBq of ^{131}I escaped into the atmosphere. It was estimated that, in the Three Mile Island accident (1979), about 1.6 PBq of ^{85}Kr and 740 GBq of ^{131}I were released to the environment. In the largest reactor accident, which happened in 1986 at Chernobyl, vast amounts of various radionuclides were released into the atmosphere (Table 8; NEA/OECD, 2002).

4.5. Nuclear weapons

The radioactive debris from nuclear weapon explosions categorises into three groups depending on the explosion yield and the height of bursts. The first group consists of the larger particles, which fall out within hours close to the site of the explosion and are of high radioactivity. The particles of the second group are comparatively small and behave similarly to aerosols. They are dispersed into the troposphere from which they are washed out by wet (precipitation) and dry deposition. The third group of radioactive fallout products penetrates the stratosphere where their residence time is in the order of 1 year. Underground explosions can also produce fallout and result in extensive surface contamination if they vent through the ground.

The stratospheric fallout of atmospheric nuclear bomb explosions with yields[11] in the megaton range[12] is the principal source of worldwide contamination by long-lived radionuclides. More than 200 nuclides of 35 elements are produced in a nuclear burst. Most of the nuclides are radioactive with short half-lives. Table 9 compiles the main radionuclides produced by the fission reactions of nuclear explosions.

In addition to fission products, a number of induced radionuclides are produced by the detonation of nuclear devices, among them substantial

[11] Yields of nuclear weapon explosions are usually expressed in equivalent mass of TNT, the conventional explosive trinitrotoluene.

[12] The fallout of nuclear bombs with yields in the kiloton range does not penetrate the stratosphere. Tropospheric fallout contains only a small fraction of long-lived radionuclides but can be a source of heavy exposure by short-lived radionuclides such as ^{131}I.

Table 8 Estimate of total release of radionuclides during the Chernobyl accident in 6 May 1986.

Nuclide	Half-life	Activity (PBq)
^{133}Xe	5.3 days	6,500
^{131}I	8.0 days	~1,760
^{134}Cs	2.0 days	~54
^{137}Cs	30.0 days	~85
^{132}Te	78.0 h	~1,150
^{89}Sr	52.0 days	~115
^{90}Sr	28.0 years	~ 10
^{140}Ba	12.8 days	~240
^{95}Zr	1.4 h	196
^{99}Mo	67.0 h	>168
^{103}Ru	39.6 days	>168
^{106}Ru	1.0 year	>73
^{141}Ce	33.0 days	196
^{144}Ce	285.0 days	~116
^{239}Np	2.4 days	~95
^{238}Pu	86.0 years	0.035
^{239}Pu	24,400.0 years	0.03
^{240}Pu	6,580.0 years	0.042
^{241}Pu	13.2 years	~6
^{242}Cm	163.0 days	~0.9

Table 9 Fission and production yields of radionuclides in weapons testing (UNSCEAR, 1982).

Nuclide	Half-life	Representative fission yield (%)	Normalized production (PBq per Mt fission energy)
^{89}Sr	50.5 days	2.56	590
^{90}Sr	28.8 a	3.50	3.9
^{95}Zr	64 days	5.07	920
^{103}Ru	39.4 days	5.20	1,500
^{106}Ru	368 days	2.44	78
^{131}I	8.04 days	2.90	4,200
^{136}Cs	13.2 days	0.036	32
^{137}Cs	30.2 a	5.57	5.9
^{140}Ba	12.8 days	5.18	4,700
^{141}Ce	32.5 days	4.58	1,600
^{144}Ce	284 days	4.69	190

amounts of ^{239}Pu. Other radionuclides are produced in neutron-induced reactions in non-radioactive elements of the nuclear device, the atmosphere, soil or seawater. These reactions were sufficient to produce a significant increase above the natural backgrounds of radiocarbon and tritium. Typical yields per megaton were 3.2×10^{23} atoms ^{14}C and 1×10^{19} atoms ^{3}H. The neutrons produced in nuclear bursts, similar to the neutrons produced in cosmic-ray interactions, are capable of producing a variety of radionuclides in the atmosphere (Table 3).

Studies conducted in the early years of weapon tests identified ^{90}Sr as one of the most hazardous radionuclides. Essentially, all of the ^{90}Sr injected into the atmosphere from weapons testing prior to the test-ban agreement in 1963 was deposited on the Earth's surface by 1970. The global deposition reached a maximum of about 450 PBq at the end of 1967 (Volchok and Kleinman, 1971). The deposition in the northern hemisphere was three times higher than in the southern. Due to radioactive decay, the inventory of deposited ^{90}Sr decreased by 2.5% per year. The deposition at any given locality was influenced mainly by the meteorological factors that influence troposphere–stratosphere exchange including the amount of precipitation. ^{89}Sr is produced in much larger amounts but, due to its short half-life, it nearly vanishes after 1 year.

The distribution of ^{137}Cs in the stratosphere and its terrestrial deposition are similar to those of ^{90}Sr.[13] On average, the accumulation by the end of 1965 ranged between 2.2 and 3.7 GBq km^{-2}. Observations by Gustafson et al. (1970) have shown that ^{137}Cs deposition reached a peak of about 4.8 GBq km^{-2} during 1965–1966.

Prior to 1954, cosmogenic ^{14}C had been in secular equilibrium in the biosphere at a concentration of $\sim$0.275 Bq gC^{-1}. This natural equilibrium was disturbed by the additional ^{14}C produced by detonation of thermonuclear devices. The most precise estimates of the amounts of ^{14}C produced in weapons tests are based on measurements of the ^{14}C activity in stratospheric air using high-flying aircraft and balloons. A total inventory of 355 PBq was estimated (NCRP, 1985). In the mid-1960s, the ^{14}C content increased by almost 100% of the natural level, and thereafter it gradually decreased. By the end of 1967, the tropospheric ^{14}C content in the northern hemisphere was still about 60% higher (Nydal, 1968) and at present, it is about 10%.[14]

In nuclear weapon tests, significant amounts of plutonium isotopes were also produced and injected into the atmosphere. These isotopes include ^{238}Pu (half-life 87.7 years), ^{239}Pu (half-life 24,310 years), ^{240}Pu (half-life 6,570 years) and ^{241}Pu (half-life 14 years). It has been estimated that about 12 PBq plutonium was distributed globally, 0.15 PBq remaining in the

[13] The terrestrial distribution and applications of ^{137}Cs are discussed in more detail in Chapter 7.

[14] The production of ^{14}C and its distribution in the global carbon cycle are discussed in more detail in Chapter 4.

stratosphere. The deposition patterns and accumulation rates of plutonium are similar to those for strontium. Therefore, the plutonium deposition could be derived from the strontium data; a value of about 52 MBq km^{-2} was estimated. Between 1963 and 1964, the maximum atmospheric concentration of plutonium was about 6.29×10^{-5} Bq m^{-3} (Bennett, 1978).

The total amount of ^{131}I released to the atmosphere was estimated to be about 500 EBq.[15] A major part of ^{131}I was injected into the stratosphere, where it decayed substantially before reaching the troposphere.

The natural tritium inventory prior to the nuclear bomb tests was about 4 kg. The various thermonuclear tests carried out until the test-ban agreement in 1963 produced an additional amount of about 800 kg. The tritium concentration in precipitation of the northern hemisphere rose from about 1 Bq l^{-1} prior to the bomb tests to a maximum of several 100 Bq l^{-1} in 1963. After the moratorium on atmospheric bomb tests, the tritium level in precipitation gradually decreased, and during the last decade, the natural level has been reached again. The bomb tritium has been used in numerous hydrological studies during the last five decades as a powerful transient tracer (Chapter 5).

Other radionuclides produced in weapon tests were either short lived or produced in insignificant amounts.

REFERENCES

Anderson, E. C., W. F. Libby, S. Weinhouse, A. F. Reid, A. D. Kirshenbaum, and A. V. Grosse. (1947). Natural radiocarbon from cosmic radiation. *Physical Review*, 72, 931–936.

Beer, J., F. Joos, Ch. Lukasczyk, W. Mendel, J. Rodriguez, U. Siegenthaler, and R. Stellmacher. (1994). ^{10}Be as an indicator of solar variability and climate. In: *The Solar Engine and its Influence on Terrestrial Atmosphere and Climate* (Ed. E. Nesme-Ribes). Springer, Berlin, Heidelberg, pp. 221–233.

Bennett, B. G. (1978). *Environmental Aspects of Americium*. EML Report No. 348. USDOE, New York.

Blinov, A. (1988). The dependence of cosmogenic isotope production rate on solar activity and geomagnetic field variations. In: *Secular Solar and Geomagnetic Variations in the last 10,000 Years* (Eds F. R. Stephenson and A. W. Wolfendale). Kluwer, Dordrecht, pp. 329–340.

Damon, P. E., J. C. Lerman, and A. Long. (1978). Temporal fluctuations of atmospheric ^{14}C: Causal factors and applications. *Annual Review of Earth and Planetary Sciences*, 6, 457–494.

Eisenbud, M., and T. Gesell. (1997). *Environmental Radioactivity*. Academic Press Inc., Orlando, FL, 656 pp.

Faul, H. (1954). *Nuclear Geology*. Wiley, New York.

Gosse, J. C., and F. M. Phillips. (2001). Terrestrial *in situ* cosmogenic nuclides: Theory and application. *Quaternary Science Reviews*, 20, 1475–1560.

[15] 1 EBq = 1×10^{18} Bq.

Gustafson, P. F., D. M. Nelson, S. S. Brar, and S. E. Muniak (1970). *Recent Trends in Radioactive Fallout.* Report ANL-7760, Part III. Argonne National Laboratory, Argonne, IL, p. 246.

Guyodo, Y., and J.-P. Valet. (1996). Relative variations in geomagnetic intensity from sedimentary records: The past 200,000 years. *Earth and Planetary Science Letters*, 143, 23–36.

Hess, C. T., R. E. Casparius, S. A. Norton, and W. F. Brutsaert. (1980). Investigation of natural levels of radon-222 in groundwater in Maine for assessment of related health effects. In: *The Natural Radiation Environment III* (Eds T. S. Gesell and W. M. Lowder). NTIS, Springfield, VA. USDOE CONF-708422, pp. 529–546.

Hess, C. T., J. Michel, T. R. Horton, H. M. Prichard, and W. A. Coniglio. (1985). The occurrence of radioactivity in public water supplies in the United States. *Health Physics*, 48, 553–586.

Hess, W. N., E. H. Canfield, and R. E. Lingenfelter. (1961). Cosmic ray demography. *Journal of Geophysical Research*, 66, 665–677.

Holaday, D. A. (1959). The nature of wastes produced in the mining and milling of ores. In: *Industrial Radioactive Waste Disposal.* Hearings before Joint Committee on Atomic Energy, USGPO, Washington, DC.

Kohman, T., and N. Saito. (1954). Radioactivity in geology and cosmology. *Annual Review of Nuclear Science*, 4.

Lal, D. (1988). Theoretically expected variations in the terrestrial cosmic-ray production of isotopes. In: *Solar–Terrestrial Relationships.* Societa Italiana di Fisica, Bologna, Italy, XCV corso, pp. 216–233.

Lal, D. (1991). Cosmic ray labelling of erosion surfaces: *In situ* nuclide production rates and erosion models. *Earth and Planetary Science Letters*, 104, 424–439.

Lal, D., and B. Peters (1967). Cosmic ray produced radioactivity on the Earth. In: *Handbuch der Physik, XLVI/2*, Springer, Berlin, pp. 551–612.

Libby, W. F. (1946). Atmospheric helium three and radiocarbon from atmospheric radiation. *Physical Review*, 69, 671–672.

Light, E. S., M. Merker, H. J. Vershell, R. B. Mendel, and S. A. Korff. (1973). Time dependent worldwide distribution of atmospheric neutrons and of their products, 2: Calculations. *Journal of Geophysical Research*, 78, 2741–2762.

Lingenfelter, R. E. (1963). Production of carbon-14 by cosmic-ray neutrons. *Reviews of Geophysics*, 1, 35–55.

Magno, P. J., P. R. Groulx, and J. C. Apidianakis. (1970). Lead-210 on air and total diets in US. *Health Physics*, 18, 383–388.

Masarik, J., and J. Beer. (1999). Simulation of particle fluxes and cosmogenic nuclides production in the Earth's atmosphere. *Journal of Geophysical Research*, 104, 12099–12112.

Masarik, J., and R. C. Reedy. (1994). Effects of bulk chemical composition on nuclide production processes in meteorites. *Geochimica et Cosmochimica Acta*, 58, 5307–5317.

Masarik, J., and R. C. Reedy. (1995). Terrestrial cosmogenic-nuclide production systematic calculated from numerical simulations. *Earth and Planetary Science Letters*, 136, 381–395.

Mauchline, J., and W. L. Templeton. (1964). Artificial and natural radioisotopes in the marine environment. *Annual Review of Oceanography and Marine Biology*, 2, 229–279.

Monaghan, M. C., S. Krishnaswami, and K. K. Turekian. (1985/86). The global-average production rate of ^{10}Be. *Earth and Planetary Science Letters*, 76, 279–287.

Muller, R. A. (1977). Radioisotope dating with a cyclotron. *Science*, 196, 489–494.

National Council on Radiation Protection and Measurements. (1985). *Carbon-14 in the Environment.* NCRP Report No. 81. Bethesda, MD.

NEA/OECD. (2002). *Chernobyl, Assessment of Radiological and Health Impacts – 2002 Update of Chernobyl.* NEA/OECD, Paris.

Newkirk, L. L. (1963). Calculation of low-energy neutron flux in the atmosphere by the Sn method. *Journal of Geophysical Research*, 68, 1825–1839.

Nir, A., S. T. Kruger, R. E. Lingenfelter, and E. J. Flamm. (1966). Natural tritium. *Reviews of Geophysics*, 4, 441–456.

Nishiizumi, K., C. P. Kohl, J. R. Arnold, R. Dorn, J. Kelin, D. Fink, R. Middelton, and D. Lal. (1993). Role of *in situ* cosmogenic nuclides ^{10}Be and ^{26}Al in the study of diverse geomorphic processes. *Earth Surface Processes and Landforms*, 18, 407–425.

Nydal, R. (1968). Further investigation in the transfer of radiocarbon in nature. *Journal of Geophysical Research*, 73, 3617–3675.

O'Brien, K. (1979). Secular variations in the production of cosmogenic isotopes in the Earth's atmosphere. *Journal of Geophysical Research*, 84, 423–431.

Oeschger, H., J. Houtermans, H. Loosli, and M. Wahlen. (1969). The constancy of cosmic radiation from isotope studies in meteorites and on the Earth. In: *Radiocarbon Variations and Absolute Chronology* (Ed. I. U. Olsen). John Wiley, Stockholm. Nobel symposium.

Pertsov, L. A. (1964). *The Natural Radioactivity of Biosphere*. Atomizdat, Moscow, (in Russion).

Preiss, N., M. A. Melieres, and M. Pourchet. (1996). A compilation of data on ^{210}Pb concentrations in surface air and fluxes at the air surface and water-sediment interfaces. *Journal of Geophysical Research*, 101(D22), 28847–28862.

Rankama, K., and T. G. Sahama. (1950). *Geochemistry*. University of Chicago Press, Chicago.

Reyss, J.-L., Y. Yokoyama, and F. Guichard. (1981). Production cross sections of ^{26}Al, ^{22}Na, ^{7}Be from argon and of ^{10}Be, ^{7}Be from nitrogen: implications for production rates of ^{26}Al and ^{10}Be in the atmosphere. *Earth and Planetary Science Letters*, 53, 203–210.

Schiger K. J. (1986). Disposal of uranium mill tailings. *Proceedings of the Twenty First Annual Meeting of the NCRP*, National Council of Radiation Protection and Measurement, Bethesda, MD.

Shea, M. A., and D. F. Smart. (1992). Recent and historical solar proton events. *Radiocarbon*, 34, 255–262.

Simpson, J. A. (1983). Elemental and isotopic composition of the galactic cosmic rays. *Annual Review of Nuclear Science*, 33, 323–381.

Taylor, S. R., and S. M. McClennan. (1985). *The Continental Crust: Its Composition and Evolution; An Examination of the Geochemical Record Preserved in Sedimentary Rocks*. Blackwell Scientific Publications, Oxford, 308 pp.

UNSCEAR. (1982). *Ionizing Radiation: Sources and Biological Effects*. 1982 Report. United Nations, New York.

Vogt, S., G. F. Herzog, and R. C. Reedy. (1990). Cosmogenic nuclides in extra-terrestrial materials. *Reviews of Geophysics*, 28, 253–275.

Volchok, H. L., and M. T. Kleinman (1971). *Worldwide Deposition of Sr-90 Through 1970*. Report HSL-243. USAEC, New York.

CHAPTER 2

Radionuclides as Tracers and Timers of Processes in the Continental Environment – Basic Concepts and Methodologies

Klaus Froehlich[1] *and* Jozef Masarik[2,*]

Contents

1. Introduction

Applications of radionuclides in environmental studies are based on the general concept of 'tracing', in which either intentionally introduced

*Corresponding author. Tel.: +421 2 602 95 456; Fax: +421 2 654 25 882
E-mail address: masarik@fmph.uniba.sk

[1] Viktor-Wittner-Gasse 36/7, 1220 Vienna, Austria
[2] Department of Nuclear Physics and Biophysics, Comenius University, Mlynská dolina, 842 48 Bratislava, Slovakia

Radioactivity in the Environment, Volume 16
ISSN 1569-4860, DOI 10.1016/S1569-4860(09)01602-7

radionuclides or naturally occurring (environmental) radionuclides are employed. Environmental radionuclides have the distinct advantage over injected (artificial) tracers in that they facilitate the study of various processes on a much larger temporal and spatial scale through their natural distribution in environmental systems. Thus, environmental radionuclides are unique tools in regional studies to investigate the time- and space-integrated characteristics of environmental systems. The use of injected artificial tracers is generally effective for site-specific local applications.

This chapter deals with the basic principles underlying the various applications of environmental radionuclides in tracer studies of terrestrial systems. There are two major groups of applications: (1) providing timescales of past and present processes; (2) tracing substances involved in terrestrial transport, exchange and mixing processes. The applications of both groups are based on the phenomenon of radioactivity.

2. Radioactive Decay – Fundamentals of Radiometric Dating

2.1. Radioactivity

Radioactivity is a phenomenon related to the spontaneous transmutation (decay) of an unstable atomic nucleus accompanied by emission of radiation. In this definition, a number of terms is used that need to be explained. In the following, a brief summary of the definitions and fundamentals is given. For more details, the reader may refer to a suitable textbook (e.g. Turner, 2007; Magill and Galy, 2005).

The *nucleus* of an atom consists of *protons* and *neutrons* and is surrounded by a cloud of orbiting *electrons* forming the *shell* of the atom (Figure 1). The mass of a proton is about 2,000 times heavier than that of an electron. The proton carries a positive and the electron a negative elementary charge. The neutron is electrically neutral and only slightly heavier than the proton (Figure 1). In a neutral atom, the number of electrons is equal to the number of protons, which is called the *atomic number* Z. While the mass of an atom is nearly equal to the mass of its nucleus (about 2×10^{-27} to 400×10^{-27} kg), the size (diameter) of the atom is controlled by the atomic shell (about a fraction of a nanometre), that is about 10,000 times larger than the nucleus.

The term *nuclide* defines an atomic species with a given number of protons and neutrons in the nucleus. The number of protons and neutrons is called the *atomic mass number* $A = N+Z$, where N is the number of neutrons and Z the number of protons (atomic number). The common name of proton and neutron is *nucleon*, that is A is the number of nucleons. A nuclide is characterised by the symbol ${}^{A}_{Z}X$, where X stands for the

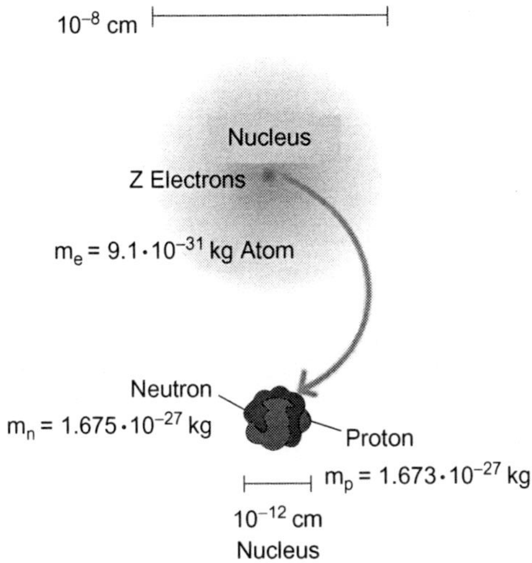

Figure 1 Structure of the atom.

chemical element to which the nuclide belongs. For example the nuclide of carbon with 12 nucleons is described by ${}^{12}_{6}C$.

Nuclides belonging to the same chemical element, that is having the same atomic number but different atomic mass numbers, are called *isotopes*. For example the following nuclides are isotopes of the element carbon: ${}^{12}_{6}C$ (most abundant carbon isotope), ${}^{13}_{6}C$ (heavy stable carbon isotope) and ${}^{14}_{6}C$ (heavy unstable carbon isotope). Since the chemical element is defined by the atomic number, the latter can be omitted in the nuclide symbol. Hence, the above carbon isotopes can be written as ${}^{12}C$, ${}^{13}C$ and ${}^{14}C$.

Nucleons are bound in the nucleus by *nuclear forces*, also called strong forces. These are short-range attractive forces that exist between protons and neutrons and are strong enough to overcome the electrostatic repulsion which exists between the charged protons. Like atomic electrons of the shell, nucleons 'occupy' discrete energy levels of the nucleus. The situation in the nucleus is, however, complicated by the fact that two types of nucleons, protons and neutrons, are to be accommodated. The energy levels involved are of the order of megaelectron volt, much higher than the energies of atomic electron levels. Because the charged protons experience an electrostatic repulsion between them, which the uncharged neutrons do not, the proton levels appear at slightly higher energies than the neutron levels. As in the case of atomic electrons, each nuclear level can only hold a certain number of protons or neutrons. Because of the higher energy of the proton levels, stable nuclei tend to have slightly more neutrons than protons. There are possible transitions between neutron and proton energy levels. These transitions lead to the emission of particles and transformation of one kind of nucleons to another. If the nucleus has, for example a single proton in a high energy level with a vacancy for a neutron in a lower level, that proton can transmute into a neutron and emit a particle with positive charge, called a positron, and a gamma quantum that takes the excess energy. This process is called *beta plus* decay. The opposite process, called *beta minus* decay, occurs if a nucleus has too many neutrons and thus transmutes a neutron into a proton.

The *gamma quantum* is a photon that has no rest mass and no charge. Therefore, the gamma transition changes neither Z nor A of the emitting nucleus. Like X-rays, the gamma quantum is equivalent to electromagnetic radiation. Thus, the alternative term is *gamma ray* (radiation). The energy of the gamma quantum E_γ is given by the energy difference between the corresponding energy states E_2 (higher state) and E_1 (lower state):

$$E_\gamma \equiv hf = E_2 - E_1$$

where h is Planck's constant $= 6.626068 \times 10^{-34}\,\mathrm{m^2\,kg^{-1}\,s^{-1}}$ and f the frequency of the equivalent electromagnetic radiation. In most cases, the nucleus remains in the higher energy state for less than 10^{-14} s.

The basic condition for the existence of *stable nuclides* is the following: the total energy of a stable nucleus must be lower than the total energy of any other combination of the same number of neutrons and protons. If this condition is not fulfilled, the nucleus is unstable and will spontaneously decay into another configuration. The latter process is called *radioactive decay* and the decaying nuclide is called a *radionuclide*. There are radionuclides that can decay in two or three different ways, but the majority of radionuclides decay in a single way. In either case, radioactive decay changes Z and N of the nucleus. The product nuclide, also called the *daughter nuclide*, may also be radioactive and transmute spontaneously, as happens in the natural decay series (Chapter 1). In the following sections, a short overview of the various radioactive decay types will be given.

2.1.1. Beta minus decay

A nuclide with too many neutrons can get closer to stability by converting a neutron into a proton. The simultaneous emission of an electron e^- (in this process also called beta minus particle β^-) and an antineutrino $\bar{\nu}$ (a particle that carries energy but no charge and nearly no mass) ensures the conservation of charge and energy, respectively. This process

$$n \rightarrow p + e^- + \bar{\nu}$$

increases Z by 1 and decreases N by 1 so that A remains unchanged. The decay product (daughter nucleus) can initially be in an excited state and emit the excess energy in the form of one or more gamma quantum (gamma rays). A typical example is the decay of the radionuclide ^{131}I into its daughter nuclide ^{131}Xe:

$$^{131}_{53}I \rightarrow ^{131}_{54}Xe + e^- + \bar{\nu} + \gamma$$

In this process, the atomic number increased by 1 and therefore the iodine nucleus changed to a xenon nucleus under emission of a beta minus particle β^-, an antineutrino $\bar{\nu}$ and a gamma ray.

Details of a radioactive decay process are given by the decay scheme. In Figure 2a, it is shown that ^{131}I decays with a half-life of 8 days under emission of a beta minus particle to one of two states of ^{131}Xe. The decay may either emit a beta particle with maximum energy of 606 keV, resulting in an excited state of the ^{131}Xe nucleus 364 keV above the ground state, or a beta particle with a maximum energy of 333 keV, resulting in a state at 637 keV above the ground state. These are the maximum beta particle energies. In reality, the beta particles may have any energy between zero and the maximum value, with a typical average value being about one-third of the maximum energy. The remaining energy is carried away by the neutrino. The excited states of ^{131}Xe immediately decay to the ground state. In 82% of the decay processes, a gamma ray of 364 keV energy is emitted and in 7% of decays, a gamma ray of 637 keV is emitted. For the sake of clarity, the remaining 11% of decay processes have not been specified in this simplified version of the decay scheme. Figure 2b demonstrates what information is included in the chart of the nuclides. The box representing ^{131}I shows the half-life of this radionuclide (8 days), its decay type (beta minus decay), the maximum beta particle energies (0.61 and 0.33 MeV) and the associated gamma ray energies (0.36 and 0.64 MeV). Since beta minus decay increases Z by 1 and decreases N by 1, the daughter nuclide ^{131}Xe is represented by a box that is shifted one unit

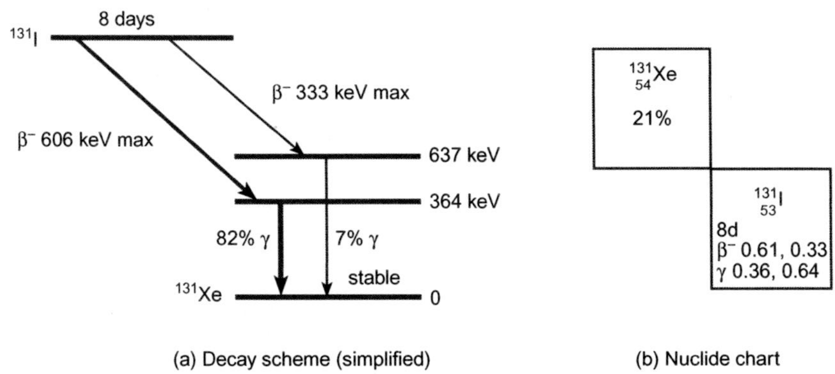

Figure 2 Radioactive decay of ^{131}I into ^{131}Xe. (a) Simplified decay scheme; (b) Part of the nuclide chart characterising parent and daughter nuclide of this decay process.

above and one unit left of the ^{131}I box. The ^{131}Xe box shows that this nuclide is stable and occurs naturally with an abundance of 21%, that is it accounts for 21% of all naturally occurring Xe isotopes.

2.1.2. Beta plus decay

Beta plus decay occurs when a nucleus consists of too many protons; then a proton changes into a neutron plus a positron e^+ and a neutrino ν:

$$p \rightarrow n + e^+ + \nu$$

This decay process is similar to the beta minus decay except that the roles of proton and neutron are reversed and the emitted particle is positively charged instead of negatively. A positron has the same mass as an electron but a positive elementary charge instead of a negative one. In this decay process, Z decreases by 1 and N increases by 1, and hence there is also no change in A. The emitted positron usually travels only a short distance; it will almost immediately collide with an electron of the surrounding material. Since electron and positron are antiparticles, they disappear, and their energy equivalent (each particle 511 keV) is converted into two gamma rays that are emitted in opposite directions (annihilation radiation).

An alternative way to lower the number of protons involves the capture of an orbiting electron. The electron from an inner orbit combines with one of the protons in the nucleus to form a neutron and a neutrino according to

$$p + e^- \rightarrow n + \nu$$

This process is called electron capture (EC). The vacancy in the inner shell is immediately filled by an electron from the outer shell, and the difference in the atomic binding energy is emitted as a characteristic X-ray. Furthermore, the energy difference between the parent and daughter nucleus is emitted as a gamma ray.

Often, beta plus decay and EC are competing for the decay of a radionuclide with proton excess. A typical example is the decay of ^{124}I into ^{124}Te. The decay through beta plus emission is described by

$$^{124}_{53}I \rightarrow ^{124}_{52}Te + e^+ + \nu + \gamma$$

and through EC by

$$^{124}_{53}I + e^- \rightarrow ^{124}_{52}T + \nu + \gamma + X$$

The decay scheme (Figure 3) shows that ^{124}I decays with a half-life of 4.2 days into either an excited state of ^{124}Te, followed by emission of a 603 keV gamma ray, or directly to the ground state. Twenty-four percent of the decays are by EC to the 603 keV level and 23% to the ground state of

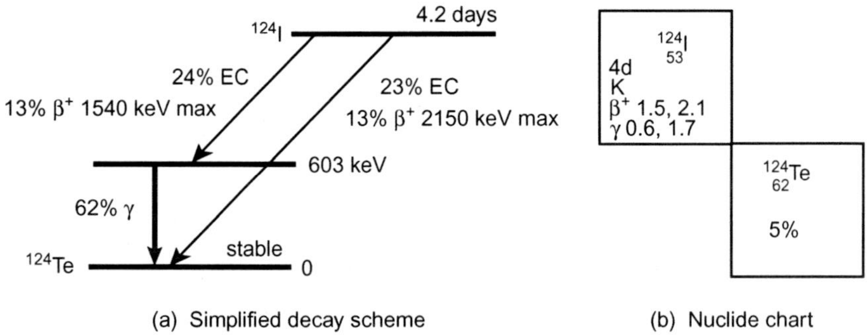

Figure 3 Radioactive decay of ^{124}I into ^{124}Te. (a) Simplified decay scheme; (b) Part of the nuclide chart characterising parent and daughter nuclide of this decay process.

^{124}Te. Further 13% of the decays are through positron emission with a maximum energy of 1,540 keV and another 13% with a maximum energy of 2,150 keV. The remaining 27% are by EC to higher energy levels not shown in this simplified diagram, but most of these eventually reach the 603 keV level. Therefore, in total, 62% of the decays result in emission of a 603 keV gamma ray, and there are also other less abundant gamma ray emissions from the higher levels which are not shown. In total, 26% of decays are by positron emission, each of which results in two annihilation gamma rays.

2.1.3. Alpha decay and nuclear fission

The heaviest nuclides are unstable. The nuclear force that holds the nucleus together is of very short range. If the nucleus contains too many nucleons, the distance between the nucleons becomes too far for the nuclear force to be able to hold all nucleons together. The repulsive electrostatic force between the protons overcomes nuclear forces and the nucleus can spontaneously decay. There are two modes for this decay.

The nucleus emits an alpha particle that consist of two protons and two neutrons which are bound together by the nuclear force in a very stable configuration that is identical to a helium nucleus $\alpha = {}^4_2\mathrm{He}$. Thus, alpha decay decreases both Z and N by 2 and shifts, that is transmutes, the parent nuclide into another nuclide with mass number 4 less and atomic number 2 less:

$$^A_ZX \rightarrow {}^{A-4}_{Z-2}Y + {}^4_2\mathrm{He}$$

In the chart of nuclides, the box of the daughter nuclide is shifted two places to the left and two rows lower.

An example of alpha decay involves ^{228}Th:

$$^{228}_{90}\text{Th} \rightarrow {}^{224}_{88}\text{Ra} + {}^{4}_{2}\text{He} + 5,421\ \text{MeV}$$

The energy difference between the ground states of parent and daughter nuclide is 5,421 MeV, in which 71% of this alpha decay process is taken by the alpha particle. There are also transitions with lower energy particles where the difference from the total energy is emitted by gamma rays (Figure 4).

Nuclear fission is an alternative decay mode for very heavy nuclides in which the nucleus splits up into two fragments with a mass ratio of about 2:3. Fission can be induced by neutron capture or it can happen spontaneously. The latter mode of radioactive decay is energetically feasible for atomic mass numbers above 230 including thorium and uranium. In the case of ^{238}U, the spontaneous fission probability per decay of ^{238}U is 5.4×10^{-7}. The recoil of the fission fragments causes trails of damage in uranium-containing minerals. The density of fission tracks in minerals increases with time and thus can be used for age determination of minerals (fission track dating).

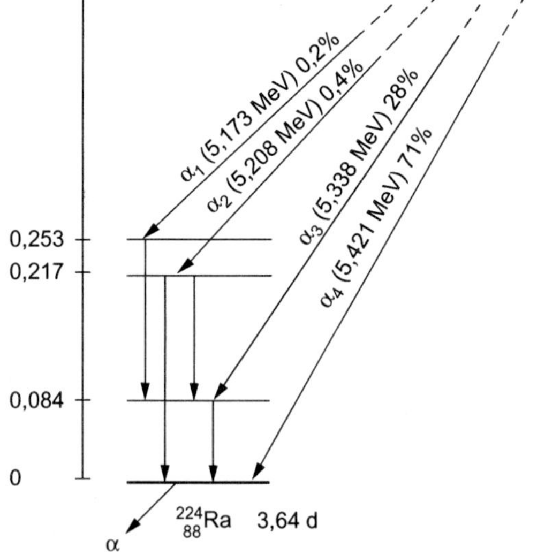

Figure 4 Scheme of decay of $^{228}_{90}$Th.

2.2. Exponential decay law

As discussed above, the spontaneous decay of an unstable nucleus is entirely random, that is it is impossible to predict when a particular nucleus will decay. The probability that a given unstable nucleus decays is neither dependent on its history (age) nor on external chemical or physical conditions. Given a sample of N atoms of a particular radionuclide, the probability of an atom decaying in a small interval of time dt is $-dN/N$ and depends only on the length of this time interval:

$$\frac{-dN}{N} \propto dt$$

that is the decay probability per unit of time $-dN/(N \cdot dt)$ is constant. It is called the decay constant λ

$$\frac{-dN}{N \bullet dt} = \lambda \tag{1}$$

Each radionuclide has its own decay constant (also called decay rate). The solution of this differential equation is the following exponential function:

$$N = N_0 \bullet e^{-\lambda t} \tag{2}$$

N_0 and N are the number of atoms at time $t = 0$ and at time t, respectively. Instead of the decay constant λ, an alternative parameter is often used in the decay equation, which is the *half-life* $T_{1/2}$. The time necessary for the number of atoms to fall to half of its original value is called the half-life of the given radionuclide. With this definition and Equation (2), it follows that

$$\frac{N_0}{2} = N_0 \bullet e^{-\lambda T_{1/2}} \tag{3}$$

and by transforming this equation, the relationship between decay constant and half-life is obtained (Figure 5):

$$T_{1/2} = \frac{\ln(2)}{\lambda} \tag{4}$$

The *activity* A of a sample containing N atoms of a particular radionuclide is defined as the number of atoms decaying per unit of time. Via Equation (2), it follows that:

$$A \equiv \frac{-dN}{dt} = \lambda N \tag{5}$$

According to this definition, the *activity unit* is $[A] = 1\,s^{-1}$, which is also called 1 becquerel (Bq), thus

$$1\,Bq = 1\,s^{-1}$$

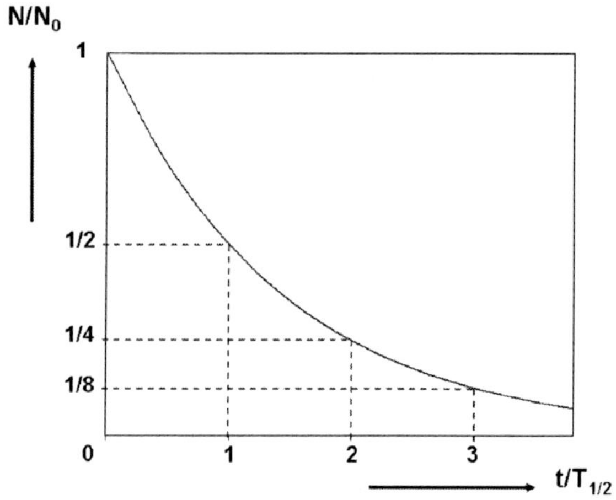

Figure 5 Exponential decrease of the number of atoms of a radionuclide with time.

An obsolete activity unit is the *curie* (Ci), where $1\ \text{Ci} = 3.7 \times 10^{10}\ \text{Bq}$.

The relationship between the activity A and the mass m of a radionuclide in a given sample is, according to Equation (5),

$$A = \lambda N = \lambda \frac{m}{A_r} N_A \tag{6}$$

where N_A = Avogadro's number = $6.022 \times 10^{23}\ \text{mol}^{-1}$ and A_r = molar mass of the radionuclide.

2.3. Fundamentals of radiometric dating

2.3.1. Radioactive parent–stable daughter nuclide

The age of rocks, minerals, groundwater and other terrestrial and marine substances is defined as the time elapsed since their formation. Environmental radionuclides occurring in these substances can be used as an internal clock to determine the age of such material. Assuming that the activity A of a particular radionuclide included in the sample changed only due to radioactive decay (no admixture or removal by other physical and chemical processes) and that the initial activity A_0 is known, the age t of the sample can be determined by the following expression derived from Equations (2), (4) and (5):

$$t = \frac{T_{1/2}}{\ln(2)} \ln\left(\frac{A_0}{A}\right) \tag{7}$$

This relationship shows that the age range that can be covered by a particular radionuclide depends on the half-life. The half-lives of environmental radionuclides suitable for radiometric dating range from less than days to billions of years (Chapter 1). Many dating methods using cosmic-ray-produced radionuclides (Chapter 1) are based on Equation (7). For example cosmogenic radiocarbon with its half-life of 5,730 years allows dating of geological and groundwater samples in a range from less than 1,000 to about 40,000 years (Chapters 4 and 5).

Often it is difficult or even impossible to determine/reconstruct the initial activity of a sample which has to be dated. In such cases, the age can still be determined radiometrically if the daughter product of the radioactive decay is a stable nuclide, remains in the sample and can be measured. If N_0 is the number of atoms of the parent radionuclide in the sample, N the remaining number and D the number of atoms of the daughter nuclide, it follows that $N_0 = N + D$, and via Equations (2) and (4), the age t of the sample is given by the expression:

$$t = \frac{T_{1/2}}{\ln 2} \ln\left(1 + \frac{D}{N}\right) \tag{8}$$

A prominent application of this equation is the ^{3}H–^{3}He dating method for young groundwater, described in more detail in Chapter 5. On the basis of this model, a number of dating techniques have been developed, especially for dating ancient rocks and minerals using long-lived radionuclides (Table 1).

The parent nuclides in Table 1 represent primordial radionuclides (Chapter 1) which survived since synthesis of the elements because of their long half-lives. Here, only two of the most frequently used methods are briefly discussed. A comprehensive description of all the methods listed in Table 1 is given by Dickin (2005).

Table 1 Primordial radionuclides used for dating of ancient rocks and minerals.

Parent	Stable daughter	Currently accepted half-life $T_{1/2}$ in 10^9 years	Natural abundance in %
^{238}U	^{206}Pb	4.468	99.27
^{235}U	^{207}Pb	0.704	0.72
^{232}Th	^{208}Pb	14.05	100
^{40}K	^{40}Ar, ^{40}Ca	1.277	0.0117
^{87}Rb	^{87}Sr	47.5	27.835
^{138}La	^{138}Ce, ^{138}Ba	105	0.0902
^{147}Sm	^{143}Nd	106	15.02
^{176}Lu	^{176}Hf	37.8	2.59
^{187}Re	^{187}Os	43.5	62.60
^{190}Pt	^{186}Os	37.8	2.59

2.3.2. Potassium–argon method

One of the most widely used methods of radiometric dating is the potassium–argon method. Potassium is one of the 10 most abundant elements in the Earth's crust and can be found in most rock-forming minerals. The method covers a wide age range and it can be used to date young rocks of a few thousand years as well as the oldest rocks found on Earth. There are three decay modes of ^{40}K: beta minus emission, positron emission and EC

$$^{40}_{19}K \rightarrow ^{40}_{20}Ca + e^{-} + \overline{\nu}$$
$$^{40}_{19}K \rightarrow ^{40}_{18}Ar + e^{+} + \nu$$
$$^{40}_{19}K + e^{-} \rightarrow ^{40}_{18}Ar + \nu$$

When ^{40}K undergoes beta minus decay, it transmutes into ^{40}Ca, the most abundant isotope of calcium. Since calcium is also very common in minerals, it is not possible to distinguish the ^{40}Ca produced from the decay of ^{40}K from the ^{40}Ca present in the rock in question since its formation. However, when ^{40}K undergoes positron emission or EC, it transmutes into the inert gas atom ^{40}Ar. When the rock melts, ^{40}Ar escapes. Usually, after solidification of magma, the rock crystals are free of ^{40}Ar. With progressing time, ^{40}Ar is produced by the decay of ^{40}K and accumulates in the crystals. However, if magma cools quickly on the surface of the Earth, some of the Ar remains trapped so that the age derived from the ^{40}Ar measurements appears to be older than in reality. Another source of incorrect dating is the partial loss of argon if the rock is heated above 300°C, for instance by metamorphism. If all the ^{40}Ar escapes during metamorphism, the apparent age derived from the ^{40}Ar measurement is just the time elapsed since the metamorphic event. Minerals suitable for potassium–argon dating include muscovite, biotite, hornblende, glauconite and K-feldspar.

2.3.3. Rubidium–strontium method

The age of minerals and rocks can be determined precisely by measurements of the radionuclide rubidium-87 and its decay product strontium-87 (Table 1). Rearranging Equation (8) and replacing D and N by ^{87}Sr and ^{87}Rb, respectively, the following expression is obtained:

$$^{87}Sr = ^{87}Rb(e^{\lambda t} - 1) \tag{9}$$

where ^{87}Sr is the amount of strontium-87 produced by the rubidium-87 decay in the sample (radiogenic strontium-87) and ^{87}Rb the measured rubidium amount in the sample. Taking an initial concentration of strontium-87 at formation of the mineral $^{87}Sr_0$ into account, the total amount of strontium-87 in the sample is given by

$$^{87}Sr_t = ^{87}Sr_0 + ^{87}Sr \tag{10}$$

Combining Equations (9) and (10), it follows that:

$$^{87}\mathrm{Sr}_t = {}^{87}\mathrm{Sr}_0 + {}^{87}\mathrm{Rb}(e^{\lambda t} - 1) \qquad (11)$$

In Equation (11), the amount of $^{87}Sr_0$ is unknown, but can be determined by measuring the amount of ^{86}Sr in the sample. This stable strontium isotope is not of radiogenic origin and thus does not change with time. Dividing Equation (11) by ^{86}Sr, the following expression is obtained:

$$\left(\frac{^{87}\mathrm{Sr}}{^{86}\mathrm{Sr}}\right)_t = \left(\frac{^{87}\mathrm{Sr}}{^{86}\mathrm{Sr}}\right)_0 + \left(\frac{^{87}\mathrm{Rb}}{^{86}\mathrm{Sr}}\right)_t (e^{\lambda t} - 1) \qquad (12)$$

The quantities that can be measured directly with a mass spectrometer are $(^{87}Sr/^{86}Sr)_t$ and $(^{87}Rb/^{86}Sr)_t$. The two unknowns are the ratio $(^{87}Sr/^{86}Sr)_0$ and t. The different minerals within a given rock have the same age t. Both rubidium and strontium are trace elements (the average abundance of Sr is about 10 ppm and that of Rb is 3 ppm) and do not form their own minerals. Before solidification of the rock, the distribution of the rubidium isotopes in the magma was uniform, that is all the minerals of the rock had the same initial ratio $(^{87}Sr/^{86}Sr)_0$. Thus, by plotting the measured $(^{87}Sr/^{86}Sr)_t$ ratio against the measured $(^{87}Rb/^{86}Sr)_t$ ratio of the various minerals of a given rock, according to Equation (12), a straight line should be obtained for which the intercept is $(^{87}Sr/^{86}Sr)_0$ and the slope $(e^{\lambda t}-1)$. Since the age t is constant, this line is called an isochrone. An example of the use of this method is demonstrated in Figure 6. Minerals suitable for

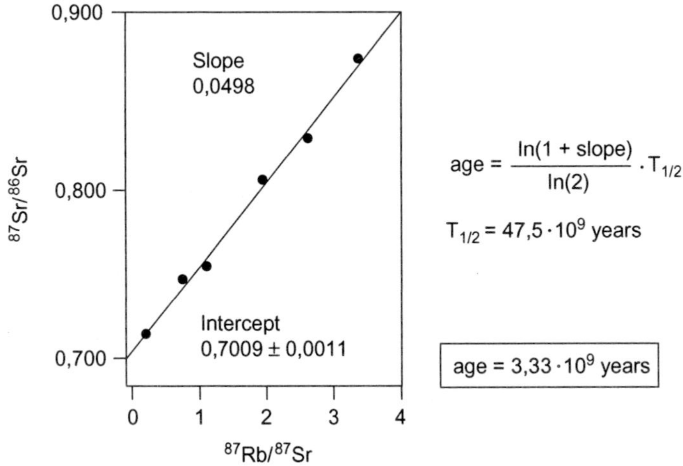

Figure 6 Example of the application of the rubidium–strontium method to determine the age of a rock sample.

rubidium–strontium dating include K-micas, K-feldspars, biotite, metamorphic rock and glauconite.

2.3.4. Parent–daughter pairs of natural decay series

A number of dating methods are based on measuring the radioactivity of parent–daughter pairs of the natural decay series (Chapter 1). Considering closed systems, the temporal change of the number of atoms of the daughter radionuclide (dN_2/dt) of the sample is given by the difference between its production due to radioactive decay of the parent radionuclide ($\lambda_1 N_1$) and radioactive decay of the daughter nuclide ($-\lambda_2 N_2$):

$$\frac{dN_2}{dt} = \lambda_1 N_1 - \lambda_2 N_2 \tag{13}$$

If at the formation of the sample ($t = 0$) the parent radionuclide was separated from its precursor (or if this radionuclide has no precursor, such as ^{238}U, ^{235}U and ^{232}Th), it follows that

$$N_1 = N_1^0 e^{-\lambda_1 t}$$

and after integration of Equation (13),

$$N_2 = \frac{\lambda_1}{\lambda_2 - \lambda_1} N_1^0 (e^{-\lambda_1 t} - e^{-\lambda_2 t}) + N_2^0 e^{-\lambda_2 t} \tag{14}$$

Often systems are considered in which the half-life of the parent radionuclide is orders of magnitudes longer than that of the daughter, that is $T_{1/2}(2) \ll T_{1/2}(1)$ or $\lambda_2 \gg \lambda_1$, and then it follows, via Equation (14), that

$$N_2 = \frac{\lambda_1 N_1}{\lambda_2} + \left(N_2^0 - \frac{\lambda_1 N_1}{\lambda_2} \right) e^{-\lambda_2 t} \tag{15}$$

where N_1 = constant. After six half-lives, N_2 approaches the constant value $N_2^S = \lambda_1 N_1 / \lambda_2$ or $\lambda_2 N_2^S = \lambda_1 N_1$, that is the daughter nuclide activity A_2^S becomes equal to that of the parent nuclide A_1 (Figure 7). This state is called secular equilibrium (superscript S).

A typical example relevant to studies of terrestrial systems is the relationship between ^{226}Ra ($T_{1/2} = 1{,}600$ years) and ^{222}Rn ($T_{1/2} = 3.82$ days) in a closed system; after six half-lives, that is about 20 days, the system is in secular equilibrium.

2.3.5. $^{230}Th/^{234}U$ dating method

An important method for dating terrestrial samples is based on the determination of the activity ratios $^{234}U/^{238}U$ and $^{230}Th/^{234}U$. If such a sample represents a closed system (no exchange of these radionuclides with the environment), a relationship can be derived – in analogy to the derivation

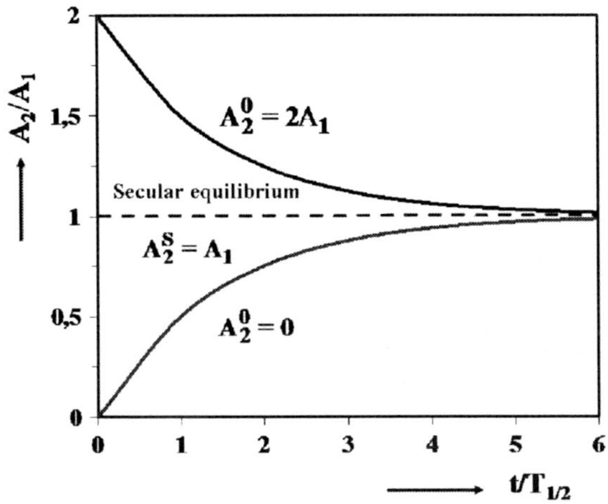

Figure 7 Relationship between long-lived parent radionuclide and relatively short-lived daughter radionuclide.

of Equation (14) – for the three consecutive members of the ^{238}U series: ^{238}U, ^{234}U and ^{230}Th. Assuming that, at the time of formation of the sample ($t = 0$) the ^{230}Th content was zero, the following expression is obtained:

$$\frac{^{230}\mathrm{Th}}{^{234}\mathrm{U}} = \frac{^{234}\mathrm{U}}{^{238}\mathrm{U}}(1 - e^{-\lambda_{230}t}) + \frac{\lambda_{230}}{\lambda_{230} - \lambda_{234}}\left(1 - \frac{^{234}\mathrm{U}}{^{238}\mathrm{U}}\right)(1 - e^{-(\lambda_{230}-\lambda_{234})t}) \quad (16)$$

where ^{230}Th, ^{234}U and ^{238}U stand for the activity concentration of these radionuclides in the given sample and the subscripts and superscripts '230', '234' and '238' are related to these radionuclides. This method is used for samples that can retain uranium and thorium, such as carbonate sediments, speleothems, travertine, bones and teeth, and also volcanic rocks and minerals, peat and pedogenic carbonate. The accuracy of the age determination is between ± 1 and ± 10 ka for sample ages up to about 150 ka, and about ± 50 ka for ages up to 350 ka.

3. The Concept of Residence Time in Environmental Systems

3.1. Introduction

In terrestrial studies, one of the principal uses of environmental radionuclides is for age determinations, that is determination of the time elapsed since formation of substances such as rocks, minerals and groundwater, and since occurrence of events such as volcanic eruptions. Radionuclides enclosed in these substances can serve as internal timers that count down from a specified time interval like an hourglass. If the radioactive decay dominates the changes in radionuclide concentrations, these changes can be used to infer timescales of processes such as water exchange in lakes (turnover time), groundwater recharge and flow regime (infiltration time, travel time, transit time and residence time) or atmospheric circulation processes (atmospheric residence time of water and aerosols). In the following section, the concept of the residence time distribution in terrestrial systems and the information that can be derived from this approach will briefly be discussed.

3.2. The concept of residence time

For radiometric dating, it is essential to define clearly the various concepts of time. In the following, the definitions given by Nir (1986) are adopted, and only systems under steady-state conditions are considered, in which the mass inflow is equal to the mass outflow of the system and these are constant in time. Three different measures of time are used (Figure 8):

- t – the chronological time (calendar time).
- T – the time elapsed since a particle entered the system up to the instant of observation t. T is commonly called age in the system.
- τ – the time a particle spends in the system from its entry until it leaves at the boundary. It is commonly called transit time (often the term residence time is used).

In Figure 8, t is the chronological time at which the system is observed (observation time); its value is determined by an arbitrary chronological timescale (e.g. present date and hour of the day). T is the age of a particle that entered the system at the chronological time $t-T$. The value of T starts at 0 for each particle entering the system at any chronological time. It advances with t on the same timescale, that is $\Delta T = \Delta t$. The age T cannot exceed the transit time of the system, that is $T \leq \tau$. Thus, it is important to distinguish between the age T and the transit time τ of a particle in the system. Each particle has its own τ value which may not be known precisely, but an effort should be made to know its probability distribution $h(\tau)$, also termed as the *system response* function.

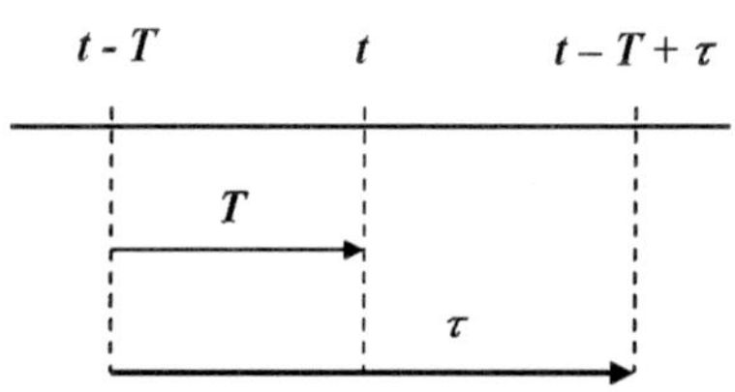

Figure 8 Time relations for the steady state (from Nir, 1986).

Before we discuss specific examples of the various available response functions, some general properties should be considered. As $h(\tau)$ is a normalised probability density function, it follows that

$$\int_0^\infty h(\tau)d\tau = 1 \tag{17}$$

and for the mean flux transit time τ_m (often called mean residence time, MRT):

$$\tau_{\mathrm{m}} = \int_0^\infty \tau h(\tau)d\tau \tag{18}$$

In the population statistics, q is the constant birth and death rate, $h(\tau)$ the distribution of life times, and τ_m is equivalent to the *life expectancy* or the *average age at death*.

In addition to the flux-related distribution $h(\tau)$, the mass-related distribution $M(\tau)$ also has to be considered. If $\mathrm{d}M(\tau)$ is the portion of the total mass in the system having transit times between τ and $\tau+\mathrm{d}\tau$, the following relationship can be found (Nir, 1986):

$$\mathrm{d}M(\tau) = \tau q h(\tau)\mathrm{d}\tau \tag{19}$$

and by integration

$$M = q\int_0^\infty \tau h(\tau)d\tau = q\tau_m \tag{20}$$

where M is the total mass of the traced component in the system. Therefore, without any assumption about the flow and mixing pattern in the system, the relation for the mean transit time has been found as

$$\tau_m = \frac{M}{q} \tag{21}$$

In a similar way, the average age T_m can be calculated (Nir, 1986) by

$$T_m = \frac{1}{2}\frac{\int_0^\infty \tau^2 h(\tau)d\tau}{\int_0^\tau \tau h(\tau)d\tau} = \frac{q}{2M}\int_0^\infty \tau^2 h(\tau)d\tau \tag{22}$$

In the population analogy, T_m corresponds to the *average age* of the *total population*. In contrast to the mean transit time, the average age depends on the specific form of the response function.

From the above equation, it can be concluded that the response function is the primary target of 'age' determination of terrestrial systems, for example of groundwater dating. The system is described by a box model (Figure 9), also called *lumped parameter model* (Maloszewski and Zuber, 1996), in which $h(\tau)$ represents the dynamics of the system (e.g. groundwater flow and recharge regime, lake water balance and air mass transport). It should be

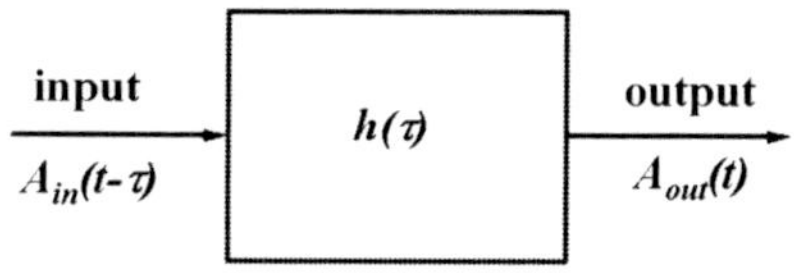

Figure 9 Box model of a terrestrial system.

emphasised that the response of the system to an input pulse depends only on τ. For example the concentration of a tracer (radionuclide) at the outflow of a well-mixed lake depends only on the time but not on the place of sampling. For a steady mass flow through the system, the tracer output $A_{\rm out}(t)$ (in the case of a radionuclide as tracer the activity concentration of the radionuclide with a decay rate λ) at the observation time t can be related to its input activity concentration $A_{\rm in}(t-\tau)$ by the convolution integral:

$$A_{\rm out}(t) = \int_0^{\infty} A_{\rm in}(t-\tau)e^{-\lambda\tau}h(\tau)d\tau \tag{19'}$$

The main parameter of any $h(\tau)$ function is the mean transit or residence time of the given tracer (radionuclide) in the system; it depends on specific system parameters, for example flow rate along flow path in an aquifer or rate of inflow into a system like a lake or an unconfined aquifer. Some of the often used system response functions $h(\tau)$ and their practical relevance will be discussed in the following section.

3.2.1. Piston flow model

If the tracer flow through the system is like a piston flow (without mixing and dispersion), the response function is represented by a *Dirac delta* function (Figure 10):

$$h(\tau) = \delta(\tau - \tau_{\rm PFM}) \tag{20'}$$

with $h(\tau) = 0$ for $\tau \neq \tau_{\rm PFM}$ and $h(\tau) = 1$ for $\tau = \tau_{\rm PFM}$.

The residence time $\tau_{\rm PFM}$ is the average tracer (radionuclide) transit time and is related to the system parameter M (total mass of the traced component in the system) and q (mass flux, flow of this component through the system) according to Equation (21)

$$\tau_{\rm PFM} = \tau_m = \frac{M}{q} \tag{21a}$$

If the system is represented by a confined aquifer with length x, cross-section A, aquifer porosity p, groundwater flow velocity v and water density ρ, it follows that $M = Ap\rho x$ and $q = Ap\rho v$ and via Equation (21a) that

$$\tau_{\rm PFM} = \frac{x}{v} \tag{21b}$$

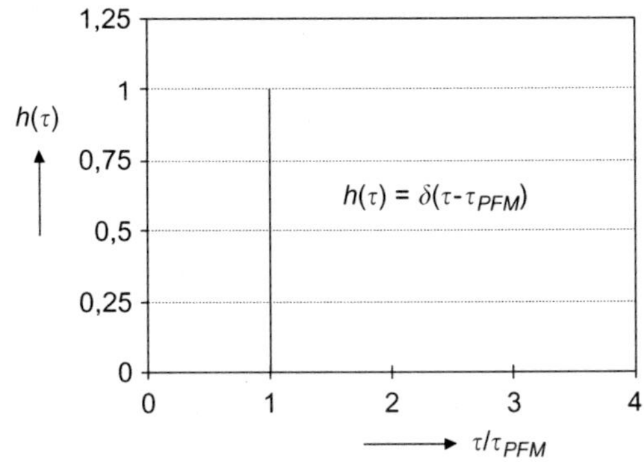

Figure 10 The piston flow model.

Combining Equations (19′) and (20′), the relation

$$A_{\text{out}}(t) = A_{\text{in}}(t - \tau_{\text{PFM}})e^{-\lambda\tau_{\text{PFM}}} \tag{22′}$$

between input and output activity concentrations is obtained. For the application of Equation (22′), the input concentration $A_{\text{in}}(t-\tau_{\text{PFM}})$ has to be known or reconstructed. Measuring the output concentration, the average transit time of the tracer can be determined, and according to Equations (21a) and (21b), the groundwater flow rate in a confined aquifer and the accumulation and/or erosion rate of soils and sediment (Chapters 5 and 7) can be determined.

3.2.2. Exponential model

If the system is well mixed, $h(\tau)$ represents an exponential function (Figure 11):

$$h(\tau) = \frac{e^{-\tau/\tau_{\text{EM}}}}{\tau_{\text{EM}}} \tag{23}$$

where the average tracer transit time is $\tau_{\text{EM}} = \tau_m = M/q$. Combining Equation (23) with Equation (19′), it follows that

$$A_{\text{out}}(t) = \frac{1}{\tau_{\text{EM}}}\int_0^{\infty} A_{\text{in}}(t - \tau)e^{-(\lambda+1/\tau_{\text{EM}})\tau}d\tau \tag{24}$$

If the tracer input function is known, the average tracer transit time τ_{EM} (in the case of reservoirs, it is also called the *turnover time*) can be derived from Equation (24). For constant input function, Equation (24) becomes

$$A_{\text{out}}(t) = \frac{A_{\text{in}}}{1 + \lambda\tau_{\text{EM}}} \tag{25}$$

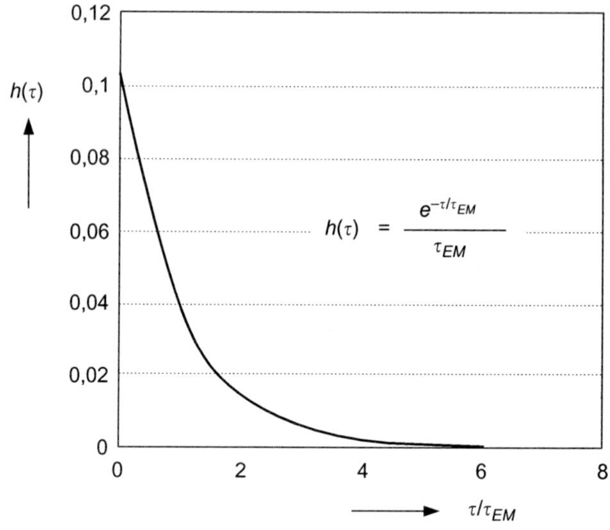

Figure 11 The exponential model.

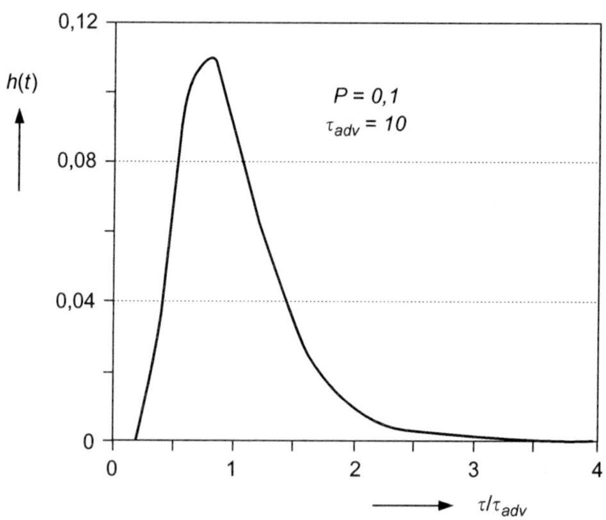

Figure 12 Dispersion model.

3.2.3. Dispersion model

Assuming that a mass flow along a flow path (e.g. groundwater in a confined aquifer) is subject to (hydro)dynamic dispersion, the system response function $h(\tau)$ is given by the expression:

$$h(\tau) = \frac{1}{\tau}\left(\frac{4\pi P\tau}{\tau_{\mathrm{adv}}}\right)^{-1/2} \exp\left[-\frac{(1-\tau/\tau_{\mathrm{adv}})^2}{4P\tau/\tau_{\mathrm{adv}}}\right] \tag{26}$$

where P is the apparent *dispersion parameter* (reciprocal of the *Peclet number*) and τ_{adv} the (average) advective travel time ($= x/v$). In Figure 12, the function $h(\tau)$ is shown for typical values of these parameters. For higher P values, the distribution of the transit time is wider and more asymmetrical. For constant radionuclide input concentration, the following expression can be derived from Equations (19′) and (26):

$$A_{\mathrm{out}}(t) = A_{\mathrm{in}} \exp\{(2P)^{-1}(1-(1+4\lambda P\tau_{\mathrm{adv}})^{1/2})\} \tag{27}$$

3.3. Conclusion

The concept of residence time distribution in terrestrial systems has shown wide application, especially in studies of hydrological systems (e.g. Maloszewski and Zuber, 1996). The given examples of lumped parameter models represent basic approaches in using radionuclides as tracers for time determinations of terrestrial systems. Further models include the linear model, the combined exponential–piston flow model (EPM) and the combined linear–Piston flow model (LPM) (Maloszewski and Zuber, 1996). Lumped parameter models are also available for variable flow conditions (Zuber et al., 1986). Furthermore, a variety of software packages have been developed for the use of these models, for example MULTIS (Richter et al., 1993) and BOXmodel (Zoellmann et al., 2001).

In all these models, the studied system is treated as a whole and the variations of parameters within the system are neglected. It has been emphasised (Zuber, 1994) that for constant radionuclide input, only single parameter models can provide unique solutions (PFM, EM). In general, the models combine de-convolution of radionuclide input with a system response function (also called weighting function) $h(\tau)$ to calculate the activity concentration of the radionuclide in the output of the system. Adjusting the response function to optimise the fit between measured and computed output concentration provides the values of the model parameters from which the values of the system-specific quantities can be calculated (e.g. groundwater flow rate, reservoir turnover time). It should be emphasised that the derived average transit time and average age values represent those of the particular radionuclide used in the study. Differences between the average transit time of the tracer and the traced substance ('radionuclide age' and water age, in the case of groundwater dating) may occur because of interaction of the tracer with matter in the system, for example with rocks of the aquifer, in the case of groundwater dating. For groundwater systems, a detailed discussion of the strengths and limitations of the residence time distribution concept has been given by Zuber (1994).

4. On the Definition of Environmental Tracers

For five decades, isotopes have been used as tracers to study environmental systems. While in the initial phase, the application of artificial radionuclides dominated, later the emphasis shifted to naturally occurring stable and radioactive nuclides, also known as environmental nuclides. They are broadly distributed in the environment and variations in their abundances can be used to determine pathways of mass transport and timescales of environmental processes.

An *ideal tracer* is a substance that behaves in the system exactly as the traced matter in the particular process to be studied. More precisely, a tracer is a minute amount of substance having similar physical and dynamic characteristics (to a satisfactory degree) as those of one of the major or minor components. Changes in its concentration do not affect the behaviour of the other components of the system and are measurable (Nir, 1986). The tracer must have at least one property that distinguishes it from the traced material (e.g. radioactive decay). This definition implies that an ideal tracer should have neither sources nor sinks in the system other than those to be studied. In practice, however, substances that have other sources or sinks are still regarded as tracers if these sources and sinks can be properly accounted for, or if their influence is negligible within the measurement accuracy (Zuber, 1994). An ideal tracer should move with the same velocity in the system (air, water, soil and sediment) as the material to be traced. For example, the tracer movement should not be retarded significantly by interaction with other substances (e.g. sorption, desorption processes in groundwater, soil and sediments). Tracers should not be confused with trace elements, which have no corresponding major or minor components; thus, the behaviour of trace elements depends on their own concentration and properties.

In general, environmental tracers that satisfy these criteria are environmental radionuclides and less abundant heavy isotopes of common elements in environmental substances. Two typical examples of radioactive environmental tracers for mass transport are the following:

(1) $^{14}CO_2$ satisfies the requirement of a tracer for the CO_2 dynamics in the atmosphere if it can be corrected for (or if it is justified to neglect) radioactive decay and possible isotope effects. These latter effects may appear at phase transitions and at physical boundaries such as the air–water interface. If CO_2 closely follows the air motion, $^{14}CO_2$ can readily be used to trace atmospheric air circulation.

Table 2 Environmental processes that lead to mass-dependent isotope fractionation of elements involved in these processes.

Physical processes	Chemical processes	Biological processes
Evaporation	Chemical reactions	Assimilation
Condensation	Sorption/desorption	Respiration
Diffusion	Ion exchange	Denitrification
Ultra-filtration		

(2) Tritium (T) as HTO has been used as a tracer of atmospheric moisture. In the period following atmospheric nuclear weapon tests, the mixing of higher T concentration stratospheric moisture with that evaporating from surface water bodies could be traced. Tritium can also trace trajectories of constant moisture air parcels, if these are not affected by mixing with other air masses. Furthermore, tritium is an excellent tracer of groundwaters.

There are a number of physical, chemical and biological processes in the environment that differentiate on a mass basis between the isotopes of the involved elements or compounds and thus lead to measurable changes in their isotopic compositions (Table 2). The elements experiencing measurable isotope fractionation and commonly used in hydrological studies include H, Li, B, C, N, O, S and Cl. In Chapter 5, the isotope fractionation of the elements of the water molecule is briefly discussed. Examples of the application of oxygen and hydrogen isotopic ratios to determine the origin and mixing of ground and surface water are given in Chapter 5, and in Chapter 8, the use of these isotopes as climate indicators is discussed.

In this book, emphasis is placed on the use of environmental radionuclides (Chapter 1) as tracers of environmental systems. The following chapters focus on the use of environmental radionuclides as tracers of processes in terrestrial systems, such as Chapters 3 and 4 (atmosphere), Chapter 5 (hydrosphere) and Chapters 6 and 7 (lithosphere). Their use as time-markers of climate changes in combination with stable isotopes as climate indicators is presented in Chapter 8.

REFERENCES

Dickin, A. P. (2005). *Radiogenic Isotope Geology*. Second ed., Cambridge University Press, Cambridge, UK.

Magill, J., and J. Galy. (2005). *Radioactivity Radionuclides Radiation (including the Universal Nuclide Chart on CD-ROM)*. Springer-Verlag, Berlin; Heidelberg, NY, ISBN 3-540-21116-0. p. 259.

Maloszewski, P. and A. Zuber. (1996). Lumped parameter models for interpretation of environmental tracer data. In: *Manual on Mathematical Models in Isotope Hydrogeology*. IAEA-TECDOC-910. IAEA, Vienna, pp. 9–50.

Nir, A. (1986). Role of tracer methods in hydrology as a source of physical information. Basic concepts and definitions. Time relationship in dynamic systems. In: *Mathematical Models for Interpretation of Tracer Data in Groundwater Hydrology*. IAEA-TECDOC-381. IAEA, Vienna, pp. 7–44.

Richter, J., P. Szymczak, and P. Jordan. (1993). Use of combination of lumped parameter models to interpret groundwater isotopic data. *Journal of Contaminant Hydrology*, 14, 1–13.

Turner, J. E. (2007). *Atoms, Radiation and Radiation Protection (Physics Textbook)*. Wiley-VCH, Berlin, ISBN-13: 978-3-527-40606-7. p. 586.

Zoellmann, K., W. Kinzelbach, and C. Fulda. (2001). Environmental tracer transport (^3H and SF_6) in the saturated and unsaturated zones and its use in nitrate pollution management. *Journal of Hydrology*, 240(3–4), 187–205.

Zuber, A. (1994). On calibration and validation of mathematical models for the interpretation of environmental tracer data in aquifers. In: *Mathematical Models and Their Applications to Isotope Studies in Groundwater Hydrology*. IAEA-TECDOC 777. IAEA, Vienna, pp. 11–41.

Zuber, A., P. Maloszewski, A. Herrmann, and W. Stichler. (1986). Tracer relations in variable flow. In: The 5th International Symposium on Underground Water Tracing, IGME (Institute of Geology and Mineral Exploration), Athens, pp. 355–360.

CHAPTER 3

Radionuclides as Tracers of Atmospheric Processes

Ivan Sykora[1,*] *and* Klaus Froehlich[2]

Contents

*Corresponding author. Tel.: +421-2-60295-458
E-mail address: Ivan.Sykora@fmph.uniba.sk

[1] Department of Nuclear Physics and Biophysics, Faculty of Mathematics, Physics and Informatics, Comenius University, Mlynská dolina, 842 48 Bratislava, Slovakia
[2] Viktor-Wittner-Gasse 36/7, 1220 Vienna, Austria

Radioactivity in the Environment, Volume 16
ISSN 1569-4860, DOI 10.1016/S1569-4860(09)01603-9

1. Introduction

For more than three decades, atmospheric radionuclides have been used to trace atmospheric transport and exchange processes (Turekian and Graustein, 2003; Gäggeler, 1995; Baeza et al., 1996; Arimoto et al., 1999). There are three principal sources of atmospheric radioactivity: radon isotopes and their progeny, cosmic-ray-produced radionuclides (cosmogenic radionuclides) and anthropogenic radionuclides. Chapter 1 of this volume extensively discusses the production of these radionuclides. This chapter is focused on their atmospheric behaviour, distribution and applications.

Radon isotopes are members of the natural decay series (Chapter 1), that is the ^{238}U decay series (^{222}Rn, half-life 3.8 days), the ^{232}Th decay series (^{220}Rn, also called thoron, half-life 56 s) and the ^{235}U decay series (^{219}Rn, also called actinon, half-life 3.9 s). Almost all radon in the atmosphere is produced in soils and rocks by radioactive decay of the respective precursor (Table 1), from which it is released and transported to the atmosphere by diffusion. In atmospheric studies using radon isotopes, ^{222}Rn plays a dominant role, as its longer half-life facilitates a greater diffusive transport and an atmospheric concentration that is 100 times higher than that of ^{220}Rn. The atmospheric activity concentration of ^{219}Rn is relatively negligible.

Table 2 is a compilation of the cosmogenic radionuclides most relevant to atmospheric studies. Given the particularly wide spectrum of ^{14}C applications in atmospheric and other studies, this radionuclide is discussed in a dedicated chapter of this volume (Chapter 4).

Anthropogenic radionuclides contributing to atmospheric radioactivity include the gaseous radionuclides tritium (^{3}H), radiocarbon (^{14}C, incorporated in $^{14}CO_2$), ^{85}Kr, ^{37}Ar, ^{133}Xe and a number of others such as ^{137}Cs and ^{90}Sr, ^{131}I, ^{144}Ce, ^{106}Ru, and radioactive isotopes of plutonium. The sources, production and release to the atmosphere of these radionuclides are discussed in Chapter 1 of this volume. Here, we focus on their atmospheric distributions. Further applications of these radionuclides are described in Chapters 5 and 7 of this volume.

2. Atmospheric Radionuclides

2.1. Radon and its decay products

About 99% of ^{222}Rn in the atmosphere originates in soils and rocks, where it is produced by radioactive decay of its parent nuclide ^{226}Ra. Once released to the atmosphere, it remains there until its radioactive decay. As it

Table 1 Parts of the natural decay series starting with the respective radium isotopes (see also Table 2 in Chapter 1).

Nuclide	Half-life	Radiation
Uranium series		
^{226}Ra	1,622 years	α,γ
^{222}Rn	3.8 days	α,γ
^{218}Po	3.05 min	α
^{214}Pb	26.8 min	β,γ
^{218}At	1.5–2.0 s	α
^{214}Bi	19.7 min	β,α
^{214}Po	1.64×10^{-4} s	α,γ
^{210}Tl	1.3 min	β,γ
^{210}Pb	22 years	β,γ
^{210}Bi	5.0 days	β,α
^{210}Po	138 days	α,γ
^{206}Tl	4.2 min	β
^{206}Pb	Stable	
Thorium series		
^{224}Ra	3.64 days	α,γ
^{220}Rn	55.6 s	α,γ
^{216}Po	0.145 s	α
^{212}Pb	10.6 h	β,γ
^{212}Bi	60.5 min	β,α,γ
^{212}Po	3.04×10^{-7} s	α
^{208}Tl	3.1 min	β,γ
^{208}Pb	Stable	
Actinium series		
^{223}Ra	11.4 days	α,γ
^{219}Rn	4.0 s	α,γ
^{215}Po	1.77×10^{-3} s	α,β
^{211}Pb	36.1 min	β,γ
^{215}At	$\sim 10^{-4}$ s	α
^{211}Bi	2.16 min	α,β,γ
^{211}Po	0.52 s	α,γ
^{207}Tl	4.79 min	β,γ
^{207}Pb	Stable	

is a chemically inert gas, physical or chemical processes cannot remove it from the atmosphere. Therefore, it is suitable to trace atmospheric mass transport and to identify air masses derived from the continental boundary layer. Furthermore, atmospheric ^{222}Rn is the source of its decay products (^{214}Bi, ^{214}Pb, ^{214}Po, ^{210}Pb, ^{210}Bi and ^{210}Po) in the atmosphere.

Table 2 Production rates and global inventory of cosmogenic radionuclides (from Turekian and Graustein, 2003).

Nuclide	Half-life	Production rate (atoms $cm^{-2} s^{-1}$)		Global inventory
		Troposphere	Total atmosphere	
^{3}He	Stable	6.7×10^{-2}	0.2	3.2×10^{3} t
^{10}Be	1.5×10^{6} years	1.5×10^{-2}	4.5×10^{-2}	260 t
^{26}Al	7.1×10^{5} years	3.8×10^{-5}	1.4×10^{-4}	1.1 t
^{81}Kr	2.3×10^{5} years	5.2×10^{-7}	1.2×10^{-5}	8.5 kg
^{36}Cl	3.0×10^{5} years	4×10^{-4}	1.1×10^{-3}	15 t
^{14}C	5,730 years	1.1	2.5	75 t
^{39}Ar	268 years	4.5×10^{-3}	1.3×10^{-2}	52 kg
^{32}Si	140 years	5.4×10^{-5}	1.6×10^{-4}	0.3 kg
^{3}H	12.36 years	8.4×10^{-2}	0.25	3.5 kg
^{22}Na	2.6 years	2.4×10^{-5}	8.6×10^{-5}	1.9 g
^{35}S	87 days	4.9×10^{-4}	1.4×10^{-3}	4.5 g
^{7}Be	53 days	2.7×10^{-2}	8.1×10^{-2}	3.2 g
^{37}Ar	35 days	2.8×10^{-4}	8.3×10^{-4}	1.1 g
^{33}P	25.3 days	2.2×10^{-4}	6.8×10^{-4}	0.6 g
^{32}P	14.3 days	2.7×10^{-4}	8.1×10^{-4}	0.4 g

2.1.1. Flux of radon from soil to the atmosphere

In most soils, ^{226}Ra is nearly in secular equilibrium with its precursor ^{238}U, that is their specific activities are nearly equal. The specific activities of ^{238}U in soils range from about 10 to 50 Bq kg^{-1}. The emission of an alpha particle during decay of a ^{226}Ra atom causes a slight displacement of the product ^{222}Rn atom in the soil particle. Although this displacement is only a few tens of a nanometre, ^{222}Rn atoms close to the surface can escape into the pore space of the soil. The fraction of the total produced ^{222}Rn escaping into soil air depends on the specific surface of the soil particles and thus on the grain size distribution of the soil. Typical values found for this emanating fraction are between 1% and about 25% (Graustein and Turekian, 1990). Radon in the soil air of the top layer is transported by molecular diffusion to the surface where it escapes to the atmosphere.

Numerous determinations of the ^{222}Rn flux from the continents have been made (Wilkening et al., 1975). It has been found that the ^{222}Rn flux at the soil surface depends on the ^{226}Ra content of the soil, the distribution of the soil grain size and the diffusion resistance in the soil zone. The ^{222}Rn flux can be determined directly by measuring the ^{222}Rn concentration increase with time under an inverted cup placed above the soil surface. Alternatively, it can be derived from measured profiles of ^{222}Rn in the soil (e.g. Dörr and Münnich, 1989). A simple solution of the diffusion equation for ^{222}Rn in soil air is given by the expression:

$$C = C_{\infty}(1 - e^{-z/\bar{z}}) \tag{1}$$

where $\bar{z} = \sqrt{D/\lambda}$ is the relaxation depth, D the diffusion coefficient in soil, λ the ^{222}Rn decay constant, and C_{∞} the equilibrium concentration ($z \gg \bar{z}$). Figure 1 shows ^{222}Rn concentration profiles measured in sandy forest soil and clayey soil. With the given values for relaxation depth and equilibrium concentration (Figure 1), the theoretical curves satisfactorily reproduce the measurements. For the clayey soil, the radon flux to the atmosphere has been calculated to be 4,200 dpm m^{-2} h^{-1}, which corresponds to 9.2×10^3 atoms m^{-2} s^{-1} (Dörr and Münnich, 1989).

Measurements of soils from different sites in Germany gave values between about 2×10^3 and 1.3×10^4 atoms m^{-2} s^{-1}. Clayey soil showed higher values than sandy soils. Furthermore, the authors showed that seasonal variations of the soil moisture content give rise to seasonal variations in the ^{222}Rn flux (higher soil moisture corresponds to lower radon flux). For US soils, the mean ^{222}Rn flux has been found to be between 1.5×10^4 and 2×10^4 atoms m^{-2} s^{-1} (Turekian and Graustein, 2003). The frequency distribution of mean ^{222}Rn flux values around the

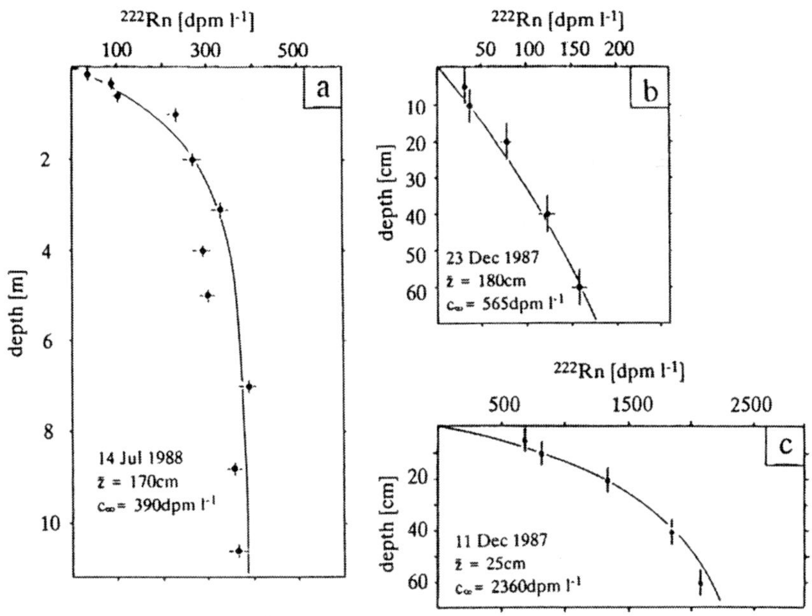

Figure 1 ^{222}Rn concentration profiles in different soils and depth resolution. Samples for profiles a and b have been taken from sandy soil, profile c is from clayey soil. (Adapted from Dörr and Münnich, 1989 with permission from Elsevier.)

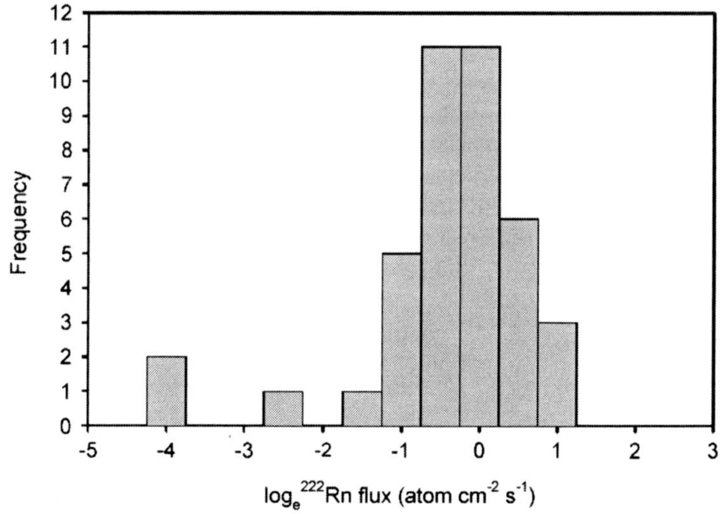

Figure 2 Frequency distribution of mean ^{222}Rn flux values for 40 studies from around the world using direct flux measurement methods. (Adapted from Conen, 2004 with permission from WMO/GAW.)

world (Figure 2) reveals a large variability. Considering the use of the ^{222}Rn flux in global atmospheric chemistry models, it has been suggested that a value of 1×10^4 atoms cm^{-2} s^{-1} be adopted for the latitude range from 60°S to 30°N, with a linear decrease from 1×10^4 to 0.2×10^4 atoms cm^{-2} s^{-1} for the range from 30°N to 70°N (Conen, 2004). The atmospheric ^{222}Rn activity concentration ranges from about 1 to 100 Bq m^{-3} (corresponding to about 10^4–10^6 atoms ^{222}Rn per m^3 air).

The concentration of ^{226}Ra in surface water of the oceans is at least three orders of magnitude lower than in soils. Since the ^{222}Rn flux to the atmosphere is proportional to the ^{226}Ra concentration, it can be concluded that the oceans are not an important source of atmospheric ^{222}Rn (Turekian and Graustein, 2003).

2.1.2. Short-lived decay products of ^{222}Rn

Table 1 shows the short-lived radionuclides following the decay of ^{222}Rn through the long-lived ^{210}Pb. The half-life of ^{218}Po, the first of these radionuclides, is long enough that newly formed ^{218}Po atoms can encounter the surface of aerosol particles to which they attach. If the residence time of these aerosol particles is sufficiently longer than the half-life of the ^{218}Po decay products, ^{214}Pb and ^{214}Bi, they will reach radioactive equilibrium. In fact, measurements of ^{214}Pb and ^{214}Bi in boundary layer air indicated radioactive equilibrium and, thus, aerosol residence time longer than a few hours (Turekian et al., 1999).

2.1.3. ^{210}Pb and its decay products

Atmospheric ^{210}Pb is mainly produced within the atmosphere by decay of ^{222}Rn; its direct precursor is ^{214}Po (Table 1). The produced ^{210}Pb atoms are attached to aerosols in the diameter size range between 0.1 and 0.5 μm, the so-called 'accumulation mode' (see Section 3.2). For particles of this size, precipitation is the main mechanism of their removal from the atmosphere. ^{210}Pb is a minor constituent of this aerosol type (one ^{210}Pb-containing aerosol particle in about 10^4 aerosol particles) and, thus, it is useful to trace their atmospheric transport and to determine their atmospheric residence time.

2.1.4. ^{220}Rn and its decay products

Like ^{222}Rn, thoron (^{220}Rn) diffuses from topsoil into the atmosphere. Typical ^{220}Rn flux values are at 100 atoms $m^{-2} s^{-1}$, which is about 100-fold lower than for ^{222}Rn. The decay of ^{220}Rn produces additional airborne radioactive atoms, including ^{216}Po, ^{212}Pb and ^{212}Bi. Most of the activity of the radon progeny is associated with small diameter particles from 6 to 200 nm. Because of its very short half-life (0.145 s), the activity concentration of ^{216}Po closely tracks the activity concentration of its immediate parent radionuclide thoron (secular equilibrium). Thoron has a much shorter half-life than ^{212}Pb (55.6 s vs. 10.6 h), so thoron does not mix to as high in the atmosphere as does ^{212}Pb. While ^{220}Rn and ^{216}Po are limited to heights above ground of some tens of metres, their decay products ^{212}Pb and ^{212}Bi reach heights up to a few thousand metres (Gat and Assaf, 1968). On the other hand, since they are concentrated in a much smaller volume of space, ^{220}Rn and ^{216}Po have a much higher activity concentration near the ground, which is between 0.01 and 1 Bq m^{-3}.

2.2. Cosmogenic radionuclides

Lal et al. (1958), Lal and Peters (1962, 1967), Masarik and Beer (1999) and Nagai et al. (2000) have published reviews on the production and distribution of cosmogenic radionuclides. Table 2 is a compilation of the cosmogenic radionuclides, of which ^{14}C (Chapter 4), ^{3}H, ^{7}Be and ^{10}Be are the most frequently used in atmospheric studies.

2.2.1. Tritium

The activity concentration of tritium ($^{3}H = T$) is often expressed in tritium units (TU), where 1 TU = one tritium atom per 10^{18} hydrogen atoms; this corresponds to a tritium activity concentration of 0.12 Bq l^{-1} water. In the atmosphere, except the upper stratosphere, the dominating species containing tritium is HTO, mainly in the form of water vapour (e.g. Chamberlin, 1991; NCRP, 1979). Most atmospheric tritium is washed out by precipitation. Due to isotopic enrichment, the T/H ratio in liquid water is 1.1 times higher than in water vapour. A minor part of tritium occurs as HT; the ratio of HT/HTO is about 10^{-4} (Chamberlin, 1991). The natural tritium concentration in precipitation depends on latitude; in Polar Regions, it is about $5\,Bq\,l^{-1}$ and at latitudes of 45° about $0.6\,Bq\,l^{-1}$. This distribution is governed by the stratosphere–troposphere exchange (STE) of the air masses (Gäggeler, 1995). The atmospheric tritium distribution shows a so-called continental effect, characterised by an increase of the tritium concentration with distance from the coast. At coastal stations, the tritium values are close to that of sea surface water.

2.2.2. Beryllium-7

The decay of ^{7}Be (half-life 53 days) to ^{7}Li by electron capture is associated with the emission of a 478 keV gamma ray that facilitates measurement of its activity by low-level germanium detector (Chapter 9). Lal and Peters (1962, 1967) have shown that the ^{7}Be production rate decreases with atmospheric depth (Figure 2). Most ^{7}Be resides in the stratosphere. Its production rate reaches a maximum in the upper stratosphere at about 20 km (Masarik and Beer, 1999) and decreases with decreasing altitude down to ground level due to STE (e.g. Stohl et al., 2000). The relatively high production rates of ^{7}Be in the upper troposphere, combined with its transport from the lower stratosphere to the upper troposphere, normally maintain a steep vertical concentration gradient between the upper and lower troposphere (Feely et al., 1989).

The source functions of cosmogenic radionuclides depend mainly on latitude and altitude (Benioff, 1956; Lal et al., 1958; Lal and Peters, 1962; O'Brien, 1979; Masarik and Beer, 1999). Considering ^{7}Be, only 33% is produced in the troposphere, particularly in the upper troposphere, while the rest is produced in the stratosphere (Figure 3). There is an anticorrelation between the long-term variation of the ^{7}Be air concentration and variation of solar activity (Cannizzaro et al., 2004; Al-Azmi et al., 2001; Durana et al., 1996; Ioannidou et al., 2005). Figure 4 shows these variations at ground level in Palermo, which is located at 38°7′N and 13°21′E (Cannizzaro et al., 2004). Altogether more than 4,600 air samples have been analysed, from which a mean value of the daily ^{7}Be concentrations of $5.06\,mBq\,m^{-3}$ has been derived. The monthly mean ^{7}Be concentration ranged from $2.16\,mBq\,m^{-3}$ in December 1990 to $10.57\,mBq\,m^{-3}$ in July 1998. Lower values are observed during summer; they have been attributed to reduce washout of beryllium from the atmosphere (lower rainfall in the summer season) and seasonal variability of vertical exchange. However, measurements of the fluxes of galactic cosmic-ray particles using balloons, satellites and surface-based neutron monitors

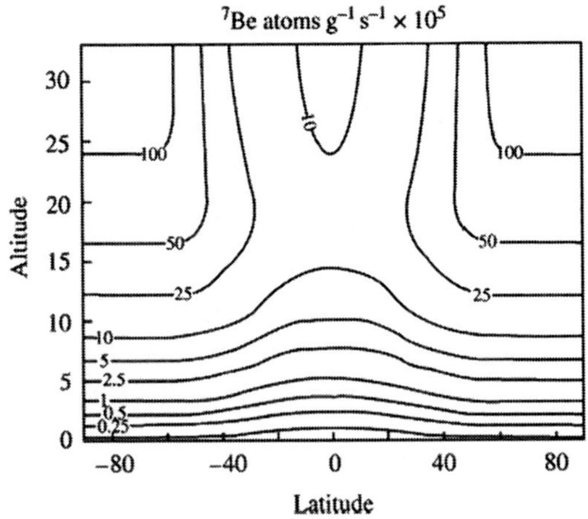

Figure 3 Production of ^{7}Be in the atmosphere as a function of latitude and altitude. (Adapted from Turekian and Graustein, 2003 with permission from Elsevier.)

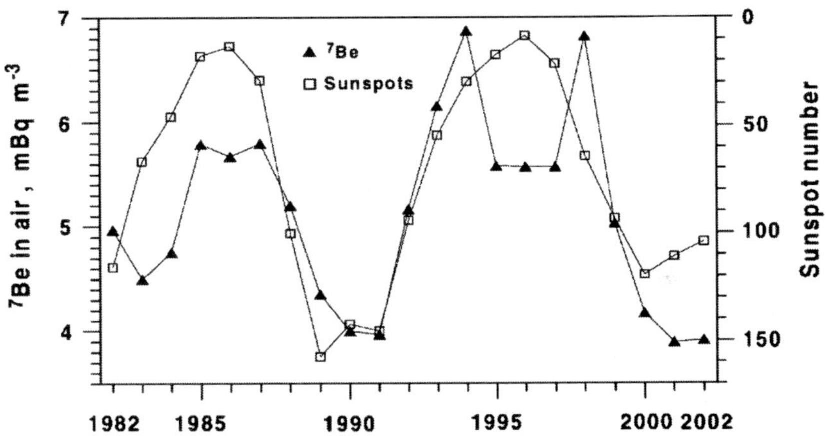

Figure 4 Annual averages of ^{7}Be concentration and sunspot number for the period 1982–2002 (ordinate scales in opposite direction). (Adapted from Cannizzaro et al., 2004 with permission from Elsevier.)

have shown that modulation processes are more complicated and cannot be described simply by an anticorrelation with solar activity.

2.2.3. Beryllium-10

^{10}Be (half-life 1.5×10^6 years) decays by internal conversion to ^{10}B emitting a high-energy beta particle. Its concentration is measured by accelerator mass spectrometry (Chapter 9).

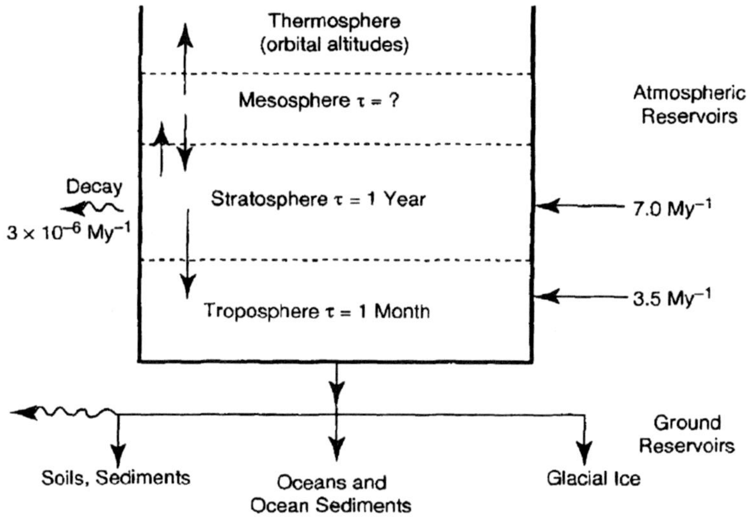

Figure 5 Cycling of cosmogenic ^{10}Be in the atmospheric layers (adapted from McHargue and Damon, 1991). Inventories are given in moles per year ($mol\,yr^{-1}$), and τ is the residence time. (Adapted from Gregory, 1996 with permission from Elsevier.)

The rate of ^{10}Be deposition to the Earth's surface is the same as the rate of its production in the atmosphere (4.5×10^2 atoms $m^{-2}\,s^{-1}$). The ratio of the production rates $^{10}Be/^7Be$ is approximately 0.5 (O'Brien, 1979). About 70% of the atmospheric production of both isotopes occurs in the stratosphere. The atmospheric distribution of these beryllium isotopes is controlled by their intrusion into the troposphere at about 40–50° latitude. Thus, their deposition to the Earth's surface increases during (or immediately following) seasonal weather disturbances of the stability of the tropopause.

Figure 5 shows the cycling of cosmogenic ^{10}Be in the Earth's surface layers. As the residence time of aerosols in the stratosphere is more than 1 year and in the troposphere less than 1 month (Figure 4), the $^{10}Be/^7Be$ ratio in the stratosphere is higher than in the troposphere. The $^{10}Be/^7Be$ ratio is therefore useful to indicate stratospheric air injected to the troposphere and to distinguish between stratospheric ozone and pollution-based tropospheric ozone.

2.3. Anthropogenic radionuclides

The main sources of production and release to the atmosphere of anthropogenic radionuclides are nuclear bomb explosions and nuclear facilities (reactors, reprocessing plants).

2.3.1. Nuclear weapons tests

Between 1945 and 1980, more than 500 nuclear tests were carried out that injected radioactive debris into the atmosphere on a global scale. Their atmospheric input was higher in the northern than in the southern hemisphere, because most of the atmospheric tests took place at northern latitudes (Chapter 1). Non-gaseous radionuclides attach to aerosols and participate with them in atmospheric processes. Radionuclides with half-lives shorter than the stratospheric residence time of aerosols (about 2 years) are decayed before reaching the troposphere. Concerning atmospheric injection and deposition, the radioactive debris can be divided into three groups: (1) deposition near the explosion, (2) injected into the troposphere and (3) injected into the stratosphere. Local deposition is mainly due to surface explosions (80% of the radioactive debris) and explosions at the sea surface (20%); merely 10% are deposited near atmospheric explosions. The fraction of radionuclides injected by atmospheric explosions into the troposphere is only about 5%, but it is of significance for the deposition of short-lived radionuclides such as ^{131}I and their human impact. The stratospheric component stems from high-yield (megaton) atmospheric nuclear explosions and represents the principal source of global contamination with long-lived radionuclides. The contribution to the stratospheric inventory of explosions in the kiloton range is insignificant, because in this case the nuclear debris do not penetrate into the stratosphere.

2.3.2. Operation of nuclear reactors

In the early years of nuclear reactor operation, some of the nuclear waste was directly deposited in the environment. Volatile gases from chemical processing were released into the atmosphere, sometimes without filtering. As an example, through the US nuclear programme, about 3.7×10^{16} Bq ($>99\%$ ^{131}I) were released into the atmosphere by reprocessing plants and about 2.8×10^{18} Bq ($>99\%$ ^{41}Ar) by production reactors (U.S. Department of Energy, 2001). In modern reactors (boiling water reactors), gases pass through refrigerated activated charcoal beds that remove the remaining radionuclides and noble gases and store them to allow for radioactive decay. Another possibility is to compress the gases in tanks and store them accordingly (Eisenbud and Gesell, 1997). After 3 months hold-up, ^{131}I is transmuted into a stable nuclide. The remaining radioactivity is mainly represented by ^{85}Kr, which is either directly released into the atmosphere or cryogenically condensed for long-term storage.

2.3.3. Nuclear reactor accidents

Accidents of nuclear reactors can be an important source of atmospheric radioactivity. In the past, three nuclear accidents with substantial releases of

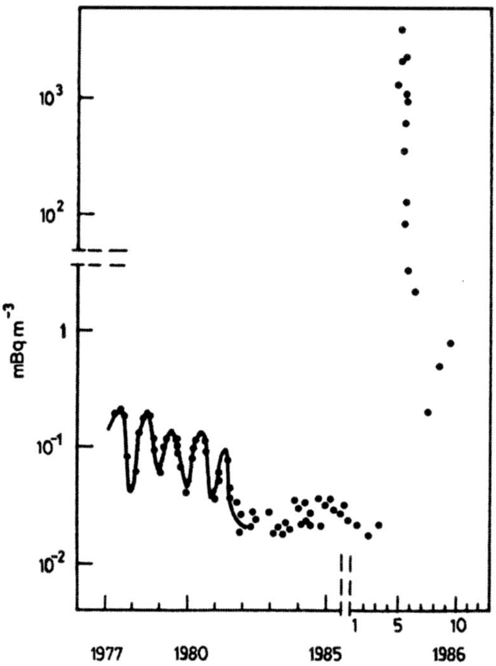

Figure 6 Increase of ^{137}Cs concentration in the atmosphere in Bratislava (Slovakia) after the Chernobyl accident (Povinec et al., 1988).

radionuclides into the atmosphere happened: Windscale, England (1957); Three Mile Island, PA, USA (1979) and, the largest accident, Chernobyl (Ukraine, former Soviet Union, 1986) (Chapter 1 of this volume). During the Chernobyl accident, the reactor core was destroyed and all the inventory of gaseous fission products released into the atmosphere (OECD, 1995). Subsequently, after the Chernobyl accident, in Bratislava, Slovakia for example, the concentration of ^{137}Cs reached a level 10 times higher than the maximum in 1963 caused by atmospheric nuclear bomb tests (Figure 6). In addition, the concentrations of other long-lived radionuclides, particularly ^{90}Sr and ^{85}Kr, greatly increased.

2.3.4. Fuel reprocessing

The spent fuel from nuclear reactors is chemically treated in nuclear reprocessing plants in order to recover the remaining ^{235}U and the amount of ^{239}Pu that has been produced during burning the fuel in the reactor. During the reprocessing, radionuclides are released into the environment including the atmosphere. The atmospheric releases include ^{129}I, ^{131}I and the radioactive gases ^{85}Kr and ^{3}H. Further details on these releases are discussed in Section.4.3 of Chapter 1 and by NRC (1996).

2.3.5. Man-made tritium

After the release of tritium in a nuclear bomb explosion (Chapter 1), it is rapidly incorporated into water molecules and enters with the water vapour into the hydrological cycle. Due to nuclear bomb tests in the late 1950s and early 1960s, the tritium content of precipitation and surface waters in temperate zones of the northern hemisphere increased substantially; the transfer of bomb-tritium into the southern hemisphere was comparatively low. The atmospheric tritium input due to thermonuclear explosions was estimated to be about 1.9×10^{20} Bq for the northern hemisphere and about 0.5×10^{20} Bq for the southern hemisphere (UNSCEAR 2000). Since the atmospheric test ban in 1963, tritium decreased in precipitation and, since the 1990s, reached nearly the natural level (cosmogenic tritium). The French and Chinese tests between 1968 and the early 1980s modified the global decrease of atmospheric tritium slightly. Other sources of atmospheric tritium include nuclear facilities and industrial applications of tritium, especially the production of luminous watches (Gäggeler, 1995).

3. Behaviour of Radionuclides in the Atmosphere

The following focuses on the distribution of the most frequently used radionuclides including ^{222}Rn and its long-lived decay products ^{210}Pb and ^{7}Be. The distribution of other airborne radionuclides is considered in the context of discussion of the behaviour of atmospheric aerosols to which the radionuclides attach.

3.1. Distribution of ^{222}Rn in the atmosphere

The vertical distribution of radon over the continents depends on its supply from soils, upward transport (mainly turbulent diffusion) and radioactive decay. The atmospheric ^{222}Rn concentration reaches its highest value near the land surface and decreases with both altitude and distance from land (Turekian and Graustein, 2003). Jacobi and Andre (1963) have developed a one-dimensional turbulent-diffusion model which reproduces fairly well the measured vertical ^{222}Rn profiles over mid-continental regions, where vertical transport dominates (Figure 7). Turekian et al. (1977) modelled the large-scale atmospheric circulation for ^{222}Rn and its long-lived daughters (Figure 8). The model takes into account the sharp discontinuity of the ^{222}Rn flux between the continents and oceans as well as the air-mass transit time and the mean residence time of aerosols.

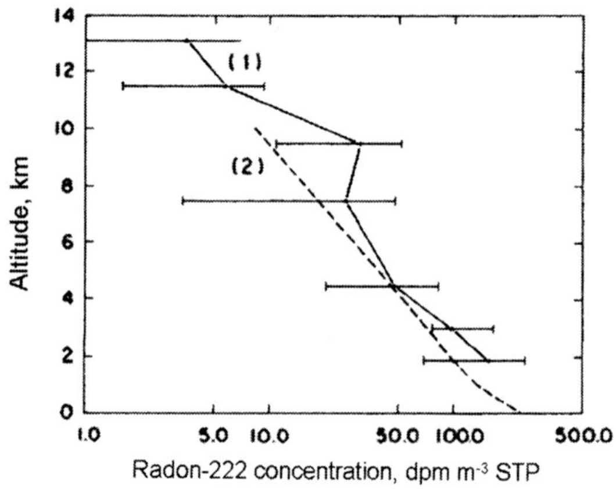

Figure 7 Vertical profile of ^{222}Rn in the atmosphere: (1) the mean of six profiles over west-central United States, (2) profile calculated by Jacobi and Andre (1963). The error bars represent one standard deviation from the mean. (Adapted from Moore et al., 1973 with permission from Elsevier.)

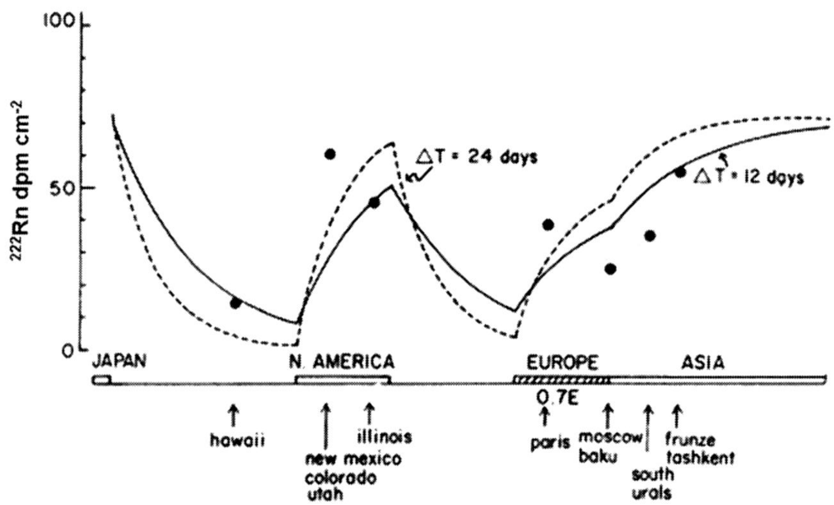

Figure 8 Distribution of ^{222}Rn in the atmosphere of the mid-northern hemisphere, calculated for a ^{222}Rn soil flux of 1.2×10^4 atoms cm^{-2} s^{-1}. $\Delta T =$ transit time across the North Pacific Ocean (9,000 km). (Adapted from Turekian et al., 1977 with permission from Elsevier.)

(The residence time of an aerosol particle is defined as the time the particle will be retained in a region of interest before disappearance or removal by any process.) The parameters ΔT (transit time across the North Pacific Ocean) and τ_R (aerosol mean residence time) are coupled. For $\Delta T = 12$

days, the aerosol mean residence time is $\tau_R = 5$ days, while $\Delta T = 24$ days is compatible with $\tau_R = 18$ days. The combination of $\Delta T = 12$ days and $\tau_R = 5$ days appeared to be reasonable, as confirmed by the air trajectories determined by other means (Turekian et al., 1977).

3.2. Distribution of ^{210}Pb in the atmosphere

Numerous vertical profiles have also been measured for ^{210}Pb and its relatively long-lived decay products ^{210}Bi (half-life 5 days) and ^{210}Po (half-life 138 days). Evaluating these profiles, it must be taken into account that – in contrast to ^{222}Rn as an inert gas – ^{210}Bi and ^{210}Po are attached to aerosol particles and thus subject to removal from the atmosphere by wet and dry deposition. Figure 9 shows the measured and the theoretical vertical distributions of ^{210}Pb for the same region as for Figure 7. The theoretical curve has been calculated with the same model used to evaluate the ^{222}Rn

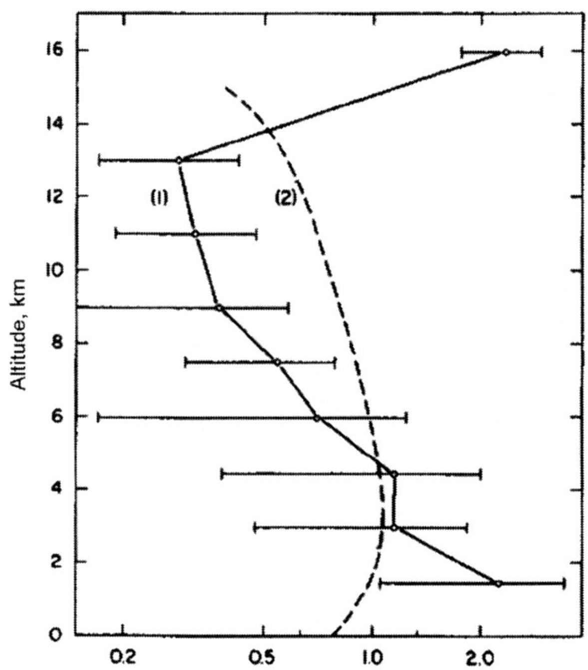

Figure 9 Vertical distribution of ^{210}Pb in the atmosphere in mid-continental region of the United States. (1) the mean over 12 profiles, (2) profile calculated by Jacobi and Andre (1963). The error bars represent one standard deviation from the mean. (Adapted from Moore et al., 1973 with permission from Elsevier.)

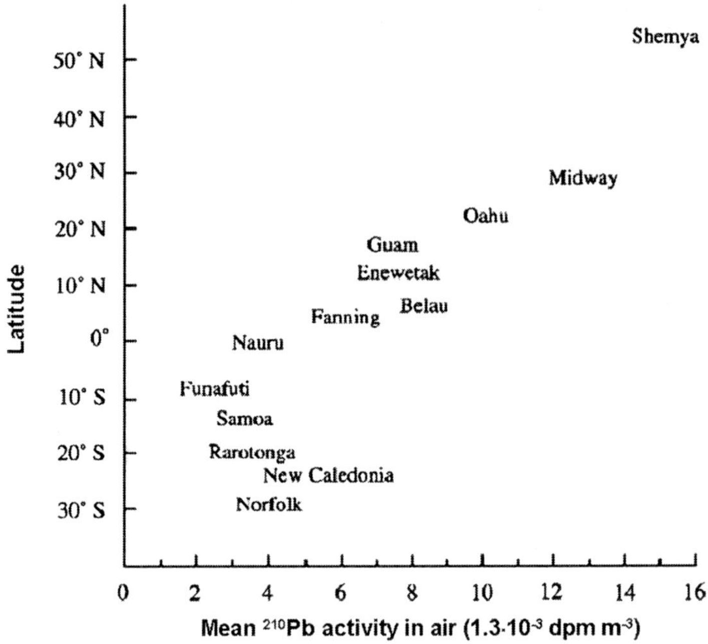

Figure 10 Distribution of ^{210}Pb in air at stations of the SEAREX network. The site name is centred over the point representing the data. The length of the name approximates to the error of the mean. (Adapted from Turekian and Graustein, 2003 with permission from Elsevier.)

data except that two additional terms are included in the turbulent-diffusion equation, which account for the removal of aerosol-attached ^{210}Pb and its production by decay of ^{222}Rn (Jacobi and Andre, 1963). The discontinuity in the measured ^{210}Pb concentration above the tropopause points to an abrupt decrease of the removal rate at this altitude. The higher ^{210}Pb concentration near the ground is due to 'recycling' of ^{210}Pb from the continental surface.

Figure 10 shows an example of the spatial distribution of ^{210}Pb in air over the Pacific. The sampling sites belong to the network of the SEAREX (sea/air exchange) programme. The increase of the concentration with latitude north correlates with the size of the land-mass upwind from the sampling sites (Turekian and Graustein, 2003).

3.3. Distribution of ^7Be in the atmosphere

Measurements at different locations carried out by Papastefanou and Ioannidou (1995), Takayuki et al. (1996), El-Hussein et al. (2001) and Gerasopoulos et al. (2001) have shown that the ^7Be concentration in air at ground level depends on latitude, altitude and on local meteorological conditions. At middle latitude, the ^{7}Be values are generally higher than at high latitudes (Al-Azmi et al., 2001; Aldahan et al., 2001; Baeza et al., 1996; Feely et al., 1989; Todorovic et al., 1999). In Bratislava (latitude about 48°N), measurements from 2001 to 2005 yielded a monthly average ^{7}Be activity concentration in ground air of $2.85\,mBq\,m^{-3}$ (Sykora et al., 2005); during the period from 1981 to 1995, the monthly average was $3.12\,mBq\,m^{-3}$ at nearly the same location (Durana et al., 1996). These values are remarkably lower than the value of $5.06\,mBq\,m^{-3}$ derived from daily measurements during the period 1982 to 2002 in Palermo, Italy (39°N), and the value of $5.21\,mBq\,m^{-3}$ obtained in Kuwait (29°N) during 1994–1998 by Al-Azmi et al. (2001). The larger values at the latter two stations, which are located close to the sea, may be attributed to enhance vertical air mass exchange at coastal stations compared to inland stations such as Bratislava. Figure 11 illustrates the latitudinal effect of ^{7}Be measured at various ground-level stations.

The weekly concentrations of ^{7}Be show clear seasonal changes with high values during spring–summer and low values in autumn–winter. The amplitude of the seasonal pattern may also vary from year to year at the same latitude. The average monthly values were found to be generally higher at southern stations (at about 56°S) than at the same latitude of northern stations (about 68°N). Therefore, the atmospheric distribution of ^{10}Be and ^{7}Be can be used to trace the global transport of air masses. For example Dibb et al. (1997) used ^{7}Be data to evaluate the outflow of continental air into the atmosphere of the western Pacific region. From 1987 to 2001, Ioannidou et al. (2005) measured ^{7}Be concentrations in

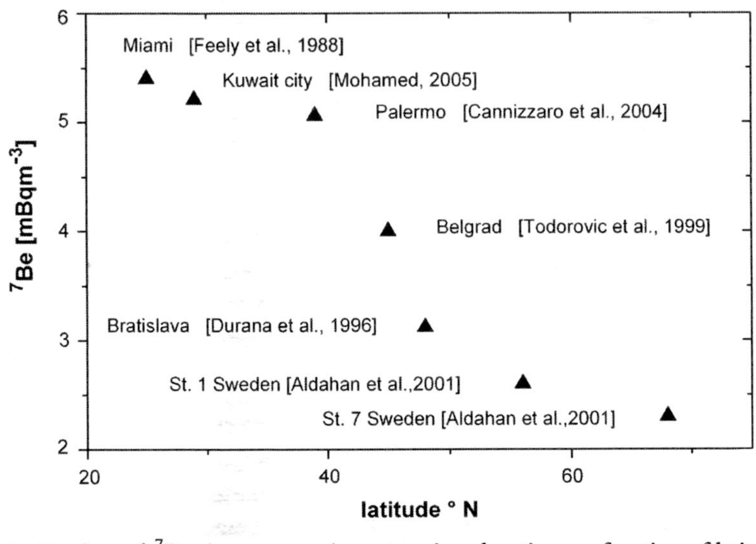

Figure 11 Annual ^{7}Be air concentrations at various locations as function of latitude.

ground-level air at Thessaloniki, Northern Greece (40°38′N, 22°58′E), and found values ranging from 0.47 to 12.7 mBq m^{-3} and calculated a mean activity concentration of 5.02 ± 2.49 mBq m^{-3} which is characteristic for the latitude of 40°N. At Miami (25°N), the annual ^{7}Be concentration during 1994 to 1998 was found to be 5.4 mBq m^{-3} (Hernández Alonso et al., 2004). The global average tropospheric ^{7}Be is 12.5 mBq m^{-3} (UNSCEAR, 2000).

Apart from the effects of latitude and altitude, the meteorological conditions are of significant influence on ^{7}Be in ground-level air. The highest ^{7}Be concentrations reached during the warm season appear to be caused by the stronger vertical transport of air masses during this season. Furthermore, in mid-latitudes (above 40°N), the tropopause is higher in summer (Gerasopoulos et al., 2001) than in winter. During July and August, the tropopause rises to maximum heights of 18 km (average values is about 12 km) (Ioannidou et al., 2005). In winter, the atmosphere is more stable and the tropopause is lower, which causes lower ^{7}Be values in surface air. During the spring period, the ^{7}Be concentration increases significantly due to STE processes. Figure 12 shows the mean monthly atmospheric concentrations of ^{7}Be versus temperature, and Figure 13 gives an example of the seasonal variation of the ^{7}Be concentration and meteorological parameters.

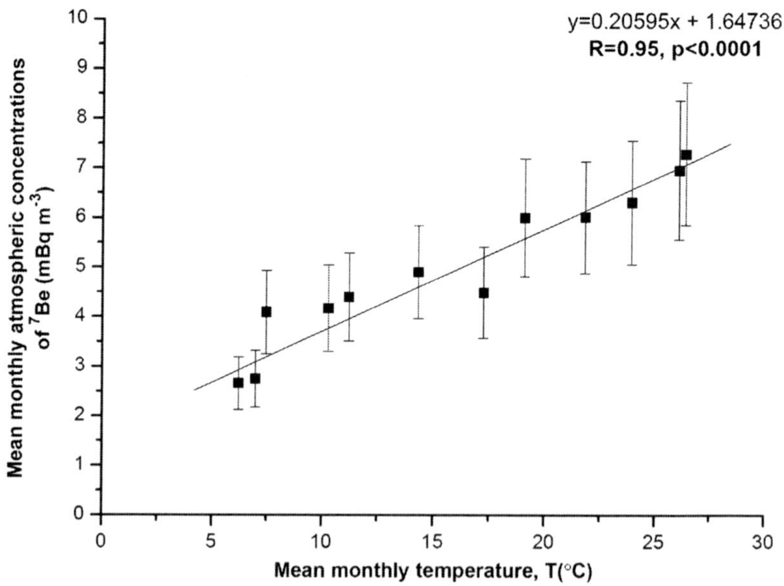

Figure 12 Mean monthly atmospheric concentrations of ^{7}Be versus temperature. The error bars represent the standard deviation of the monthly values. (Adapted from Ioannidou et al., 2005 with permission from Elsevier.)

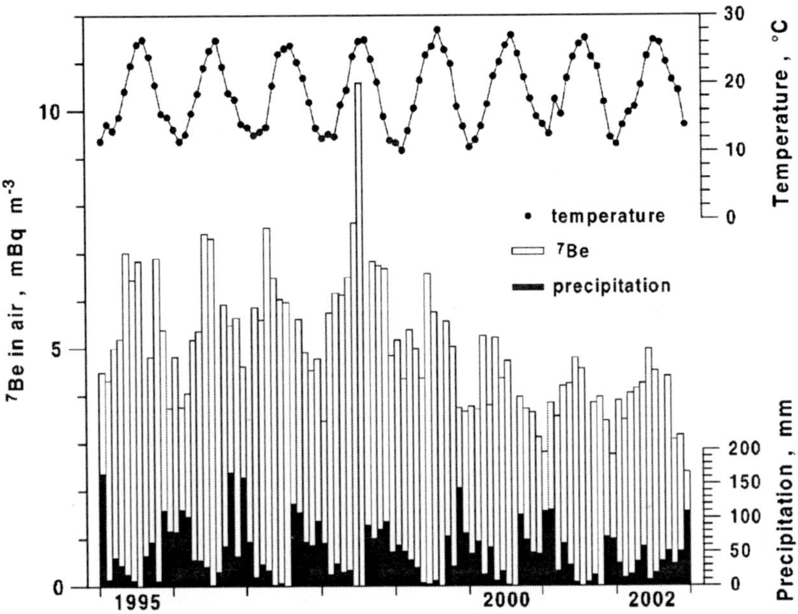

Figure 13 Average monthly concentration of ^7Be, total monthly precipitation and monthly average temperature for the period 1995–2002 in Palermo, Italy. (Adapted from Cannizzaro et al., 2004 with permission from Elsevier.)

At high altitudes, the changes of the ^7Be concentration due to meteorological conditions are more pronounced. Gerasopoulos et al. (2001) measured the ^7Be activity concentrations from 1996 to 1998 at four high-altitude stations: Jungfraujoch (3,580 masl), Switzerland; Zugspitze (2,962 masl), Germany; Sonnblick (3,106 masl), Austria and Mt. Cimone (2,165 masl), Italy. The ^7Be time series were analysed in combination with a set of meteorological and atmospheric parameters (tropopause height, relative and specific humidity and 3D back trajectories). At the three Alpine stations, a positive correlation between tropopause height and ^7Be concentration was found. The ^7Be values appeared to be negatively correlated with the relative humidity, indicating the influence of wet scavenging. Using composite maps of the geopotential height at 500 mbar, Zanis et al. (1999) showed that ^7Be concentrations higher than 8 mBq m^{-3} at Jungfraujoch were associated with an upper ridge over Switzerland. Trajectory statistics showed that low ^7Be concentrations typically originate from lower-altitude subtropical ocean areas, while high concentrations arrive from the north and high altitudes where stratospheric intrusions occur. Although the ^7Be activity concentrations were highly episodic, the monthly means indicate an annual cycle with a late-summer maximum at all stations, when the tropopause is higher and the vertical transport from upper tropospheric levels to the lower troposphere enhanced.

3.4. Behaviour of aerosols in the atmosphere

3.4.1. Activity size distribution of aerosols

As briefly discussed in Section 3.1, most of the airborne radionuclides are carried by aerosol particles that absorb them on their surface. Therefore, these radionuclides in principle can be used as tracers of aerosol particles. Their activity concentration in the atmosphere is influenced by their production rate, transport processes, radioactive decay and, for the longer-lived radionuclides (half-lives > 1 day), removal processes (dry and wet deposition) from the atmosphere to the ground. During longer residence in the atmosphere, the aerosol coagulation process significantly changes the activity size distribution of the original radioactive aerosol. This effect has been used to estimate the residence time of radioactive aerosol particles in the atmosphere.

In general, atmospheric aerosol particles follow a trimodal distribution (NRC, 1979): (1) the nucleation mode (from 3 to 70 nm, average 15 nm), (2) the accumulation mode (from 70 to 2,000 nm, average 300 nm) and (3) the coarse mode (from 2,000 to 36,000 nm, average > 10,000 nm). ^{7}Be and ^{210}Pb are preferentially attached to accumulation-mode aerosols (Figure 14) for which they are a useful tracer of their transport and

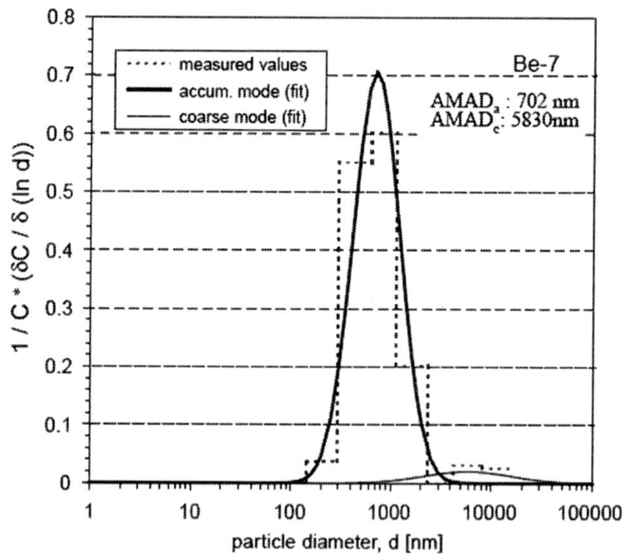

Figure 14 Typical relative activity size distribution of ^{7}Be in outdoor air of Göttingen, Germany. Measurement period: 4 weeks. (Adapted from Gründel and Porstendörfer, 2004 with permission from Elsevier.)

deposition (Papastefanou and Ioannidou, 1995). Cascade impactors with various numbers of stages (Mohamed, El-Hussein, 2005; Lee et al., 2004) are used to measure the size distribution of aerosol-attached radionuclides.

Gründel and Porstendörfer (2004) used online impactors over a period of 28 days to collect ^{7}Be and radon and thoron decay products in outdoor air near Göttingen, Germany.

Various radionuclides were preferentially attached to various sizes of aerosols.

- For short-lived radon and thoron decay products (^{218}Po, ^{214}Po, ^{210}Po and ^{212}Po) the following size distribution was obtained: 12–19% in the nucleation mode, 81–88% in the accumulation mode and nil in the coarse mode. The activity median aerodynamic diameters of the accumulation mode (AMADa) varied between 332 nm (^{218}Po) and 347 nm (^{214}Po) for the short-lived radon progeny and between 382 nm (^{212}Po) and 421 nm (^{212}Pb) for the thoron decay products.
- Between 93% and 96% of the long-lived radon decay products, ^{210}Pb and ^{210}Po were adsorbed on aerosol particles in the accumulation size range and only 4–7% on nuclei with diameters <60 nm.

It can be concluded from this and other similar studies that the surface area of the aerosol particles controls the attachment of natural radionuclides on aerosols. Furthermore, the longer the atmospheric residence of radionuclides with greater half-lives (^{210}Pb, ^{210}Po, ^{212}Pb, ^{212}Bi/^{212}Po and ^{7}Be), the higher is the coagulation with non-active aerosol particles and thus the greater are the AMADa values. In the case of ^{7}Be, the bigger particle size may be a consequence of the formation in the upper region of the atmosphere.

Winkler et al. (1998) studied the seasonal influence on the size distribution of cosmogenic ^{7}Be and the long-lived radon progeny ^{210}Pb in ambient aerosols sampled continuously from December 1994 to end of March 1996 in ground-level air. The variation of atmospheric processes during the study period resulted in changes of the activity median diameter from 440 to 740 nm for ^{7}Be and from 280 to 740 nm for ^{210}Pb. Although in the winter months (October to April), the difference between ^{210}Pb (AMD: 595 nm) and ^{7}Be (AMD: 590 nm) was negligible, in summer, ^{210}Pb was associated with significantly smaller aerosols (AMD: 430 nm) than ^{7}Be (AMD: 520 nm). The generally smaller values in summer point to a shorter residence time in that season. Furthermore, the difference in the activity size distribution between the long-lived radon decay products and ^{7}Be indicates that ^{7}Be and ^{7}Be aerosols are preferentially formed in the upper region of the atmosphere, while ^{210}Pb aerosols are formed in the lower atmosphere.

Mohamed (2005) measured, in addition to ^{7}Be and ^{212}Pb, ^{40}K in outdoor and indoor air in El-Minia city, Egypt. This ^{40}K stems from mechanical disintegration of dust in the atmosphere. Most of the aerosol particles were found to be in the size range between 200 and 2,200 nm, and

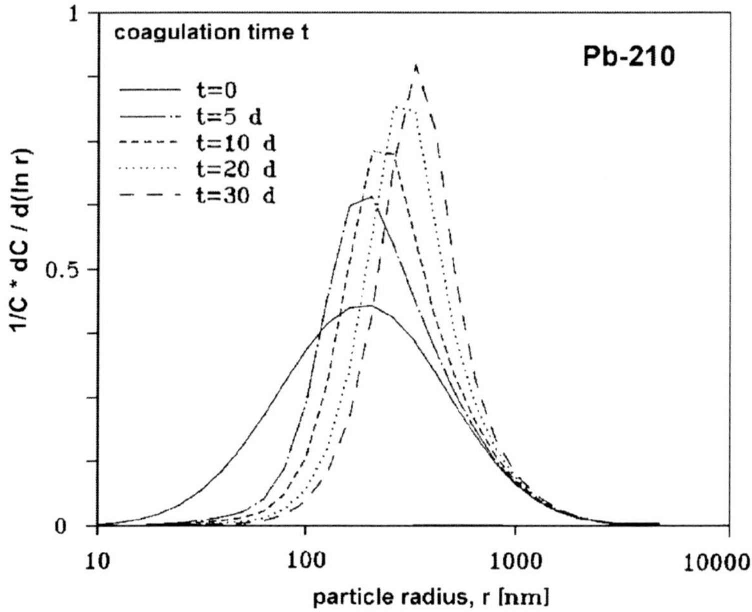

Figure 15 Change of the aerosol size distribution of ^{210}Pb by coagulation with the non-active particles of the atmospheric aerosol obtained by model calculations. (Adapted from Gründel and Porstendörfer, 2004 with permission from Elsevier.)

the aerosol particle concentration ranged between 1.1×10^3 and 3×10^3 cm^{-3} with a mean value of $1.8 \times 10^3 \pm 200$ cm^{-3}. Among other things, the mean activity concentration of unattached ^{212}Pb was found to be about 19 ± 3 mBq m^{-3}. It decreased with increasing aerosol particle concentration, while the attached activity concentration of ^{212}Pb increased.

Model calculations by Butterweck (1991) suggest that the increase in size of radioactive aerosol during their residence time in the atmosphere is due to coagulation with the non-active aerosol particles. The coagulation of the radioactive aerosol with the inactive aerosol was, in this model, numerically calculated for the example of ^{210}Pb (Figure 15). The real ^{210}Po aerosol size distribution can be obtained after a coagulation time of about 15–20 days. This time is about the average residence time of an aerosol of that size distribution in the atmosphere (Poet et al., 1972).

3.4.2. Health effects of radioactive aerosols

The UNSCEAR Report (UNSCEAR, 1982) concluded that inhalation of thoron, radon and their decay products is the most important component of man's exposure to natural radiation (see also Steinhausler, 1993 and Pillai et al., 1994). The health effects are mainly due to internal irradiation from alpha particles. About 90% of the total alpha-radiation energy of the ^{220}Rn decay chain is contributed by the alpha radiation of ^{212}Pb (ICRP, 1981; Shimo et al., 1990; Guo et al., 1992; Doi and Kobayashi, 1994). Due to the relatively long half-life of ^{212}Pb, a considerable fraction of ^{212}Pb deposited in the human respiratory tract can be transferred to other tissues. The deposition probability of unattached and aerosol-attached activities in the human respiratory tract during inhalation depends primarily on the particle size (between 0.5 and 5 nm for the unattached fraction). The unattached atoms are deposited in the respiratory tract during inhalation, whereas about 80% of attached atoms are exhaled without deposition (James et al., 1991). The size of short-lived radon and thoron decay products unattached on aerosols is very small (below 4 nm).

In conclusion, the activity size distribution is a very important parameter for the estimation of radiation dose by inhalation.

3.4.3. Aerosol composition

The major constituents of atmospheric aerosol are inorganic ions (Appel et al., 1980). Aerosols can be of natural or anthropogenic origin. Sea salt, soil dust and oceanic sulphate constitute a major part of the global natural aerosol during volcanically quiet periods. Some part of the atmospheric dust is also due to anthropogenic activities (Satheesh and Krishna Moorthy, 2005; Sokolik et al., 1998; Haywood et al. 2003). Soot is the main absorbing fraction of aerosol (Babu and Moorthy, 2002) and is one of the most complex aerosol types. Both natural and anthropogenic processes such as forest fires, combustion of fuel and transport (Schwartz, 1995) produce soot. The aerosol composition over land is different from that over the sea. After production, dust aerosols are often transported long distances from their sources (Arimoto, 2001). Examples are dust transport from the Sahara across the Atlantic Ocean, Arabian dust transport across the Arabian Sea and dust from China across the Pacific. Mineral dust plays an important role in marine biological processes (Falkowski et al., 1998). For example iron occurring in oceanic dust acts as a nutrient for phytoplankton. This, in turn, emits dimethyl sulphide (DMS) and thus forms sulphate aerosols over the ocean. These natural sulphate aerosols over oceans act as condensation nuclei for the formation of clouds. Sea salt aerosols, which are hygroscopic, also give rise to cloud formation (Satheesh and Krishna Moorthy, 2005).

3.4.4. Deposition of aerosols

Aerosol deposition to the Earth's surface that occurs without enhancement due to rain or snow is called dry deposition. Dry deposition rates are generally slower than those wet deposition.

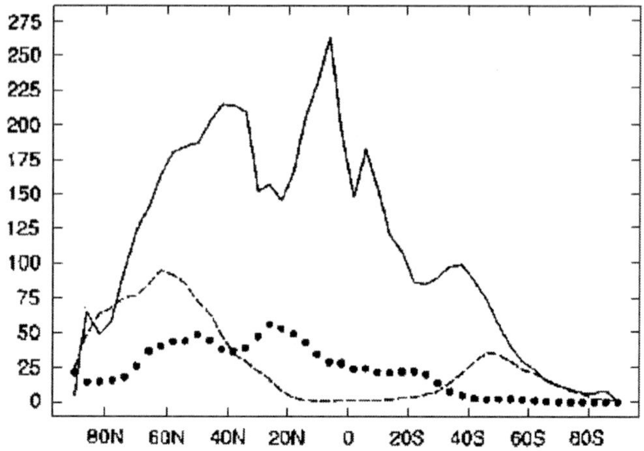

Figure 16 Zonal mean depositional fluxes of ^{210}Pb contributed by convective precipitation (solid line), large-scale precipitation (dashed line) and dry processes (dotted line). Values are annual averages. (Adapted from Balkanski et al., 1993 with permission from Elsevier.)

However, dry deposition is a continuous process, while wet deposition occurs only intermittently during rain- and snowfalls. Thus, in regions with low rainfall, dry deposition may dominate the scavenging of aerosols. On the other hand, rain is able to sweep through large portions of the troposphere at high speed carrying aerosol particles along with it. In addition, snow, hail and even fog can remove aerosol particles from the atmosphere.[1] Thus, with the exception of arid regions, wet deposition dominates the washout of aerosol particles from the atmosphere. Balkanski et al. (1993) suppose that wet deposition accounts for 86% of the global deposition flux, the remaining 14% being due to dry deposition (Figure 16). Major factors controlling deposition of aerosols are concentrations of aerosols in air and precipitation amount (Baskaran, 1995; Baskaran et al., 1993; McNeary and Baskaran, 2003). Wet deposition of aerosol-attached radionuclides (e.g. ^{7}Be) is also influenced by climatic conditions. While under dry conditions the influence can be small (Ioannidou et al., 2005), high precipitation amounts can cause a remarkable decrease of ^{7}Be in air (Papastefanou and Ioannidou, 1991). Figure 17 compares the seasonal change of the ^{7}Be concentration in air with those of precipitation and temperature at Palermo, Italy. At this station, low precipitation corresponds to high ^{7}Be concentration in air and vice versa. Megumi et al. (2000) found that high ^{7}Be concentration in air corresponds to high ^{7}Be concentration in precipitation.

[1] In the stratosphere, however, there is no rain and thus the removal rate for aerosol particles is much smaller.

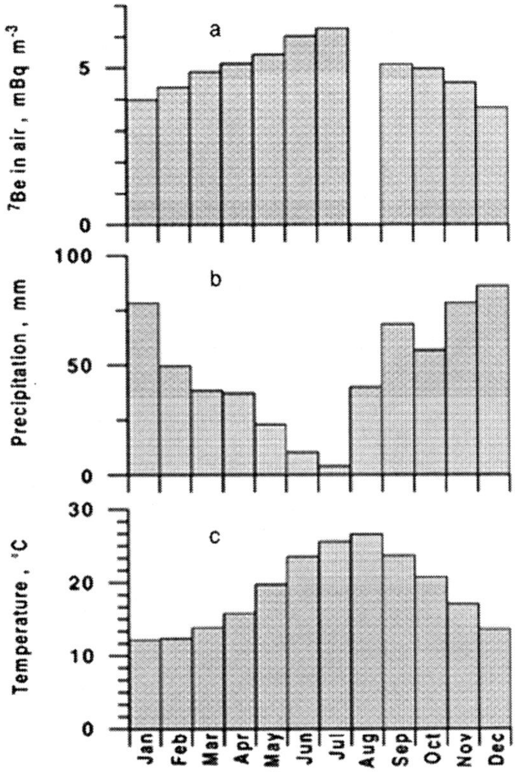

Figure 17 Mean values of (a) average monthly ^{7}Be concentrations in air, (b) total monthly precipitation, (c) monthly average temperatures, for the period 1995–2002 in Palermo, Italy. (Adapted from Cannizzaro et al., 2004 with permission from Elsevier.)

4. Application of Radionuclides in Atmospheric Studies

Natural radionuclides from terrestrial and upper atmospheric sources (^{222}Rn, ^{220}Rn, ^{212}Pb, ^{210}Pb, ^{7}Be, ^{10}Be, etc.) and of anthropogenic origin are widely used as tracers to examine atmospheric processes relevant to air quality and climate and to validate atmospheric models simulating transport, transformation and removal processes of gases and aerosols. The validation and calibration of such models require accurate experimental data on source functions and temporal and spatial variation of relevant radionuclides. Yet, the effective use of such data is still limited, especially due to the insufficient accuracy of radionuclide source functions. As ^{222}Rn is the parent nuclide of a number of daughter products relevant to atmospheric studies, the knowledge of its sources is essential. So far, only patchy data are available on

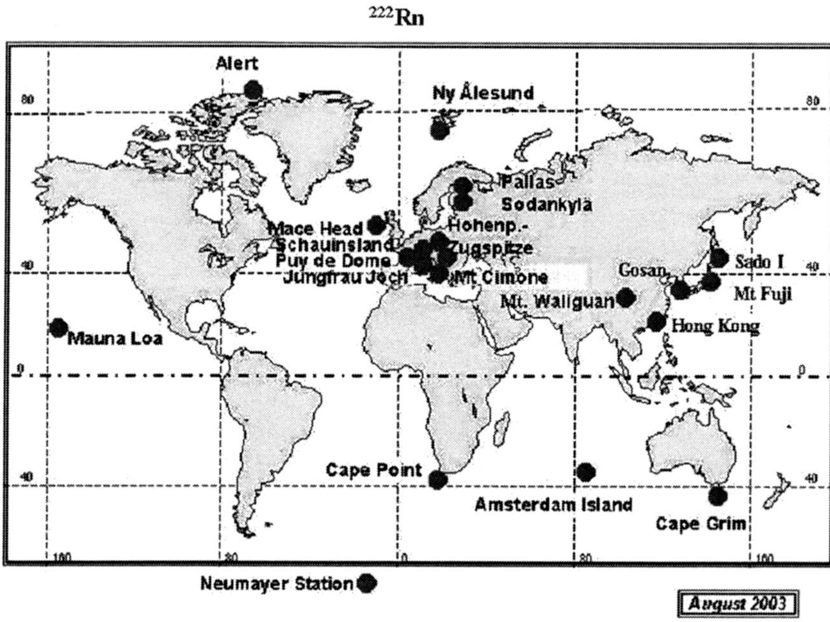

Figure 18 Routine observations of ^{222}Rn at GAW global, regional and contributing stations as of 2003. (Adapted from WMO/GAW, 2004 with permission.)

the magnitude and global distribution of the ^{222}Rn flux to the atmosphere. To improve this situation, a global atmospheric monitoring network for ^{222}Rn flux measurements has been established (WMO/GAW, 2004). Figure 18 shows the global distribution of the ^{222}Rn stations. Currently surface air concentrations of ^{210}Pb and ^{7}Be are also measured routinely at the Global Atmosphere Watch (GAW) stations.

Table 3 compiles key radionuclides and relevant applications in atmospheric modelling. Some of these applications include the following:

- Vertical profiles of ^{222}Rn are useful in constraining vertical mixing in the troposphere. ^{222}Rn has been proved as a tracer of small-scale vertical transport processes in the atmosphere.
- ^{222}Rn measurements at surface sites, from ships or aircraft, are used to calibrate models on boundary layer transport as well as on exchange between the boundary layer and free troposphere.
- Radionuclides like ^{210}Pb, ^{7}Be, ^{10}Be and ^{90}Sr attached on particle surfaces serve as tracers of aerosols in physical processes including wet and dry deposition.
- Some of these radionuclides also serve as surrogates for other atmospheric components. ^{210}Pb, for example is a useful surrogate for atmospheric trace constituents and pollutants such as sulphate, lead and mercury

Table 3 Radionuclides useful for atmospheric modelling (from WMO/GAW, 2004).

Species	Application	Source
^{222}Rn	Turbulent transport	Soil
	Zonal transport	
	Transcontinental transport	
^{210}Pb	Scavenging	^{222}Rn decay
^{212}Pb	Scavenging	^{222}Rn decay
	Vertical transport	
^{14}CO	Downward transport (STE)	Cosmogenic
	Tropospheric OH	
$^{14}CO_2$	Carbon cycle	Cosmogenic
^{7}Be	Scavenging	Cosmogenic
	Downward transport (STE)	
^{10}Be	Scavenging	Cosmogenic
	Downward transport (STE)	
^{3}H	Water vapor transport	Cosmogenic
^{36}Cl	Palaeoclimate	Cosmogenic
	Geomagnetic field	
^{85}Kr	Transcontinental transport	Nuclear processing
^{137}Cs	Wind erosion	Bomb tests
	Resuspension	Nuclear accident
^{90}Sr	Downward transport	Bomb tests

(Turekian and Graustein, 2003). Furthermore, combined ^{210}Pb and ^{7}Be measurements are able to identify sources of ozone.

- Nuclear facilities releasing ^{85}Kr are mostly located at mid-latitudes of the northern hemisphere, which makes it a good tracer for inter-hemispheric transport of pollutants. From ^{85}Kr latitudinal profiles, an inter-hemispheric exchange time of approximately 1.1 years has been derived.
- ^{14}CO turned out to be a very good tracer of STE, and it can be used to assess the tropospheric OH abundance. It is mainly produced in the atmosphere by cosmic radiation; its only sink is oxidation by OH.

The application of radionuclides produced in the atmosphere by cosmic rays has substantially benefitted from recent progress in numerical simulation of their atmospheric production rates as functions of altitude, latitude, and solar and geomagnetic field intensity. New measurement techniques for particle flux have been developed which allow testing of the quality of numerical models (Masarik, 2004).

A concise review of these achievements and trends in the application of radionuclides in atmospheric studies as well as lists of relevant references can be found in WMO/GAW (2003). In the following section, some methodological approaches and selected applications are presented.

4.1. Atmospheric residence time of aerosols – use of ^{222}Rn daughters

The atmospheric residence time of an aerosol particle is defined as the time the particle will be retained in the atmosphere before disappearance or removal by any process. In the context of residence time determinations by airborne radionuclides, the so-called apparent residence time is represented by the reciprocal of the removal rate of a given radionuclide, which is defined by the following relationship for ^{222}Rn decay products (Moore et al., 1973):

$$\frac{dN_2}{dt} = N_1\lambda_1 - N_2(\lambda_2 + \lambda_R) \tag{2}$$

where N_1 and N_2 are the number of atoms, λ_1 and λ_2 the radioactive decay constants of the parent and daughter radionuclides, respectively, and λ_R the first-order rate constant for the removal of aerosols from the atmosphere by all processes. This removal rate is zero for ^{222}Rn and is assumed equal for all of its decay products. The apparent mean residence time is

$$\tau_R = \frac{1}{\lambda_R}$$

For the ^{222}Rn decay products and steady-state conditions ($dN_2/dt = 0$), the following expressions can be derived from Equation (2):

$$\tau_R\left(\frac{\text{Rn}}{\text{Pb}}\right) = \frac{A_{\text{Pb}}}{A_{\text{Rn}}}\frac{1}{\lambda_{\text{Pb}}}$$

$$\tau_R\left(\frac{\text{Bi}}{\text{Pb}}\right) = \frac{A_{\text{Bi}}}{A_{\text{Pb}} - A_{\text{Bi}}}\frac{1}{\lambda_{\text{Bi}}}$$

A corresponding expression follows for τ_R(Po/Pb) (Moore et al., 1973). The experimental data given by Moore et al. (1973) show that, for a given radionuclide ratio, these apparent residence times are relatively constant but differ from the values obtained for other radionuclide ratios. Furthermore, the apparent residence times are nearly constant within the troposphere but increase abruptly above the tropopause. The wide variations found in residence time values between the three ratios ($\tau_{\text{Rn/Pb}} = 3.4$ days; $\tau_{\text{Bi/Pb}} = 7.8$ days; $\tau_{\text{Po/Pb}} = 20$ days) suggest that the real atmosphere does not conform to the model assumptions (steady-state and 'closed box' model). Reviewing the data of Moore et al. (1973), Turekian and Graustein (2003) assumed that the air samples represent a mixture of individual small parcels that behave as closed systems. With this assumption, they found, for most of the stratospheric samples, residence times between about 100 and 180 days, which are consistent with the behaviour of the lower stratosphere characterized by suppressed vertical mixing and non-existing precipitation. The tropospheric samples were found to be consistent with mixing of aerosols of varying residence times, the values ranged from 1–2 days to about 50 days. This mixing approach does not contradict with the interpretation of Moore et al. (1973) which said that the residence time in the lower troposphere is about 4 days and increases by up to a factor of 3 towards the top of the troposphere. Modelling by Balkanski et al. (1993) confirmed the increase of the mean residence time of ^{210}Pb with altitude. The residence time of ^{210}Pb produced in the lowest 500 m of the atmosphere was found to be on average four times shorter than that of ^{210}Pb produced in the upper atmosphere. Moreover, the model estimated the variation of the residence time with latitude for January and July (Figure 19). In the southern tropics, where ^{210}Pb is largely produced at high altitudes, high values of the mean residence time have been found. Residence times at southern mid-latitudes with frequent rainfall are shorter (about 5 days). At northern mid-latitudes, the residence time changes from 5 days in winter to 10 days in summer, reflecting the seasonality in convection.

There are also many studies using cosmic-ray-produced ^{7}Be to determine tropospheric residence times. For example Durana et al. (1996) derived the mean residence time from measured ^{7}Be tropospheric

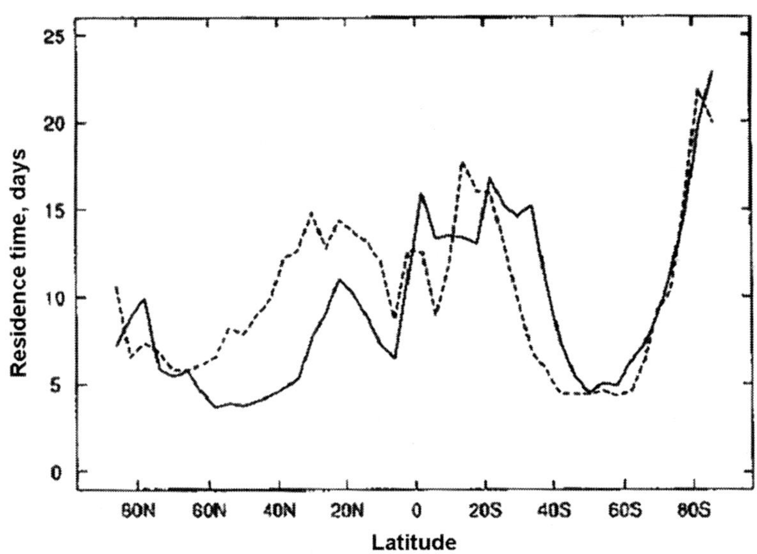

Figure 19 Zonal mean tropospheric residence time of ^{210}Pb as a function of latitude for January (solid line) and July (dashed line). (Adapted from Balkanski et al., 1993 with permission from Elsevier.)

deposition rate and average tropospheric concentration. The average value of 30 days obtained for a tropospheric height of 9 km appeared higher than expected for ^{7}Be of tropospheric origin. This indicates that the tropospheric ^{7}Be contained an admixture of ^{7}Be produced in the stratosphere. Liu et al. (2001) found good agreement between simulated and experimental values of the ^{7}Be and ^{210}Pb atmospheric concentrations using global mean tropospheric lifetimes of ^{7}Be and ^{210}Pb against deposition of 21 and 9 days, respectively.

4.2. Atmospheric mixing and transport: use of ^{7}Be and ^{210}Pb

Useful information about transport, removal and residence time of aerosols in the atmosphere can be obtained from combined measurements of ^{7}Be and ^{210}Pb. ^{7}Be is produced in the stratosphere and upper troposphere, and ^{210}Pb in the lowermost metre of the atmosphere over continents and islands. While ^{7}Be is mixed downward, ^{210}Pb is mixed upward, but both are removed from the atmosphere by precipitation. Thus, changes in the ^{7}Be/^{210}Pb ratio in space and time reflect both vertical and horizontal transport in the atmosphere. In particular, seasonal variations in the stability of the atmosphere over land have a dominant influence on the ^{7}Be/^{210}Pb ratio. In summer, surface heating generates convective vertical transport which reduces ^{210}Pb in surface air by mixing it through a larger volume. Simultaneously, the enhanced transport of ^{7}Be from the upper troposphere to the surface creates a summer maximum in ^{7}Be. In winter, the stability of the atmosphere keeps ^{210}Pb at the surface with an accordingly higher concentration. At maritime stations, the relationship between the two radionuclides is different, mainly because of the missing local source (^{222}Rn) for ^{210}Pb. Virtually, all ^{210}Pb at these sites is transported from the continent, which makes the ^{7}Be/^{210}Pb ratio an excellent indicator of continental sources of local aerosols: low values of the ratio due to high ^{210}Pb reflect high continental influence, while low values point to relative isolation from continental sources. This behaviour is reflected in the long-term records of the near-surface ^{7}Be/^{210}Pb ratio at a continental station (Champaign, IL) of the USA, and a maritime station (Bermuda Island) (Figure 20). The seasonal cycle of ^{7}Be/^{210}Pb at the island site is 6 months out of phase with the continental site. A strong seasonal cycle has also been observed at the high elevation site Izania (2,367 masl) on Tenerife (Canary Islands). This site represents free troposphere conditions and indicates that the long-distance transport is especially efficient in the free troposphere. The high ^{7}Be/^{210}Pb values in winter are due to the atmospheric stability over the continents in that season which hinders transfer of surface aerosols to the free troposphere and limits the lateral transport distance. During spring and summer, the troposphere is mixed by convective processes giving rise to an effective exchange between boundary layer and free troposphere and

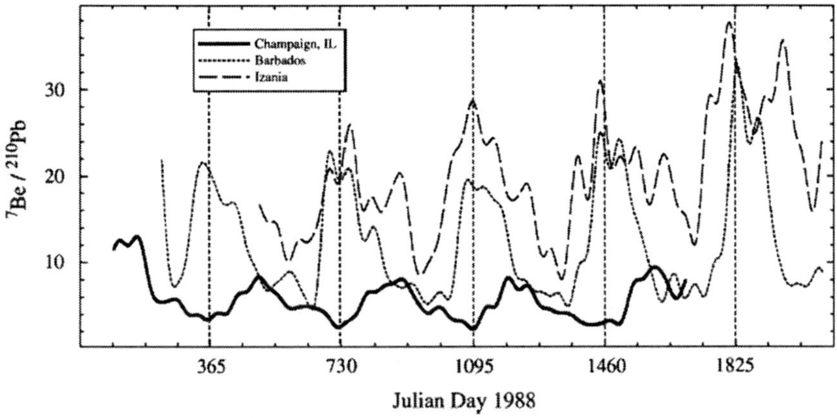

Figure 20 Time series of $^7Be/^{210}Pb$ in surface air at a continental (Champaign, IL) and an oceanic (Barbados Island) site, and in oceanic free troposphere (Izania, Tenerife). (Adapted from Turekian and Graustein, 2003 with permission from Elsevier.)

enhancing long-distance transport of aerosols. This leads to a uniform distribution of aerosols between the continental and maritime boundary layers (Turekian and Graustein, 2003).

Hirosea et al. (2004) carried out simultaneous measurements of atmospheric deposition of various radionuclides (^{210}Pb, ^7Be and thorium isotopes) at Tsukuba (2000 and 2001) and Nagasaki (2000), Japan. ^{210}Pb and ^7Be concentrations in rainwater showed a typical seasonal variation with maximum from winter to spring and minimum in summer. This variation was controlled by the seasonal exchange between continental (the northern free troposphere) and oceanic air masses. The low concentrations in summer may also be caused by the strong vertical mixing during that season which replaces ^{210}Pb-enriched air of the lower troposphere with ^{210}Pb-depleted air of the upper troposphere. Such behaviour is typical for Asian regions that are influenced by the monsoon, contrary to Europe and North America where the maximum concentration of ^7Be is reached in summer and minimum from winter to spring. The significant difference found between ^{210}Pb and ^7Be concentrations during spring months indicates intrusion of stratospheric ^7Be because of the STE at this time. The measured concentrations of the thorium isotopes in precipitation revealed that the ratio $^{230}Th/^{232}Th$ is a useful tracer of mineral dust originating from the East Asian continents.

The pair ^{210}Pb and ^7Be is also useful to test global atmospheric models for wet and dry deposition (Liu et al., 2001). Long-term data records of ^{210}Pb (Preiss et al., 1996) and ^7Be (Feely et al., 1989; Larsen and Sanderson, 1995) are available from worldwide networks and provide good constraints for model validation. ^7Be and ^{210}Pb have also been used to validate global numerical models describing other processes. These processes include transport of continental air over the oceans (Turekian et al., 1989; Balkanski et al., 1993), transport from the stratosphere (Rangarajan and Gopalakrishnan, 1970; Viezee and Singh, 1980; Sanak et al., 1985; Dibb et al., 1992, 1994; Rehfeld and Heimann, 1995) and subsidence in the troposphere (Feely et al., 1989; Koch et al., 1996).

4.3. Identification of ozone sources

The ratio $^{7}Be/^{210}Pb$ can also be used to determine the sources of chemical species in the lower free troposphere such as ozone and nitrate. Graustein and Turekian (1990) have correlated the ozone concentration with a function of ^{7}Be and ^{210}Pb measured at Izania, Tenerife, and found that the primary source of ozone in the lower free troposphere of this eastern Atlantic region was the upper troposphere.

Long-term records of monthly mean surface air concentration of ^{210}Pb, ^{7}Be and O_3 measured at New York City, USA (Lee, 2004), indicate that the summer maximum of ozone corresponds well with the summer minimum of ^{210}Pb and maximum of ^{7}Be (Figure 21).

Figure 22 shows ^{7}Be and ozone concentrations measured during summer campaigns in free tropospheric air at Mt. Fuji, Japan. The results of

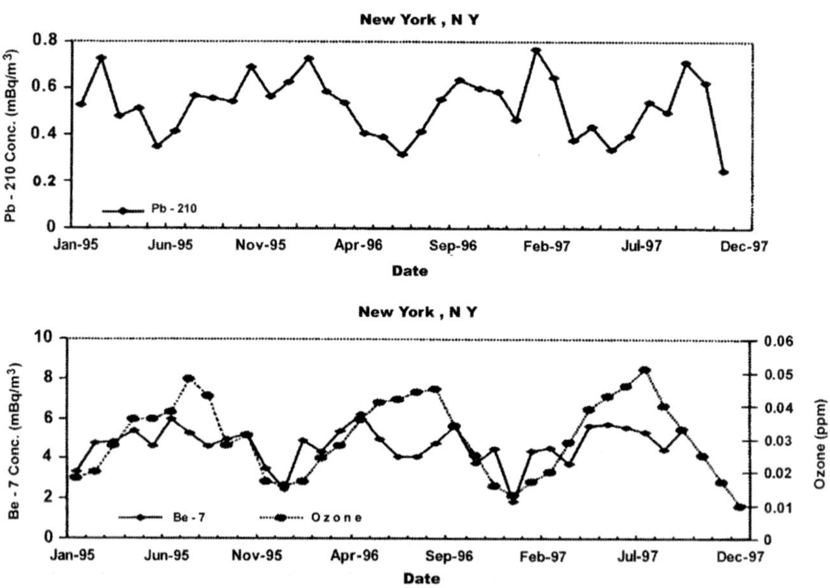

Figure 21 Monthly surface air measurements of ^{210}Pb (top) and ^{7}Be (bottom) corresponded with the surface ozone measurements at New York City, NY, USA. (Adapted from Lee, 2004 with permission from WMO/GAW.)

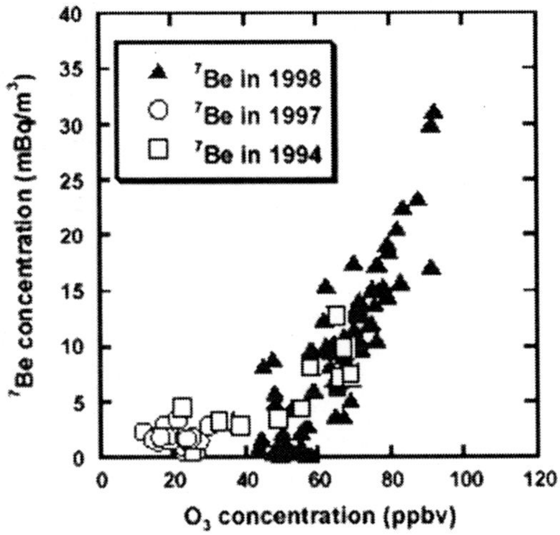

Figure 22 Plot of ^{7}Be and ozone concentrations measured in free tropospheric air at Mt. Fuji, Japan, during summer campaigns. (Adapted from Lee, 2004 with permission from WMO/GAW.)

the 1998 campaign and part of the values of the 1994 campaign show a clear correlation between ^{7}Be and ozone, which indicates transport from the lowermost stratosphere and uppermost troposphere. The ozone values at low ^{7}Be concentration represent a photochemical component of ozone produced in the troposphere. Based on these measurements, the ozone influx from the stratosphere has been estimated to be about 0.9×10^{11} molecules cm^{-2} s^{-1} (Lee, 2004).

4.4. ^{14}CO research

Airborne ^{14}CO is a rather novel radioactive tracer complementing CO in its role as tracer in atmospheric chemistry (Jöckel and Brenninkmeijer, 2004). Its atmospheric abundance is mainly controlled by cosmic-ray production of ^{14}C, which is rapidly oxidised to ^{14}CO. The average cosmic-ray source strength is 1.6–2 molecules cm^{-2} s^{-1}, which corresponds to a total production of approximately 13–16 kg ^{14}CO per year. The cosmogenic ^{14}CO production rate oscillates with a period of 11 years (solar modulation of the cosmic-ray intensity); the production rates are higher during times of low solar activity. Some 20% of the total atmospheric ^{14}CO abundance is recycled ^{14}CO produced by oxidation of biogenic methane and higher hydrocarbons as well as burning of biomass. Fossil fuel is ^{14}C free and thus its use does not contribute to the ^{14}CO inventory.

The chemical lifetime of ^{14}CO is relatively short, about 1–2 months. The principal sink for atmospheric ^{14}CO is oxidation to ^{14}CO2 by the hydroxyl radical (OH) according to $^{14}CO + 2OH = {}^{14}CO_2 + H_2O$. Thus,

^{14}CO is useful to constrain the distribution and seasonality of this pivotal atmospheric oxidant hydroxyl, which is essential to understand tropospheric chemistry, the distribution of reactive trace gases and their rates of removal from the atmosphere (Krol et al., 2007).

REFERENCES

Al-Azmi, D., A. M. Sayed, and H. A. Yatim. (2001). Variations in ^{7}Be concentrations in the atmosphere of Kuwait during the period of 1994 to 1998. *Applied Radiation and Isotopes*, 55, 413–417.

Aldahan, A., G. Possnert, and I. Vintersved. (2001). Atmospheric interactions at northern high latitudes from weekly Be isotopes in surface air. *Applied Radiation and Isotopes*, 54, 345–353.

Appel, B. R., E. M. Kothny, E. M. Hoffer, and J. J. Waselowski. (1980). Sulfate and nitrate data from the California Aerosol Characterisation Experiment (ACHEX). In: *The Character and Origins of Smog Aerosols* (Eds G. M. Hidy, et al.). John Wiley, New York, pp. 315–335.

Arimoto, R. (2001). Eolian dust and climate: Relationships to sources, tropospheric chemistry, transport and deposition. *Earth Science Reviews*, 54(1–3), 29–42.

Arimoto, R., J. A. Snow, W. C. Graustein, J. L. Moody, B. J. Ray, R. A. Duce, K. K. Turekian, and H. B. Maring. (1999). Influences of atmospheric transport pathways on radionuclide activities in aerosol particles from over the North Atlantic. *Journal of Geophysical Research*, 104(D17), 301–321.

Babu, S. S., and K. K. Moorthy. (2002). Aerosol black carbon over a tropical coastal station in India. *Geophysical Research Letters*, 29(23), 2098, doi: 10.1029/2002GL015662.

Baeza, A., L. M. Delrio, A. Jimenez, C. Miro, J. M. Paniagua, and M. Rufo. (1996). Analysis of the temporal evolution of atmospheric beryllium as a vector of the behavior of other radionuclides in the atmosphere. *Journal of Radioanalytical and Nuclear Chemistry*, 207(2), 331–344.

Balkanski, Y. J., D. J. Jacob, G. M. Gardner, W. C. Graustein, and K. K. Turekian. (1993). Transport and residence times of tropospheric aerosols inferred from a global three-dimensional simulation of ^{210}Pb. *Journal of Geophysical Research*, 98, 20570–20586.

Baskaran, M. (1995). A search for seasonal variability on the depositional fluxes of ^{7}Be and ^{210}Pb. *Journal of Geophysical Research*, 100, 2833–2840.

Baskaran, M., C. H. Coleman, and P. H. Santschi. (1993). Atmospheric depositional fluxes of ^{7}Be and ^{210}Pb at Galveston and College Station, Texas. *Journal of Geophysical Research*, 98, 20555–20571.

Benioff, P. A. (1956). Cosmic ray production rate and mean removal time of Beryllium-7 from the atmosphere. *Physical Review*, 104, 1122–1130.

Butterweck, G. (1991). Dissertation, Faculty of Physics, Georg-August University, Göttingen, Germany.

Cannizzaro, F., G. Greco, M. Raneli, M. C. Spitale, and E. Tomarchio. (2004). Concentration measurements of ^{7}Be at ground level air at Palermo, Italy – Comparison with solar activity over a period of 21 years. *Journal of Environmental Radioactivity*, 72, 259–271.

Chamberlin, A. C. (1991) Radioactive aerosols. *Cambridge Environmental Chemistry Series*, Cambridge University Press, ISBN 0-521-40121-6.

Conen, F. (2004) Variation of ^{222}Rn flux and its implication for tracer studies. In: *1st International Expert Meeting on Sources and Measurements of Natural Radionuclides Applied to Climate and Air Quality Studies*. World Meteorological Organization Technical Document No. 1201, pp. 35–42.

Dibb, J. E., L. D. Meeker, R. C. Finkel, J. R. Southon, M. W. Caffee, and L. A. Barrie. (1994). Estimation of stratospheric input to the Arctic troposphere. ^{7}Be and ^{10}Be in aerosols at Alert, Canada. *Journal of Geophysical Research*, 99, 12855–12864.

Dibb, J., R. Talbot, B. Lefer, E. Scheuer, G. Gregory, E. Browell, J. Bradshaw, S. Sandholm, and H. Singh. (1997). Distribution of Beryllium 7 and Lead 210 and soluble aerosol-associated ionic species over the western Pacific: PEM West B, February–March 1994. *Journal of Geophysical Research*, 102, 28287–28302.

Doi, M., and S. Kobayashi. (1994). Characterization of Japanese wooden houses with enhanced radon and thoron concentrations. *Health Physics*, 66, 274–282.

Dörr, H., and K. O. Münnich. (1989). Downward movement of soil organic matter and its influence on trace-element transport (^{210}Pb, ^{137}Cs) in the soil. *Radiocarbon*, 31(3), 655–663.

Durana, L., M. Chudy, and J. Masarik. (1996). Investigation of ^{7}Be in the Bratislava atmosphere. *Journal of Radioanalytical and Nuclear Chemistry, Articles*, 207(2), 345–356.

Eisenbud, M., and T. Gesell. (Eds). (1997). *Environmental Radioactivity; From Natural, Industrial, and Military Sources*. 4th ed., Academic Press, San Diego, CA, 656pp.

El-Hussein, A., A. Mohamemed, M. A. El-Hady, A. A. Ahmed, A. E. Ali, and A. Barakat. (2001). Diurnal and seasonal variation of short-lived radon progeny concentration and atmospheric temporal variations of ^{210}Pb and ^{7}Be in Egypt. *Atmospheric Environment*, 35(25), 4305–4313.

Falkowski, P. G., R. T. Barber, and V. Smetacek. (1998). Biogeochemical controls and feedbacks on ocean, primary production. *Science*, 281(5374), 200–206.

Feely, H. W., R. J. Larsen, and C. G. Sanderson. (1989). Factors that cause seasonal variations in Beryllium-7 concentrations in surface air. *Journal of Environmental Radioactivity*, 9, 223–249.

Gäggeler, H. W. (1995). Radioactivity in the atmosphere. *Radiochimica Acta*, 70/71, 345–353.

Gat, J. R., and G. Assaf. (1968). Atmospheric bismuth-212: Measurement and geophysical applications. *Science*, 159(3818), 977–979.

Gerasopoulos, E., P. Zanis, A. Stohl, C. S. Zerefos, C. Papastefanou, W. Ringer, L. Tobler, S. Huebner, H. W. Gäggeler, H. J. Kanter, L. Tositti, and S. Sandrini. (2001). A climatology of ^{7}Be at four high-altitude stations at the Alps and the Northern Apennines. *Atmospheric Environment*, 35, 6347–6360.

Graustein, W. C., and K. K. Turekian. (1990). ^{222}Rn fluxes from soils to the atmosphere measured by ^{210}Pb–^{226}Ra disequilibrium in soils. *Geophysical Research Letters*, 17, 841–844.

Gregory, J. C. (1996). Atmospheric radioactive isotopes at orbital altitudes. *Radiation Measurements*, 26(6), 841–850.

Gründel, M. M., and J. Porstendörfer. (2004). Differences between the activity size distributions of the different natural radionuclide aerosols in outdoor air. *Atmospheric Environment*, 38, 3723–3728.

Guo, Q., M. Shimo, Y. Ikebe, and S. Minato. (1992). The study of thoron progeny and radon progeny concentrations in different kinds of dwellings in Japan. *Radiation Protection Dosimetry*, 45, 357–359.

Haywood, J., P. Francis, S. Osborne, M. Glew, N. Loeb, E. Highwood, D. Tanre, G. Myhre, P. Formenti, and E. Hirst. (2003). Radiative properties and direct radiative effect of Saharan dust measured by the C-130 aircraft during SHADE. 1. Solar spectrum. *Journal of Geophysical Research*, 108(D18), 8577.

Hernández Alonso, C. M., H. Águila Cartas, M. Asencio Díaz, and A. Caravaca Muñoz. (2004). Reconstruction of ^{137}Cs signal in Cuba using ^{7}Be as tracer of vertical transport processes in the atmosphere. *Journal of Environmental Radioactivity*, 75, 133–142.

Hirosea, K., T. Honda, S. Yagishita, Y. Igarashi, and M. Aoyama. (2004). Deposition behaviors of ^{210}Pb, ^{7}Be and thorium isotopes observed in Tsukuba and Nagasaki. *Japan Atmospheric Environment*, 38, 6601–6608.

ICRP. (1981). Limits for intakes of radionuclides by workers. *Annals of the ICRP*, 6(2/3), Publication 30, Part 3. Pergamon Press, Oxford.

Ioannidou, A., M. Manolopoulou, and C. Papastefanou. (2005). Temporal changes of ^{7}Be and ^{210}Pb concentrations in surface air at temperate latitudes (40°). *Applied Radiation and Isotopes*, 63(2), 277–284.

Jacobi, W., and K. Andre. (1963). The vertical distribution of radon 222, radon 220 and their decay products in the atmosphere. *Journal of Geophysical Research*, 68(1963), 3700–3813.

James, A. C., F. T. Cross, J. S. Durham, J. K. Briant, P. Gehr, R. Masse, R. G. Cuddihy, and A. Birchall. (1991). Dosimetry model for bronchial and extrathoracic tissues of the respiratory tract. *Radiation Protection Dosimetry*, 37, 221–230.

Jöckel, P., and C. Brenninkmeijer. (2004). ^{14}CO and its application in studies of atmospheric chemistry and Transport. In: *1st International Expert Meeting on Sources and Measurements of Natural Radionuclides Applied to Climate and Air Quality Studies*. World Meteorological Organization Technical Document No. 1201, pp. 81–83.

Koch, D. M., D. J. Jacob, and W. C. Graustein. (1996). Vertical transport of tropospheric aerosols as indicated by ^{7}Be and ^{210}Pb in a chemical tracer model. *Journal of Geophysical Research*, 101, 18651–18666.

Krol, M. C., J. Fokke-Meirink, P. Bergamaschi, J. E. Mak, D. Lowe, P. Jöckel, S. Houweling, and T. Röckmann. (2007). What do ^{14}CO measurements tell us about OH? *Atmospheric Chemistry and Physics Discussions*, 7(4), 10405–10438.

Lal, D., P. K. Malhotra, and B. Peters. (1958). On the production of radioisotopes in the atmosphere by cosmic radiation and their application to meteorology. *Journal of Atmospheric and Terrestrial Research*, 12, 306–328.

Lal, D., and B. Peters. (1962). Cosmic ray isotopes and their application to problems in geophysics, *Progress in Cosmic Ray Physics*, Vol. 6, No. 3. N. Holland Publishing Co., Amsterdam.

Lal, D., and B. Peters. (1967). Cosmic ray produced radioactivity on Earth, *Encyclopedia of Physics*, Vol. 46, No. 2, Springer, New York, pp. 551–612.

Larsen, R. J., and C. G. Sanderson (1995). *EML surface air sampling program, 1990–1993 data*. Environmental Measurement Laboratory Report EML-572. US Department of Energy, New York, USA.

Lee, H. N. (2004). Issues and challenges of using natural radionuclides as tracer for atmospheric studies. In: *1st International Expert Meeting on Sources and Measurements of Natural Radionuclides Applied to Climate and Air Quality Studies*. World Meteorological Organization Technical Document No. 1201, pp. 30–34.

Lee, L. Y. L., R. C. W. Kwok, Y. P. Cheung, and K. N. Yu. (2004). Analyses of airborne ^{7}Be concentrations in Hong Kong using back-trajectories. *Atmospheric Environment*, 38, 7033–7040.

Liu, H. Y., D. J. Jacob, I. Bey, and R. M. Yantosca. (2001). Constraints from ^{210}Pb and ^{7}Be on wet deposition and transport in a global three-dimensional chemical tracer model driven by assimilated meteorological fields. *Journal of Geophysical Research*, 106D, 12109–12128.

Masarik, J. (2004). Simulation of the Cosmogenic Nuclide Production in the Earth Atmosphere. In: *1st International Expert Meeting on Sources and Measurements of Natural Radionuclides Applied to Climate and Air Quality Studies*. World Meteorological Organization Technical Document No. 1201, pp. 65–67.

Masarik, J., and J. Beer. (1999). Simulation of particle fluxes and cosmogenic nuclide production in the Earth's atmosphere. *Journal of Geophysical Research*, 104, 12099–12111.

McHargue, L. R., and P. E. Damon. (1991). The global beryllium-10 cycle. *Reviews of Geophysics*, 29, 141–158.

McNeary, D., and M. Baskaran. (2003). Depositional characteristics of ^{7}Be and ^{210}Pb in southeastern Michigan. *Journal of Geophysical Research*, 108, 4210, doi:10.1029/202JD003021.
Megumi, K., T. Matsunami, N. Ito, S. Kiyoda, A. Mizohata, and T. Asano. (2000). Factors, especially sunspot number, causing variations in surface air concentrations and depositions of ^{7}Be in Osaka, Japan. *Geophysical Research Letters*, 27, 361–364.
Mohamed, A. (2005). Activity size distributions of some naturally occurring radionuclides ^{7}Be, ^{40}K and ^{212}Pb in indoor and outdoor environments. *Applied Radiation and Isotopes*, 62, 751–757.
Mohamed, A., and A. El-Hussein. (2005). Comparison of outdoor activity size distributions of ^{220}Rn and ^{222}Rn progeny. *Applied Radiation and Isotopes*, 62, 955–959.
Moore, H. E., S. E. Poet, and E. A. Martell. (1973). ^{222}Rn, ^{210}Pb, ^{210}Bi, and ^{210}Po profiles and aerosol residence times versus altitude. *Journal of Geophysical Research*, 78, 7065–7075.
Nagai, H., W. Tada, and T. Kobayashi. (2000). Production rates of ^{7}Be and ^{10}Be in the atmosphere. *Nuclear Instruments and Methods in Physics Research Section B: Beam Interactions with Materials and Atoms*, 172(1–4), 796–801.
NCRP. (1979). *Tritium in Environment*. National Council on Radiation Protection and Measurement, Washington, DC.
NRC (National Research Council). (1979). *Airborne Particles*. University Park Press, Baltimore, MD.
NRC (National Research Council). (1996). *Nuclear Wastes: Technologies for Separations and Transmutations*. Commission on Geosciences, Environment and Resources, Committee on Separations Technology and Transmutation Systems, 592 pp. ISBN: 0-309-05226-2.
O'Brien, K. (1979). Secular variations in the productions of cosmogenie isotopes in the atmosphere. *Journal of Geophysical Research*, 84, 42–43.
OECD. (1995). *Chernobyl Ten Years On*. Organisation for Economic Cooperation and Development, Paris.
Papastefanou, C., and A. Ioannidou. (1991). Depositional fluxes and other physical characteristics of atmospheric beryllium-7 in the temperate zones (40°N) with a dry (precipitation-free) climate. *Atmospheric Environment*, 25A(10), 2335–2343.
Papastefanou, C., and A. Ioannidou. (1995). Aerodynamic size association of ^{7}Be in ambient aerosols. *Journal of Environmental Radioactivity*, 26, 273–282.
Pillai, P. M. B., A. C. Paul, I. S. Bhat, M. R. Iyer, and K. C. Pillai. (1994). Deposition and clearance of ^{212}Pb in humans. *Health Physics*, 66, 343–345.
Poet, S. E., H. E. Moore, and E. A. Martell. (1972). Lead 210, Bismuth 210 and Polonium 210 in the atmosphere: Accurate ratio measurement and application to aerosol residence time determination. *Journal of Geophysical Research*, 77, 6515–6527.
Povinec, P., M. Chudý, I. Sýkora, J. Szarka, M. Pikna, and K. Holý. (1988). Aerosol radioactivity monitoring in Bratislava following the Chernobyl accident. *Journal of Radioanalytical Nuclear Chemistry*, 126–467.
Preiss, P. J., M.-A. Melieres, and M. Pourchet. (1996). Use of a new database of lead-210 for global aerosol model validation. *Journal of Geophysical Research*, 101(25), 325–347, 357.
Rangarajan, C., and S. S. Gopalakrishnan. (1970). Seasonal variation of beryllium^{7} relative to caesium^{137} in surface air at tropical and sub-tropical latitudes. *Tellus*, 22, 115–121.
Rehfeld, S., and M. Heimann. (1995). Three dimensional atmospheric transport simulation of the radioactive tracers ^{210}Pb, ^{7}Be, ^{10}Be, and ^{90}Sr. *Journal of Geophysical Research*, 100(D12), 26141–26161.
Sanak, J., G. Lambert, and B. Ardouin. (1985). Measurements of stratosphere-to-troposphere exchange in Antarctica by using short-lived cosmonuclides. *Tellus*, 37B(2), 109–115.
Satheesh, S. K., and K. Krishna Moorthy. (2005). Radiative effects of natural aerosols: A review. *Atmospheric Environment*, 39, 2089–2110.
Schwartz, S. E. (1995). Connections between aerosol properties and forcing climate. In: *Aerosol Forcing and Climate* (Eds R. J. Charlson and J. Heintzenberg). John Wiley, New York, pp. 251–280.

Shimo, M., Guo, Q., Ikebe, Y., and S. Minato. (1990). Tn daughters concentration in several indoor and outdoor airs and lung dose due to Rn and Tn daughters. *Proceedings of the 25th Annual Meeting of the Japanese Health Physics Society*, Tsukuba, Japan, 54pp.

Sokolik, I. N., O. B. Toon, and R. W. Bergstrom. (1998). Modeling the radiative characteristics of airborne mineral aerosols at infrared wavelengths. *Journal of Geophysical Research*, 103(D8), 8813–8826.

Steinhausler, F. (1993) Management of radiation protection in the miningand milling and downstream processing of mineral sands. *Proceedings of the International Conference on Occupational Health and Safety in the Minerals Industry*. Minesafe International. ISBN No. 1875449124. Perth, Australia. pp. 239–248.

Stohl, A., N. Spichtinger-Rakowsky, P. Bonasoni, H. Feldmann, M. Memmesheimer, H. E. Scheel, T. Trickl, S. Huebner, W. Ringer, and M. Mandl. (2000). The influence of stratospheric intrusions on alpine ozone concentrations. *Atmospheric Environment*, 34, 1323–1354.

Sykora, I., Meresova, J., and K. Holy. (2005). *Radioactivity of Bratislava atmosphere in 2004.* Progress Report, KJFB FMFI UK-144/05, May 2005, Bratislava.

Takayuki, T., Y. Kenji, H. Koh, and T. Shizuo. (1996). Seasonal variations of residence time and upper atmospheric contribution of aerosols studied with Pb-210, Bi-210, Po-210 and Be-7. *Tellus*, 48(B), 690–702.

Todorovic, D., D. Popovic, M. Radenkovic, and G. Djuric. (1999). Concentration measurements of ^{7}Be and ^{137}Cs in ground level air in Belgrade city area. *Environment International*, 25(1), 59–66.

Turekian, K. K. and W.C. Graustein (2003). Natural Radionuclides in the Atmosphere. In: *The Atmosphere* (Ed. R. F. Keeling), Vol. 4. *Treatise on Geochemistry* (Eds. H. D. Holland and K. K. Turekian) Elsevier–Pergamon, Oxford, pp. 261–279.

Turekian, K. K., W. C. Graustein, and J. K. Cochran. (1989). Lead-210 in the SEAREX program: An aerosol tracer across the Pacific. In: *Chemical Oceanography* (Ed. J. P. Riley). Academic Press, London, pp. 51–81.

Turekian, V. C., W. C. Graustein, and K. K. Turekian. (1999). The ^{214}Bi to ^{214}Pb ratio in lower boundary layer aerosols and aerosol residence times at New Haven, Connecticut. *Journal of Geophysical Research*, 104, 11593–11598.

Turekian, K. K., Y. Nozaki, and L. K. Benninger. (1977). Geochemistry of atmospheric radon and radon products. *Annual Review of Earth and Planetary Sciences*, 5, 227–255.

UNSCEAR. (1982). *Ionizing radiation sources and biological effects*. United Nations, New York.

UNSCEAR. (2000). *Sources and Effects of Ionizing Radiation*. United Nations, New York.

U.S. Department of Energy. (2001). Summary data on the radioactive waste, spent nuclear fuel, and contaminated media managed by the U.S. Department of Energy, Washington, http://cid.em.doe.gov

Viezee, W., and H. B. Singh. (1980). The distribution of Be-7 in the troposphere; implications for stratospheric/tropospheric exchange. *Geophysical Research Letters*, 7, 805–808.

Wilkening, M. H., W. E. Clements, and D. Stanley. (1975). Radon-^{222} flux in widely separated regions. In: *Natural Radiation Environment II* (Ed. J. A. S. Adams). US Energy Research and Development Administration, Oak Ridge, TN, pp. 717–730.

Winkler, R., F. Dietl, G. Frank, and J. Tschiersch. (1998). Temporal variation of ^{7}Be and ^{210}Pb size distributions in ambient aerosol. *Atmospheric Environment*, 32(6), 983–991.

WMO/GAW. (2003). *Aerosol Measurement Procedures Guidelines and Recommendations No. 153*. September, WMO TD No. 1178.

WMO/GAW. (2004). *1st International Expert Meeting on Sources and Measurements of Natural Radionuclides Applied to Climate and Air Quality Studies*. World Meteorological Organization Technical Document No. 1201.

Zanis, P., E. Schuepbach, H. W. Gaeggeler, S. Huebner, and L. Tobler. (1999). Factors controlling Beryllium-7 at Jungfraujoch in Switzerland. *Tellus*, 51B, 789–805.

CHAPTER 4

Radiocarbon as a Tracer in the Global Carbon Cycle

Gordon T. Cook[1,*], E. Marian Scott[2] *and* Douglas D. Harkness[3]

Contents

*Corresponding author. Tel.: +44-1355-270136; Fax: +44-1355-229898
E-mail address: g.cook@suerc.gla.ac.uk

[1] Scottish Universities Environmental Research Centre (SUERC), Radiocarbon Dating Laboratory, Scottish Enterprise Technology Park, East Kilbride G75 0QF, UK
[2] Statistics Department, University of Glasgow, Glasgow G12 8QW, UK
[3] NERC Radiocarbon Laboratory, SUERC, UK

Radioactivity in the Environment, Volume 16
ISSN 1569-4860, DOI 10.1016/S1569-4860(09)01604-0

1. Introduction

The earth's life-support system is contained within the global carbon cycle and therefore mankind is entirely dependent on the complex interactions between the atmosphere, oceans, land and, of course, living organisms themselves. Scientific observations over many years have established that the environmental conditions sustained by the natural biogeochemical cycling of carbon are not comfortably static but can be liable to significant and often catastrophic change. Furthermore, such changes can have a geographical distribution ranging from regional through to global scales. Much of the evidence for past environmental change points clearly to the interaction and interdependence of natural forces. However, observations and measurements reported since the mid-20th century provide irrefutable proof that man has become a major contributor to the inducement of global environmental change. A particularly pressing priority concerns the effect of modern industrialisation and land management practices on global climate. Since the late 19th century, anthropogenic practices have resulted in a continuing increase in the concentration of carbon dioxide (CO_2) in the atmosphere with its resultant spectre of global warming and the consequential aspects of climate change – the so-called 'greenhouse effect'!

Clearly, we need to improve and refine our knowledge of those natural and anthropogenic processes that determine the distribution and exchange of carbon between the major reservoirs, in order to design appropriate strategies for objective management of the global environment. For example, given our present concern over the potential for anthropogenic-induced global warming, we need to (i) research the extent, timing and causes of past episodes of climate change and (ii) develop mathematical models to quantify the interdependence of the global climate system to external forcing. In this context, it is particularly fortunate that the element carbon occurs in nature in three isotopic forms, *viz.* ^{12}C and ^{13}C, both of which are stable, and the radioactive species, ^{14}C (radiocarbon). The relative abundance of these isotopes differs among specific compartments of the carbon cycle and this pattern can be exploited to provide a quantitative measure of the crucial pathways and dynamics for carbon exchange, for example, between the ecologically sensitive atmosphere and the potential storage sinks for excess anthropogenic CO_2, that is the terrestrial biosphere and the oceans.

It is often remarked that predictions of the future are best formulated against experiences from the past. This sentiment is particularly relevant in the context of studies regarding the sensitivity of the modern global environment to changes in the natural distribution of carbon. ^{14}C provides a timeframe that is universally applicable to events pertaining to the past 50,000 years or so of earth's history. This chronological ordering of the evidence for past events covers:

(i) the last major climatic shift from full glacial to the temperate conditions that have prevailed for the past 13,000 years or so, and
(ii) the prehistoric development of human civilisation.

Timing is also important in the context of the dynamics of mass exchange between specific carbon pools and allied to their respective retention capacity. In particular, we need to assess the ability of the most sensitive regions of the natural carbon cycle (arguably the atmosphere) to absorb and then recover from rapid and violent forcing events, whether these are natural and/or anthropogenic. In this situation, the benefits of the conventional radiocarbon timescale are usefully supplemented by changes imposed by human activities on the modern environmental distribution of the heavier carbon isotopes, that is ^{13}C and ^{14}C. For example, the excess CO_2 released by the utilisation of fossil fuels is '^{14}C-free' and also carries a stable (^{13}C) isotopic signature that is significantly different from that of the pre-industrial atmosphere. However, the most significant event in terms of such carbon isotope tracer studies was the massive and, on the ^{14}C timescale, virtually instantaneous injection of ^{14}C into the upper atmosphere during the nuclear weapons test programmes of the late 1950s and early 1960s. The progressive redistribution of this "bomb ^{14}C" throughout the dynamic carbon cycle continues to afford an otherwise impossible degree of time resolution for the rates of atmospheric and biospheric processes.

Thus, in ^{14}C, we have a tracer that enables us to study the timing and duration of past changes in our environment, both prior to and during the period of man's influence on the environment, through the measurement of the natural ^{14}C abundance. While, furthermore, recent anthropogenic carbon inputs (from fossil fuel burning and bomb ^{14}C) have given us the potential to study rates of mixing and exchange within and between the major carbon-containing reservoirs.

2. Global Distribution of Carbon

2.1. Carbon inventories

The element carbon provides the key building block for all organic material, living and dead. It is also an important component of the atmosphere. Carbon occurs in significant amounts and in various molecular forms in the oceans, the continental hydrosphere and the earth's crustal rocks. The total amount of carbon contained in these pools still remains subject to considerable debate and numerical uncertainty. Bolin and Fung (1992) provide a comprehensive re-evaluation of the earth's carbon inventories and the associated rates of exchange (carbon fluxes) across their common boundaries. The authors also highlight and discuss key issues that remain outstanding for further research to improve our understanding of these parameters and their associated environmental management issues.

An objective appreciation of the geological and environmental processes that have given rise to and sustain the natural distribution of carbon is most conveniently achieved by the concept of a biogeochemical cycle. A schematic representation of the major carbon reservoirs that must be recognised in this quasi-steady-state model is depicted in Figure 1. Timing is a major consideration in evaluating the distribution of carbon in nature.

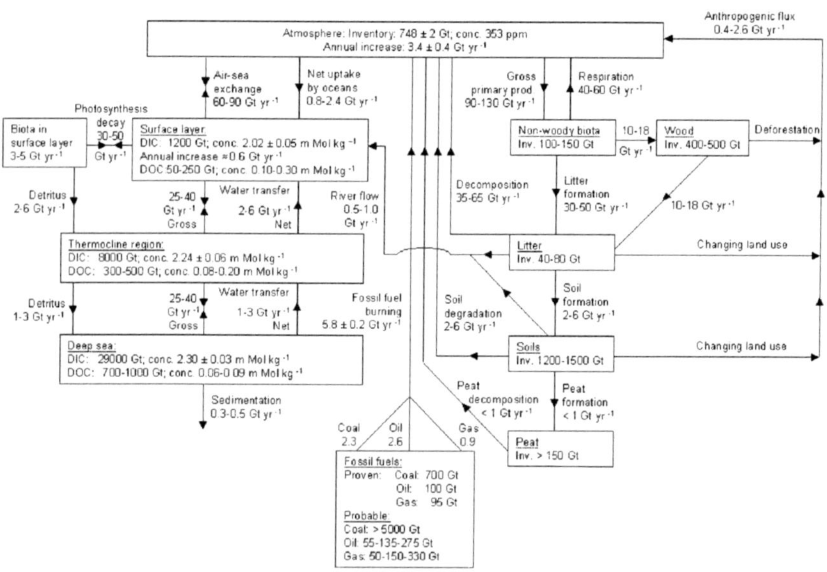

Figure 1 The carbon cycle. The magnitudes of the carbon compartments are given in gigatons (Gt $= 10^9$ ton) of C. The fluxes between these are expressed in Gt C/yr. (Adapted from Bolin and Fung, 1992 with permission from University Corporation of Atmospheric Research/Office of Inter-Disciplinary Earth Sciences.)

The present global distribution of carbon is the result of physical, chemical, biological and geological processes that have proceeded over many thousands of years. These processes proceed at widely differing rates. For example, carbon turnover in geological processes generally requires tens of thousands to millions of years whereas atmospheric mixing and/or photosynthetic fixation reactions are often characterised by timescales of 1 year or less. It is convenient therefore to differentiate between the 'slow' or 'sedimentary' and the 'dynamic' compartments within the carbon cycle model. This chapter focuses on the relatively rapid processes of natural carbon exchange, that is, the recognition of those carbon reservoirs that contribute to the dynamic system with characteristic internal turnover times of <10,000 years. The limit here is set arbitrarily to correspond with the mean age of basal post-glacial peat deposits. There is one unavoidable exception in later discussion, that is, the anthropogenic transfer of significant amounts of carbon from the essentially static carbon reservoir comprising coal, oil and gas deposits directly to the earth's atmosphere.

It is appropriate at this time to highlight some significant features of the carbon mass distribution summarised in Figure 1.

The atmosphere is by far the smallest of the dynamic reservoirs, with <2% of the total mass of dynamic carbon. It is also the most homogeneous, comprising gaseous molecules; primarily CO_2 gas (at approximately 370 ± 2 ppmv of dry air) with traces of methane. These characteristics make it the most sensitive reservoir to any perturbation of the otherwise near-equilibrium attained in the natural distribution pattern.

The world's oceans are by contrast the largest sink for dynamic carbon, containing more than 92% of the mass inventory. Most (ca. 96%) of the carbon retained by the oceans is present in dissolved inorganic (DIC) form, that is as CO_2, CO_3^{2-}, HCO_3^- and carbonic acid. Approximately 3% occurs as dissolved organic carbon (DOC), whereas marine organisms account for <1% of this total carbon inventory. Due to the relatively slow rates of mixing, particularly of deep waters, the oceanic reservoir is also the most heterogeneous in terms of its ^{14}C distribution. Near-surface waters have apparent radiocarbon ages of a few hundred years (the global average is around 400 years) whereas the carbon contained in the deep water that upwells around the Antarctic continent has ages that are typically well in excess of 1,000 years.

The bulk reservoir of carbon represented by *the terrestrial biosphere* presents the greatest diversity in terms of both the size of its component carbon pools and their respective internal turnover rates. The actual size of these various organic deposits is also least confidently recorded. The predominant molecular forms in the biospheric inventory are the hydrocarbon structures associated with the cellular structure of plants and their degradation products (Summons, 1993). An obvious distinction is between living matter and recognisable organic detritus. The best available estimates point to carbon in living structures representing <30% of the biospheric inventory (Bolin and Fung 1992). With few exceptions, the immediate source of carbon in living matter is atmospheric CO_2 via the agencies of photosynthesis and/or feeding. The mean retention or turnover time for carbon in living materials is therefore commensurate with the average rate of tissue renewal. This is a paramount assumption in conventional radiocarbon dating (see Sections 3.2 and 3.3). A small but significant amount of carbon is contained in relatively young plant litter on the soil surface; however, the bulk of organic detritus is considerably older and contained as humus in soils and peat deposits.

2.2. Carbon isotope geochemistry

The relative abundances of the three isotopes of carbon are ^{12}C – 98.89%, ^{13}C – 1.11% and ^{14}C – 1×10^{-10}%. Slight but significant variations occur in these median values due to isotopic preferences in the transfer of carbon between different geochemical phases or, in the case of ^{14}C, as the result of ongoing radioactive decay.

^{14}C is produced via the bombardment of atmospheric nitrogen atoms by a relatively narrow energy range of cosmic-ray produced neutrons (Korff, 1940; Lingenfelter, 1963; Castagnoli and Lal, 1980).[1] The neutron flux originates from spallation reactions caused by the interaction of the high-energy component of primary cosmic radiation with atoms and molecules in the upper atmosphere. A fraction of these cosmic-ray-produced neutrons escapes back into space but the energy of a significant majority is reduced to thermal levels by successive collisions, at which point they become available for capture by the abundant supply of atmospheric nitrogen atoms, forming ^{14}C with the release of a proton

$$^{14}N(n, p)^{14}C$$

Once formed, the radioactive carbon atoms are oxidised to $^{14}CO_2$ and as such become a component of the natural carbon cycle.

The rate of natural ^{14}C production is not geographically uniform (Lingenfelter, 1963). The shielding effect of the earth's geomagnetic field on the incoming cosmic rays results in the intensity of secondary cosmic neutrons at the geomagnetic poles being about five times greater than at equatorial latitudes.[2] Fortunately, the resultant latitudinal gradient in ^{14}C production rate effectively disappears as a result of rapid mixing of the gaseous $^{14}CO_2$ molecules that transfer from the stratosphere to the physically well-mixed lower atmosphere (Harkness, 1970). However, temporal variations in atmospheric ^{14}C concentration are apparent and these become significant for interpretation of the conventional radiocarbon timescale (see Section 3.3).

[1] Chapter 1 discusses production mechanisms and rates of cosmogenic radionuclides including ^{14}C in detail.
[2] See also Chapter 1.

Light et al. (1973) estimated the natural cosmogenic ^{14}C production rate to be 1.4×10^6 GBq y^{-1}, which is in reasonable agreement with the later estimates of Bush et al. (1983) and UNSCEAR (1993) of 9.2×10^5 and 1×10^6 GBq y^{-1}, respectively.

The total inventory of natural ^{14}C within the carbon cycle is determined by the equilibrium balance achieved over many thousands of years between an approximately constant global production rate and the immutable radioactive decay process. The concentration of ^{14}C at a particular location within the carbon cycle is defined by the extent to which the radioactive decay process predominates over direct isotopic exchange with the production source, that is the earth's atmosphere. The maximum natural specific activity of ^{14}C occurs in atmospheric CO_2 and in the absence of anthropogenic disturbance is approximately 13.56 dpm g^{-1} carbon (0.226 Bq g^{-1} carbon). As a result of radioactive decay, ^{14}C reverts to stable nitrogen (^{14}N) by the emission of a beta particle ($E_{max} = 156$ keV) and an anti-neutrino.

$$^{14}C \rightarrow {}^{14}N + \beta^- + \nu$$

2.3. Anthropogenic disturbance

Since the late 19th century, the relative isotopic abundances of carbon in the dynamic cycle have been altered progressively and significantly by human activities: Firstly, since the onset of the industrial era, by the combustion of fossil fuels (the *Industrial* or *Suess effect*) and then more recently by the detonation of thermonuclear devices in the upper atmosphere (the *Bomb* or *Nuclear Effect*).

2.3.1. The Suess effect

Early researchers in applied radiocarbon measurement found that the ^{14}C concentration recorded by known-age tree rings in relatively modern wood appeared to reflect a progressive dilution of the ambient atmospheric ^{14}C concentration. This phenomenon was particularly evident from the late 19th century onwards and by 1950 had culminated in an apparent 3% decrease relative to anticipated values (Cain and Suess, 1976). The primary source of the ^{14}C dilution was traced to the release into the atmosphere of increasing amounts of '^{14}C-free' CO_2 produced by the combustion of fossil fuels, that is coal, oil and natural gas. Because of their great geological age, these fossil organic derivatives are effectively free from measurable ^{14}C. With the identification of the 'Suess effect', it was soon realised that materials that had grown since the late 19th century onwards could not be used to define the zero-aged reference activity required for implementation of a conventional radiocarbon timescale. This was a major factor in the ultimate decision to prepare and employ contemporary reference standards such as the US National Bureau of Standards (NBS) oxalic acid (Section 5.3).

2.3.2. The bomb effect

The onset of nuclear weapons testing programmes in the mid-1950s resulted in the release of large fluxes of thermal neutrons to the upper atmosphere. As in the natural production reaction, this excess of subatomic particles was readily captured by atmospheric nitrogen atoms and produced significant amounts of artificial or 'bomb' ^{14}C (Rafter and Fergusson, 1957). Between 1954 and 1963, the specific activity of ^{14}C, monitored in ground-level air at mid to high latitudes of the northern hemisphere, had effectively doubled. A moratorium agreement in 1963 halted atmospheric testing by the majority of the nuclear powers and, in the absence of a further significant input of artificial ^{14}C, internal mixing and the exchange of carbon with the adjacent reservoirs of the carbon cycle resulted in a steady decline in atmospheric ^{14}C concentrations (see Figure 2). Temporal and

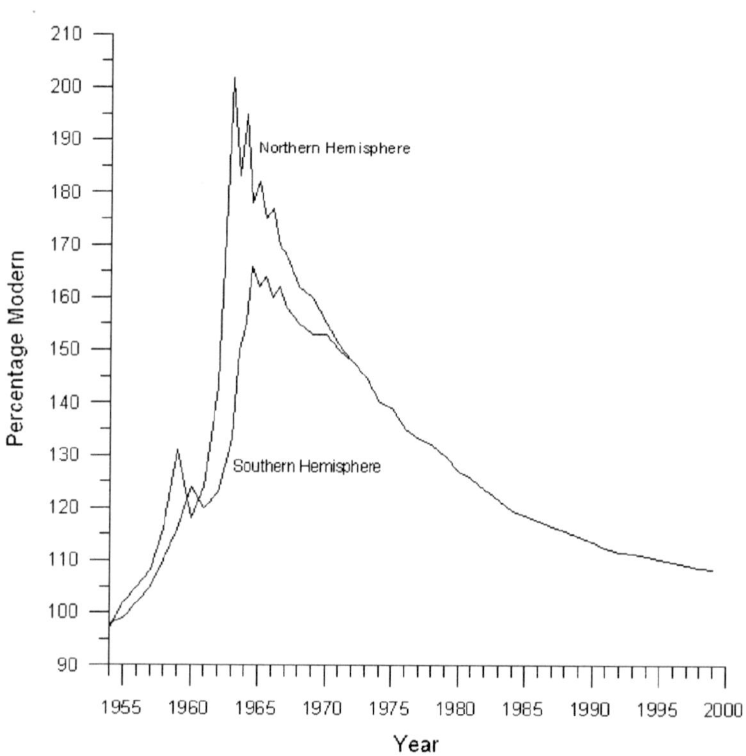

Figure 2 Long-term observation of $\Delta^{14}C$ in atmospheric CO_2 in the northern and in the southern hemisphere (after Levin and Hesshaimer, 2000).

spatial variations in the distribution of bomb ^{14}C within the atmosphere and the decline in this atmospheric inventory were closely monitored and reported by several research groups (Tauber, 1960; Willis et al., 1960; Nydal, 1963, 1966; Munnich and Vogel, 1963; Rafter, 1965; Walton et al., 1970). Since most of the 'bomb' ^{14}C was injected into the stratosphere over high northern hemisphere latitudes, it could hardly have been better designed as a tracer spike for the dispersal of natural production. Near-ideal conditions prevailed to determine contemporary mixing patterns and to quantify, with hitherto impossible resolution: (i) the rates of exchange of carbon between the major reservoirs of the carbon cycle ('turn-over times') and (ii) the storage capacity ('mean residence time') for carbon within a particular component of the natural cycle (Broecker and Peng, 1982) (see Section 6). The opportunity for 'bomb ^{14}C' tracing is of course transient since the bomb excess will be diluted eventually to an imperceptible level by near-equilibrium distribution throughout the relatively massive inventory of naturally produced radiocarbon.

As a consequence of these effects, it is widely recognised that 1890 AD represents the last year when the ^{14}C record was not significantly perturbed by anthropogenic effects. This point will be discussed in more detail in Section 5.3, in relation to ^{14}C standards.

3. The Conventional Radiocarbon Dating Technique

3.1. Brief history

In the general ambit of scientific awareness, conventional ^{14}C dating is perhaps the most widely recognised of all the radionuclide-based methods available for reconstructing a timescale for discrete episodes of earth history.

While the major site of formation of ^{14}C and its production mechanism were first postulated by Korff (1940), it was the work of Willard Libby and his co-workers at the Institute for Nuclear Studies and Department of Chemistry at the University of Chicago that led to the development of the conventional radiocarbon dating technique (e.g. Libby, 1946; Anderson et al., 1947; Libby et al., 1949). Libby (1946) postulated that ^{14}C should be present in living matter and, subsequently, Anderson et al. (1947) demonstrated its presence, following thermal diffusion enrichment, at an activity of 10.5 dpm g^{-1} C in methane recovered from the city of Baltimore sewage system. A subsequent worldwide assay of living terrestrial material (Libby et al., 1949) demonstrated apparent uniformity of ^{14}C with an average activity of 12.5 ± 0.2 cpm g^{-1} C. The value was intended to be absolute disintegrations per minute but the authors recognised that the absolute calibration of their detectors could be as much as 10% in error.

Given that their detectors were of very low efficiency (about 6%) and the background was approximately 10 cpm, their value is remarkably close to the true 1940s atmospheric value of about 13.15 dpm g^{-1} C. The group also dated wood samples from two ancient Egyptian tombs approximately 4,600 years old and obtained count rates from the samples that were consistent with the activity of their present day samples and their half-life for ^{14}C (Libby et al., 1949). This was followed by further studies on (i) dating more samples of known age, and (ii) comparing the ^{14}C inventory with predictions from cosmic ray data and estimates of total exchangeable carbon (Arnold and Libby, 1949; Anderson and Libby, 1951). Then, in 1951, Arnold and Libby (1951) published the first ever radiocarbon date list in 'Science'. Libby's book *Radiocarbon Dating* appeared a year later (Libby, 1952) and within the next 3 years a further seven radiocarbon dating laboratories had been established. Furthermore, the breadth of opportunity afforded by the dating method was emerging. For example, measurements made at the USGS laboratory in Washington, DC during 1953/1954 (Suess, 1992; Rubin and Suess, 1955) provided evidence to the effect that:

(i) The maximum extent of the North American ice sheet had occurred around 18,000 years previously.
(ii) The progressive release of ^{14}C-free CO_2 to the atmosphere from the combustion of fossil fuels had caused a measurable decrease in the natural ^{14}C activity in the atmosphere.
(iii) *Homo neanderthalensis* had survived in some regions of Europe until about 30,000 years ago.

Some of the earliest age calculations (Arnold and Libby, 1949) were carried out using a half-life of $5{,}720 \pm 47$ years, determined by Engelkemeir et al. (1949); however, a value of $5{,}568 \pm 30$ years (now known as the Libby half-life) was used for the age calculations presented in the first date list and was subsequently adopted universally. This half-life was subsequently re-evaluated and accepted at $5{,}730 \pm 40$ years (Godwin, 1962). However, since a significant number of radiocarbon age measurements had been reported and interpreted on the basis of the Libby half-life, it is an international agreement that all conventional radiocarbon age values are still calculated and reported on that basis (Stuiver and Polach, 1977; Mook and van der Plicht, 1999).

By 1967, some 80 registered laboratories worldwide had published date lists in the dedicated journal *Radiocarbon*. Their efforts had by that time resulted in approximately 2,650 English language articles referring to applied ^{14}C measurement scattered throughout some 360 scientific journals (Polach, 1988). This latter statistic highlights the fact that, within its first 20 years, applied natural ^{14}C measurement had grown into a multidisciplinary science.

A prominent feature of this development was and continues to be the mutual benefit achieved by two-way exchange between the isotope science community *per se* (the providers) and the much more diverse group of natural scientists (the users) who have a direct interest in the method's dating and/or tracing capabilities. This relationship has proved a symbiotic success: The users present the challenge to apply meaningful isotope analyses to a wide range of specific questions. The providers advance their technical capabilities to match the set challenge and in doing so can often demonstrate to existing and/or potential users an opportunity for entirely new areas of applied ^{14}C science. At the close of the 20th century, there were at least 130 specialist radiocarbon laboratories operating worldwide.

In addition to upper-atmospheric production, ^{14}C is also formed in surface minerals but at a significantly lower rate (of the order of tens of atoms of ^{14}C per gram of SiO_2 per annum). This formation occurs by a number of reactions including $^{16}O(n, 2pn)^{14}C$ and $^{17}O(n, \alpha)^{14}C$(Gosse and Phillips, 2001). Davis and Schaeffer (1955) first proposed that those cosmogenic radionuclides, formed *in situ* within surface minerals, could be used for studying rates of certain geological processes.[3] However, it was not until the late-1970s/1980s when sufficiently sensitive mass spectrometers were developed, that the possibilities of *in situ* ^{14}C dating began to be explored. Nevertheless, the ability to extract and measure *in situ* produced ^{14}C from minerals has only recently been accomplished (Handwerger et al., 1999; Lifton et al., 2001). The development of the accelerator mass spectrometry (AMS) technique is undoubtedly also the most important advance in conventional ^{14}C dating as it reduced the sample size requirement from the gram quantities of elemental carbon required for radiometric measurement (i.e. measuring the number of radioactive decays, see Section 4.2) down to milligram quantities. In turn, this enabled ^{14}C dating of much more precisely selected samples or, alternatively, the dating of specific geochemical fractions that have been isolated from samples.

3.2. Fundamental theory

The 1960 Nobel Prize for Chemistry was awarded to Willard F. Libby for, in the words of the citation, 'his method to use Carbon–14 for age determinations in archaeology, geology, geophysics and other branches of science'. With hindsight, the logic of Libby's basic dating theory was elegant in its simplicity. He reasoned that if a relatively constant cosmic ray flux had prevailed over a period equivalent to 6 or 7 ^{14}C half-life values, that is approximately 35,000 years, then an equilibrium situation would be established whereby the ongoing production of ^{14}C would be balanced by the rate of decay of those radioactive atoms stored within the various

[3] This use of *in situ* produced cosmogenic radionuclides is the main subject of Chapter 6.

reservoirs of the natural carbon cycle. Furthermore, if the relative distribution of carbon within and between the compartments of the carbon cycle had remained essentially unaltered then the carbonaceous compounds in each reservoir would exhibit a characteristic and constant isotopic signature. He recognised terrestrial life forms as a particular case.

Thus, the ^{14}C atoms that are produced in the upper atmosphere by the $^{14}N(n, p)^{14}C$ reaction are rapidly oxidised to $^{14}CO_2$ and mix with the $^{12}CO_2$ and $^{13}CO_2$. The CO_2 is taken up by green plants in the terrestrial biosphere and converted to carbohydrates by the process of photosynthesis. Subsequent consumption of plants by animals (and animals by other animals) transfers the ^{14}C throughout the entire terrestrial food chain. It is now well established that there is global uniformity in natural $^{14}C/^{12}C$ enrichment between the well-mixed atmosphere and the terrestrial flora and fauna that atmospheric carbon supports, provided that due corrections are made for the degree of isotopic fractionation that takes place during the initial uptake and subsequent metabolic fixation of atmospheric carbon by the primary producers (plant life) and, thereafter, any fractionation that occurs during transport through the food chain. In contrast, however, the oceans and the life that they support represent a rather heterogeneous reservoir that is not in equilibrium with the atmosphere. The problems encountered in dating samples of a marine origin will be discussed later.

The four fundamental assumptions in the conventional radiocarbon dating method are that:

1. The rate of formation of ^{14}C in the upper atmosphere has been constant over the entire applied ^{14}C dating timescale (approximately the last 50,000 years).
2. The activity of the atmosphere and the biosphere with which it is in equilibrium has been constant over the applied timescale.
3. The rate of transfer of ^{14}C between different reservoirs of the carbon cycle is rapid with respect to the average lifetime of ^{14}C (approximately 8,300 years).
4. The half-life is accurately known.

On the basis of the above assumptions, all living organisms, throughout the entire applied ^{14}C dating timescale, would have been labelled to the same extent with ^{14}C during life. As long as the organism was alive this isotopic equilibrium prevails. However, after death, the isotopic equilibrium with the atmospheric carbon source is no longer sustained and the concentration of ^{14}C in the detrital organic carbon decreases exponentially with passing time and in accordance with its immutable rate of radioactive decay. It follows therefore that if the half-life value of ^{14}C is known then a direct comparison of the residual ^{14}C specific activity in the detrital carbon sample with that in zero-aged (living) material can provide an index of the time elapsed since death of the organism.

There is one other rather obvious, yet often poorly considered, requirement, *viz.* that a secure relationship exists between the sample selected for isotopic analysis and the specific event or phenomenon that it is assumed to have coincided with its death.

3.3. Shortcomings in the fundamental theory

The experience of over 50 years of applied ^{14}C dating has shown that, although Libby's basic assumptions hold in a general sense, none is strictly true, and each requires careful consideration and often a degree of adjustment or fine-tuning to ensure best scientific return from the radiocarbon chronology. Examination of the detail within the four assumptions listed above reveals the following assumptions.

3.3.1. First assumption

Perhaps the most fundamental assumption of the basic dating theory is that the relative isotopic abundance of carbon atoms in a particular reservoir has remained constant over the duration of the radiocarbon timescale. For example, the ideal situation requires that plants and animals living at the present time carry an identical $^{14}C/^{13}C$ ratio to those that lived at any time during the past 50,000 years or so of earth history. This assumption was first challenged by discrepancies reported between the measured radiocarbon ages and the known calendar growth dates of relatively modern wood samples (deVries, 1959). A follow-up investigation by the dating laboratories at Cambridge, Copenhagen and Heidelberg proved that the ^{14}C ages recorded by known age tree-ring growth sequences showed a pattern of consistent discrepancies of up to 160 years over the previous 1,300 years of dendrochronologically established tree growth (Willis et al., 1960). Indeed, direct measurement of the variations in atmospheric ^{14}C concentration, recorded in annually produced tree rings, indicate that throughout post-glacial time (the last 10,000 years or so), the concentration of ^{14}C in the earth's atmosphere has varied by up to 12% due to variations in the ^{14}C production rate. There are both short- and long-term variations in natural production rate (Suess, 1965). The short-term fluctuations are usually attributed to heliomagnetic modulation of the primary cosmic ray flux (Stuiver, 1961; Damon et al., 1989), that is, to changes in solar sunspot activity in which periods of high activity result in decreased cosmic ray incidence on the earth and hence decreased ^{14}C production. The longer-term fluctuations (millennia scale) are attributed to geomagnetic modulation, that is the charged cosmic rays, which ultimately bring about ^{14}C production, are deflected to a greater or lesser degree depending on the earth's dipole moment (Elsasser et al., 1956; Sonett, 1992; Sternberg, 1992).

3.3.2. Second assumption

It follows from the above that variations in production rate will influence the activity of the atmosphere and the biosphere with which it is in equilibrium, and this is further complicated by changes in reservoir sizes (e.g. associated with glaciation/de-glaciation) and their influence on ^{14}C fluxes and reservoir activities. This is most easily demonstrated by the fact that the calibration curve is a non-linear function.

3.3.3. Third assumption

This assumption is, in the main, valid for the well-mixed atmosphere and the terrestrial flora and fauna that it supports; however, the situation is more complex for samples formed in the aquatic environment where the initial ^{14}C specific activity may differ from the contemporaneous atmosphere. The concept of reservoir effects is well established for the oceanic environment, which is a rather large (containing about 92% of the dynamic carbon) and heterogeneous reservoir with respect to radiocarbon activity. ^{14}C uptake occurs by exchange at the ocean–atmosphere interface; however, at the polar regions, where the world's main deep waters are produced, surface waters sink because of their increased density (brought about by cooling) and then travel slowly through our deep oceans (for many hundreds of years) before finally reaching the surface again. Of course, during this slow circulation of deep water, ^{14}C decay is occurring so that even after mixing with surface water, the oceans are still depleted relative to the atmosphere. The slow movement and mixing of deep-water currents, combined with their complex pattern of circulation and upwelling, result in significant heterogeneity with respect to ^{14}C, both temporally and spatially (Stuiver and Braziunas, 1993; Ascough et al., 2004).

Lakes that have "old carbon" in the dissolved inorganic component of the water column (mainly due to carbonate in the surrounding catchment) require the calculation of site-specific offsets from the atmospheric activity. These may be derived by the measurement of contemporaneous paired samples that derived their carbon from the water column or atmosphere (e.g. submergent and emergent plant macrofossils) taken from the sediment; however, there is always the question of how constant an offset has been through time. Recently, a reservoir effect of a similar magnitude to the average marine effect has been noted in the River Danube (Cook et al., 2001).

3.3.4. Fourth assumption

Radiocarbon ages are all calculated using an incorrect half-life (the Libby half-life of 5,568 years) despite the fact that the accepted physical half-life value is 5,730 $\pm$ 40 years (Godwin, 1962).

The fine-tuning required to account for problems associated with the first, second and fourth assumptions, can be achieved using (atmospheric based) calibration curves (discussed below). Marine calibration curves, which are modelled curves based on the atmospheric activities, have been established to account for the depletions in the oceans.

3.4. Calibration

The realisation that it was necessary to make a clear distinction between 'conventional ^{14}C years' and 'calendar years' prompted several laboratories to specialise in the production of a calibration curve that could be used to reconcile measured radiocarbon ages with true calendar time. The approach was to employ a direct comparison of precisely measured ^{14}C ages with those values that could be ascribed to the same carbonaceous sample using independent dating methods.

3.4.1. The calibrated timescales

Radiocarbon age measurements are always reported in years BP $\pm$ 1 sigma error, where 'BP' represents 'before present' and 'present' is the year 1950 AD. A radiocarbon age is calculated on the assumption that the atmospheric radiocarbon concentration has been constant through time (and equal to that in 1890, see Section 5.3 on ^{14}C standards). However, as we have seen, this basic assumption is not entirely valid, and as a result, radiocarbon ages must be calibrated to the calendar timescale.

Calibration of radiocarbon measurements therefore requires knowledge of the levels of atmospheric radiocarbon in the past. This has been provided through ^{14}C measurements made on long series of known-age material including annually produced tree rings, corals and varved sediments of marine origin, etc. The most important material for the Holocene is dendrochronologically dated tree rings. This record has provided material of known age that has been used by radiocarbon practitioners to produce a calibration curve that plots absolute age of the tree rings (normally in decadal or bi-decadal spans) against radiocarbon age. The first such dendro-dated series was based on the very long-lived Bristlecone pine from North America, and was created in the 1960s (Suess, 1970). Publication of the Bristlecone pine calibration curve stimulated a considerable interest in the refinement and temporal extension of the tree-ring record of past variations in atmospheric ^{14}C concentration (Damon et al., 1972; Stuiver, 1982; Pearson et al., 1986). In particular was the development of high-precision dating techniques (with the reduction of analytical uncertainties to <20 years) and the availability of dendro-dated samples of fossil wood from deposits in Ireland and Germany to supplement and compare with the record from North America. Collaborative research led to the progressive publication of a series of precise calibration curves extending back to cover the past 6 millennia (Stuiver and Pearson, 1986, 1993; Pearson and Stuiver, 1986, 1993). These data sets conformed to the features of the original Bristlecone pine record and also served to confirm a much more complex pattern of short-term variations in atmospheric ^{14}C concentration. Beyond the upper limit of the European oak chronologies, a cross-dated sequence of sub-fossil pine recovered from southern Germany and northern Italy (Becker and Kromer, 1986) allowed extension of the high-precision calibration back to 11,400 years BP (Becker, 1993; Becker and Kromer, 1993.) and finally to the current limit of 12,460 years BP (Friedrich et al., 2004). ^{14}C measurements on corals and varved marine sediments have extended the timescale back to approximately 26,000 years (Reimer et al. 2004). This is achieved by converting the marine data to the atmospheric equivalents using a site-specific marine reservoir correction. Nevertheless, there remain some issues concerning the comparability of these different materials. A number of special issues of the journal *Radiocarbon* (Vols. 28, 35, 40 and 46) have provided a compendium (and best knowledge at the time) of the calibration data.

The availability of a high-precision data set derived from the ^{14}C specific activity in these materials allows the ready conversion of conventional radiocarbon ages to calendar dates. However, complexities arise due to the complex pattern of short-term ^{14}C variations (typically of 100 to 200 years duration) in the temporal record. In conjunction with the unavoidable analytical uncertainty associated with all natural ^{14}C measurements, this can often result in a poorer resolution of real time, that is, a significant increase in the uncertainty range for the calibrated age measurement.

Once calibrated, a radiocarbon date should be expressed in terms of a cal BC or cal AD range that should correspond to historical years BC and AD. The cal prefix indicates that the dates are the result of radiocarbon calibration. The term cal BP means the number of calendar years before 1950.

3.4.2. Calibration programmes

There are a number of calibration programmes available for terrestrial samples; amongst the most widely used are Calib, BCal and OxCal. These can be found at the following web addresses: depts.washington.edu/qil, bcal.shef.ac.uk, rlaha.ox.ac.uk. The latter two have the ability to adopt a Bayesian approach. These programmes effectively use the same primary data sets although they may produce slightly different cal ranges because of the different mathematical routines employed. Marine calibration curves are employed for samples of marine origin or samples that contained a marine component within their diet. The latest (Marine04), which covers the period from 0 to 26 kyr BP, uses dendrochronologically dated tree-ring samples for the period 0–10.5 kyr. These data are converted to marine mixed-layer ages using a box diffusion model. Between 10.5 and 26 kyr, high-resolution ^{14}C data are derived from foraminifera in varved sediments and U/Th dated corals (Hughen et al., 2004). For totally marine-derived samples, site- and time-specific ΔR values (deviations from the global average depletion) are required to be input to these models for maximum chronological accuracy while mixed curves can be used for samples containing proportions of both terrestrial and marine carbon. Currently, considerable research effort is being directed towards the assessment of site-specific ΔR values and temporal and spatial variations in this parameter (Angulo et al., 2005; Ascough et al., 2004, 2005, 2006; Cage et al., 2006). Marine sample calibration is a feature of both the Calib and OxCal programmes.

For the period after 1950, there is a great deal of data on the atmospheric radiocarbon concentration, which has varied widely because of anthropogenic influences such as the burning of fossil fuels and atmospheric testing of nuclear devices (see Section 2.3). Although there is no internationally agreed calibration curve, Queen's University, Belfast has a very useful calibration curve called CALIBomb (http://calib.qub.ac.uk/CALIBomb/) that can be used for post-1950 terrestrial samples.

3.4.3. Methods of calibration

The method used to calibrate a ^{14}C age has also changed over time. Originally, the intercepts of the radiocarbon age (at ± 2 sigma) with the calibration curve were projected down onto the calendar timescale, to give a calendar age range. This range corresponded to the years in which the radiocarbon activity in tree rings was within two standard deviations of the ^{14}C measurement. More recent calibration programmes use a probabilistic approach, resulting in a calibrated age probability distribution. Commonly, probabilities are ranked and summed to find the 68.3% (1 sigma) and 95.4% (2 sigma) confidence intervals. An example using OxCal is shown below. This approach can result in small, disjointed calendar age ranges and means that the simple summary 'calibrated age (± 2 sigma)' is no longer appropriate.

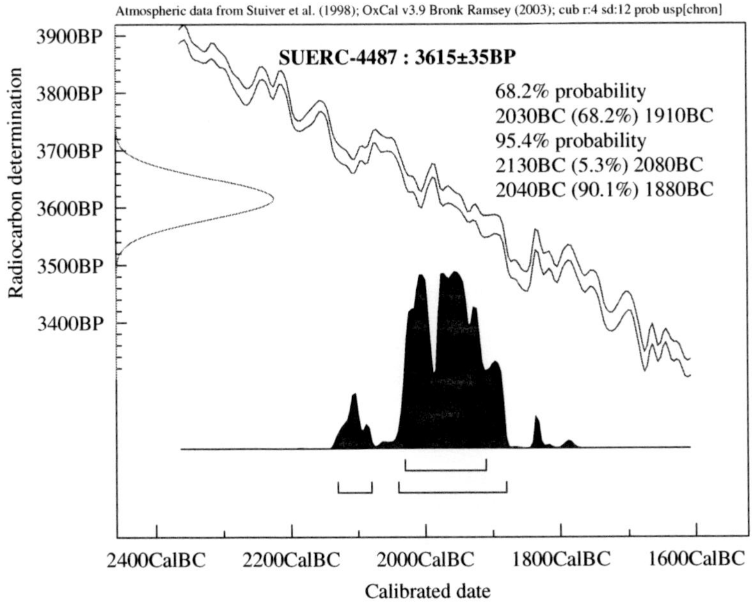

In addition, the calibration programmes also include options for calibration of multiple dates, especially useful when there is stratigraphic information and for techniques such as wiggle matching, where a floating (i.e. not fixed to the calendar timescale) chronology is matched to the master calibration curve.

3.5. Sample selection and potential pitfalls

A radiocarbon measurement can be made on any sample of material (organic or inorganic) that contains carbon. However, the two most fundamental questions that have to be asked about the sample are: (1) Does it contain sufficient carbon for the age to be measured by the available technique? (2) Does a secure relationship exist between the sample selected for isotopic analysis and the specific event or phenomenon that is assumed to have coincided with its death.

The first question can be answered fairly succinctly with the following: As a rule of thumb, 1 mg of elemental carbon (after pre-treatment) is required to make an AMS measurement; however, a sample containing >1 mg carbon would be required if an isotope ratio mass-spectrometric measurement of the $\delta^{13}C$ is also required. For radiometric measurement, a minimum of 1 g of elemental carbon would be required but for most laboratories several grams would be preferred. Of course, when a sample is first collected in the field, the quantity of carbon that it contains may not be obvious. For example: (i) Bone samples can undergo significant organic carbon (collagen) loss while buried and this may not be entirely obvious upon collection. (ii) Charcoal samples can appear significantly heavier than they actually are due to a relatively small quantity of adhering soil. (iii) Sediments can be quite deceiving and their carbon content difficult to predict under field conditions The "rule of thumb" that is applicable here is to collect as much of a sample as possible without compromising its integrity.

There are many factors that can influence the relevance of a sample to a period or event and this is probably best illustrated by taking a simple example:

> A coastal archaeological site is being excavated and a secure context containing charred material (grain, round-wood and fragments from a sizeable tree) from a fire and marine mollusc shells is uncovered. What events do these sample types best reflect?

The burnt grain and to a lesser extent the round-wood charcoal are short lived. The grain is from a single year's growth while the round-wood charcoal will represent a few years growth at most. Both are therefore likely to reflect, fairly closely, the time that has elapsed since the fire. The charcoal fragments from the sizeable tree and the mollusc shells will not. In the case of the charcoal, it must be borne in mind that the only time when carbon exchange with the atmosphere occurs is the year in which a specific ring is being laid down. After this, exchange effectively ceases. Therefore, the age of the charcoal fragments is the average time that has elapsed since the specific rings contained in the fragments were laid down, that is, there is potentially an 'old wood' effect. Therefore, using this material to estimate the time of the fire could conceivably give an age that is too old. This can be further complicated at coastal sites where drift-wood that had been in the environment for some considerable time was commonly collected and burnt or if timber was scarce and only burnt when it was no longer of use for structural purposes. The mollusc shells are relatively short-lived but will be subject to a marine reservoir effect as previously discussed. If a good estimate of the ΔR is available for the specific time and geographic location then, potentially, the mollusc shells can be used to estimate the time elapsed since the fire. Of course, they were not directly associated with the fire and so there has to be some reliance on the security of the archaeological context.

4. Methods of Isotope Measurement[4]

Useful exploitation of the naturally occurring isotopes of carbon in environmental research is entirely dependent on the levels of accuracy and precision at which they can be monitored. The challenge is considerable and this arises primarily from the need to resolve small changes in the distribution of the heavier isotopes (^{13}C and ^{14}C), which are already present in relatively low abundances. Although the stable form, ^{13}C, occurs as approximately 1% of all carbon, the magnitude of the isotopic variations that provide a proxy for the evaluation and/or quantification of natural processes are typically a factor of 10–100 times lower. With ^{14}C the situation is even less encouraging. The radioactive isotope is present in the atmosphere and living organic material at only 10^{-10}% abundance, that is, only one in a million million of these carbon atoms is radioactive. Furthermore, the beta particles that are emitted during the radioactive decay of ^{14}C have a low energy, *viz.* the characteristic E_{max} value is 156 keV. Effective radioactive monitoring therefore demands a highly sensitive arrangement with little or no physical barrier between the ^{14}C atoms and the beta particle detection device. Finally, but by no means least important, is the radioactive half-life value. The already low specific activity of zero-aged material (ca. 13.56 dpm g^{-1} C) decreases by a factor of two for each 5,730 year period elapsed since the death of the sample organism or the physiochemical removal of an inorganic sample material from active

[4] Chapter 9 is dedicated to the various measuring methods for environmental radionuclides including ^{14}C. The use of carbon isotopes as tracers in the carbon cycle and for precise absolute dating of carbonaceous material requires specific analytical approaches. This sub-chapter pays special attention to these approaches and thus is a useful complement to Chapter 9.

isotopic exchange with its source reservoir within the dynamic carbon cycle. The development of a mass-spectrometric technique in the late 1970s for the measurement of ^{14}C abundance is arguably the most significant advance in the technique. However, irrespective of the measurement technique, it is the half-life that effectively defines the practical upper limit of the conventional radiocarbon timescale. After the passage of 8–9 half-lives (approximately 50,000 years), there are insufficient residual ^{14}C atoms to provide a discernable signal above the inevitable noise (background) of the detection equipment and any contaminant ^{14}C that is added during sample preparation.

4.1. ^{13}C Measurement

^{13}C measurements are typically made on a sub-sample of CO_2 derived from the combustion of organic samples or acid hydrolysis of carbonates. In the case of carbonate sediments, the CO_2 is produced from dissolution in concentrated phosphoric acid. The gas is isolated and purified (using cryogenic separation procedures) in a vacuum system and analysed by mass spectrometry.

The relative abundance of masses 44, 45 and 46 in the gas are obtained using an Isotope Ratio Mass Spectrometer. Their relative intensities are compared with those of a working standard reference gas of known isotopic composition. In practice, this is achieved by automatic valve switching and data collection whereby reference gas and sample gas are alternately bled into the mass spectrometer, switching a number of times over a period of several minutes, thus obtaining a mean delta value for the sample with respect to the reference gas using the delta notation shown below.

$$\delta\text{sample to ref} = \left[\frac{R\text{sample}}{R\text{ref}} - 1\right] \times 10^3$$

where R is the 45/44 mass ratio (to get $\delta^{13}C$) and δ is in per mil (‰).

Similarly, the 46/44 mass ratio can be used to obtain the $\delta^{18}O$ of the sample with respect to the reference gas.

Contributions to the 45 and 46 peaks from minor isotope combinations (e.g. ^{17}O) in the CO_2 molecules are compensated for using the 'Craig corrections' (Craig, 1957), which are a feature of modern software. The latter procedure provides us with 'raw' delta values with respect to the internal reference gas. Adjusting the deltas to get values with respect to the international standards (V-PDB for ^{13}C and SMOW for ^{18}O) involves the following equation:

$$\delta_{\text{A to C}} = \delta_{\text{A to B}} + \delta_{\text{B to C}} + 10^{-3}(\delta_{\text{A to B}} \times \delta_{\text{B to C}})$$

where A is the sample gas, B the internal reference gas and C the international standard. The $\delta_{\text{B to C}}$ value is that of the internal reference gas with respect to the international standard. The latter is pre-calibrated using CO_2 sample gases of known isotopic composition produced from International Reference Materials such as NBS 19 and IAEA-CO-1 (carbonate).

Accelerator mass spectrometers that have the ability to measure all three carbon isotopes can therefore provide the data necessary to undertake 'on-line' $\delta^{13}C$ calculations. These measurements are typically of poorer precision ($\pm$1–2‰) than would be achieved by 'off-line' stable isotope mass spectrometry (approximately $\pm$0.1‰); however, there is an argument that the on-line measurements reflect potential fractionation effects in the AMS. Some AMS laboratories measure their $\delta^{13}C$ values on-line, while others prefer the off-line method.

4.2. ^{14}C Measurement

A paramount consideration in the technical design of suitably sensitive methods for the measurement of natural ^{14}C specific activities is the achievement of optimum detection efficiency allied to minimum background. 'Efficiency' is defined as the number of events that are recorded, as opposed to those that actually occur during a given period. System 'background' is the occurrence of recorded events that are spurious but otherwise indistinguishable from true ^{14}C disintegrations.

The routine measurement of natural ^{14}C concentrations is achieved by two approaches: (i) radiometric counting of beta particles from the decaying ^{14}C atoms and (ii) since the early to mid-1980s, direct ion counting using particle accelerators as high-energy mass spectrometers (AMS).

Radiometric counting systems fall into three categories. Each is characterised by the method used to prepare and present the sample carbon for measurement.

Solid carbon counting as used by Libby and his co-workers to produce the first radiocarbon ages was based on an infinitely thick source of elemental carbon (carbon black) inserted into a screen wall counter (e.g. Anderson et al., 1951). The original system required about 8.5 g C per sample, spread over an area of 400 cm^2 to give a slight excess over the 20 mg cm^{-2} required to produce an infinitely thick source. Although the background could be reduced to an acceptable 4.5 cpm by the use of passive and anti-coincidence shielding, the device was relatively inefficient since only approximately 5% of the ^{14}C decay events were detected. This poor detection efficiency defined the upper dating range at <25,000 years BP and the associated counting statistics translated to uncertainty ranges of 200–300 years on samples up to 5,000 years old. The practical demands of preparing samples were considerable and, following the introduction of trace quantities of nuclear fission products into the environment, this source of anthropogenic contamination precluded further development of the method.

Gas proportional counting (GPC) was introduced during the mid to late 1950s as a much improved approach to natural ^{14}C measurement. In particular, there was a marked improvement in detection efficiency allied with the avoidance of radioactive contamination in the prepared sample. Improvement in the signal-to-noise ratio of the detector systems led, under favourable conditions, to a maximum dating limit close to 50,000 years BP. Indeed, finite age values of as much as 75,000 years BP have been reported from carbonaceous gas samples that had been subjected to a preliminary isotopic enrichment procedure (Grootes et al., 1975).

The initial experimental requirement was near-quantitative conversion of the raw sample carbon to a suitable gas, which functioned as the counter gas in the central detector. Various detector arrangements were developed to utilise CO_2, methane (CH_4) or acetylene (C_2H_2). Irrespective of the preferred molecular form, a relatively complex vacuum system was necessary to prepare and store the preferred counting gas and to enable filling of the detector volume. Individual counters were designed to accommodate sample sizes ranging from 0.5 to 20 g C. The degree of analytical confidence achieved by the routine application of GPC technology was typically ± 50 years at best. This attainment was determined primarily by the amount of sample carbon that could be accommodated within the detector volume and the extent to which the background (noise) could be reduced and maintained at a relatively stable rate over the typical 24 h or more necessary to accumulate a reasonable number of recorded ^{14}C events. The development of GPC technology heralded the attainment of so-called 'high-precision' dating where specialised systems were employed to achieve analytical confidence limits of less than ± 20 years (1 sigma) on samples younger than approximately 8,000 years (Stuiver and Pearson, 1993).

Liquid scintillation counting (LSC) was also introduced in the 1950s and has subsequently become the preferred method for radiometric analysis of ^{14}C. Hereforth (1948) was the first to demonstrate that certain aromatic compounds (fluors) could convert the absorbed energy from nuclear radiation into photons of light. Virtually all modern day scintillation counters employed for ^{14}C dating are based on the design of Hiebert and Watts (1953) which employs two diametrically opposed photomultiplier tubes (PMTs) that convert the light output from the sample into electrical pulses, and fast coincidence circuitry (typical resolving time 20 ns approximately) to differentiate between decay events occurring in the sample vial that is situated between the PMTs and the random noise (dark current) from the PMTs.

In addition to coincidence circuitry, several other techniques are used in modern low-level scintillation counters, used for ^{14}C analysis, to reduce background. These include: (1) Various forms of pulse shape analysis to differentiate between the pulses produced by beta decays and the so-called non-quenchable component of background. (2) Active guard detectors that include PMTs in anti-coincidence with the sample PMTs. (3) Pulse amplitude comparison circuitry. (4) Large passive, graded shields comprising lead with cadmium and copper linings.

A typical scintillation cocktail that would be added to a sample would comprise one or more aromatic solutes (fluors) dissolved in an aromatic solvent. In the case of ^{14}C dating, the sample is typically converted to benzene, which is both the sample and the solvent. The effective 4π counting system gives high counting efficiency for ^{14}C (95% in an open counting window). High-precision measurements to better than ± 20 years (1 sigma) can also be achieved by this method (Pearson and Stuiver 1986). A small number of laboratories undertake direct absorption of CO_2 into a quaternary amine and then add a scintillation cocktail to the medium but this is a lower precision method due to the amount of carbon that can be accommodated within a scintillation vial (approximately 20 ml).

Accelerator mass spectrometry is a highly sensitive technique for counting individual ^{14}C atoms, as opposed to the radiometric techniques, which measure the beta particles produced by the very small percentage of ^{14}C atoms that decay during the measurement period (on average, 13.56 pm g^{-1} of 'modern' carbon). The AMS technique emerged from research carried out in low energy physics laboratories and the first ^{14}C measurements, made using a tandem van de Graaff accelerator, appeared in the late 1970s (Bennett et al., 1977; Nelson et al., 1977). Purser et al. (1980) developed the first dedicated AMS system. ^{14}C measurements are feasible by AMS because of the excellent discrimination against: (1) Molecular interferences (e.g. $^{12}CH_2$, ^{13}CH, etc), (2) mass resolution from the ^{12}C and ^{13}C isotopes and (3) isobaric interference from ^{14}N. The fundamental principles of the measurement technique are as follows:

1. A beam of Cs^+ ions is used to sputter the carbon sample (usually in the form of a carbon coating on iron powder) to produce C^- ions, which are pre-accelerated to a few tens of keV, focussed using various electrostatic devices and mass analysed by an injector magnet, which directs the ions into the tandem accelerator. The isobaric ^{14}N interference is removed because nitrogen does not form stable negative ions.
2. The mass analysed negative ions are accelerated in the first stage of the tandem accelerator to the positive high voltage terminal where electrons are stripped from the ions using a low-pressure gas such as argon. Typically, terminal voltages of 2–4.5 MV have been used to produce C^{3+} or C^{4+} ions although recently small units using terminal voltages <1 MV and producing C^+ and C^{2+} ions have become operational (Suter et al., 2000).
3. The positively charged C ions are then accelerated back to ground potential in the second stage of the tandem and following further focussing they are selected with respect to charge state and energy using

an analyser magnet and directed to a detector that determines both the atomic number and mass of the atoms.

Radiocarbon ages can be calculated using either the $^{14}C/^{13}C$ or the $^{14}C/^{12}C$ atom ratio. Fifield (1999) provides an excellent overview of the technique.

4.3. Relative advantages of the methods

Currently, three analytical techniques are routinely used for ^{14}C dating, namely, the radiometric techniques of GPC and LSC, and AMS. Screen wall counting of carbon has not been employed for many years due to the requirement for large samples, the tedious and difficult sample preparation and low counting efficiency. LSC is now the more common approach for radiometric analysis although there is little difference in accuracy and precision between this technique and GPC with both techniques being capable of high-precision measurements ($< \pm 20$ years) (Stuiver and Pearson, 1993; Pearson and Stuiver, 1986). The popularity probably relates to the fact that liquid scintillation spectrometers, suitable for radiocarbon dating, are manufactured commercially for a range of scientific disciplines and can be purchased readily. In addition, the major manufacturers have good service and repair backup. In contrast, GPC equipment is less readily available.

The major advantages of AMS over the radiometric techniques are the counting time and sample size requirements. Whereas a gram of 'modern' carbon has an activity of 0.226 Bq and would require a counting time of approximately 17 h to achieve a 1 sigma Poisson error of 1.0% (assuming a counting efficiency of about 75%), the same gram of carbon contains approximately 6×10^{10} atoms of ^{14}C. Using a typical counting rate for ^{14}C atoms in an AMS system of approximately $100\,s^{-1}$, the same precision can be obtained on 1 mg of modern carbon counted for about 15 min. The reduction in sample size has enabled researchers to be much more selective in their choice of sample. Selection can be regarded as having three levels of complexity: (1) Single entity dating, that is dating single pieces of bone, charcoal, shell, etc. (Ashmore, 1999). For example, in archaeology it was common practice, following excavation of archaeological sites, to date bulk samples of these materials by radiometric measurement; however, evidence has shown that, even in situations where the excavator was confident that the deposit represented a single event, residual material from an earlier event was often present and so the age obtained was an average for the deposit and might not even reflect a period when the site under study was occupied. AMS has enabled single entities from a deposit to be dated and it is now common for archaeologists to date several single entities from a single context or feature, which enables residuality problems to be identified. (2) The separation and measurement of particular gross fractions from a larger complex sample, for example Geyh et al. (1998) demonstrated that terrestrial plant remains separated from lake sediments provided the most reliable means of dating the sediment. (3) Measurements made on specific compounds or groups of compounds, for example, the separation and subsequent age measurement of specific lipids from potsherds (Stott et al., 2001). This type of work enables the lipids common to food residues left on pot fragments to be separated from other components that may derive from the surrounding soil, etc.

Small samples have been dated by both LSC and GPC. Otlet et al. (1986) review practical miniature GPC systems (i.e. detectors of between about 5 and 50 ml volume) for the dating of samples as small as 2.7 mg C while Haas and Trigg (1991) discuss the dating of samples in the range 200–600 mg C using a modern low-level LSC system. The major drawback in both systems is the counting time required to achieve acceptable counting statistics. For example, 2.7 mg of modern carbon measured in a miniature GPC system would require a counting time of about 7 months to produce a Poisson error of 1% (90% counting efficiency). While most miniature gas counting systems employ an array of small counters within a single anti-coincidence shield, throughput of samples is still a limiting factor. Similarly, 200 mg of modern C counted in a modern LSC system would require about a month of counting time to achieve the same precision (at 75% counting efficiency).

Inherently, there should be no difference in accuracy between an AMS and a radiometric measurement and the radiometric laboratories that undertake high-precision measurements can routinely produce errors on single analyses ($<$20 years on samples of approximately 1 half-life) that are of higher precision than can be produced by AMS on equivalent samples (25–40 years approximately by AMS). Of course, to achieve this by LSC, approximately 15–20 g of elemental carbon would be required. Based solely on counting statistics, the counting error on an AMS measurement can be reduced to a much smaller value than could be achieved on a reasonable timescale by a radiometric counter, however, other factors have to be taken into consideration in the determination of the overall error on the age. Where AMS gains is that optimum precision can be maintained for older samples without increasing the counting time to the point where throughput is limited.

5. Definition, Calculation and Reporting of C Isotope Enrichment Values

5.1. ^{14}C Half-life

Some of the earliest ^{14}C age calculations (Arnold and Libby, 1949) were carried out using a half-life value of 5,720$\pm$47 years, determined by Engelkemeir et al. (1949); however, a value of 5,568$\pm$30 years (now known as the Libby half-life) was soon adopted (e.g. Arnold and Libby, 1951) and this value continues to be used in age calculations. This half-life was the best estimate based on the work of a number of research groups, although, in fact, the half-life that Libby and his co-workers used originally is very close to the current best estimate of 5,730$\pm$40 years.

5.2. Basic ^{14}C age calculation

The radiometric ^{14}C age of a sample is calculated from a re-arranged form of the first-order decay equation by assuming that the atmospheric ^{14}C activity has been constant throughout all past time. The most basic form of the age equation is:

$$t = \frac{1}{\lambda} \ln \left[\frac{A_0}{A_t} \right]$$

where t is sample age, which is the time that has elapsed since removal of the carbonaceous material from the carbon cycle (e.g. death of an organism); $\lambda =$ a decay constant ($\ln 2/t_{1/2}$) $= 1.2449 \times 10^{-4}$, where $t_{1/2} = 5{,}568$ years (Libby half-life); $A_t =$ the activity of the carbonaceous material t years after death as measured in the radiocarbon laboratory; $A_0 =$ 'modern equilibrium living activity' of the sample, again as measured in the radiocarbon laboratory.

5.3. ^{14}C Standards

Radiocarbon age calculations require that both the 'modern equilibrium living activity' of ^{14}C (A_0 term) and the activity remaining t years after death (A_t term) are measured. The A_0 activity was originally based on the age-corrected activity of tree rings growing during 1890 AD and in equilibrium with 1890 AD atmospheric CO_2. This equated to 0.226 Bq g^{-1} C. The year 1890 AD was chosen as this represents the last time when the ^{14}C record was not significantly perturbed by anthropogenic influences. In 1958, a primary modern reference standard (SRM 4990, often termed Oxalic acid I), in the form of a batch of oxalic acid, was produced by the US National Bureau of Standards and calibrated against 1890 AD wood, producing the following internationally accepted reference value: 0.95 $\times$ the activity of SRM 4990, *in 1950 AD*, normalised to a δ^{13}C value of -19‰ (with respect to V-PDB) = activity of 1890 AD wood. Although the measurements were not made in 1950 AD, this was judged to be an appropriate reference year. Thus, when a sample is dated, its age is always presented in years BP (Before Present) where 'Present' is 1950 AD. Therefore, since all samples and the oxalic acid standard have been decaying at the same rate post-1950, there is no need for decay corrections, provided that all ages are expressed relative to 1950. Supplies of SRM 4990 were largely exhausted some time ago and the current internationally accepted modern standard is SRM 4990C, which is oxalic acid manufactured from the 1977 harvests of French sugar beet molasses. It is commonly known as Oxalic acid II and is distributed by the National Institute of Standards and Technology, MD, USA. The internationally accepted reference activity for Oxalic acid II is: 0.7459 × the specific activity normalised to a $\delta^{13}C$ value of −25‰ = 0.95 × the activity of Oxalic acid I, normalised to a $\delta^{13}C$ value of −19‰. All radiocarbon dating laboratories report their results directly (or indirectly via a secondary standard) relative to the international standard.

5.4. ^{13}C enrichment

During the transfer of carbon within and between the various reservoirs of the global carbon cycle, isotopic fractionation occurs. For example, atmospheric carbon has an enrichment of about −8‰ relative to the V-PDB carbonate standard, whereas C_3 plants, which derive their CO_2 directly from the atmosphere, are even more depleted with a typical enrichment value of about −23‰. Similarly, the dissolved inorganic component of seawater, which is directly exchanging with atmospheric CO_2, is around 0‰. In addition to natural processes, preparation of a sample in the laboratory can induce further fractionation, for example by incomplete oxidation.

Since the percentage depletion/enrichment is proportional to the difference in atomic mass, ^{14}C fractionation will be double that of ^{13}C. If fractionation were ignored, samples of different materials (e.g. wood, marine shell and bone) that are contemporaneous in age would appear to have different radiocarbon ages. To account for this, all sample activities are normalised relative to a $\delta^{13}C$ of −25‰ (theoretical value for wood) by calculation of a fractionation factor (FF), which approximates to the equation:

$$\mathrm{FF} = 1 - \left[\frac{2(25 + \delta^{13}\mathrm{C})}{1{,}000}\right]$$

If an activity is left uncorrected for fractionation, each 1‰ deviation from −25‰ will result in an age offset of 16 years Therefore, using the above figures, the plants would appear 240 radiocarbon years older than the contemporaneous atmosphere. The above correction factor is used in radiometric dating and in AMS when the $^{14}C/^{12}C$ atom ratios are employed. However, in AMS, when the $^{14}C/^{13}C$ ratio is measured, the factor of 2 is lost and the equation approximates to:

$$\mathrm{FF} = 1 - \left[\frac{(25 + \delta^{13}\mathrm{C})}{1{,}000}\right]$$

In this case, each deviation of 1‰ from −25‰ corresponds to an age offset of 8 years.

5.5. ¹⁴C enrichment

5.5.1. Radiometric ^{14}C age calculation

On the basis of the above discussion of fractionation, the normalised sample activity for radiometric dating is denoted by:

$$A_{SN} = A_S\left[1 - \frac{2(25 + \delta^{13}C_S)}{1,000}\right]$$

where A_S is the measured sample activity and A_{SN} the normalised sample activity. $\delta^{13}C_S = \delta^{13}C$ value of the sample.

Similarly, the normalised activity for Oxalic acid II is denoted by:

$$A_{ON} = 0.7459 A_{OXII}\left[1 - \frac{2(25 + \delta^{13}C_{OXII})}{1,000}\right]$$

where A_{OX} is the measured activity of Oxalic acid II and A_{ON} is 74.59% of this activity, normalised to −25‰.

Thus, the fraction modern (F) of a sample is given by:

$$F = \frac{A_S[1 - (2(25 + \delta^{13}C_S))/1,000]}{0.7459 \times A_{OXII}[1 - (2(25 + \delta^{13}C_{OXII}))/1,000]} = \frac{A_{SN}}{A_{ON}}$$

and the age (t) is given by:

$$t = \frac{1}{\lambda}\ln\left[\frac{1}{F}\right] = 8,033\ \ln\left[\frac{1}{F}\right]$$

F remains constant irrespective of the year of measurement after 1950 because the standard and the sample are decaying at the same rate. Therefore, the age is constant and independent of the year of measurement.

Stuiver and Polach (1977) and Mook and van der Plicht (1999) provide a more complete discussion of the above.

5.5.2. AMS ^{14}C age calculation

The term F is also derived when age calculations are made using isotopic ratios. When the $^{14}C/^{12}C$ ratio is measured, the calculation is:

$$F = \frac{(14/12)_S \times [1 - 2 \times (25 + \delta^{13}C_S)/1,000]}{0.7459 \times (14/12)_{OXII} \times [1 - 2 \times (25 + \delta^{13}C_{OXII})/1,000]} = \frac{A_{SN}}{A_{ON}}$$

When the $^{14}C/^{13}C$ ratio is measured, the calculation is:

$$F = \frac{(14/13)_S \times [1 - (25 + \delta^{13}C_S)/1,000]}{0.7459 \times (14/13)_{OXII} \times [1 - (25 + \delta^{13}C_{OXII})/1,000]} = \frac{A_{SN}}{A_{ON}}$$

Ages are again calculated from:

$$t = 8,033 \ \ln\left[\frac{1}{F}\right]$$

The radiometric and AMS calculations described above produce what is termed a *conventional radiocarbon age*. This implies the following:

a. The age is in years BP (before present) where present is 1950 AD.
b. The Libby half-life of 5,568 years was used to calculate the age.
c. SRM 4990, SRM 4990C or a suitable secondary standard was used in the analytical scheme.
d. The standard's activity/isotopic ratio was normalised to the appropriate $\delta^{13}C$ value (−19‰ for SRM 4990 and −25‰ for SRM 4990C).
e. All sample activities/ratios were normalised to −25‰.

Mook (1986) details the recommendations/resolutions of the radiocarbon community with respect to reporting of ages, calibration of data, etc.

The age should also be accompanied by an error quoted at the 1σ level of confidence that, in the case of radiometric measurements, should have components derived from:

(a) The total counts from the modern standard(s).
(b) The total counts from the background sample(s).
(c) The total counts from the unknown.
(d) The error on the $\delta^{13}C$ measurement.
(e) Other components such as the errors on: the degree of quenching and sample benzene dilution in LSC, the gas filling pressure and gas dilution in GPC.

The ultimate guide to a realistic error term is via the analysis of a series of replicate samples over a period of time. On a series of say 10 analyses conducted on sub-samples from a bulk homogeneous sample, the standard deviation on the 10 analyses should be approximately equal to the 1σ error on an individual analysis if all sources of error have been taken into account. If the standard deviation is larger then not all errors have been accounted for.

5.6. International quality assurance in natural ^{14}C measurement

In keeping with the principles of analytical science, radiocarbon laboratories have always been conscious of the importance of accuracy and precision for their reported results, that is, the ethic of analytical quality control (QC) which in turn is the foundation for the wider concept of quality assurance (QA). Natural ^{14}C measurements require traceability to international standards whose activities are known exactly by independent means and also to reference materials whose activities are estimated and typically accompanied by associated uncertainty statements. This concern for good quality management within the radiocarbon community is exemplified by the care and effort given to establishing and maintaining primary standards and reference materials and in the routine organisation of laboratory inter-comparisons or collaborative trials to verify comparability of measurements. ^{14}C measurement uses a number of standards and reference materials with different properties. Historically, the absolute calibration of ^{14}C measurement was tied to 1890 wood, through the 'primary' standard of Oxalic acid I, subsequently replaced by Oxalic acid II, as discussed above. These are both internationally calibrated and certified materials, whose ^{14}C activities are known absolutely. A second tier of materials, often called secondary standards or reference materials are also used routinely. These include internationally recognised materials such as ANU-sucrose (now also known as IAEA-C6), Chinese-sucrose and the IAEA C1-C6 series (Rozanski et al., 1992), augmented by additional oxalic acid samples (now IAEA C7 and C8) (Le Clercq et al., 1998). The activity of these materials has been estimated from large numbers of measurements made by many laboratories, often through a laboratory inter-comparison. Recently, further natural materials from the Third and Fourth International Radiocarbon Inter-comparisons (TIRI (Gulliksen and Scott, 1995) and FIRI (Bryant et al., 2000)) have been added to this list. The activities of these standards and reference materials span both the applied ^{14}C age range and the chemical composition range of typical samples. A summary of the inter-comparisons organised within the ^{14}C dating community can be found in Scott et al. (2003) and Radiocarbon 45(2), 2003. The first large inter-comparison was the International Collaborative Study (ICS) (Scott et al., 1990) involving more than 50 ^{14}C laboratories worldwide. Following the ICS, in 1990, the International Atomic Energy Agency distributed six new potential reference materials to over 130 laboratories for characterisation. This study was less concerned with laboratory performance but more with the suitability of the test materials for traceability of ^{14}C measurement. Results from 69 laboratories were ultimately reported and formed the basis of the characterisation of the materials through calculation of their consensus activity. Overall, there was generally good agreement in the results, but a number of difficulties were subsequently identified with some of the materials.

TIRI (Scott et al., 1992; Gulliksen and Scott, 1995) was begun in 1991 and involved a large number of laboratories (over 70). TIRI was designed to provide an independent assessment of laboratory performance, following the recently completed IAEA study and hence the materials were designed to test the full laboratory procedures and to provide new reference materials. The consensus values (Gulliksen and Scott, 1995) for the samples were estimated using an approach similar to that taken in the IAEA study

(Rozanski et al., 1992). The results from TIRI, as in the other studies, indicated that laboratories were in general providing accurate results, but pointed to variation in the results beyond that described by the quoted uncertainties. The most recent laboratory inter-comparison was completed at the end of 2001. A set of core samples was distributed to over 120 laboratories during 1999 and by the deadline of December 2000, 92 sets of results had been received. These included results from more than 75% of operational AMS facilities (Bryant et al., 2000; Boaretto et al., 2002; Scott et al., 2003) involved in ^{14}C dating. Wood, peat, bone, marine carbonate (turbidite) and grain, together with specific fractions of samples such as the cellulose fraction of wood and the humic acid fraction of peat were used, from which a set of seven core materials was selected. Three sets of duplicate samples were provided on a blind trial basis to allow an assessment of laboratory precision. Consensus values for the FIRI materials were derived. In particular, the consensus values for the four known-age, dendro-dated wood samples and the average ^{14}C ages corresponding to the master calibration curve were such that the differences all lay within the limits of the quoted errors, providing some confirmation of the validity of the consensus value estimation. On average, no evidence of significant differences between the different measurement techniques was observed. Again, there was some evidence that the quoted errors were underestimates of the laboratory precision. There was also evidence that a number of laboratories had significant but generally small offsets relative to the consensus results. All of the studies cited above served several purposes. First, they have provided valuable information to laboratories and hence to users about precision and accuracy of measurements. As a result, it is clear that collaborative trials such as TIRI and FIRI are important parts of laboratory quality assurance. Second, the studies have increased the numbers and types of reference materials. This is important, since by inclusion of a variety of reference materials in routine laboratory practice, the ^{14}C determinations can be constrained to a greater degree.

5.7. Statistical significance of a ^{14}C age

Since a ^{14}C age can be considered as a realisation of a random process (due to the nature of radioactive decay), then we can consider that there is some underlying true ^{14}C age and an uncertainty about this value. The commonly used probability model for the process is the Gaussian or normal model, the mean of which equals the true but unknown ^{14}C age. This uncertainty is quantified in the error (or standard deviation or sigma) that is reported with a measured (estimated) radiocarbon age. It is primarily based on counting statistics, but should also include assessments of the uncertainties associated with other components of the isotopic analysis (e.g. fractionation correction). It may also have a minimum value based on replicate analyses so that, even where counting statistics are not a limiting factor, the standard deviation on a series of replicates dictates this minimum value for the error term. In recent years, there have been a number of publications where the use of a laboratory error multiplier has been introduced. Such an error multiplier captures sources of variation in the result that are not accounted for in the quoted error. This has frequently been used as a measure of laboratory performance in inter-laboratory trials. Thus, the interpretation of a single measured ^{14}C age and its associated error is as follows: we assume that the measured ^{14}C age is a realisation of a random Gaussian process with a mean 'true ^{14}C age' and sigma equal to the quoted error. The best estimate of the true ^{14}C age is the measured ^{14}C age, but the true value is highly (95%) likely to have come from the range: measured radiocarbon age ± 2 sigma.

6. ^{14}C as a Tracer

The development of mathematical analogues (models) is essential to the advancement of our appreciation and successful management of the Earth's carbonaceous environment. Such models are of varying complexity and are designed to represent either the transfer of carbon on a global scale or within a specific regional and/or ecological context. However, any model designed to represent a natural system must be capable of accounting for and complying with measured scientific data. Carbon isotope geochemistry provides a near-ideal benchmark. Not only is the element virtually ubiquitous in natural systems; the chronological opportunity afforded by conventional radiocarbon dating has allowed quantification of the rates of exchange of carbon both within and among the major reservoirs of the dynamic biogeochemical cycle (see Section 2). Today, in the post-Kyoto era, nations trade responsibilities for carbon emissions and agreements are made to facilitate the overall reduction of greenhouse gas emissions, including CO_2 and methane. A key tool in the scientific endeavour to understand the global carbon cycle is ^{14}C.

In compliance with basic dating theory, and in the absence of measurable anthropogenic influence, the measured ^{14}C age of a carbonaceous component within a specific region of the natural carbon cycle is a measure of its mean residence time within that spatial and/or geochemical environment. When allied with estimates of the near steady-state mass distribution of carbon within the dynamic cycle, such ^{14}C enrichment values can be interpreted in terms of the rates and fluxes of carbon transfer within and/or between the major carbon reservoirs. A prime example here is provided by the world's oceans where variations in the spatial distribution of the residual ^{14}C concentration in DIC, is sufficient to delineate discrete water bodies and to provide a measure of both mass transfer rates and physical mixing patterns. However, time resolution is generally constrained at >100 years by a combination of the relatively long radioactive half-life of ^{14}C, its low isotopic abundance and allied to the limited experimental precision that can be achieved in natural ^{14}C measurement.

A considerable, and totally fortuitous, opportunity for much improved time resolution is provided by progressive anthropogenic-induced changes to the natural isotopic abundances of carbon within some regions of the dynamic cycle. These isotopic disturbances are first noted from the late 19th century when the increasing use of fossil fuels started the effective transfer of carbon from the environmentally inert sedimentary system directly into the Earth's atmosphere as gaseous CO_2 (see Section 2.3). Because this increase to the steady-state inventory of the dynamic cycle comprised geologically old carbon, it carried the very convenient quantitative label of being '^{14}C-free'. The progressive build up of fossil-derived CO_2 in the atmosphere is accurately documented in the annual growth increments of trees and these record a ca. 2–3% dilution of the previous natural ^{14}C abundance by the mid-20th century. More recently, the value of ^{14}C as an environmental tracer was increased greatly by the rapid and relatively short-lived injection into the upper atmosphere of radiocarbon atoms produced as a consequence of nuclear weapons test programmes (see Section 2.3). The progressive mixing of this artificial excess through the dynamic carbon cycle from its virtual point source injection region (the upper atmosphere of the high latitude northern hemisphere) is of course superimposed on the natural isotopic abundance distribution but with no alteration to the mass distribution of elemental carbon. The principal advantage gained from the ongoing dispersal of the "bomb ^{14}C" pulse is the opportunity for much improved time resolution for dynamic transfers in most regions of the dynamic cycle (Oeschger et al., 1975; Broecker et al., 1980). Radioactive decay can be discounted and the conservative properties of the 'bomb spike' allow transfer and/or residence times to be assessed in terms of years through decades. The value of this transient tool has been most evident in respect of establishing atmospheric mixing rates and the quantitative determination of carbon fluxes between this most environmentally sensitive reservoir and its principal sources and/or sinks, *viz.* the oceans and the terrestrial biosphere.

In essence, the use of environmental ^{14}C measurement as a tracer parameter for carbon transfer in natural systems must be recognised as a two-component opportunity. These natural and artificially sourced tracer components are mutually supportive although in some instances difficult, if not impossible, to resolve. In any ^{14}C-based model, it is necessary therefore, to consider (i) a long-term time index derived from natural (cosmic ray) production which will persist, balanced by radioactive decay, into the foreseeable future and (ii) the transient and relatively short-term index provided by the 'bomb ^{14}C' pulse which will eventually be dampened to insignificant levels by natural dispersal throughout the dynamic carbon cycle. Where it is appropriate to do so, it may also become increasingly necessary to consider the mass imbalance incurred by the release of fossil-derived (^{14}C-free) CO_2 into the natural system.

A detailed discussion of the uses of ^{14}C as a tracer is well beyond the scope of this chapter and so what follows is a brief description of some of the applications together with references to more detailed discussions.

6.1. Atmospheric studies

^{14}C has been instrumental in elucidating a number of processes related to the atmospheric carbon reservoir.

The first major anthropogenic perturbation of the carbon cycle was the burning of fossil fuels, which added $^{12}CO_2$ and $^{13}CO_2$ to the atmosphere, but of course, no $^{14}CO_2$. The effect on the ^{14}C:^{12}C isotopic ratio is observable in tree rings from the latter half of the 19th century onwards, until the early 1950s when it is masked by addition of ^{14}C from atmospheric nuclear weapons tests (Stuiver and Quay, 1981). It had been postulated that the increase in the atmospheric CO_2 concentration could be due to an ongoing natural perturbation of the carbon fluxes between the atmosphere, oceans and biosphere; however, Suess (1955) was able to demonstrate conclusively, through ^{14}C analyses of 19th and 20th century tree rings, that there was in fact a reduction in ^{14}C in the 20th century and that the increase in atmospheric CO_2 concentration was due to fossil fuel combustion (Suess effect). He further concluded that, after allowing for decay in the older wood, the mid-20th century reduction in atmospheric ^{14}C activity was probably $<1\%$. Subsequent research indicates that this decrease was in the region of 2.3–3% (Stuiver and Quay, 1981), although in more densely populated areas with greater pollution, regional effects of significantly greater magnitude (10% anthropogenic contamination level) have been observed (Levin et al., 1989).

Nuclear weapons tests, which commenced in 1945, produce ^{14}C by the same ^{14}N(n, p)^{14}C reaction that brings about natural ^{14}C production. The main periods of testing were 1954–1958 and 1961–1962 (Carter and Moghissi, 1977), after which a partial test ban treaty brought about a significant reduction in tests. It has been estimated that of the order of 10^{17} Bq of ^{14}C were produced by nuclear weapons testing (Cambray et al., 1985; Hesshaimer et al., 1994) up to 1980 (the year of the last major test). The majority of the tests took place in the northern hemisphere; hence the ^{14}C levels were significantly higher than in the southern hemisphere. In the northern hemisphere, the maximum tropospheric activity (an approximate doubling of the natural level) was measured in 1963/1964. However, since the maximum ^{14}C production occurred around 1961 and the majority of the ^{14}C was injected into the stratosphere of the northern hemisphere, the lag time can be attributed to the finite residence time of carbon in the stratosphere (Walton et al., 1970).

In the southern hemisphere, the maximum was significantly less at approximately 68% above the natural level and was reached in spring 1965 (Gupta and Polach, 1985). Therefore, the time taken for the activities in the two hemispheres to equilibrate has allowed the finite mixing time between them to be estimated.

Figure 2 illustrates the atmospheric records for both the northern and southern hemispheres. Levin and Hesshaimer (2000) present a very useful review of several atmospheric and oceanic studies involving ^{14}C as a tracer.

6.2. Ocean studies

The oceans represent a much larger and rather more heterogeneous reservoir for ^{14}C than the atmosphere, therefore, the best estimate of maximum nuclear weapons ^{14}C activity in surface waters would be the late 1960s to very early 1970s. A maximum increase of approximately 20% was recorded for temperate northern latitudes and a few percent lower for southern latitudes (Nydal et al., 1984; Nydal, 1967; Nydal and Lovseth, 1965). This nuclear weapons signal has been used widely as a tracer for the exchange of CO_2 between the atmosphere and the surface oceans while both natural and weapons testing ^{14}C have been used for studying deeper water circulation patterns.

The first large-scale programme that included ^{14}C measurements was begun in 1971 and called GEOSECS (Geochemical Ocean Section Study). GEOSECS was a global survey of the three-dimensional distribution of chemical, isotopic and radiochemical tracers in the oceans, carried out partly as a baseline study for future chemical changes, and partly to investigate large-scale ocean water transport and mixing processes. The expeditions took place in the Atlantic Ocean from July 1972 to May 1973, in the Pacific Ocean from August 1973 to June 1974, and in the Indian Ocean from December 1977 to March 1978. Amongst the main conclusions was the fact that the northern Pacific Ocean does not exhibit the same deep convection and deep-water renewal that is characteristic of the Atlantic Ocean. GEOSECS also enabled a more realistic estimate of the timescale of the renewal and circulation of the bulk of the old Pacific deep water (Ostlund and Stuiver, 1980).

The success of the GEOSECS programme led to further large-scale ocean studies including TTO (Transient Tracers in the Ocean), WOCE (World Ocean Circulation Experiment), SAVE (South Atlantic Ventilation Experiment) and JGOFS (Joint Global Ocean Flux Studies). The TTO programme, which was limited primarily to the Atlantic Ocean, was underway by 1978 and was divided into two sub-projects: TTO/NAS-North Atlantic Study; TTO/TAS-Tropical Atlantic Study. The programme was a broad attempt to study ocean circulation using several hydrographic parameters and tracers. The programme was stimulated by growing concern about future climate and the consequences of increasing CO_2 in the atmosphere and the possibility of using bomb ^{14}C to calibrate ocean models for CO_2 uptake. The fact that the TTO programme was a decade after GEOSECS also allowed comparison of the depth of penetration of the weapons testing ^{14}C signal.

The WOCE programme ran between 1990 and 2002 and was a component of the World Climate Research Programme (WCRP). The overall aims of the programme were to obtain a detailed description of the physical properties and circulation of the global ocean, and the role of the ocean in global climate. Sophisticated numerical ocean models were developed to provide a framework within which the observations could be interpreted and predictions for the future made. Parts of the programme were carried out in collaboration with JGOFS. A main objective in the programme was to separate nuclear weapons tests ^{14}C from the natural background so that data could be used to calibrate numerical global ocean models, determine upwelling and thermocline ventilation rates and estimate the transfer of CO_2 from the atmosphere to the ocean (Key, 1996). *Radiocarbon* 38(3) (1996) is given over largely to ^{14}C tracer studies in the oceans while Nydal (2000) presents a good overview of the various marine programmes and a fairly comprehensive reference list.

Further details may also be obtained at a range of websites including:

http://www.soc.soton.ac.uk/OTHERS/woceipo/ipo.html for WOCE publications, etc.

http://markov.ldgo.columbia.edu:81/SOURCES/.GEOSECS/ for GEOSECS data, etc.

6.3. Soil studies

In a recent estimate, Jobbagy and Jackson, 2000 calculate a global inventory of ca. 1,500 Pg, that is 1.5×10^{18} g for carbon stored in the upper 1 m of mineral soil. This is more than twice the amount of carbon present in the Earth's atmosphere as CO_2 ($\sim$725 Pg) or as living material within the terrestrial biomass ($\sim$660 Pg) (Anderson, 1992). Unfortunately, in terms of global modelling, there is an immediate problem with the definition and quantification of soil organic carbon (SOC) *per se*. This arises from a considerable diversity in the geographical distribution of soil organic matter (SOM), its biogeochemical composition and allied to the fact that measured ^{14}C age values for specific components of the SOM pool can range between 1year and several millennia. Consequently, there is a well-accepted consensus that in any carbon cycle model, the SOC pool cannot be readily characterised by a single age or mean residence time (Jenkinson and Raynor, 1977; O'Brien, 1984; Parton et al., 1987). Empirical evaluations indicate that at least three component carbon pools are necessary to describe soil carbon dynamics, *viz.* fast, active and recalcitrant. Each of these SOC sub-pools is then defined in terms of the mean time required for its component organic detritus to either mineralise to CO_2 and return to the atmosphere or accumulate within the mineral soil regime. As a first approximation, 'fast carbon' can be characterised by turnover times of < 10 years, 'active carbon' is typically resident for decades through centuries and the remaining 'recalcitrant' carbon persists in the mineral soil on timescales of centuries to millennia. The most recently available estimates (Matthews, 1997; Schimel, 1995; Potter and Klooster, 1997) indicate that only 20–30% of the global SOC pool falls in the 'fast' and 'active' categories, that is, the majority of the carbon present in mineral soils is held effectively in long-term (recalcitrant) natural storage.

Conventional radiocarbon dating has been applied in attempts to determine the ages of fossil soils and of older organic components in contemporary soil horizons (Paul et al., 1964; Campbell et al., 1967a, 1967b; Sharpenseel, 1972; Guillet, 1982). However, there is a general recognition that the measured ages are inevitably an underestimate due to the fact that SOM is the product of the ongoing processes of acquisition and loss (Wang et al., 1996). It is necessary therefore to employ time-dependent models to equate the ^{14}C enrichment monitored in SOC with its true age or turnover rate of SOM within a particular soil system. Several approaches to modelling soil carbon turnover rates are described and discussed in the scientific literature (e.g. O'Brien and Stout, 1978; Harkness et al., 1986; Jenkinson et al., 1992; Harrison and Harkness, 1993; Harrison, 1996; Harrison et al., 2000; Gaudinski et al., 2000; Hahn and Buchmann, 2004).

There are numerous reports in the scientific literature that evidence the important contribution that ^{14}C measurement can provide for investigations of the flux of carbon through the SOM regime. Several themes predominate.

An early and continuing interest focuses on the extent to which changing plant/soil relationships, either natural or man managed, for example forest clearance, the abandonment of agricultural land or afforestation, can influence the productivity and/or carbon storage capacity of a particular soil type (Jenkinson, 1990; Harkness and Harrison, 1989; Pessenda et al., 1996; Rutberg et al., 1996).

When compared on a year-to-year basis, some global-scale models of mass transfer within the carbon cycle invoke the influence of an apparent 'missing sink' for fossil fuel derived CO_2 released to the atmosphere. The soil carbon reservoir would appear to offer the most likely solution to this enigma. It has been estimated (Raich and Potter, 1995) that 60–80 Pg of soil-derived carbon is respired as CO_2 each year and this is ca. 15 times the calculated rate for anthropogenic emissions of fossil carbon. The apparent 'missing sink' may well be a consequence of intermediate-term storage of photosynthesised carbon, that is on a decal timescale (Fung et al., 1997).

The anticipated rise in mean global temperature in response to increasing concentrations of atmospheric CO_2 has stimulated a considerable interest in the sensitivity of SOM status to prevailing climate. Jenkinson et al. (1991) have estimated that, by the mid-21st century, the additional amount of CO_2 respired from soils as a direct consequence of an anticipated 0.03°C per year mean temperature rise will have contributed an additional 19% to the total produced from the combustion of fossil fuels. Of particular present day concern is the question of whether any future global warming, 'kick started' by industrial activity, may lead to an accelerated release of CO_2 from the large stores of recalcitrant carbon that comprise the peaty soils that cover vast areas in northern latitudes, for example Canada and Siberia (Davidson et al., 2000). This aspect of soil carbon stability is being addressed by a series of ongoing field experiments conducted on organic-rich upland soils in the UK (Huang et al., 1996; Bol et al., 1996; Bol et al., 1999).

Can soils be managed to provide a significant net sink for excess CO_2 (Bolin, 1998)? There is a general consensus that, in temperate regions, forest clearance is associated with a net loss of SOC whereas afforestation of grassland or previously established agricultural soils enhances the content of SOM. Bomb ^{14}C signatures are invaluable in affording direct inter-comparison of age/depth profiles for selected soil profiles which in conjunction with measures of their relative mass distributions of SOC can be used to equate and compare the net rates of input and loss of SOC as a result of changing land management (Ladyman and Harkness, 1980; Harkness and Harrison, 1989).

Considerable progress has been achieved in quantifying the dynamics of carbon transfer through soils. However, the confidence that can be ascribed to many modelled estimates is limited by the physical and/or chemical complexity of the typical SOM sample available for isotope analyses. In particular is the intimate association of fast and slow cycling fractions. Several approaches in the application of ^{14}C measurement have employed physical and/or chemical procedures as pre-treatment to isolate and recover major fractions from bulk SOM (Sharpenseel, 1976; Trumbore and Zheng, 1996; Leavitt et al., 1996). In most instances, however, the minimum amount of carbon required for ^{14}C measurement by radiometric counting methods (ca. 1 g) remained a practical constraint. The development and increasing availability during the past decade of AMS for direct measurement of ^{14}C enrichment has provided the opportunity to examine soil carbon dynamics in a much wider range of experimental situations than was realistically possible before. Without doubt, the major advantage from AMS technology comes with the ca. 1,000-fold reduction in sample size requirement and allied with a significant increase in analytical throughput. Examples of hitherto impossible investigations include: (i) the use of AMS coupled with sophisticated chromatographic separation techniques to investigate soil processes at the molecular level (Huang et al., 1996; Bol et al., 1996) and (ii) the direct measurement of the ^{14}C signatures in CO_2 respired at various depths within the soil profile (Bol and Harkness, 1995; Tegan and Dorr, 1996).

6.4. Local effects

^{14}C is produced via a range of reactions within nuclear fuel cycle operations. Table 1 lists the major production pathways.

Of all the stages in the nuclear fuel cycle, the running of reactors and reprocessing of spent fuel contribute most to ^{14}C releases to the environment. Briggs and Hart (1988) estimated that the total ^{14}C production between 1985 and 2000 would be 5.87×10^6 GBq. However, despite this input, McCartney et al. (1988a) estimate that the global ^{14}C activity of the atmosphere will continue to fall until at least AD 2015 because of the greater effect of fossil fuel burning. These localised discharges can be used to study biogeochemical processes going on in the immediate environment. We have chosen to use the Sellafield nuclear fuel reprocessing plant in Cumbria, England as a case study of what can be accomplished. The Sellafield plant is operated by British Nuclear Fuels plc, and currently represents the largest anthropogenic source of ^{14}C to the environment. Sellafield (then Windscale) commenced discharges in the early 1950s and since the mid-1960s, annual discharges (aerial and aquatic to the NE Irish Sea) have been of the order of 3–20 TBq per annum approximately. The aquatic discharges were typically around 1–2 TBq per annum until the mid-1990s when there was a change in discharge policy from mainly atmospheric to mainly aquatic.

Most of the early reported discharges were estimates and there have been several attempts to reconstruct the annual aerial discharges (McCartney et al.,

Table 1 ^{14}C production pathways and sources within the nuclear fuel cycle.

Reaction	Main source
^{13}C (n, γ) ^{14}C	Graphite moderated reactors and CO_2 coolant
^{17}O (n, α) ^{14}C	Water moderated reactors and oxide fuels
^{16}O (n, ^3He) ^{14}C	Not significant except in fast reactor neutron spectra
^{15}N (n, d) ^{14}C	Not significant except in fast reactor neutron spectra
Ternary fission	Few percent of the total
^{14}N (n, p) ^{14}C	Principal contributor is from nitrogen impurities in the fuel, but also in the moderator, coolant and structural materials

Source: From Otlet et al. (1992).

1988b; Otlet et al., 1989; McNamara and McCartney, 1998) using ^{14}C measurements made on tree rings. These studies rely on the accurate determination of a calibration factor that relates the observed excess ^{14}C specific activity in a ring to the releases for that year. However, early studies were hampered by the paucity of known discharge data on which to base the calibration factor. Isogai et al. (2002) produced a complete estimated record of the discharges based on ^{14}C measurements on rings from an oak tree growing close to the Sellafield site. They used a calibration factor based on 6 years of detailed measured discharges and wind direction data. Also, Cook et al. (2004a) reconstructed the aquatic discharge record by applying simple empirical models to ^{14}C data for annually collected seaweed samples.

Work carried out in the NE Irish Sea since the early 1990s on the four biogeochemical fractions of the water column reflects the fact that the ^{14}C discharges are primarily in the form of DIC. The ^{14}C is subsequently transferred to the particulate organic carbon (POC) and, to a lesser extent, the dissolve organic carbon (DOC) and finally, the particulate inorganic carbon (PIC). Also, the water column results appear to reflect the change in discharge policy that took effect in 1994. After this time, the activities in the four biogeochemical fractions are typically greater than pre-1994 (Begg et al., 1992; Cook et al., 1995, 2004b). Analysis of different size fractions of the inorganic component of the inter-tidal sediment suggests that there is likely to be a gradual increase in the specific ^{14}C activity of the sediment in the general area as Sellafield-contaminated organisms die and their shells are ground down by natural processes (Cook et al., 1995, 2004b). Cook et al. (2004b) and Begg et al. (1992) also studied the inter-tidal biota close to Sellafield and found all species analysed to be enriched in ^{14}C relative to ambient background. The degree of enrichment was mussels >limpets->anemones = winkles>seaweed. This trend appears to reflect the positions of the biota in the food chain once the dilution in seaweed from atmospheric uptake is taken into account. This type of study could easily be extended to determine the ratio of uptake by seaweed of inorganic carbon from the aquatic and atmospheric reservoirs. This could be accomplished by studying the relative activities in seaweed at different positions on a beach relative to the high and low tide marks, using ^{14}C as a tracer, because the atmospheric activity where the original study was conducted is relatively unaffected by the aerial discharges.

7. Radiocarbon Literature

The journal 'Radiocarbon' is an ideal starting point for both advances in measurement techniques and applications. In addition, the website for the journal: www.radiocarbon.org has details of all the radiocarbon laboratories currently in operation and links to the laboratory websites.

Also, the proceedings of the AMS conferences are a useful source of information. These are published in *Nuclear Instruments and Methods in Physics Research B.*

8. A Look Forward

The future of applied ^{14}C measurements would seem to be assured by ongoing developments in the design of new generations of AMS instrumentation. Hitherto, AMS ^{14}C measurements have been undertaken using tandem accelerators operating at terminal voltages of 2.5 MV and higher, with a preference for the production of 3+ and 4+ charge states to achieve the required precision. The obvious disadvantages are that these instruments are large, expensive and operationally demanding. More recently, design features have tended towards smaller instruments, for example, using a 0.5 MV tandem to utilise the 1+ charge state at levels of precision comparable to that of the higher voltage systems (Suter et al., 1999). Currently, ongoing design features are focussing on single-stage accelerator mass spectrometers (SSAMS) (Schroeder et al., 2004). These systems would appear to promise the relative simplicity of a low voltage (250 kV) open air insulated accelerator as opposed to the complexities of a high pressure gas insulated tandem, with little loss in analytical resolution. Allied to these developments in AMS technology are significant advancements in the preparation and presentation of sample carbon to the accelerator. For example, the continuing development of gas sources which allow the sample to be presented to the accelerator as CO_2 gas (sometimes using an inert carrier gas) (Xu et al., 2004), rather than a graphite pellet. This development allows much smaller samples to be analysed and leads directly to the concept of on-line gas chromatography-accelerator mass spectrometry (GC-AMS) (Bronk Ramsey et al., 2004). The realisation of these considerable technical developments would seem set to open the floodgates of applied ^{14}C measurement across both established and novel areas of science.

REFERENCES

Anderson, E. C., J. R. Arnold, and W. F. Libby. (1951). Measurement of low level radiocarbon. *The Review of Scientific Instruments*, 22, 225–230.

Anderson, E. C., and W. F. Libby. (1951). World-wide distribution of natural radiocarbon. *The Physical Review*, 81, 64–69.

Anderson, E. C., W. F. Libby, S. Weinhouse, A. F. Reid, A. D. Kirshenbaum, and A. V. Grosse. (1947). Natural radiocarbon from cosmic radiation. *The Physical Review*, 72, 931–936.

Anderson, J. M. (1992). Responses of soils to climate change. *Advances in Ecological Research*, 22, 163–210.

Angulo, R. J., M. C. de Souza, P. J. Reimer, and S. K. Sasaoka. (2005). Reservoir effect of the southern and southeastern Brazilian coast. *Radiocarbon*, 47, 67–73.

Arnold, J. R., and W. F. Libby. (1949). Age determination by radiocarbon content; checks with samples of known age. *Science*, 110, 675–680.

Arnold, J. R., and W. F. Libby. (1951). Radiocarbon dates. *Science*, 113, 111–120.

Ascough, P., G. T. Cook, and A. J. Dugmore. (2005). Methodological approaches to determining the marine radiocarbon reservoir effect. *Progress in Physical Geography*, 29, 532–547.

Ascough, P. L., G. T. Cook, M. J. Church, A. J. Dugmore, S. V. Arge, and T. H. McGovern. (2006). Variability in North Atlantic marine radiocarbon reservoir effects at c. 1000 AD. *The Holocene*, 16, 131–136.

Ascough, P. L., G. T. Cook, A. J. Dugmore, J. Barber, E. Higney, and E. M. Scott. (2004). Holocene variations in the Scottish marine radiocarbon reservoir effect. *Radiocarbon*, 46, 611–620.

Ashmore, P. J. (1999). Radiocarbon dating: Avoiding errors in dating by avoiding mixed samples. *Antiquity*, 73, 124–130.

Becker, B. (1993). An 11,000-year German oak and pine dendrochronology for radiocarbon calibration. *Radiocarbon*, 35, 201–213.

Becker, B., and B. Kromer. (1986). Extension of the Holocene dendrochronology by the preboreal pine series, 8000 to 10,000 BP. *Radiocarbon*, 28, 961–968.

Becker, B., and B. Kromer. (1993). The continental tree-ring record absolute chronology, ^{14}C calibration and climatic change at 1 ka BP. *Palaeogeography, Palaeoclimatology, Palaeoecology*, 103, 67–721.

Begg, F. H., G. T. Cook, M. S. Baxter, E. M. Scott, and M. McCartney. (1992). Anthropogenic radiocarbon in the eastern Irish Sea and Scottish coastal waters. *Radiocarbon*, 34, 707–716.

Bennett, C. L., R. P. Beukens, M. R. Clover, H. E. Gove, R. B. Liebert, A. E. Litherland, K. H. Purser, and W. E. Sondheim. (1977). Radiocarbon dating using accelerators; negative ions provide the key. *Science*, 198, 508–510.

Boaretto, E., C. Bryant, I. Carmi, G. T. Cook, S. Gulliksen, D. D. Harkness, J. Heinemeier, E. McGee, P. Naysmith, G. Possnert, M. Scott, J. van der Plicht, and M. J. van Strydonck. (2002). Summary findings of the fourth international radiocarbon intercomparison (FIRI), 1998–2001. *Quaternary Science*, 17, 633–639.

Bol, R. A., and D. D. Harkness. (1995). The use of zeolite molecular sieves for trapping low concentrations of CO_2 from environmental atmospheres. *Radiocarbon*, 37, 643–647.

Bol, R. A., D. D. Harkness, Y. Haung, and D. M. Howard. (1999). The influence of soil processes on carbon isotope distribution and turnover in the British Uplands. *European Journal of Soil Science*, 50, 41–51.

Bol, R. A., Y. Haung, J. Meredith, G. Eglinton, D. D. Harkness, and P. Ineson. (1996). The ^{14}C age and residence time of organic matter and its lipid constituents in a stagnohumic gley soil. *European Journal of Soil Science*, 47, 215–222.

Bolin, B. (1998). The Kyoto negotiations on climate change: A science perspective. *Science*, 279, 330–331.

Bolin, B., and I. Fung. (1992). Report: The carbon cycle revisited. In: *Modelling the Earth System* (Ed. D. Ojima). UCAR/Office of Interdisciplinary Earth Studies, Boulder, CO.

Briggs, A., and D. Hart. (1988). *The Management of Carbon-14 and Iodine-129 Wastes, a Site Specific Survey of Current and Future Arisings, Possible Management Options and Potential Impact with Respect to the United Kingdom*. DOE Report DOE/RW/89.031, UKAEA, Harwell.

Broecker, W. S., and T. Peng. (1982). *Tracers in the Sea.* Lamont-Doherty Geological Observatory, Columbia University, Palisades, New York.

Broecker, W. S., T. H. Peng, and R. Engh. (1980). Modeling the carbon system. *Radiocarbon,* 22, 565–598.

Bronk Ramsey, C., P. Ditchfield, and M. Humm. (2004). Using a gas ion source for radiocarbon AMS and GC-AMS. *Radiocarbon,* 46, 25–32.

Bryant, C., I. Carmi, G. T. Cook, S. Gulliksen, D. D. Harkness, J. Heinemeier, E. McGee, P. Naysmith, G. Possnert, E. M. Scott, J. van der Plicht, and M. van Strydonck. (2000). Sample requirements and design of an inter-laboratory trial for radiocarbon laboratories. *Nuclear Instruments and Methods in Physics Research B,* 172, 355–358.

Bush, R. P., R. F. White, and G. M. Smith. (1983). Carbon-14 Waste Management. AERE-R10543, United Kingdom Atomic Energy Authority, Harwell, Didcot, UK.

Cage, A. G., J. Heinemeier, and W. E. N. Austin. (2006). Marine radiocarbon reservoir ages in Scottish coastal and fjordic waters. *Radiocarbon,* 48, 31–43.

Cain, W. F., and H. E. Suess. (1976). Carbon 14 in tree-rings. *Journal of Geophysical Research,* 81, 3688–3694.

Cambray, R. S., G. N. J. Lewin, and K. Playford. (1985). *Radioactive fallout in air and rain: Results to the end of 1984.* UK Atomic Energy Authority, Harwell, AERE-R11915.

Campbell, C. A., E. A. Paul, D. A. Rennie, and K. J. McCullum. (1967a). Factors affecting the accuracy of the carbon – dating method in soil humus studies. *Soil Science,* 104, 81–83.

Campbell, C. A., E. A. Paul, D. A. Rennie, and K. J. McCullum. (1967b). Factors affecting the accuracy of the carbon dating method of analysis to soil humus studies. *Soil Science,* 104, 217–224.

Carter, M. W., and A. A. Moghissi. (1977). Three decades of nuclear testing. *Health Physics,* 33, 55–71.

Castagnoli, G., and D. Lal. (1980). Solar modulation effects in terrestrial production of carbon-14. *Radiocarbon,* 22, 133–158.

Cook, G. T., F. H. Begg, P. Naysmith, E. M. Scott, and M. McCartney. (1995). Anthropogenic ^{14}C marine geochemistry in the vicinity of a nuclear fuel reprocessing plant. *Radiocarbon,* 37, 459–467.

Cook, G. T., C. Bonsall, R. E. M. Hedges, K. McSweeney, V. Boroneant, and P. B. Pettitt. (2001). A freshwater diet-derived reservoir effect at the stone age sites in the Iron Gates gorge. *Radiocarbon,* 43, 453–460.

Cook, G. T., A. B. MacKenzie, G. K. P. Muir, G. Mackie, and P. Gulliver. (2004a). Sellafield-derived anthropogenic ^{14}C in the marine intertidal environment of the NE Irish Sea. *Radiocarbon,* 46, 877–884.

Cook, G. T., E. M. Scott, A. B. MacKenzie, F. H. Naysmith, K. Isogai, P. J. Kershaw, R. Anderson, and P. Naysmith. (2004b). Reconstructing the history of ^{14}C discharges from Sellafield. Part 2. Aquatic discharges. *Journal of Radioanalytical and Nuclear Chemistry,* 260, 239–247.

Craig, H. (1957). Isotopic standards for carbon and oxygen and correction factors for mass-spectrometric analysis of carbon dioxide. *Geochimica et Cosmochimica Acta,* 12, 133–140.

Damon, P. E., S. Cheng, and T. W. Linick. (1989). Fine and hyperfine structure in the spectrum of secular variations of atmospheric ^{14}C. *Radiocarbon,* 31, 704–718.

Damon, P. E., A. Long, and E. J. Wallick. (1972). Dendrochronologic calibration of the Carbon-14 time scale. *Proceedings of the 8th International Conference on Radiocarbon Dating,* Royal Society of New Zealand, Wellington, pp. 44–59.

Davidson, E. A., S. E. Trumbore, and R. Amundson. (2000). Soil warming and organic carbon content. *Nature,* 408, 789–790.

Davis, R. J., and O. A. Schaeffer. (1955). Chlorine-36 in nature. *Annals of the New York Academy of Science,* 62, 105–122.

deVries, H. (1959). Measurement and use of natural radiocarbon. In: *Research in Geochemistry* (Ed. P. H. Abelson). Wiley, New York.

Elsasser, W., E. O. Ney, and J. R. Winckler. (1956). Cosmic-ray intensity and geomagnetism. *Nature*, 178, 1226–1227.

Engelkemeir, A. G., W. H. Hamill, M. G. Inghram, and W. F. Libby. (1949). The half-life of radiocarbon. *Physical Review*, 75, 1825–1833.

Fifield, L. K. (1999). Accelerator mass spectrometry and its applications. *Reports on Progress in Physics*, 62, 1223–1274.

Friedrich, M., S. Remmele, B. Kromer, J. Hofmann, M. Spurk, K. F. Kaiser, C. Orcel, and M. Küppers. (2004). The 12,460-year oak and pine tree-ring chronology from Central Europe – a unique annual record for radiocarbon calibration and paleoenvironment reconstructions. *Radiocarbon*, 46, 1111–1122.

Fung, I., C. B. Field, J. A. Berry, M. V. Thompson, J. T. Randerson, C. M. Malmstrom, P. M. Vitousek, G. J. Collatz, P. J. Sellers, D. A. Randall, A. S. Denning, F. Badeck, and J. John. (1997). Carbon 13 exchanges between the atmosphere and biosphere. *Global Biogeochemical Cycles*, 4, 507–533.

Gaudinski, J. B., S. E. Trumbore, E. A. Davidson, and S. H. Zheng. (2000). Soil carbon cycling in a temperate forest: Radiocarbon-based estimates of residence times, sequestration rates and partitioning of fluxes. *Biogeochemistry*, 51, 33–69.

Geyh, M. A., U. Schotterer, and M. Grosjean. (1998). Temporal changes of the ^{14}C reservoir effect in lakes. *Radiocarbon*, 40, 921–931.

Godwin, H. (1962). Half-life of radiocarbon. *Nature*, 195, 944.

Gosse, J. C., and F. M. Phillips. (2001). Terrestrial in situ cosmogenic nuclides: Theory and application. *Quaternary Science Reviews*, 20, 1475–1560.

Grootes, P. M., W. G. Mook, J. C. Vogel, A. E. de Vries, A. Haring, and J. Kistemaker. (1975). Enrichment of radiocarbon for dating samples up to 75,000 years. *Zeischrift für Naturforschung*, 30, 1–14.

Guillet, B. (1982). Study of the turnover of soil organic matter using radioisotopes (^{14}C). In: *Constituents and Properties of Soils* (Eds M. Bonneau and B. Souchier). Academic Press, London, UK, pp. 230–257.

Gulliksen, S., and E. M. Scott. (1995). Report of the TIRI workshop. *Radiocarbon*, 37, 820–821.

Gupta, S. K., and H. A. Polach. (1985). *Radiocarbon Dating Practices at ANU*. ANU, Canberra, Australia.

Haas, H., and V. Trigg. (1991). Low-level scintillation counting with a LKB Quantulus counter establishing optimal parameter settings. In: *Liquid Scintillation Counting and Organic Scintillators* (Eds H. Ross, J. E. Noakes, and J. D. Spaulding). Lewis Publishers, Lansing, MI, pp. 669–675.

Hahn, V., and N. Buchmann. (2004). A new model for soil organic turnover using bomb carbon. *Global Biogeochemical Cycles*, 18, GB 1019.

Handwerger, D. A., T. E. Cerling, and R. L. Bruhn. (1999). Cosmogenic ^{14}C in carbonate rocks. *Geomorphology*, 27, 13–24.

Harkness, D. D. (1970). Artificial carbon-14, a tracer for carbon in the atmosphere and biosphere. PhD Thesis, University of Glasgow.

Harkness, D. D., and A. F. Harrison. (1989). The influence of afforestation on upland soils; the use of 'bomb ^{14}C' enrichment as a quantitative tracer for changes in organic status. *Radiocarbon*, 31, 637–643.

Harkness, D. D., A. F. Harrison, and P. Bacon. (1986). The temporal distribution of bomb – ^{14}C in a forest soil. *Radiocarbon*, 28, 328–337.

Harrison, A. F., and D. D. Harkness. (1993). Potential for estimating carbon fluxes in forest soils using ^{14}C techniques. *New Zealand Journal of Forestry Science*, 23, 367–379.

Harrison, A. F., D. D. Harkness, A. P. Rowland, J. S. Garnett, and P. J. Bacon. (2000). Annual carbon and nitrogen fluxes in soils along the European forest transect, determined using

^{14}C-bomb. In: *Carbon and Nitrogen Cycling in European Forest Ecosystems. Ecological Studies* (Ed. E. D. Scultz). Springer-Verlag, New York, pp. 237–256.

Harrison, K. G. (1996). Using bulk soil radiocarbon measurements to estimate soil organic matter turnover times: Implications for atmospheric CO_2 levels. *Radiocarbon*, 38(3), 181–190.

Hereforth, L. (1948). Die Fluoreszensanregung Organischer Substanzen mit Alphateilchen, Schnellen Elektronen und Gammastrahlen. Thesis presented at Technical University Berlin-Charlottenburg.

Hesshaimer, V., M. Heimann, and I. Levin. (1994). Radiocarbon evidence for a smaller oceanic carbon dioxide sink than previously believed. *Nature*, 370, 201–203.

Hiebert, R. D., and R. J. Watts. (1953). Fast coincidence circuits for H3 and C14 measurements. *Nucleonics*, 11, 38–41.

Huang, Y., R. Bol, D. D. Harkness, P. Ineson, and G. Eglinton. (1996). Post-glacial variation in distributions, ^{13}C and ^{14}C contents of aliphatic hydrocarbons and bulk organic matter in three types of British acid upland soils. *Organic Geochemistry*, 24, 273–287.

Hughen, K. A., M. G. L. Baillie, E. Bard, et al. (2004). Marine04 marine radiocarbon calibration, 0–26Cal kyr BP. *Radiocarbon*, 46, 1059–1086.

Isogai, K., G. T. Cook, and R. Anderson. (2002). Reconstructing the history of ^{14}C discharges from Sellafield: 1. Atmospheric discharges. *Journal of Environmental Radioactivity*, 59, 207–222.

Jenkinson, D. E. (1990). The turnover of organic carbon and nitrogen in soil. *Philosophical Transactions of the Royal Society of London B*, 329, 361–368.

Jenkinson, D. S., D. E. Adams, and A. Wild. (1991). Model estimates of CO_2 emissions from soil in response to global warming. *Nature*, 351, 304–306.

Jenkinson, D. S., D. D. Harkness, D. E. Vance, D. E. Adams, and A. F. Harrison. (1992). Calculating net primary production and annual input of organic matter to soil from the amount and radiocarbon content of soil organic matter. *Soil Biology and Biochemistry*, 24, 295–308.

Jenkinson, D. S., and J. H. Raynor. (1977). The turnover of soil organic matter in some of the Rothamstead classical experiments. *Soil Science*, 123, 298–305.

Jobbagy, E. G., and R. B. Jackson. (2000). The vertical distribution of soil organic carbon and its relation to climate and vegetation. *Ecological Applications*, 10, 423–436.

Key, R. (1996). WOCE Pacific Ocean radiocarbon program. *Radiocarbon*, 38, 415–423.

Korff, S. A. (1940). *Terrestrial Magnetism and Atmospheric Electricity*, 45, 133.

Ladyman, S. J., and D. D. Harkness. (1980). Carbon isotope measurement as an index of soil development. *Radiocarbon*, 22, 885–891.

Leavitt, S. W., R. F. Follett, and E. A. Paul. (1996). Estimation of slow- and fast-cycling soil organic carbon pools from 6N HCl hydrolysis. *Radiocarbon*, 38, 231–239.

Le Clercq, M., J. van der Plicht, and M. Gröning. (1998). New 14C reference materials with activities of 15 and 50 pMC. *Radiocarbon*, 40, 295–297.

Levin, I., and V. Hesshaimer. (2000). Radiocarbon – a unique tracer of global carbon cycle dynamics. *Radiocarbon*, 42, 69–80.

Levin, I., J. Schuchard, B. Kromer, and K. O. Münnich. (1989). The continental European Suess effect. *Radiocarbon*, 31, 431–440.

Libby, W. F. (1946). Atmospheric helium three and radiocarbon from cosmic radiation. *The Physical Review*, 69, 671–672.

Libby, W. F. (1952). *Radiocarbon Dating*. Chicago, University of Chicago Press.

Libby, W. F., E. C. Anderson, and J. R. Arnold. (1949). Age determination by radiocarbon content: World-wide assay of natural radiocarbon. *Science*, 109, 227–228.

Lifton, N. A., A. J. T. Jull, and J. Quade. (2001). A new extraction technique and production rate estimate for in situ cosmogenic 14C in quartz. *Geochimica et Cosmochimica Acta*, 65, 1953–1969.

Light, E. S., M. Merker, H. J. Verschell, R. B. Mendell, and S. A. Korff. (1973). Time dependence worldwide distribution of atmospheric neutrons and of their products. 2. Calculation. *Journal of Geophysical Research*, 78, 2741–2762.

Lingenfelter, R. E. (1963). Production of Carbon 14 by cosmic-ray neutrons. *Review of Geophysics*, 1, 35–55.

Matthews, E. (1997). Global litter production, pools and turnover times: Estimates from measured data and regeneration models. *Journal of Geophysical Research*, 102, 18771–18800.

McCartney, M., M. Baxter, and E. M. Scott. (1988a). Carbon-14 discharges from the nuclear fuel cycle: 1. Global effects. *Journal of Environmental Radioactivity*, 8, 143–155.

McCartney, M., M. Baxter, and E. M. Scott. (1988b). Carbon-14 discharges from the nuclear fuel cycle: 2. Local effects. *Journal of Environmental Radioactivity*, 8, 157–171.

McNamara, N., and M. McCartney. (1998). A new estimate of atmospheric ^{14}C discharges from Sellafield. *Journal of Environmental Radioactivity*, 41, 1–10.

Mook, W. G. (1986). Recommendations/resolutions adopted by the Twelfth International Radiocarbon Conference. *Radiocarbon*, 28, 799.

Mook, W. G., and J. van der Plicht. (1999). Reporting ^{14}C activities and concentrations. *Radiocarbon*, 41, 227–239.

Munnich, K. O., and J.C. Vogel. (1963). Investigations of meridional transport in the troposphere by means of carbon-14. *Radioactive Dating. Proceedings of a Symposium on Radioactive Dating* held in Athens, 19–23 November 1962. Vienna, IAEA, 1963 (Proceedings Series), pp. 189–197.

Nelson, D. E., R. G. Korteling, and W. R. Stott. (1977). Radiocarbon measurements using the McMaster tandem accelerator. *Science*, 198, 506–508.

Nydal, R. (1963). Increase in radiocarbon from the most recent series of thermonuclear tests. *Nature*, 200, 212–214.

Nydal, R. (1966). Variations in C-14 concentration in the atmosphere during the last several years. *Tellus*, 18, 271–279.

Nydal, R. (1967). On the transfer of radiocarbon in nature. *Radioactive Dating and Methods of Low-Level Counting. Proceedings of IAEA Symposium* held in Monaco March 1967, Vienna, IAEA (Proceedings Series), pp. 119–128.

Nydal, R. (2000). Radiocarbon in the ocean. *Radiocarbon*, 42, 81–98.

Nydal, R., S. Gulliksen, K. Lövseth, and F. H. Skogseth. (1984). Bomb ^{14}C in the ocean surface 1966–1981. *Radiocarbon*, 26, 7–45.

Nydal, R., and K. Lovseth. (1965). Distribution of carbon from nuclear tests. *Nature*, 206, 1029–1031.

O'Brien, B. J. (1984). Soil organic matter fluxes and turnover rates estimated from radiocarbon enrichments. *Soil Biology and Biochemistry*, 16, 115–120.

O'Brien, B. J., and J. D. Stout. (1978). Movement and turnover of soil organic matter as indicated by carbon isotope measurements. *Soil Biology and Biochemistry*, 10, 309–317.

Oeschger, H. U., U. Siegenthaler, U. Schotterer, and A. Gugelmann. (1975). A box diffusion model to study the carbon dioxide exchange in nature. *Tellus*, 27, 168–192.

Ostlund, H. G., and M. Stuiver. (1980). GEOSECS pacific radiocarbon. *Radiocarbon*, 22, 2553.

Otlet, R. L., M. J. Fulker, and A. J. Walker. (1992). Environmental impact of atmospheric Carbon-14 emissions resulting from the nuclear fuel cycle. In: *Radiocarbon After Four Decades* (Eds R. E. Taylor, A. Long, and R. S. Kra). Springer-Verlag, New York, pp. 519–534.

Otlet, R. L., G. Huxtable, and D. C. W. Sanderson. (1986). The development of practical systems for ^{14}C measurement in small samples using miniature counters. *Radiocarbon*, 28, 603–614.

Otlet, R. L., H. Longley, and A. J. Walker. (1989). Measurements of carbon-14 in tree-rings growing in the Sellafield area. In: *Studies of Environmental Radioactivity in Cumbria, Part 13*, AERE-R12362. United Kingdom Atomic Energy Authority, Harwell, Didcot, UK.

Parton, W. J., D. S. Schimel, C. V. Cole, and D. S. Ojima. (1987). Analysis of factors controlling soil organic matter levels in Great Plain Grasslands. *Soil Science Society of America Journal*, 51, 1173–1179.

Paul, E. A., C. A. Campbell, D. A. Rennie, and K. J. McCullum. (1964). Investigation on the dynamics of soil humus utilizing carbon dating techniques. *Transcriptions of the 8th International Congress of Soil Science,* Section 111, Bucharest, pp. 201–208.

Pearson, G. W., J. R. Pilcher, M. G. A. Baillie, D. M. Corbett, and F. Qua. (1986). High-precision ^{14}C measurement of Irish oaks to show the natural ^{14}C variations from AD 1840 to 5210 BC. *Radiocarbon*, 28, 911–934.

Pearson, G. W., and M. Stuiver. (1986). High- precision calibration of the radiocarbon timescale, 500–2500BC. *Radiocarbon*, 28, 839–862.

Pearson, G. W., and M. Stuiver. (1993). High-precision bidecadal calibration of the radiocarbon timescale, 500–2500BC. *Radiocarbon*, 35, 25–34.

Pessenda, L. C. R., R. Aravena, A. J. Melfi, E. C. C. Telles, R. Boulet, E. P. E. Valencia, and M. Tomazello. (1996). The use of carbon isotopes (^{13}C, ^{14}C) in soil to evaluate vegetation changes during the Holocene in Central Brazil. *Radiocarbon*, 38, 191–201.

Polach, D. (1988). *Radiocarbon dating Literature: The first 21 years, 1947–1968.* Academic Press, London, UK.

Potter, C. S., and S. A. Klooster. (1997). Global model estimates of carbon and nitrogen storage in litter and soil pools in response to changes in vegetation quality and biomass allocation. *Tellus Series B – Chemical and Physical Meteorology*, 49B, 1–17.

Purser, K. H., R. B. Liebert, and C. J. Russo. (1980). MACS: An accelerator based radioisotope measuring system. *Radiocarbon*, 27, 794–806.

Rafter, T. A. (1965). Carbon-14 variations in nature, part 2, increase in C-14 activities in the atmosphere of the southern hemisphere from the testing of nuclear weapons. *New Zealand Journal of Science*, 8, 472–493.

Rafter, T. A., and G. J. Fergusson. (1957). The atom bomb effect: Recent increase in the ^{14}C content of the atmosphere, biosphere, and surface waters of the oceans. *New Zealand Journal of Science and Technology*, B38, 871–883.

Raich, J. W., and C. S. Potter. (1995). Global patterns of carbon dioxide emissions from soils. *Global Biogeochemical Cycles*, 9, 23–36.

Reimer, P. J., M. G. L. Baillie, E. Bard, et al. (2004). Intcal04 terrestrial radiocarbon age calibration, 0–26 kyr BP. *Radiocarbon*, 46, 1029–1058.

Rozanski, K., W. Stichler, R. Gonfiantini, E. M. Scott, R. P. Beukens, B. Kromer, and J. van der Plicht. (1992). The IAEA ^{14}C intercomparison exercise 1990. *Radiocarbon*, 34, 506–519.

Rubin, M., and H. E. Suess. (1955). US Geological Survey radiocarbon Dates 111. *Science*, 123, 444–448.

Rutberg, R. L., D. S. Schimel, I. Hajdas, and W. S. Broecker. (1996). The effect of tillage on soil organic matter using ^{14}C: A case study. *Radiocarbon*, 38, 209–217.

Schimel, D. S. (1995). Terrestrial ecosystems and the carbon cycle. *Global Change Biology*, 1, 77–91.

Schroeder, J. B., T. M. Hauser, G. M. Klody, and G. A. Norton. (2004). Initial results with low energy single stage AMS. *Radiocarbon*, 46, 1–4.

Scott, E. M., T. C. Aitchison, D. D. Harkness, G. T. Cook, and M. S. Baxter. (1990). An overview of all three stages of the international radiocarbon intercomparison. *Radiocarbon*, 32, 309–319.

Scott, E. M., C. Bryant, G. T. Cook, and P. Naysmith. (2003). Is there a fifth radiocarbon intercomparison (VIRI)? *Radiocarbon*, 45(2), 493–495.

Scott, E. M., D. D. Harkness, B. F. Miller, G. T. Cook, and M. S. Baxter. (1992). Announcement of a further intercomparison exercise. *Radiocarbon*, 34, 528–532.

Sharpenseel, H. W. (1972). Natural radiocarbon measurements on soil organic matter fractions and on soil profiles of different pedogenesis. *Proceedings of the 8th International Conference on Radiocarbon Dating*, Lower Hutt, Royal Society of New Zealand, pp. 892–896.

Sharpenseel, H. W. (1976). The search for biologically inert and lithogenic carbon in recent soil organic matter. In: *Proceedings of the Symposium on Soil Organic Matter Studies*. Brunswick, FRG, September 1976, Vol. 2. Vienna, IAEA, pp. 193–200.

Sonett, C. P. (1992). The present status of understanding of the long-period spectrum of radiocarbon. In: *Radiocarbon After Four Decades* (Eds R. E. Taylor, A. Long, and R. S. Kra). Springer-Verlag, New York, pp. 50–61.

Sternberg, R. S. (1992). Radiocarbon fluctuations and the geomagnetic field. In: *Radiocarbon After Four Decades* (Eds R. E. Taylor, A. Long, and R. R. Kra). Springer-Verlag, New York, pp. 93–116.

Stott, A. W., R. Berstan, P. Evershed, R. E. M. Hedges, C. Bronk Ramsey, and M. J. Humm. (2001). Radiocarbon dating of single compounds isolated from pottery cooking vessel residues. *Radiocarbon*, 43, 191–197.

Stuiver, M. (1961). Variations in radiocarbon concentration and sunspot activity. *Journal of Geophysical Research*, 66, 273–276.

Stuiver, M. (1982). A high precision calibration of the AD radiocarbon timescale. *Radiocarbon*, 24, 1–26.

Stuiver, M., and T. F. Braziunas. (1993). Modeling ^{14}C atmospheric influences and ^{14}C ages of marine samples to 10,000 BC. *Radiocarbon*, 35, 137–189.

Stuiver, M., and G. W. Pearson. (1986). High-precision calibration of the radiocarbon timescale, AD1950 – 500BC. *Radiocarbon*, 28, 805–838.

Stuiver, M., and G. W. Pearson. (1993). High- precision bidecadal calibration of the radiocarbon timescale, AD1950–500BC and 2500–6000 BC. *Radiocarbon*, 35, 1–24.

Stuiver, M., and H. A. Polach. (1977). Reporting of ^{14}C data. *Radiocarbon*, 19, 355–363.

Stuiver, M., and P. D. Quay. (1981). Atmospheric ^{14}C changes resulting from fossil fuel CO_2 release and cosmic ray flux variability. *Earth and Planetary Science Letters*, 53, 349–362.

Suess, H. E. (1955). Radiocarbon concentration in modern wood. *Science*, 122, 415–416.

Suess, H. E. (1965). Secular variations of the cosmic-ray-produced carbon – 14 in the atmosphere and their interpretations. *Journal of Geophysical Research*, 70, 5937–5952.

Suess, H. E. (1970). Bristlecone-pine calibration of the radiocarbon time-scale 5200BC to the present. In: *Nobel Symposium 12: Radiocarbon Variations and Absolute Chronology* (Ed. I. U. Olsson). Wiley, Uppsala, Sweden.

Suess, H. E. (1992). The early radiocarbon years: personal reflections. In: *Radiocarbon After Four Decades* (Eds R. E. Taylor, A. Long, and R. S. Kra). Springer-Verlag, New York, pp. 11–16.

Summons, R. E. (1993). Biogeochemical cycles: A review of fundamental aspects of organic matter formation, preservation and composition. In: *Organic Geochemistry, Principles and Applications* (Eds M. H. Engel and S. A. Macko). Plenum Press, New York.

Suter, M., S. A. W. Jacob, and H.-A. Synal. (2000). Tandem AMS at sub-MeV energies – Status and prospects. *Nuclear Instruments and Methods in Physics Research*, B172, 144–151.

Suter, M., R. Huber, S. A. W. Jacob, and H.-A. Synal. (1999). A new small accelerator for radiocarbon dating. In: *15th International Conference on Application of Accelerators in Research and Industry*, 4–7 November 1998, American Institute of Physics, Denton, TX.

Tauber, H. (1960). Post-bomb rise in radiocarbon activity in Denmark. *Science*, 131, 921–922.

Tegan, I., and H. Dorr. (1996). ^{14}C measurements of soil organic matter, soil CO_2 and dissolved organic carbon (1987–1992). *Radiocarbon*, 38, 247–251.

Trumbore, S. E., and S. Zheng. (1996). Comparison of fractionation methods for soil organic matter ^{14}C analysis. *Radiocarbon*, 38, 219–229.

UNSCEAR. (1993). Sources, effects and risks of ionising radiation. Report to the General Assembly, United Nations, New York.

Walton, A., M. Ergin, and D. D. Harkness. (1970). Carbon 14 concentrations in the atmosphere and carbon dioxide exchange rates. *Journal of Geophysical Research*, 75, 3089–3098.

Wang, Y., R. Amundsen, and S. E. Trunbore. (1996). Radiocarbon dating of soil organic matter. *Quaternary Research*, 45, 282–288.

Willis, E. H., H. Tauber, and K. O. Munnich. (1960). Variations in the atmospheric radiocarbon concentration over the past 1300 years. *American Journal of Science, Radiocarbon Supplement*, 2, 1–4.

Xu, S., R. Anderson, C. Bryant, G. T. Cook, A. Dougans, S. Freeman, P. Naysmith, C. Schnabel, and E. M. Scott. (2004). Capabilities of the new SUERC 5MV AMS Facility for ^{14}C Dating. *Radiocarbon*, 46, 59–64.

CHAPTER 5

Radionuclides as Tracers and Timers in Surface and Groundwater

Contents

The hydrosphere consists of 1,386 million cubic kilometers water, of which 97.5% is saline and 2.5% freshwater. Surface water (lakes, rivers, reservoirs, etc.) is only 0.26% of the total amount of freshwater. Ice and permanent snow cover in the Arctic, the Antarctic, and in mountainous continental regions form 68.7% of this freshwater and 29.9% exists as fresh groundwater (Shiklomanov, 1998). Environmental radionuclides, in combination with stable isotopes, geochemistry, and other hydrological techniques, are a

Radioactivity in the Environment, Volume 16

ISSN 1569-4860, DOI 10.1016/S1569-4860(09)01605-2

powerful tool in studying the cycling of water in continental hydrological systems. The use of environmental radionuclides in surface water is the subject of the first two subchapters here; the third subchapter is devoted to groundwater and geothermal water studies using these techniques. Taking into consideration that most applications in groundwater and geothermal water studies require the combined use of radioactive and stable isotopes, a brief summary on stable isotopes is added to the third subchapter.

5.1. Surface Water, Unsaturated Zone, and Glacial Systems

Robert L. Michel[1], Thomas Kraemer[2] *and* L. DeWayne Cecil[3]

[1]*US Geological Survey, MS 434, Menlo Park, CA, USA*
[2]*US Geological Survey, 430 National Center, Reston, VA, USA*
[3]*US Geological Survey, INL, Idaho Falls, ID, USA*

5.1.1. Introduction

A wide range of radionuclides has been used to study movement of water, solutes, and gases in catchments. Among the hydrologic systems studied have been the unsaturated zone, glacial systems, and surface waters ranging in size from small streams to continental river basins. There are several sources of radionuclides in the environment, and the sources control the ways in which isotopes can be applied to hydrologic systems. One widely used group of nuclides is the bomb-produced species whose atmospheric concentrations peaked in the late 1950s and early 1960s. It was soon realized that these nuclides, frequently labeled transient tracers, represented a global tracer experiment in both hydrology and oceanography (Dockins et al., 1967; Suess, 1969). The bomb-produced nuclides have been primarily used to obtain timescales of physical and chemical processes occurring in the watershed. These nuclides have also been used to study water and/or solute sources and biochemical reactions occurring in the environment. Another group of radionuclides that can be utilized are those produced by cosmic-ray spallation in the atmosphere or near-surface lithosphere. Many of these nuclides, such as carbon-14 (^{14}C) and tritium (^{3}H), were also produced by nuclear weapons testing, and it is necessary to separate the two source functions when using them. A third set of nuclides that has been applied to hydrology are the nuclides produced by the uranium and thorium decay chains. The primary nuclides used from this source are radium and radon. A fourth group is composed of various nuclides produced by nuclear processes in the subsurface derived from processes such as fission and neutron capture (Andrews and Kay, 1982). Nuclides from this source (tritium, chlorine-36, etc.) require very specific conditions to be useful and are rarely used for surface water studies.

Bomb-produced nuclides are among the most important tracers used for hydrologic studies. As noted, some of these have a natural source due to cosmic-ray spallation (Lal and Peters, 1967) which has been overwhelmed by their production through atmospheric nuclear weapons testing. For instance, the total global inventory of tritium was increased from about 3.5 kg to about 800 kg (Miskel, 1973) by 1963. For tritium, almost all hydrologic studies rely on the use of the bomb pulse. Surface water concentrations of tritium are time sensitive with highest concentrations usually occurring during the peak of the bomb transient. Surface water tritium concentrations are also functions of latitude and distance from the ocean. Another bomb-produced nuclide is ^{14}C, whose atmospheric concentration was approximately doubled by nuclear weapons testing. Even prior to nuclear testing, its concentration in the atmosphere was known to undergo both short- and long-term variations (Suess, 1986; Bard et al., 1990). For most groundwater studies and some lake and ocean studies, only the prebomb concentration of ^{14}C is important. However, for studies in the unsaturated zone and surface waters, the bomb-pulse carbon-14 is frequently utilized. Chlorine-36 (^{36}Cl), another nuclide that has a natural and bomb-produced source, has also had variations in production over the past millennia. There was a major increase in ^{36}Cl during the nuclear era but, unlike tritium and ^{14}C, its greatest production occurred during the surface ocean tests of the mid-1950s. The increase in ^{36}Cl was the result of neutrons bombarding seawater chloride during sea-surface bursts at the Pacific islands, most of which occurred in the 1950s. Later tests, although larger, were atmospheric, and the ^{36}Cl production was much smaller. Chlorine-36 has been used frequently for unsaturated zone studies, and less commonly for surface water studies. Strontium-90 (^{90}Sr) and cesium-137 (^{137}Cs) were also produced in large quantities by nuclear testing. They tend to be very reactive, and the cesium absorbs very strongly to soils, so they have not received major use in hydrologic studies. Some other radioactive tracers such as krypton-85 (^{85}Kr) and radioisotopes of iodine are produced by the nuclear industry and have received limited applications to hydrologic studies of surface waters.

Cosmogenic and lithogenic nuclides whose concentrations were not impacted by nuclear testing have also found uses in certain areas of hydrology as steady-state tracers. The most important tracers with a strictly cosmogenic source are the short-lived nuclides sulfur-35 (^{35}S), phosphorus-32 (^{32}P), and beryllium-7 (^{7}Be). If there was a bomb-produced pulse for these tracers, it has already decayed away and is no longer a factor in present day systems. The lithogenic nuclides most commonly used are from the uranium and thorium decay chains. Uranium isotopes have been used to

Table 1.1 Some radionuclides commonly used in surface water studies.

Nuclide	Source	Half-life
^{3}H	Cosmic ray, nuclear testing	12.41 yr
^{7}Be	Cosmic ray	53.6 d
^{14}C	Cosmic ray, nuclear testing	5720 yr
^{32}Si	Cosmic ray	140 yr
^{32}P	Cosmic ray	14.3 d
^{33}P	Cosmic ray	25 d
^{35}S	Cosmic ray	87 d
^{36}Cl	Cosmic ray, nuclear testing	3.0×10^{5} yr
^{210}Pb	Lithogenic	22 yr
^{222}Rn	Lithogenic	3.82 d
^{224}Ra	Lithogenic	3.64 d
^{226}Ra	Lithogenic	1622 yr
^{228}Ra	Lithogenic	6.7 yr
^{234}U	Lithogenic	2.48×10^{5} yr
^{235}U	Lithogenic	7.13×10^{8} yr
^{238}U	Lithogenic	4.15×10^{9} yr

differentiate uranium from anthropogenic and natural sources and the radium and radon isotopes have been used in a large variety of hydrologic systems. Table 1.1 lists the nuclides that will be discussed in this chapter along with their half-lives and sources.

5.1.2. Catchment and river basin hydrology

Physical and chemical processes in small catchments and/or river basins involve a wide range of timescales, and radiochemicals have been applied extensively. These studies can be aimed either at determining chemical processes within the catchment or trying to determine physical parameters like the age of the water flowing out of the watershed. Conservative tracers like tritium and ^{36}Cl are generally used to follow simple physical processes while other radionuclides, such as ^{35}S and ^{32}P, may be best used to follow chemical cycling within the watershed.

Large river basins represent extremely complex systems that derive water and chemicals from a variety of sources. Chemical concentrations in water are impacted by processes such as impoundment of waters, agricultural drainage and *in situ* chemical reactions. There are continuing efforts to relate changes in water quality and isotopic data to changes in climate and land use in drainage basins (Gibson et al., 2002). However, to understand these changes, it is important to have an estimate of the timescales involved in water drainage from river basins.

5.1.2.1. Natural and bomb-produced nuclides tritium, carbon-14, and chlorine-36

5.1.2.1.1. Tritium. There are a series of nuclides that have been produced both naturally and by nuclear weapons tests. Tritium has been one of the most widely used of these tracers in hydrology, due to the fact that it exists as part of the water molecule, and physically follows the movement of water throughout the environment (Murphy, 1993). Tritium concentrations in precipitation vary both temporally and spatially. Concentrations tend to be highest in the northern latitudes, particularly between 30°N and 60°N, and at midcontinental locations (Schell et al., 1970). In the period from the beginning of the nuclear weapon test era until recently the geographic variations were caused both by the locations of the weapons tests and atmospheric circulation factors. Most atmospheric testing was in the Northern Hemisphere (Figure 1.1), which released tritium both in the northern stratosphere and troposphere. The tritium rapidly oxidized to water (HTO) and the tritium in the troposphere was deposited on the earth's surface within 1–2 months, mostly in the same hemisphere where it was produced. Tritium initially injected into the stratosphere primarily entered the troposphere during the spring of each year when continental warming resulted in the breaking up of the northern tropopause (spring leak) between 30°N and 60°N (Martell, 1959). Most stratospheric tritium, regardless of the latitude of origin, was deposited within these latitudes. Only the French and British tests were conducted in the Southern Hemisphere and these tests were much smaller then those of the USA and USSR. The amounts of tritium released to the troposphere of the Southern Hemisphere were much lower and atmospheric weapons testing in the

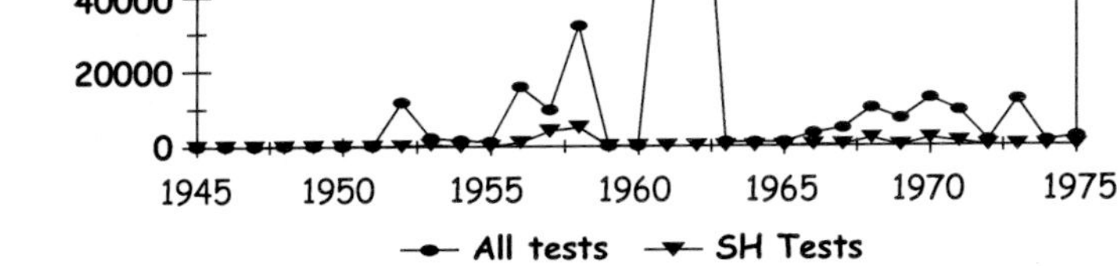

Figure 1.1 History of atmospheric nuclear testing to 1975 (derived from Carter and Moghissi, 1977). The largest detonations were predominately in the Northern Hemisphere, which is part of the reason for the global fallout patterns that developed.

Southern Hemisphere peaked in the mid-1950s. There is no southern equivalent of the spring leak so a substantial quantity of the tritium released to the stratosphere by the southern tests ended up eventually being deposited in the north. One of the major sources of tritium for the Southern Hemisphere after the major tests of the early 1960s was through interhemisphere transfer of water vapor. The delayed interhemisphere transfer of tritium as water vapor meant that for many Southern Hemisphere sites, the tritium concentrations in precipitation tended to peak in the mid-1960s as opposed to 1963–4 (Figure 1.2). This undoubtedly resulted in a similar response in the tritium concentrations of Southern Hemisphere surface waters.

The pattern of tritium fallout in precipitation determines the tritium concentrations in surface waters. The concentrations in water flowing out of river basins reflect the concentrations in precipitation falling within the basin. Figure 1.3 shows the response to the tritium transient of three river basins from about 1960 to 1987: (1) The Kissimmee River in Central Florida, which drains a small watershed at about 25°N; (2) The Mississippi River at Anoka, Minnesota, which drains the upper Mississippi River above 45°N; (3) The Mississippi River at Luling Ferry, MS, which is located at about the same latitude as the Kissimmee River (27°N) but drains a midcontinental basin including high-latitude areas. All basins show the same pattern, with highest tritium concentrations being attained in 1963, followed by a rapid drop in concentrations in the years following. The seasonal signal is much larger for the two Mississippi River sites than the Kissimmee site. The Kissimmee River, which drains a basin that is located south of the other sites, and is subject to a strong oceanic influence, has relatively low concentrations (maximum concentration of 500 TU) throughout the bomb era when compared to the other locations. This site is typical of surface water systems close to the coast in low northern latitudes. The Mississippi site at Anoka is typical of runoff at midcontinental sites in areas where tritium fallout was highest. Tritium concentrations in surface water in these areas reached the 2,000 TU range, and concentrations stayed above 100 TU for almost a decade. The Luling Ferry site drains a midcontinental area with a basin mostly in the 30–50°N latitude range. Thus, despite its relatively low-latitude location, the water in the river at this site is derived from the more northern parts of the continent. Surface waters in the Southern Hemisphere had tritium concentrations that were much lower, reflecting the lower tritium concentrations found in precipitation. No sites exist with long data sets similar to those in Figure 1.3, but most surface water concentrations were much <100 TU throughout the bomb transient. Also, concentrations did not show the sharp peak found in 1963, but tended to increase throughout the 1960s as more tritium migrated to the Southern Hemisphere by atmospheric processes (Taylor, 1966; Stewart and Taylor, 1981).

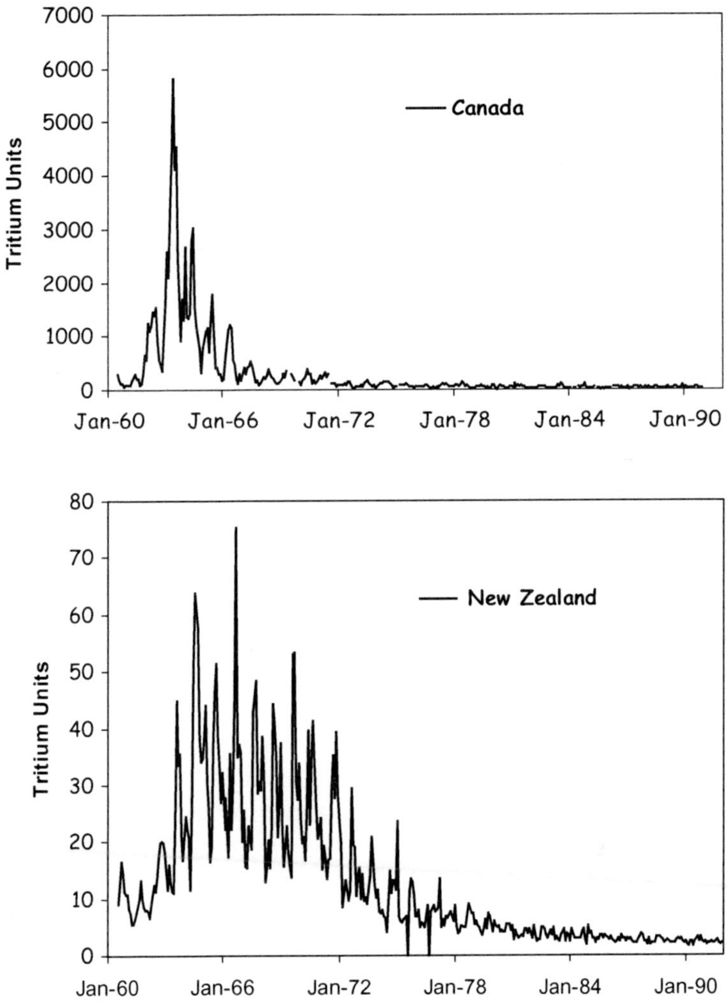

Figure 1.2 Monthly tritium concentrations in precipitation at two sites, one in Canada and the other in New Zealand, during the bomb transient. The concentrations at the Northern Hemisphere site are roughly two orders of magnitude higher than the southern sites during the 1960s and demonstrate a seasonal cycle. The bomb peak is slightly delayed in the Southern Hemisphere. The main reasons for these differences are the location of atmospheric nuclear testing, the atmospheric circulation patterns including the "spring leak" in the Northern Hemisphere, and the greater dilution by oceanic water in the Southern Hemisphere.

The changes in tritium concentrations in surface water with time have provided an opportunity to study physical processes in hydrology. One of the most common methods of applying tritium (as well as other nuclides) in small watersheds is to determine the extent of "new" versus "old" water

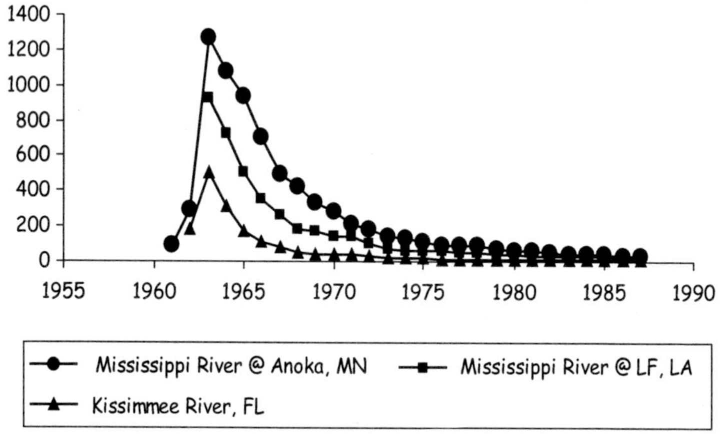

Figure 1.3 Comparison of tritium concentrations in water from three rivers: (1) Mississippi River at Anoka, MN. (2) Mississippi River at Luling Ferry, LA. (3) Kissimmee River in Central Florida. Tritium concentrations are highest at the Anoka, MN site, which receives its flow from precipitation in the midcontinental United States at about 45°N. The lowest concentrations are at Kissimmee whose flow is derived from precipitation in mid-Florida. The location at Luling Ferry, despite having a similar latitude to Kissimmee, receives its flow from higher latitudes in the midcontinent, which results in tritium concentrations similar to rivers from that area.

flowing out of a catchment after storm events (Kennedy et al., 1986). To be most useful, there should be a significant difference between the new and old water and the use of tritium in this manner has decreased as the transient has progressed (Turner and Barnes, 1998). Rose (1993) showed possible scenarios in nature where tritium could be useful in understanding ages of source waters in small springs. Figure 1.4 is a modified version of the scenarios outlined by Rose (1993) showing the five most likely responses of springs to seasonal precipitation: (A) Water exclusively from recent precipitation. (B) Only old water of prebomb origin. (C) Only water from precipitation deposited when tritium concentrations were higher than present precipitation. (D) Prebomb water during baseflow, with mostly recent precipitation at high flow. (E). Water from a period of higher tritium concentrations in precipitation at baseflow with a predominance of recent precipitation during high flow.

Many small springs are composed only of recent precipitation at all times and have seasonal tritium distributions that are similar to scenario A in Figure 1.4. For small springs where groundwater is the only source of water at all times, the tritium concentrations will also be constant throughout the seasonal cycle exhibiting the pattern in scenarios B and C of Figure 1.4. The tritium concentrations reflect "old" water with ages from a few years (scenario C) to the years prior to nuclear testing (scenario B). Scenarios D

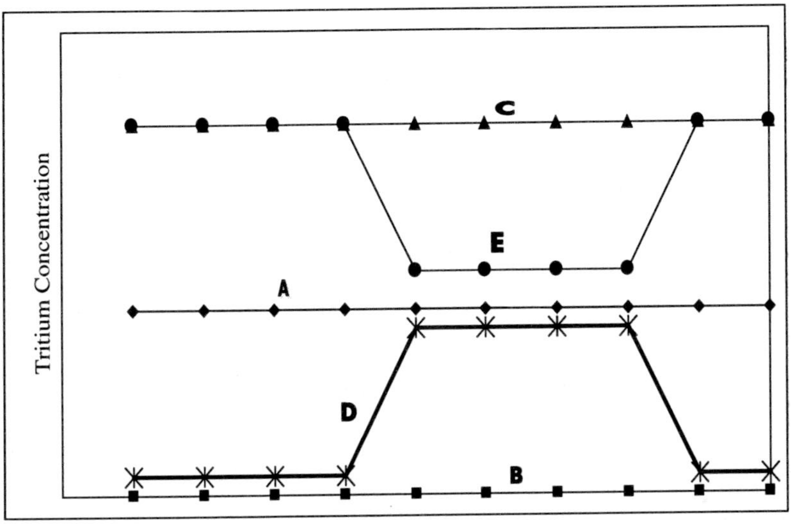

Figure 1.4 Potential response of streams to seasonal fluctuations in the source water for flow modified from the work of Rose (1993). (A) Flow derived from average recent precipitation. (B) Flow composed of waters deposited prior to the advent of atmospheric nuclear testing. (C) Flow entirely derived from water recharged in the years during higher tritium concentrations resulting from nuclear testing. (D) Water derived from pre-1950s water during low flow and predominately recent water during high flow. (E) Water derived from waters with an age of a decade or greater during low flow, but primarily from recent precipitation during high flow.

and E are two of the more common tritium responses found in watersheds. Tritium concentrations are close to recent precipitation during periods of high flow when most water is derived from this source. Tritium concentrations then change to concentrations representative of older water during low-flow periods. The concentrations during periods of low flow reflect the age of groundwater that feeds the spring during this period. McGlynn et al. (2003) studied the processes that cause differences in the tritium concentrations found in small streams at catchments in New Zealand. They determined that tritium concentrations (and residence times within the catchment) were not simply related to catchment size. They did find a correlation between the way the landscape was organized (i.e., number and size of subcatchments) and suggested that this may determine residence time.

In favorable situations, it is possible to use tritium, in conjunction with other tracers, to estimate the age of spring waters that are entering larger rivers. Katz et al. (2001) used tritium to estimate the ages of spring water feeding into the Suwannee River in Florida. Many of these springs were supplied with groundwater that had not been in contact with the surface in 1–2 decades. These waters contained nitrate concentrations resulting from agricultural practices that had been used many years earlier, so a delayed nitrate signal was being delivered to the river by these streams. This sort of information has significant implications for water management and water quality issues.

Tritium has received extensive use as a tracer for determining the timescales for the drainage of water out of large river basins. Usually, a time series of measurements is used, preferably over a period of several decades. The measurements are made at one location and the basin above that sampling point is the system being studied. The flow out of a river basin is a mixture of recent precipitation and drainage from lake and groundwater reservoirs of various ages. This mixture of waters results in a response to the tritium transient in the river that differs from that found in precipitation.

Maloszewski et al. (1983) reported on one of the early applications of ^{3}H in water of small rivers for estimating the mean residence time in the catchment area. They evaluated long-term measurements of tritium concentration in the runoff of the Lainbach creek, Germany (Figure 1.5) with conceptual hydrological models that included the mean transit time as parameter and obtained values of 1.6 and 1.8 years, respectively. Such timescale information allows predicting how long a river would need to flush out a contaminant spill of the catchment. Given the conservative nature of tritium, contaminants will not leave the river faster then the tritium peak did.

Figure 1.6 shows the response of the Mississippi River at Anoka, Minnesota to the tritium transient, compared to that of basin precipitation (Michel, 1989). Tritium concentrations in precipitation are much higher

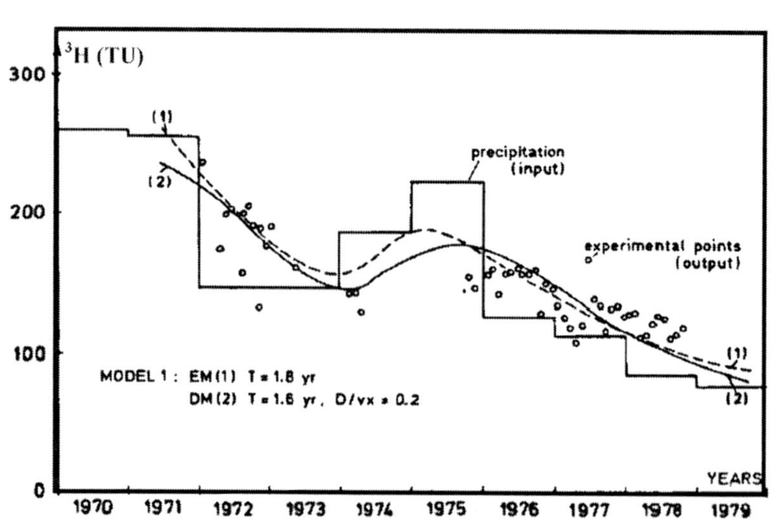

Figure 1.5 Tritium in precipitation and runoff in the catchment of the Lainbach creek in the Bavarian Alps at 1,800 m above sea level (Adapted from Maloszewski et al., 1983 with permission from Elsevier).

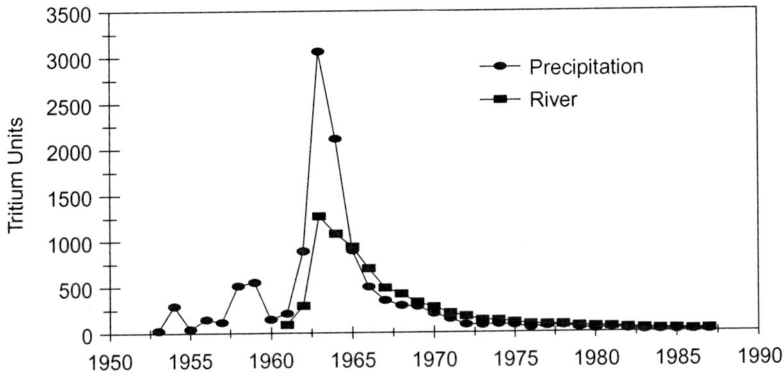

Figure 1.6 Tritium concentrations in river water and basin precipitation (Michel, 1989) for the Mississippi River at Anoka, MN. Tritium concentrations in precipitation are measured at Madison, WI, with correlations used to estimate tritium concentrations during periods when measurements were not available. Tritium concentrations in the river are measured and never reach the concentrations in precipitation during the bomb peak. However, tritium concentrations do retain the bomb-peak signal and concentrations in river water exceed those in basin precipitation by the mid-1960s.

during the peak of the transient than concentrations ever measured in river water. This is because, during the transient peak, the river water is composed of a mixture of older water, which has a relatively low tritium concentration with recent precipitation with its very high tritium concentration, damping the signal. The tritium concentration in precipitation decreases very rapidly due to the short residence time of water in the atmosphere. The residence time of waters in the river basin is much longer and many reservoirs retain a memory of the high tritium concentrations reached during the transient peak. The mixing of precipitation with older water eventually results in the tritium concentration in river water exceeding that in precipitation a few years after the peak of the transient. This is a pattern that is seen in almost all major Northern Hemisphere rivers. The differences between tritium concentrations in river water and precipitation are a function of the influence of recent precipitation on river outflow and the residence times of waters within the basin. The response of river water to the tritium transient can be used to quantify both of these processes in river basins (Weiss and Roether, 1975).

A typical example is given in Figure 1.7 that shows the tritium record of the river Danube and of precipitation at Vienna, Austria. Yurtsever (1999) evaluated these data and obtained the following results: (1) the mean transit time of the surface runoff from the catchment area upstream of the sampling site is 10 months; (2) the mean transit time of the groundwater runoff is 140 months, (3) the fraction of groundwater in the total runoff is

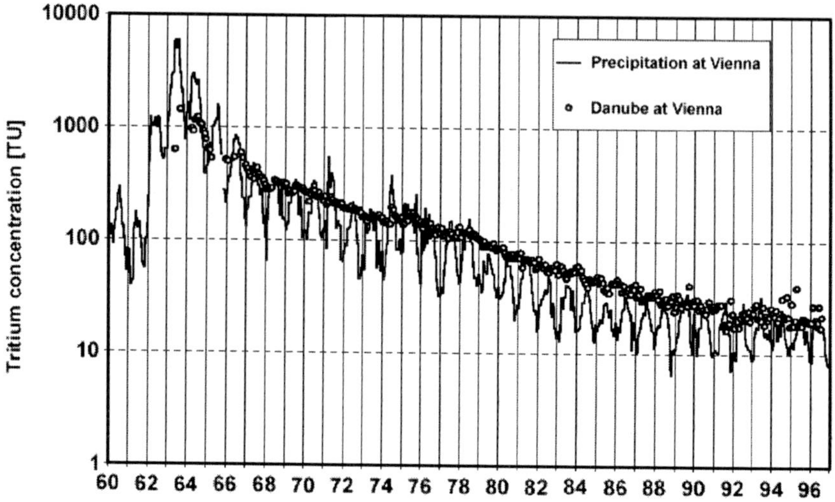

Figure 1.7 Tritium in precipitation and water of the river Danube at Vienna (from Rank et al., 1998). At Vienna, the Danube drains an area of 101,700 km^2, the mean low discharge is 804 $m^3 s^{-1}$ and the long-term average is 1,943 $m^3 s^{-1}$. Assuming that the low discharge represents the average subsurface component, the contribution to the total runoff would be 41.4%.

36%, and (4) the average travel time (flux weighted mean transit time) of the water in the whole basin is about 5 years. For the subsurface flow contribution, a value of 36% has been estimated, which agrees quite well with the value of 41.4% derived from the observed flow regime of the river.

Figure 1.8 shows the response of the Colorado River above Cisco, UT to the tritium transient in the years immediately after the bomb peak. The Colorado River at this location has not been impacted by major dams and the basic hydrologic event is the response to snowmelt in the spring of each year. Flow rates are lowest during the winter and can increase up to one order of magnitude during the height of the runoff of snowmelt. Tritium concentrations in the runoff basin reached several thousand tritium units during 1963–1964 and declined rapidly afterwards. The low tritium concentrations during the winter in this timespan are concentrations found in baseflow, which has not yet been impacted in a major way by the bomb-peak tritium. The increase in tritium concentrations during the seasonal runoff is produced by melting snow, which has received the bomb tritium. It is possible to calculate the percentage of water that is seasonally derived versus the percentage that is derived from the baseflow using tritium concentrations in precipitation and river water collected over the year. From the tritium data, it is estimated that about 30% of the flow is derived from water that had resided in the basin for <1 year. After the first

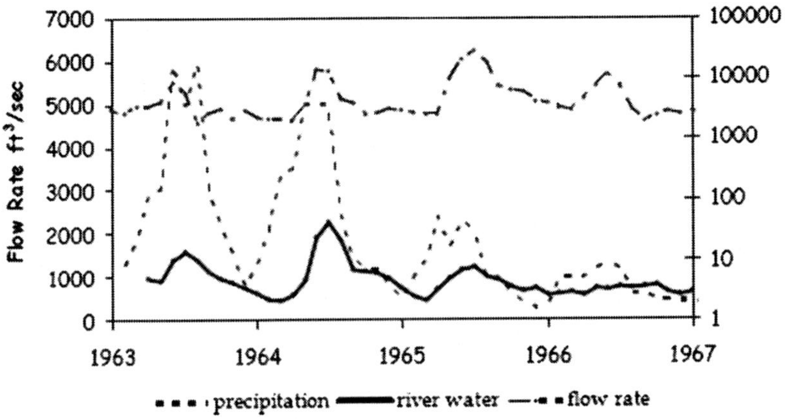

Figure 1.8 Tritium concentrations in river water and basin precipitation versus flow rate during the period immediately following atmospheric testing for the Colorado River at Cisco, UT. During periods of high flow resulting from snowmelt, the bomb-tritium concentrations are evident in the years immediately after the 1962 testing. However, the snowmelt signal disappears within about four years. Tritium concentrations in precipitation taken from Salt Lake City, UT.

3–4 years, the tritium concentrations in precipitation drop below the concentrations found in river water and this method can no longer be applied.

For watersheds where intensive data were not collected during the 1962–1966 timespan, similar parameters can be calculated using long-term data sets. In the 1950s, Brown (1961) used tritium to estimate timescales for physical processes occurring in a river basin. He used a time series of tritium measurements in the Ottawa River basin made during the 1950s to estimate timescales for water retention in the basin. The river basin had a series of small lakes and adjoining wetlands that slowed runoff from the basin. His tritium balance included decay, evaporation, runoff, and loss of water to what he referred to as the stored water reservoir of the valley. The valley reservoir was assumed to be reasonably well mixed. Residence times and the approximate size of the reservoir for the valley were calculated for the whole system as well as for the main tributaries. Eriksson (1963) later expanded on this approach in an attempt to obtain a general model. The constraint of requiring that the entire reservoir be well mixed was avoided by treating the water from each year separately.

Michel (1992, 2004) used tritium data spanning 2–3 decades from several rivers in the United States to calculate the response of the river basins to the tritium transient. A two-member model was used with one component being recent precipitation and the other component being older basin waters. It was possible to estimate the fraction of runoff that was composed of water retained in the basin for <1 year (prompt flow), and to estimate the residence time of older river basin waters. A percentage of 40% was calculated by this approach for prompt flow in the upper Colorado River which is close to the minimum percentage of 30% calculated using the seasonal variations as discussed above. The residence time obtained is a function of many different residence times that could not be separated with these data sets, so no effort was made to determine the size of the reservoir. However, the timescale represented the response of the basin to water entering the system and is useful in studies on contaminant and climatic issues. It was also possible to estimate the tritium concentration for years in which no tritium measurements were made, typically prior to about 1960. Often, rivers are the main source of water in many municipal and agricultural settings. For studies of the hydrology of these systems using the tritium transient, the tritium concentration of river water is more important then the tritium concentration of precipitation.

At the present time, precipitation is virtually free of bomb-tritium worldwide and, consequently, the tritium concentrations in rivers decrease slowly and approach the natural prebomb level. Nevertheless, the conditions for the application of tritium in catchment studies are still favorable where previous measurements of tritium in the given river basin are available from earlier studies and can be combined with present day data. Also in cases where no previous data are available, tritium monitoring may still provide useful information on the residence time of a river catchment although the measured differences in tritium concentration would be rather small and thus the uncertainties relatively high. Also in systems where the tritium concentrations are very different from the present level in local precipitation, tritium monitoring can be useful.

5.1.2.1.2. Carbon-14. Carbon-14 in dissolved inorganic carbon (DIC) has rarely been used to study timescales for river drainage. There are a large number of carbonate reactions that can occur both in the groundwater reservoirs draining into the river and occasionally within the river itself. These interactions make obtaining timescales for physical movements within the river very difficult. Carbon dioxide in the river exchanges with atmospheric carbon dioxide and that can result in a flux of atmospheric ^{14}C into river DIC. However, the rate of exchange between atmospheric carbon dioxide and DIC in water is relatively slow (Revelle and Suess, 1957) when compared to flow rates, and this exchange is usually found to have a minimal impact on ^{14}C concentrations within the river (Aravena and Suzuki, 1990). For many rivers, the DIC is derived from geologic sources such as carbonate minerals, which typically have no ^{14}C, and these sources will control the concentrations in river water. For other small streams and some rivers and lakes, the DIC is derived from volcanic or geothermal sources, which also generally have no radiocarbon. Decaying organic matter will generally have ^{14}C usually close to the ambient concentrations found in the atmosphere as noted below. Thus, the ^{14}C content of the DIC in small streams and lakes is a function of the carbon sources within the watershed and the rate of exchange of DIC with atmospheric CO_2. These concentrations can be used to determine what processes are important in determining CO_2 fluxes within the system under study (Rose et al., 1996).

The radiocarbon content of the organic fraction in rivers can be used to provide information on sources of carbon within the basin and timescales for removal of the various carbon pools. This type of research has been made possible by the development of the accelerator mass spectrometer (AMS) starting in the 1970s, which facilitates the measurement of ^{14}C in small carbon samples (Muller, 1977; Jull et al., 2003). These types of studies have both local and global implications as the export of organic carbon from the world's river basins has a major impact on the carbon flux in the world oceans (Williams and Druffel, 1987). Small mountainous rivers have been considered to have a major impact on sediment and particulate organic flux to the world oceans due to the nature of their sporadic high flow (Milliman and Syvitski, 1992). Kao and Liu (1996) studied the particulate flux, including the carbon component, from such a system, the Lanyang Hsi River in Taiwan. They analyzed the ^{14}C concentration of the particulate organic carbon (POC) component and obtained ages of $>10{,}000$ years. As there was no possibility of industrial contamination from petrochemicals, they concluded that most of the POC in this stream was derived from the bedrock in the watershed, and not from a recent biological component. Carbon-14 was measured in the various organic fractions of a similar small river, the Santa Clara River in Southern California (Masiello and Druffel, 2001). They found very low ^{14}C for the POC and black carbon and even the dissolved organic carbon (DOC) had prebomb concentrations. Again, they were able to rule out petrochemical contamination, and determined that the low ^{14}C content was from organic carbon derived from old soils. Another possibility is that the younger component of organic carbon in the rivers, particularly the DOC, may be more labile and be degraded prior to its release to the oceans. Raymond and Bauer (2001a, 2001b) investigated this problem using both ^{14}C and ^{13}C in several rivers draining into the Atlantic Ocean. POC was found to be generally older than DOC in the rivers. They found that bacteria tend to use the younger (^{14}C enriched) fraction of organic matter in the river, particularly for DOC, resulting in an export of old organic matter to the oceans. Similar results were seen for organic carbon exported by the McKenzie River (Goni et al., 2005).

In the early 1980s, samples of dissolved humic and fulvic acids and particulate organic matter were collected in the Amazon River and analyzed for ^{14}C. All samples contained ^{14}C concentrations indicative of the influence of nuclear tests, although concentrations varied between organic classes with fine particulate organic matter containing the lowest concentration and fulvic acids having the highest concentration (Hedges et al., 1986). The fulvic acid was assumed to be derived from decaying leaf litter that passed through the soil system more rapidly because of its less hydrophobic nature.

Carbon-14 in organic carbon can also be used to discriminate between natural carbon and anthropogenic carbon in pollution studies (Rosen and Rubin, 1964). Many anthropogenic carbon sources are derived from petroleum or coal and have no ^{14}C. Thus, the measurement of ^{14}C in the organic matter can distinguish between a petrochemical origin and a natural source. Spieker and Rubin (1975) developed a simple two-component model to apply to both river water and groundwater in the United States. They indicated that ^{14}C could be used as a simple environmental monitor for many types of pollution found in surface water. This approach has received less use as new developments in mass spectrometry and analytical chemistry have made it possible to identify individual carbon compounds at low concentrations. However, for individual compounds that could have either a petrochemical or natural source (e.g., methane), this approach still has value.

5.1.2.1.3. Chlorine-36. Chlorine-36 has also been used as a tracer in hydrologic processes. Chlorine-36, a beta-particle emitter, is produced in the atmosphere by cosmic rays in two major processes: (1) spallation (cosmic-ray interaction with ^{40}Ar), and (2) neutron activation of ^{36}Ar according to the following reactions (Andrews and Fontes, 1992):

^{40}Ar (p, n, α) ^{36}Cl (67% of total natural atmospheric production), and
^{36}Ar (n, p) ^{36}Cl (33% of total natural atmospheric production).

Another significant source of ^{36}Cl in the environment is the neutron activation of stable ^{35}Cl:

^{35}Cl (n, γ) ^{36}Cl,

which can occur in environments where significant chloride is available such as the sea surface. This reaction is the source of ^{36}Cl produced during near-surface weapons tests conducted by the United States and Great Britain over the Pacific Ocean during 1952–1958 (Schaeffer et al., 1960). This reaction may also produce significant ^{36}Cl *in situ* in certain subsurface environments that have a neutron source in reasonably close proximity to stable ^{35}Cl.

Bentley et al. (1986) predicted preweapons test $^{36}Cl/Cl$ ratios for the continental United States. These predictions were based on long-term deposition of both wet precipitation and dry fallout and represent integrated ratios expected for surface water that has not been exposed to anthropogenic or significant *in situ* produced ^{36}Cl. This model assumes that ET processes increase the absolute concentration of Cl^- isotopes in surface water but do not affect meteorically derived ratios. In many surface water environments, the Cl^- concentration increases along a flow path (Davis et al., 1998) and the meteoric input of ^{36}Cl may be diluted by the addition of Cl^- containing no measurable ^{36}Cl from the aquifer matrix or from the unsaturated zone that recharge must travel through. This type of total Cl^- would have a $^{36}Cl/Cl$ ratio that is in equilibrium with the *in situ* activated stable ^{35}Cl and would not fit the integrated box model postulated by Bentley et al. (1986). Moysey et al. (2003) reevaluated the model results published by Bentley et al. (1986) in view of newly collected data and the predictions of two models, by Hainsworth et al. (1994) and Phillips (2000) (Figure 1.9). The $^{36}Cl/Cl$ distribution over the continental United States as predicted by the three modeling efforts of Bentley et al. (1986), Hainsworth et al. (1994), and Phillips (2000), were compared to the results of a large continental sampling effort to determine the $^{36}Cl/Cl$ distribution in well characterized groundwater systems. The model proposed by Hainsworth et al. (1994) provided the best overall fit to the observed $^{36}Cl/Cl$ distribution as obtained from field data. It was further concluded that the model of Bentley et al. (1986) underestimates the magnitude and variability of the observed data and the model of Phillips (2000) generally overestimates the $^{36}Cl/Cl$ distribution in most parts of the United States.

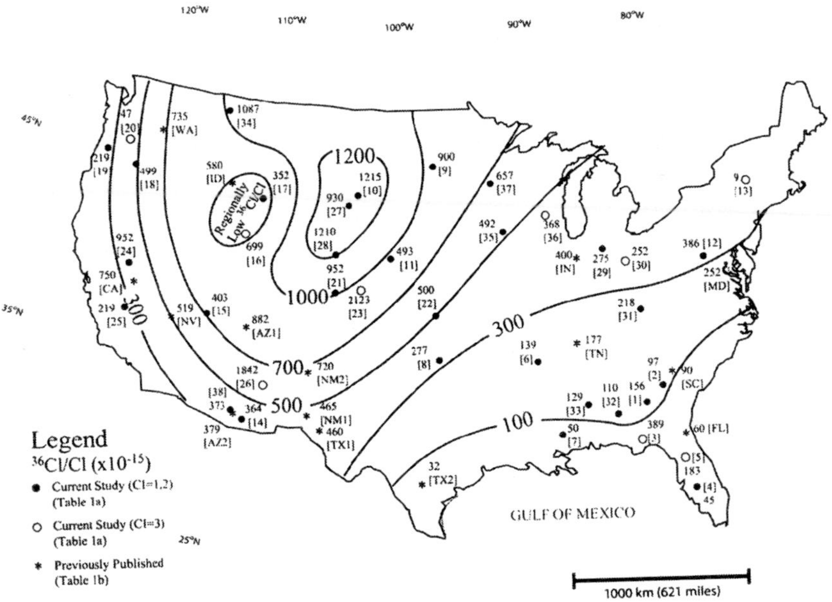

Figure 1.9 Natural levels of the chlorine-36/chlorine ($\times 10^{-15}$) ratio in precipitation and dry fallout over the United States, calculated from groundwater measurements. The numbers in parenthesis represent the sampling site ID (Adapted from Moysey et al., 2003 with permission from Springer).

The longer half-life of ^{36}Cl means that it is most useful for extremely old waters as well as waters from the bomb period. It lacks the sensitivity of tritium for younger waters, and for waters that are a few decades or centuries old which have no tritium but are too young to date by ^{36}Cl decay. Its concentration can be influenced by dissolution or input of salt from a variety of sources (Davis et al., 1998). Davis et al. (2001) used ^{36}Cl and bromine to study ages and sources of water of a series of springs in the United States. The Cl/Br ratios were used to determine what waters had sources of chloride derived from anthropogenic chemicals, ancient seawater, dissolution of halite or other processes. Young waters that had a significant anthropogenic influence would have measurable ^{36}Cl with high Cl/Br ratios. Other waters showed Cl/Br ratios similar to the ocean with little ^{36}Cl, suggesting the influence of relic seawater.

5.1.2.2. Cosmogenic phosphorus isotopes, sulfur-35, and beryllium-7

These nuclides are very short-lived and are produced by cosmic-ray spallation in the troposphere and stratosphere (Lal et al., 1960). Because of their very short half-lives (Table 1.1), they will usually be restricted to use in small watersheds with an emphasis on surface water and biogeochemical cycling.

5.1.2.2.1. Phosphorus isotopes. The use of the phosphorus isotopes (phosphorus-32, phosphorus-33) has been limited by the analytical efforts needed to analyze them (Benitez-Nelson and Buesseler, 1998). Recently, they have been used to study cycling in surface ocean water (Benitez-Nelson and Karl, 2002). In addition, these isotopes have the potential of being a valuable tracer of phosphorus in small watersheds and lakes.

5.1.2.2.2. Sulfur-35. Sulfur-35 has received a much wider use in catchment studies. Cooper et al. (1991) used both stable and radioactive nuclides to study spring runoff in a small watershed in the Arctic. They determined that, in their catchments, uptake by the biota and soils restricts the use of ^{35}S as a hydrologic tracer. In many watersheds, the ^{35}S does not leave the soil zone before it decays, resulting in a limited range of uses for this nuclide (Novak et al., 2004). This is likely also true for nuclides such as ^{7}Be and the radioactive phosphorus isotopes. However, there are many watersheds, particularly in alpine areas, where some ^{35}S can leave the soil zone and enter into the local hydrologic cycle.

Natural ^{35}S was used in the Rocky Mountains to study sulfur cycling at several alpine watersheds. Sueker et al. (1999) employed ^{35}S to study the hydrology of three small basins in the Front Range of the Rocky Mountains. Two of the basins were steep and had hydrographs with large peaks during the melt season and very low flow afterwards. The other basin had a much gentler slope, a larger debris field and a smaller flow increase in the melt season, with some level of flow occurring throughout most of the summer season. The ages of water in the two steep basins was on the order of 100–200 days, versus an age approximately double that range in the other basin. Thus, a large fraction of the water flowing out of the basin with the gentler slope is on the order of a year or more old, due to water being retained in the debris field. The two steep basins have a very limited capability to retain meltwater, and most water is drained out within one summer season.

Natural ^{35}S has also been extensively employed in the Loch Vale watershed in the Front Range of the Rocky Mountains. This watershed has poor soil development and the main hydrologic event is runoff during the snowmelt season. It is typical of watersheds in the Colorado Front Range which could be negatively impacted by an increase in sulfate concentrations resulting from increased power plant emissions (Baron, 1992). Michel et al. (2000) conducted a study of sulfate cycling over several years using natural ^{35}S as a tracer (Figure 1.10). Measurements were made of ^{35}S in snowpack and in runoff from several streams. In one stream they were able to determine that most of the sulfate came either from old depositional sulfate or was leached out by water–rock interactions. They estimated that more

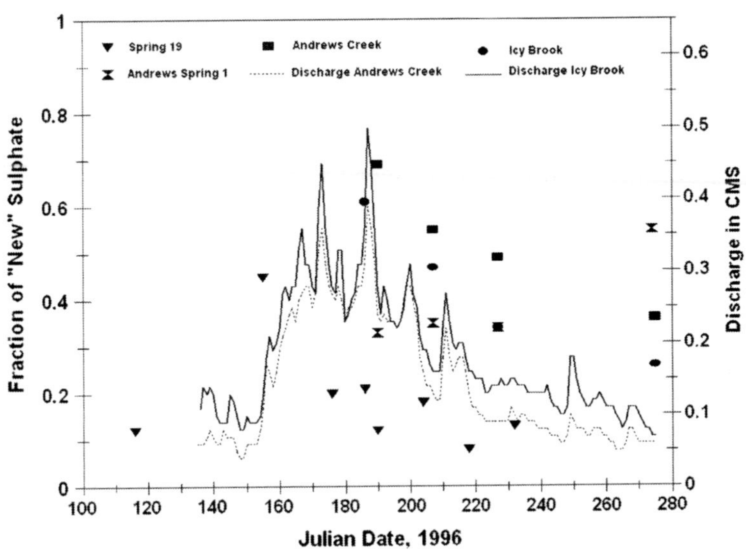

Figure 1.10 Percentage of "new" sulfate in outflow from springs in Loch Vale, CO (Adapted from Michel et al., 2000 with permission from AGU). "New" sulfate is defined as sulfate that has been deposited during or after the previous winter. Discharge rates in cubic meters/second are given for Icy Brook (solid) and Andrews Creek (dashed).

than half of the sulfate deposited in the watershed was removed every year, indicating a sensitivity of the hydrologic system to increased sulfate deposition. Further work by Kester et al. (2003) indicates that some of the atmospheric sulfate moves through the local wetlands at Loch Vale without interaction, unlike many other watersheds.

A promising technique is now being developed to use both ^{35}S and the ^{17}O anomaly to study sulfate cycling in watersheds. The ^{17}O anomaly is produced by mass-independent fractionation in the atmosphere and is only present in sulfate that has been produced by atmospheric processes (Lee and Thiemens, 2001). Once atmospheric sulfate undergoes reactions in the biosphere or lithosphere that breaks the sulfate-oxygen bond, this anomaly is destroyed (Johnson et al., 2001). Thus, the ^{17}O anomaly can give the fraction of unaltered atmospheric sulfate in streams draining a watershed. The age of this atmospheric sulfate can then be determined using ^{35}S.

5.1.2.2.3. Beryllium-7. Beryllium-7 has received limited use in surface water hydrology as it tends to be removed from the dissolved phase very quickly (Cooper et al., 1991). It has been used in studies on the partitioning of metals in natural waters (Bloom and Crecelius, 1983; Hawley et al., 1986). Work by Steinmann et al. (1999) has shown that ^{7}Be tends to partition into the colloidal phase more than the particulate or dissolved phase. This partitioning suggests that colloids are important in metal scavenging in some waters, although this tendency is partially offset by the higher concentrations of particulates in many systems.

5.1.2.3. Uranium and thorium decay series radionuclides

5.1.2.3.1. Radon and radium. Radon (more specifically ^{222}Rn), a decay product of uranium, is ubiquitous in nature, and as a noble gas, tends to accumulate in the pore spaces of rocks and sediment. Since these environments are usually filled with air or water, which can be continuously or periodically flushed, radon can be used to monitor the movement of air and water from “confined” to surface water environments.

One of the earliest uses of radon as an environmental tracer was that of Rogers (1958), who used it to locate subaqueous spring discharges to streams in the Wasatch Mountains of Utah by monitoring radon activity dissolved in the stream at various locations. He was able to estimate groundwater inflow to the streams by making the following assumptions:

(a) stream water and groundwater are well mixed downstream of the recharge point;
(b) no other source of radon or water exists except that already in the stream and that discharged to the stream by the inflowing groundwater;

(c) the rate of radon degassing to the atmosphere is constant between stream sampling points upstream and downstream of the groundwater discharge areas;
(d) the spring water radon content is representative of the radon content of the water entering below the stream bed; and
(e) the stream does not lose or gain in volume between the points of radon anomalies.

Elsinger and Moore (1983) used radon in the coastal, freshwater section of the Pee Dee River of South Carolina to estimate the gas evasion rate from surface water to the atmosphere using a stagnant film model. The river contained radon in excess of dissolved ^{226}Ra that probably was the result of radon-rich inflow of groundwater from the underlying Pee Dee formation. By monitoring the excess radon decrease down a reach of the river and accounting for minor radioactive decay during transit, they calculated a stagnant film thicknesses between 19 and 48 μm, and mass transfer coefficients from water to atmosphere from 2.1 to $4.1\,\mathrm{m\,d^{-1}}$, based on a molecular diffusivity of $(1.14 \pm 0.07) \times 10^{-5}\,\mathrm{cm^2\,s^{-1}}$ at 18°C.

Lee and Hollyday (1987) determined the occurrence of radon in groundwater and surface water in Carter's Creek, a small catchment in central Tennessee, in an attempt to quantify groundwater inflow to the stream during high and baseflow conditions. During baseflow conditions the creek consists of long, nonturbulent shallow pools connected by more turbulent riffles. During higher flow periods the stream is more continuous and more uniform in flow regime.

In the base-flow survey they found a steady decrease in radon activity in the upper section of the stream reach, with a sharp rise in radon accompanying a decrease in temperature of the surface water at the lower part of the stream reach in the catchment. This corresponded to an increase in the measured flow of the stream where higher radon values were detected. A measurement of radon from a spring in the area close to the stream was used as the groundwater radon content to calculate that the stream received an additional flow of water from a groundwater source of $0.17\,\mathrm{m^3\,s^{-1}}$. This value plus the upstream flow agreed well with the measured downstream flow.

The high-flow survey revealed overall a lower radon activity in Carter's Creek water as compared to the low-flow survey, probably the result of the greater quantity of water in the stream, thereby minimizing the initial radon values and, due to turbulence, increasing the evasion of the radon gas to the atmosphere. Two locations of moderately enhanced radon activity were observed which were attributed to groundwater inflow, increasing the flow by about $0.007\,\mathrm{m^3\,s^{-1}}$, based on downstream flow measurements, again in agreement with measured flow increases.

In neither survey of Carter's Creek did Lee and Hollyday (1987) use any adjustments to correct for noncompliance with the five assumptions already presented. Undoubtedly, evasion of radon to the atmosphere took place over the approximately 0.75 km reach of the stream, and the assumption of a constant groundwater radon activity is brought into question by the observation that the radon content of the spring changed from 489 to 615 dpm L^{-1} between the base-flow and high-flow surveys (5 months). As a result, the agreement between calculated groundwater gain and measured values may be, to some extent, fortuitous.

In an attempt to correct for the radon evasion to the atmosphere that occurs when examining larger scale reaches of more significant fluvial systems, Ellins et al. (1990) used a stagnant film model to estimate radon loss from the water's surface of the Rio Grande de Manati in Puerto Rico. This method estimates the loss of radon occurring to the atmosphere and the corrected values can give a more accurate estimation of groundwater inflow along a reach. The model assumes well-mixed water and air bodies separated by a film of water, which contains a radon gradient from the undisturbed water activity to the air activity. Transfer across this boundary is controlled by molecular diffusion, and the thickness of the film is determined by turbulence in the water. The relationship for the stagnant film thickness was given by them as:

$$z = \frac{XD}{\ln(C_u hv/C_d)} \tag{1.1}$$

where z is the thickness of the stagnant film, C_u activity of Rn in the reach upstream sampling site, C_d the activity at the downstream sampling site, X the distance between sampling sites, v the stream velocity, h the average depth of the stream, and D the molecular diffusivity of radon in water. By estimating z in the reaches of the Manati where no groundwater was thought to be entering through the bed, they were able to apply a correction to the radon in Manati water under similar conditions where groundwater input was occurring to estimate the amount of inflow to the river through its bed. Inputs of groundwater to the Manati River between 0.03 and 2.2 $m^3 s^{-1}$ were calculated.

Another method used to correct for radon degassing is to use an artificial mixing tracer injected into the surface water and monitor its behavior in the same reach as that being examined for groundwater input. This avoids the potential pitfall of the previous method, which used "control" sections of streams to examine radon loss characteristics and the need to apply the results to the "experimental" section. One volatile gas that has been used for this purpose is sulfur hexafluoride (SF_6). Wannakhof et al. (1990) used this gas to examine water to air gas transfer at Walker Branch, in Tennessee, and correct for radon evasion to quantify groundwater introduction along the reach, even in reaches where no sharp "spikes" of radon in the stream water occurred (the most obvious evidence of water addition). In this experiment, radon-free SF_6-laden water was injected into the stream water for several hours before a sampling event was carried out along a length of 292 m of Walker Branch below the injection point. Four springs provided an estimate of ^{222}Rn activity of the groundwater. Three equations were solved simultaneously and sequentially downstream to arrive at a groundwater contribution to the stream over the study reach. One stream flow measurement is required for quantitative results, in this case at the initial upstream site. These equations are:

$$[\mathrm{Rn}]_u\, Q_u = [\mathrm{Rn}]_d\, Q_d + k\,A\,[\mathrm{Rn}]_{avg} - [\mathrm{Rn}]_g Q_g \tag{1.2}$$

$$[\mathrm{SF_6}]_u\, Q_u = [\mathrm{SF_6}]_d\, Q_d + k\,A(\mathrm{SF_6})]_{avg} \tag{1.3}$$

$$Q_d = Q_u + Q_g \tag{1.4}$$

where quantities in brackets represent stream water concentration, subscripts u, d, and g stand for the upstream, downstream, and groundwater components, avg designates the average concentration, Q is flow rate, A the stream-reach surface area, and k the transfer velocity of the gases.

Genereux and Hemond (1990) quantified water input to the surface stream in Bickford watershed in Massachusetts using radon, propane, and conductivity measurements of surface and groundwaters. Radon was measured in the stream, saturated zone, and vadose zones waters, while propane (a volatile tracer to correct for gas loss from surface water to atmosphere) and sodium chloride (a conservative tracer to monitor amount of water inflow into the study reach) were injected into the stream. This information was used to calculate the fraction of vadose zone and saturated zone water making up the stream water at the sampling points along the study reach. They found that >70% of the water entering the stream-reach surface water came from the vadose zone as opposed to the saturated zone, suggesting that lateral flow through the unsaturated zone of the stream bank was more significant than vertical flow up through the saturated zone to the stream bed.

Genereux et al. (1993) extended radon studies to the previously mentioned Walker Branch watershed in Tennessee (Figure 1.11). For this study they measured radon and calcium in surface and subsurface (vadose zone, soil groundwater, and bedrock groundwater) water and constructed a three end-member mixing model to produce realistic streamflow generation estimates over varying ranges of flow conditions in the stream. Calcium was used to distinguish bedrock groundwater from soil water, and radon was used to distinguish between vadose zone water (low radon) and soil water (high radon).

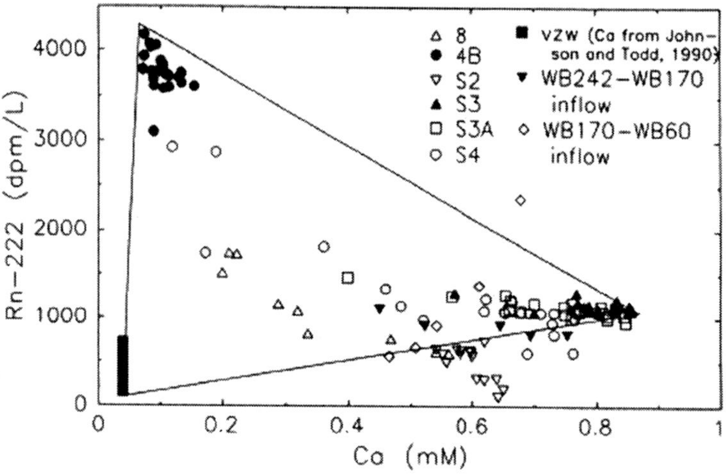

Figure 1.11 Plot of ^{222}Rn concentrations versus Ca concentrations for a variety of water samples (vadose zone water, soil groundwater, spring water, and lateral inflow to two stream reaches) from the West Fork (Genereux et al., 1993). The mixing triangle bounds the space that could be occupied by samples that are mixtures of the solutions represented by the corners of the triangle (end members).

A two-prong approach was used to estimate the contribution from the saturated and unsaturated zones. They used volatile and conservative tracers to estimate the amount of radon (and water) flowing into the stream along a length of reach through its banks and bottom. Since this water would represent flow from both the saturated and unsaturated zones, they measured the radon contents of water from both zones near the stream reach to use as end-member values. Since water from the unsaturated zone would be exceedingly difficult to obtain for radon measurement they collected the coexisting gas phase from this zone and, assuming equilibrium between the water and gas phases, used Henry's aw to calculate the radon content of this *in situ* water. Finally, calcium concentration of the end-member waters was measured for use as a non-volatile tracer to distinguish between bedrock (Ca – rich) and soil (Ca – poor) water sources. With these values (Rn and Ca content of vadose zone, saturated zone, and upstream waters) the contribution to the stream water at downstream points of each end member were calculated. Limitations and difficulties with the method were discussed, including temporal and regional variation of one or more of the end-member concentrations and difficulty in accurately measuring each end-member value.

Other members of the natural decay series have been used to infer origin and trace water movement through hydrologic systems. Eikenberg et al. (2001) used radium isotopes (^{226}Ra, ^{228}Ra, and ^{224}Ra) and strontium isotopes as geochemical tracers to distinguish between groundwater and surface water in the upper Rhine River valley of France. They found that three sources contribute to the surface Rhine River water in this area: (1) water from a granite crystalline source (which is picked up from surface streams as the river flows near the Vosges Mountains), (2) a regional groundwater source, and (3) water from a parallel river, the Ill, that seeps through intervening alluvium to emerge into the topographically lower, Rhine River.

By using graphical techniques and mixing calculations, they estimated mixing relationships of several tributaries with the river Ill using Ra isotopes and Sr isotopes, which they claim agreed well between isotopic data sets (e.g., at the confluence of the Ill with the tributary Thur, radium isotopes estimated a Thur volume fraction of 0.6 ± 0.2 below the confluence as compared to 0.43 ± 0.04 based on strontium isotopes).

5.1.2.3.2. ***Uranium.*** Uranium in surface waters results from the weathering of rocks and soil. During the process of chemical weathering the ^{234}U isotope is preferentially removed from the solid phase relative to its parent, ^{238}U. This results in isotopic fractionation of this element, which is unique among the heavy elements, and has been useful in identifying waters, tracing them in the hydrologic cycle and estimating mixing ratios when surface waters merge.

Early work on the uranium isotopic composition of river waters and geologic material was carried out to a large extent by researchers in the former Soviet Union, and this material was summarized by Cherdyntsev (1971), who tabulated much of the Russian data for rivers, lakes and groundwater, as well as rocks and surficial material analyzed in the Soviet Union (and the rest of the world) to 1969, from which much of this section is condensed.

Tuzova (1986) reported results of a study of uranium isotopes in river waters of the Sarydzhaz River basin that used uranium activity ratio (UAR) values to calculate mixing proportions of various tributaries and main-stem rivers. She also observed that both uranium concentration and UAR values are lowest in rivers fed by glaciers, and concluded this was the result of the low contact time of meltwater with rock and sediment. The results of the mixing calculation based on uranium isotopes compared well, with one exception, to the results based on hydrological estimates. She was also able to detect groundwater flow into the Sarydzhaz channel as a result of the high uranium content of the groundwater. She estimated the groundwater input to be $<3\%$ of the total river flow. In addition she estimated that the total runoff from the basin is $4\,km^3\,yr^{-1}$; the error of this value is at least 30%, as compared to hydraulically based flow estimates.

Tuzova and Filkin (1991) used UAR values of surface water and groundwater to calculate mixing ratios of small mountain rivers in the Issyk-Kul Artesian Basin. They found that these rivers contained a mix of groundwater and upland sources with respect to their uranium content and UAR values. Using these differences they calculated the proportion of each source making up the river water at various locations along the reaches, and found that groundwater addition to the rivers made up between 10 and 50% of the total water at these locations. These values agreed, in general, with the calculations from stream gauge measurements. Tuzova and Novikov (1991) also used uranium isotope data for uranium dissolved in river waters of the Pyandzh River basin in the eastern Pamir mountains to calculate mixing ratios as various tributaries merged with mountain rivers, thus gaining a greater understanding of the hydrologic dynamics and budget of the basin with only limited gauging sites.

Riotte and Chabaux (1999) examined the rivers of the Strengbach watershed in France using uranium and strontium isotopes to determine the sources of uranium in the water. They were able to differentiate sources of groundwater based on the rock-type from which the groundwater originated and noted a lack of mixing of stream water with the underlying aquifer for several miles once the stream opened out onto an alluvial plain. They also found that they could differentiate between waters from soil and deeper bedrock sources based on uranium isotopes, and follow the relative proportions of each during flood events, when more water comes from the soil zone.

Snow and Spaulding (1994) examined the uranium isotopes dissolved in the Platte River of Nebraska, USA, to estimate the tributary-main stem mixing ratios. Isotopically distinct waters were found in the Elkhorn River, Salt Creek, Shell Creek, and Loup River tributaries, which feed the Platte and, in some cases, were shown to heavily influence the uranium isotopic signature of the Platte water. Using mixing calculation and discharge measurements they showed that, where the major tributary (the Loup River) entered the Platte River, the waters did not mix for several kilometers downstream, as more Loup River water was evident on the north side of the channel (where the Loup enters) than on the south side of the channel. The differences between north and south decreased with distance downstream, indicating increasing lateral mixing. They also noticed that increased discharge from the Loup River resulted in decreased lateral mixing downstream in the Platte. A similar situation was observed for the Elkhorn River tributary as well.

They also examined a mixing situation where Salt Creek, which enters the Platte just downstream from the Elkhorn River (but from the opposite bank), and Elkhorn River discharge into the Platte. They estimated from uranium isotopic data that, at a point several kilometers downstream from these entries, the Platte consisted of ~15% Salt Creek water, ~50% Elkhorn River water, and ~35% upstream Platte River water. Data from discharge measurements suggest the water proportions should be ~10% Salt Creek, ~35% Elkhorn River, and ~55% upstream Platte River water.

The differences between uranium and flow derived data were attributed to incomplete lateral mixing within the Platte, with possible errors resulting from temporal and spatial differences of sampling sites between uranium and flow sampling, and unaccounted-for sources of water (and possibly uranium).

Kraemer et al. (2002) applied inductively coupled plasma/mass spectrometry to the measurement of uranium isotopic activity ratios in natural waters (more precise compared to alpha spectrometric techniques previously used) to study mixing in the North Platte River from the Horse Creek and Sheep Creek tributaries. Due to the $\pm 0.5\%$ precision achieved by the ICP/MS method, they were easily able to resolve binary (North Platte River and Horse Creek) mixing and ternary (North Platte, Horse Creek, and Sheep Creek) mixing in the North Platte main stem during different periods throughout 1999. These interpretations agreed qualitatively with flow measurements of the surface water bodies made at times of sampling.

Phosphate fertilizer added to cropland has been suspected of supplying uranium to surface waters through leaching from the fields into groundwater streams and rivers, especially in intensely agricultural areas. Mangini et al. (1979) examined uranium in the heavily fertilized agricultural area around the Sudkreis and Mannheim areas, Germany and concluded that, despite extensive fertilizer application to croplands (at an experimental agricultural site), anthropogenic addition of uranium to the surface waters (and hence the ocean) was unlikely to be adding significantly to the uranium burden of the world ocean. They based this conclusion on the fact that, while phosphate fertilizer contains abundant uranium, its $^{234}U/^{238}U$ activity ratio is very close to unity, while groundwater and river water draining the agricultural site was between 1.52 and 1.97 and actually higher than a control, nonfertilized forest area nearby. Major rivers in Germany that receive groundwater input from the study area carried uranium with activity ratios in roughly the same range. They concluded that uranium added by phosphate fertilizer applications is tightly bound to soil particles ($\sim$95%) and is not available for solute transport processes in their system.

Zielinski et al. (1997) examined uranium isotopes in Sixmile Creek, a tributary of the Arkansas River in the state of Colorado, USA to see if there was evidence of uranium addition to surface waters due to agricultural application of phosphate fertilizer within a drainage basin. Results of this study indicated minimal impact of fertilizer uranium compared to natural uranium backed from soil, which was not the case for nitrogen, which showed dramatic increases in water as a result of fertilizer use. They concluded, based on field data and laboratory experiments, that uranium combines with calcium and phosphorus to form insoluble products that build up in the soil becoming less available for leaching than the uranium in the local soil material.

Zielinski et al. (2000) examined uranium isotopes in the Florida Everglades to evaluate whether fertilizer application to agricultural fields was influencing the uranium content of water flowing through this large "river of grass" area. Results of the study indicate that uranium is supplied to parts of the Everglades by flow of surface water into the area from agricultural areas. But this uranium is removed from water by contact with the underlying peat in the substrate resulting in low uranium concentration in the water concomitant with a buildup in the highly organic zone of uranium with a UAR of about 1.05 in both the peat and the overlying water.

Another source of anthropogenic uranium to the environment is from depleted uranium (DU) commonly found in certain types of armor-piercing ordnance (Bleise et al., 2003). DU is thought to initially enter the environment as dust and then may move through the hydrologic- and ecosystem. Work by Jia et al. (2006) found that the distinct isotopic signature of DU could be traced in the environment. Isotopic ratios indicative of an influence from DU were found in water samples, but concentrations were too low to be of a health concern at this time. However, it is possible that this distinctive ratio could be used as a tracer for some processes.

5.1.3. Unsaturated zone

The unsaturated zone is the primary mode of recharge for many large aquifers, and radiotracers have been used to study both recharge rates and pathways for water pollutants to enter the groundwater system. Tritium is the most extensively used radioactive tracer, although other radiotracers have also been used to study mechanisms and pathways in the unsaturated zone.

5.1.3.1. Tritium and ^{36}Cl

As with surface waters, it is the pulse of ^{3}H and ^{36}Cl produced by nuclear testing in the 1950s and 1960s that is generally applied to studies of the unsaturated zone. Tritium was initially found to move through the unsaturated zone in a piston-like fashion, reflecting the same signal as that found in precipitation (Zimmermann et al., 1966). In areas with high moisture content and relatively homogeneous soil, these types of profiles were observed. It has been noted that tritium does exchange with hydrogen atoms in clays which can result in a modification of the simple piston flow (Stewart, 1966; Rat'ko et al., 2001). Andersen and Sevel (1974) were able to follow the tritium transient into the water table at a site in Denmark. The site was composed of primarily sandy sediments and a tritium peak was observed to move through the system over a period of years. Even in an ideal system such as this one, dispersion effects caused a loss of the fine structure in the tritium profile and only one large peak was observed. Estimates were made of the recharge rate and the fraction of precipitation that recharged the water table at that site. Engesgaard et al. (1996) was able to follow this pulse of tritium as it migrated into the saturated zone and calculated its flow rate once it entered the groundwater. There have also been efforts to use the tritium inventory in the unsaturated zone to calculate recharge rates at some locations. This approach requires a knowledge of tritium entering the unsaturated zone from the soil zone and a water table deep enough so that all tritium lies above it. The tritium in the soil layer is a complex function of tritium in precipitation, monthly soil moisture deficits, and soil moisture capacity (Foster and Smith-Carrington, 1980). This approach has received limited value because of the need for the large amount of auxiliary data.

Measurements comparing the movement of ^{36}Cl versus tritium have also been made at various locations. As the peak production of ^{36}Cl occurred in the mid-1950s due to neutrons released in sea level weapons tests, it was expected that the ^{36}Cl pulse would occur deeper in the soil zone than the tritium maximum. However, measurements in the southwestern United States found that the tritium pulse had advanced deeper than the ^{36}Cl pulse (Phillips et al., 1988). The bomb tritium was able to migrate deeper in the unsaturated zone because it can move as a gas as well as a liquid while ^{36}Cl can only move in the liquid phase. This is only possible in arid regions where movement of liquid in the unsaturated zone is low or nonexistent. Cook and Walker (1996) compared recharge rates estimated by chloride deposition versus those estimated through ^{36}Cl and tritium at sites where data for at least two of the tracers were available. Their work indicated that all three tracers gave similar estimates for recharge rates above about 10 mm yr^{-1}. However, for lower rates, it appears that the two bomb tracers give recharge rates that are too high (Figure 1.12). Their conclusion was that, for low recharge rates, the chloride-accumulation method was more accurate than the radioactive tracer method.

There is also evidence of differences between fluxes of ^{36}Cl and tritium in the soil and unsaturated zones in wet climates. Typically, it is assumed that chloride moves as a conservative tracer through these environments in a manner similar to tritium. Chloride is frequently used to estimate evapotranspiration in a watershed under the assumption that no interactions occur and the increase in chloride concentrations reflects the extent of water loss through evapotranspiration (Claassen and Halm, 1996). However, work by Milton et al. (2003) using bomb ^{36}Cl as a tracer indicates that chloride may be retained by interactions in the soil, particularly in clay-rich zones. They also suggest that some chloride is lost to the atmosphere as a gas, probably in organic form. Thus, ^{36}Cl data suggest that chloride may not always be a conservative tracer in many unsaturated and soil zones.

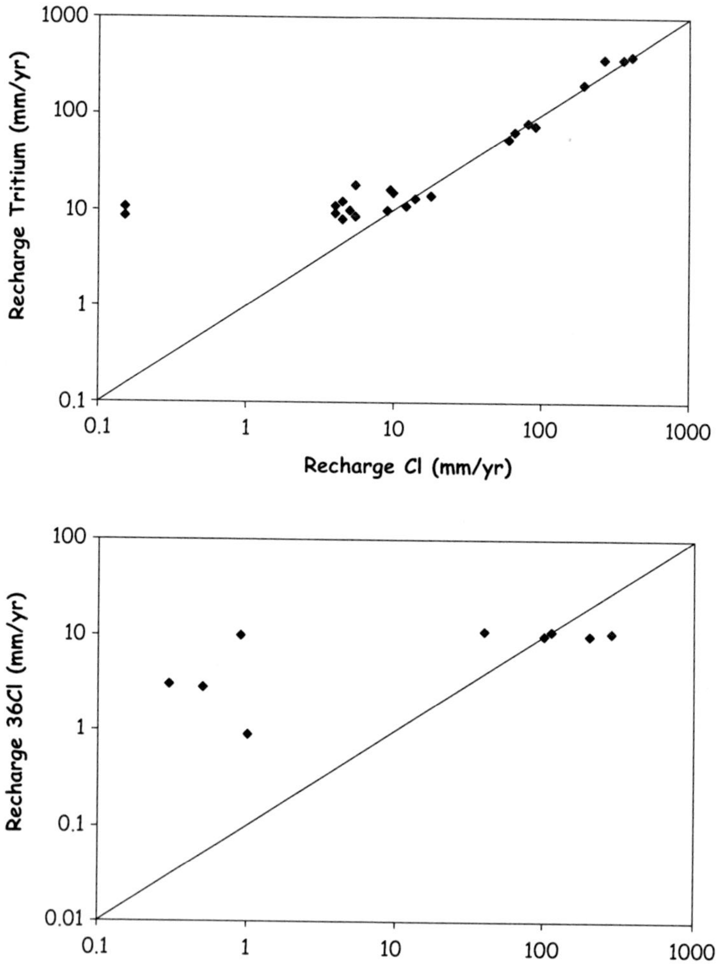

Figure 1.12 Comparison of tritium (9a) and ^{36}Cl (9b) with chloride as a tracer of recharge in arid areas modified from Cook and Walker (1996). The two transient tracers are reliable where recharge rates are 10 mm yr^{-1} or greater. However, for lower recharge rates, both tracers tend to overestimate recharge and chloride is preferred.

At the present time, the tritium pulse has moved through the unsaturated zones in many areas and interpreting tritium data from soil cores at such locations is difficult. The locations where tritium is most useful in the unsaturated zone are arid regions and locations with deep water tables (Cook et al., 1994; Gaye and Edmunds, 1996; Lin and Wei, 2006). There have been several studies in arid zones and recharge has been found to be a complex process in such areas. Specifically, tracer studies have shown that recharge frequently is focused in very limited areas. Likely sites for recharge tend to be small ephemeral streams or ponds where water can accumulate after precipitation events. These sites can be more porous and water will sink through the unsaturated zone to the water table at these locations (Wood and Sanford, 1995). Thus, areal recharge estimates can be very difficult to make, and depend on the number and density of such recharge sites. Work by Izbicki et al. (2000) on the Oro Grande Wash, an ephemeral stream in the upper Mojave Desert in California, has shown that, even at these sites, recharge can be a complex process involving significant lateral flow around small impermeable layers. The tritium and chloride profiles at the Lower Oro Grande site contained a salinity maximum at 25 m, a depth at which no tritium was present (Figure 1.13). Below the chloride maximum, bomb-tritium was found, indicating that water flowed around the layer of high salinity, and that horizontal flow occurred in the streambed. Work in fractured systems such as the proposed nuclear repository at Yucca Mountain, NV, has also indicated that preferential flow can occur, resulting in a rapid movement of water through the unsaturated zone (Guerin, 2001; Campbell et al., 2003). Bomb concentrations of ^{36}Cl were found deep in the unsaturated zone at this site, which indicates that water flows through fractures to great depths.

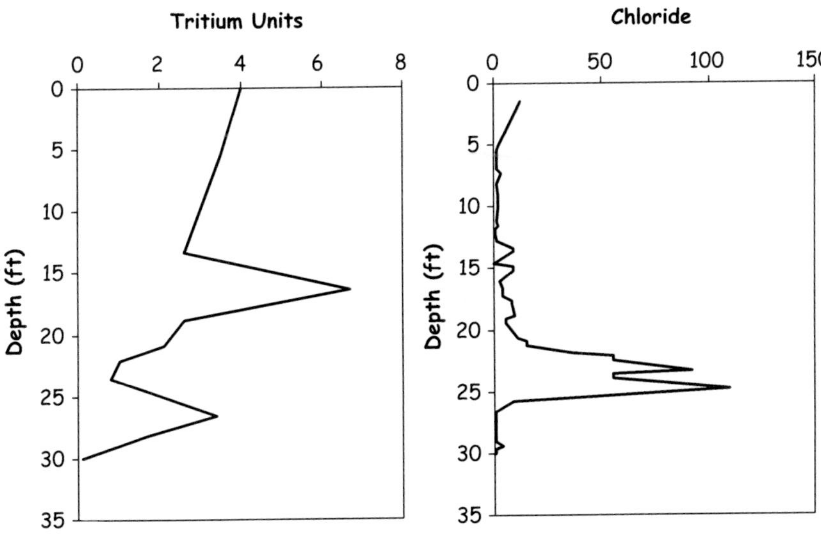

Figure 1.13 Tritium (left) and chloride (right) concentrations from the Lower Oro Grande Wash, Upper Mojave Desert, CA (Izbicki et al., 2000). A subsurface chloride peak occurs at about 10 m and tritium concentrations are below the detection level in this layer. Below this depth, tritium is detected indicating that lateral flow is an important component in this area.

5.1.3.1.1. Carbon-14. Limited use has been made of carbon-14 in unsaturated zone studies. In the liquid phase, it is a very difficult measurement because it is difficult to extract sufficient DIC out of the liquid phase without contaminating it with carbonate or organic carbon from the mineral phase. Yang (2002) squeezed water from rock samples to obtain ^{14}C values for DIC in pore waters from Yucca Mountain. From these data, he was able to calculate a water flux in the unsaturated zone. Even with this system, one has to account for reactions of DIC with carbon dioxide in the unsaturated zone air and mineral phase to try and estimate ages. If ^{14}C concentrations in soil gas were only controlled by diffusion from the atmosphere, there would be very little change of concentration with depth. However, a large number of reactions occur in the unsaturated zone including root respiration, degradation of organic matter, exchange of $CO_2(g)$ with DIC in pore water and exchange with soil carbonate (Haas et al., 1983). Root respiration can increase the CO_2 concentration in the root zone and this CO_2 typically has ^{14}C concentrations similar to those found in the atmosphere. Below the soil zone, degradation of older organic matter and the reaction of CO_2 (g) with soil DIC can lower the ^{14}C concentration. Thus, even though one-dimensional diffusion calculations indicate that bomb ^{14}C should be present throughout the unsaturated zone, measured concentrations are frequently much lower (Thorstenson et al., 1983). The same reactions make it difficult to estimate the initial ^{14}C concentration at the surface of the aquifer so that ages can be calculated for groundwater in the system. These reactions need to be taken into account when using ^{14}C to date groundwater (Plummer et al., 1994).

5.1.3.2. Hydrology of glacial systems

5.1.3.2.1. Bomb-produced radionuclides. Tritium has been extensively used to characterize glacial systems, both in terms of deposition and glacial hydrology. In polar regions and some high-altitude temperate-zone glaciers where no melting occurs, tritium and other bomb-produced isotopes can be used to determine deposition rates over the last 50 years (Merlivat et al., 1977; Koide et al., 1982). In temperate zones, tritium can be used to determine deposition rates if the strata in ice cores maintain their integrity or if melting results in movement of water vertically, followed by refreezing at other depths within the core. Work by Miller et al. (1965) on tritium in a Mount Everest glacier detailed the existence of three peaks representing the three major periods of atmospheric nuclear testing (Figure 1.14). This work demonstrated that, under proper conditions, tritium and deposition records are preserved in midlatitude ice cores. Work in the Wind River Range of Wyoming demonstrated the need to measure tritium in temperate glaciers to determine whether ice core data are compromised by refreezing of water from higher strata (Naftz et al., 1993, 1996). The ice core record, if tritium

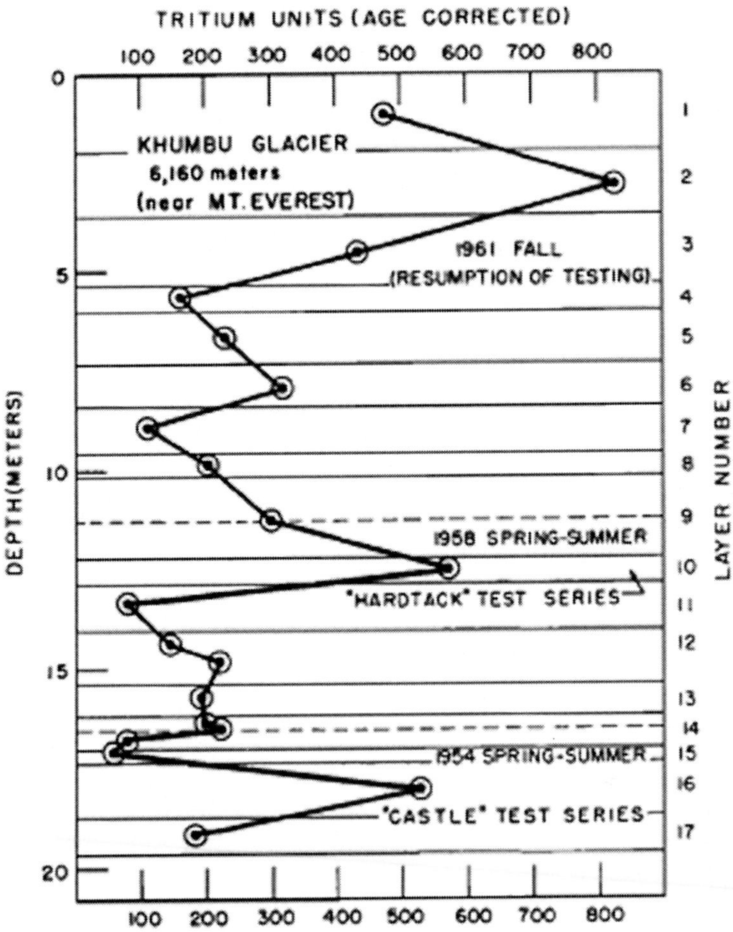

Figure 1.14 Tritium concentrations in Khumbu Glacier in the Himalayas collected in 1963 (Miller et al., 1965). Distinct bands of tritium from three periods of nuclear testing starting in the early 1950s are evident. There are approximately two layers per year deposited.

data confirm that the core is undisturbed, can then be used to study changes in depositional histories of other chemicals throughout the past. The chemical deposition history of midcontinental glaciers is frequently of more relevance than polar records in that it reflects regional and continental histories which are difficult to obtain by other methods (Schuster et al., 2002).

Other radionuclides produced by nuclear events have also been used as a time marker, particularly in the polar glaciers. In temperate glaciers, tritium has been preferred for such work due to its ease of use and the fact that the particulate and solute radionuclides may be subject to preferential transport out of the ice matrix. Furthermore, tritium can be measured with a few milliliters of water whereas other radionuclides (^{90}Sr, etc.) require larger volumes of water, which are not always easily obtained. For older cores, ^{14}C in plant and insect remains in the ice core can be dated by AMS to help determine timescales. Thompson et al. (1998) used ^{14}C and tritium to study changes in climate and dust deposition in a South American glacier. Their core showed a different distribution in dust concentrations with time compared to polar glaciers due to differences in the regional climate.

5.1.3.3. Lithogenic and cosmogenic radionuclides

Nuclides from lithogenic or cosmogenic sources have also been used to determine time lines in glacial cores, with ^{210}Pb being preferred due to its relatively constant input over time, and its ease of measurement (Picciotto et al., 1967). Another possible choice is the cosmogenically produced radionuclide ^{22}Na, but this has received little use due to analytical issues.

With its half-life of ca. 140 years, cosmogenic ^{32}Si is an excellent candidate to provide a timescale in the range from 100 to 1,000 years. Morgenstern et al. (2000) have published first ^{32}Si AMS measurements for rainwater, glacier ice, and snow. Based on their results, about 1 kg of ice is sufficient to cover a time period of seven half-lives, that is, the ^{32}Si dating limit for glacier ice is assumed to be about 1,000 years. For the Fox Glacier ice (New Zealand, 43°S) an ice age on the bottom of the glacier of more than 800 years was obtained (Morgenstern et al., 2000).

Some short-lived cosmogenic nuclides (^{7}Be, ^{32,33}P, ^{35}S) can be used in temperate-zone glaciers to determine whether the water draining from the glacier is recent precipitation or meltwater of ice older than about 1 year. Measurements were carried out on Galena Creek Rock Glacier and Upper Fremont Glacier in Wyoming, on ^{35}S and the bomb-produced nuclides tritium and ^{36}Cl to estimate the range of ages of waters draining off these glacial catchments (Naftz et al., 1996; Cecil et al., 1998). The data from Galena Creek Rock Glacier indicated that most of the water flowing out of the watershed was derived from precipitation that had been deposited within the past year. The water flowing off the Upper Fremont Glacier showed a wide range of ages from recent precipitation to water deposited prior to nuclear testing (Michel and Naftz, 1995).

5.1.3.4. Artificial and contaminant tracer studies

Public concerns and legal impediments have limited the use of artificial radioactivity in hydrologic studies. Large numbers of primary productivity studies are carried out by microbiologists every year, usually under laboratory conditions, and are not the focus of this chapter. However, many hydrologic processes can be studied with artificial radioactivity, released intentionally or accidentally. Incubation or laboratory studies of field materials have been substituted for *in situ* studies for some cases. Laboratory studies are easier to carry out than *in situ* studies, both due to legal issues and because it is easier to control all parameters in laboratories. Direct translation of laboratory studies to field sites is not always possible, so in many cases, field studies are conducted instead of/or in conjunction with laboratory studies.

One area where artificial radioactivity has been used is in the study of the movement of water and solutes through the unsaturated zone. Watwood and Fitzgerald (1988) conducted *in situ* studies to determine the fate of sulfate in soils. Using ^{35}S as a tracer, they found that sulfate could be rapidly taken up in the organic matrix of the soil and remained in that state. Later work with ^{35}S in soil incubations has suggested that some mineralization of organic S is possible and it may even be lost from the unsaturated zone as a gas (Houle et al., 2001, 2004; Novak et al., 2004). Carbon-14 can also be used for tracing the movement of contaminants through the unsaturated zone. Wietersen et al. (1993) used radiocarbon-labeled herbicides to study their migration through different soil systems. They found that the herbicides decreased in mobility in the order atrazine > metolachlor > alachlor and that mobility decreased with increasing clay and organic carbon in the soil.

In streams and rivers, it has occasionally been possible to use radioactive tracers to study uptake of some chemicals, travel times, and dilution of river water by other sources. Dincer (1967) reviewed early work and discussed issues related to using these techniques including safety issues and proper injection. He concluded that it was not a method that could be applied routinely, and he specifically was concerned that the use of long-life tracers like tritium could make future work with environmental concentrations of tritium difficult for contaminated watersheds. However, in certain well-chosen cases, he concluded that these techniques could be of benefit (Dincer et al., 1983).

Tate et al. (1995) used ^{32}P to study the uptake of phosphorus in a small mountain stream contaminated with acid-mine drainage. They concluded that adsorption on iron oxides and uptake by algae were capable of removing most phosphorous in the stream with the algal removal being the more important process. Radiotracers with different properties can also be used simultaneously to differentiate between processes in a stream reach. Jonnson et al. (2003) compared the movement of tritium and ^{51}Cr released to a small stream in Sweden. Tritium provided a conservative tracer that could be compared to the more reactive chromium. The bulk of the chromium was retained in the stream channel far longer than tritium, presumably because of interactions within the sediment bed.

Bedmar et al. (1992) released tritium and ^{82}Br into the Magdalena River in Columbia at both low and high flow to develop a model describing flow down the river. They determined that the tritium pulse had a non-Gaussian shape, skewed due to the delay of some of the tritium in areas of the river where flow could be trapped. It would leak slowly out of these areas resulting in a tailing effect.

Sites where radiochemicals are released into the environment either operationally or by accident can also provide opportunities for tracer studies. Releases of radioactivity from nuclear plants are becoming more common and can be used to study the dynamics and timescales for the transport of river water downstream from the release point (Baeza et al., 2001). Studies have also used the radioactivity released from nuclear processing plants to model the movement of river water into lakes and oceans (Lam and Durham, 1984; Bush, 1988). Tritium in leachates from domestic landfills has been used to trace the migration of contaminant plumes in groundwater around such landfill sites (Rank et al., 1992; Robinson and Gronow, 2007).

All these releases are important locally and, for some radionuclides like ^{85}Kr, releases from these operations will have an impact on the global inventory. However, for tritium and many other radionuclides released from these sources, there has not yet been an impact on the global inventories. The Savannah River, which has the highest tritium concentrations of any major river in North America, transports about 1 g of tritium to the Atlantic Ocean each year and has transported approximately 150 g out of the basin since 1966. This is about 30 times greater than the expected tritium transport for a river with a basin located at similar latitude. However, compared to the continental and global flux, this quantity is not important. The tritium transported out to the Gulf of Mexico by the Mississippi River is about a factor of 3 greater than the Savannah River transport, both yearly and during the years of the bomb transient. Present inventories of tritium in the North Atlantic Ocean are probably on the order of 10–15 kg, well in excess of any flux off the continents. Thus, while tritium from contaminant sources may be a factor in local hydrology, it is unimportant on a larger scale (Shiklomanov, 1998).

REFERENCES

Andersen, L. J., and T. Sevel. (1974). Six years of environmental tritium profiles in the unsaturated and saturated zones, Gronhoj, Denmark. In: *Symposium on Isotope Techniques in Groundwater Hydrology II*. International Energy Agency, Vienna, Austria, pp. 3–20.

Andrews, J. N., and Fontes, J. -C. (1992). Importance of the *in situ* production of ^{36}Cl, ^{36}Ar and ^{14}C in hydrology and hydrogeochemistry. In: *Isotope Techniques in Water Resources Development*. International Atomic Energy Agency: Vienna, Austria, pp. 245–269 (*IAEA-SM-319/12*).

Andrews, J. N., and R. L. F. Kay. (1982). Natural production of tritium in permeable rocks. *Nature*, 298, 361–363.

Aravena, R., and O. Suzuki. (1990). Isotopic evolution of river water in the northern Chile region. *Water Resources Research*, 26, 2887–2895.

Baeza, A., A. M. Brogueira, M. C. V. Carreiro, E. Garcia, J. M. Gil, C. Miro, M. M. Sequeira, and M. M. R. Teixeira. (2001). Spatial and temporal evolution of the levels of tritium in the Tagus River in its passage through Caceres (Spain) and the Alentejo (Portugal). *Water Research*, 35, 705–714.

Bard, E., B. Hamelin, R. G. Fairbanks, and A. Zindler. (1990). Calibration of the ^{14}C timescale over the past 30000 years using mass spectrometric U-Th ages from Barbados corals. *Nature*, 345, 405–410.

Baron, J. (Ed.) (1992). *Biogeochemistry of a Subalpine Ecosystem: Loch Vale Watershed Ecological Study*. Springer-Verlag, New York, 247pp.

Bedmar, A. P., L. Araguas, E. Torres, E. Obando, J. Jimenez, and A. Zapata. (1992). Hydraulic parameters of the Magdalena River (Columbia) derived from large scale tracer experiments. In: *Isotope Techniques in Water Resources Development*. International Atomic Energy Agency, Vienna, Austria, pp. 85–105.

Benitez-Nelson, C. R., and K. O. Buesseler. (1998). Measurement of cosmogenic ^{32}P and ^{33}P activities in rainwater and seawater. *Analytical Chemistry*, 70, 64–72.

Benitez-Nelson, C. R., and D. M. Karl. (2002). Phosphorus cycling in the North Pacific Subtropical Gyre using cosmogenic ^{32}P and ^{33}P. *Limnology and Oceanography*, 47, 762–770.

Bentley, H. W., F. W. Phillips, and S. N. Davis. (1986). Chlorine-36 in the terrestrial environment. In: *Handbook of Environmental Isotopes* (Eds P. Fritz and J.-C. Fontes). Elsevier, New York, pp. 422–480.

Bleise, A., P. R. Danesi, and W. Burkart. (2003). Properties, use and health effects of depleted uranium (DU): A general overview. *Journal of Environmental Radioactivity*, 64, 93–112.

Bloom, N., and E. A. Crecelius. (1983). Solubility behaviour of atmospheric beryllium-7 in the marine environment. *Marine Chemistry*, 12, 323–331.

Brown, R. M. (1961). Hydrology of tritium in the Ottawa Valley. *Geochimica et Cosmochimica Acta*, 21, 199–216.

Bush, K. A. (1988). Identification of Savannah River water on the South Atlantic Bight Shelf using tritium. *Journal of Geophysical Research*, 93, 9315–9331.

Campbell, K., A. Wolfsberg, J. Fabryka-Martin, and D. Sweetkind. (2003). Chlorine-36 data at Yucca Mountain: Statistical tests of conceptual models for unsaturated-zone flow. *Journal of Contaminant Hydrology*, 62, 43–61.

Carter, M. W., and A. A. Moghissi. (1977). Three decades of nuclear testing. *Health Physics*, 33, 55–71.

Cecil, L. D., J. R. Green, S. Vogt, R. Michel, and G. Cottrell. (1998). Isotopic composition of ice cores and meltwater from Upper Fremont Glacier and Galena Creek Rock Glacier, Wyoming. *Geografiska Annaler*, 80, 287–292.

Cherdyntsev, V. V. (1971). Uranium-234. Translated from Russian by Israel Program for Scientific Translations, printed by Keter Press, Jerusalem, 234p. (Translated from Uran-234, Atomizdat, Moskva, 1969).

Claassen, H. C., and D. R. Halm. (1996). Estimates of evapotranspiration or effective moisture in Rocky Mountain watersheds from chloride ion concentrations in stream baseflow. *Water Resources Research*, 32, 363–372.

Cook, P. G., I. D. Jolly, F. W. Leaney, G. R. Walker, G. L. Allan, L. K. Fifield, and G. B. Allison. (1994). Unsaturated zone tritium and chlorine 36 profiles from southern Australia: Their use as tracers of soil water movement. *Water Resources Research*, 30, 1709–1719.

Cook, P.G., and G. Walker. (1996). Evaluation of the use of ^{3}H and ^{36}Cl to estimate groundwater recharge in arid and semi-arid environments. In: *Isotopes in Water Resources Management*. International Atomic Energy Agency, Vienna, Austria, pp. 397–403.

Cooper, L. W., C. R. Olsen., D. K. Solomon, L. I. Larsen, R. B. Cook, and J. M. Grebmeier. (1991). Stable isotopes of oxygen and natural and fallout radionuclides used for tracing runoff during snowmelt in an Arctic watershed. *Water Resources Research*, 27, 2171–2180.

Davis, S. N., L. D. Cecil, M. Zreda, and S. Moysey. (2001). Chlorine-36, bromide, and the origin of spring water. *Chemical Geology*, 179, 3–16.

Davis, S. N., L. D. Cecil, M. Zreda, and P. Sharma. (1998). Chlorine-36 and the initial value problem. *Journal of Hydrogeology*, 6, 04–114.

Dincer, T. (1967). Application of radiotracer methods in streamflow measurements. In: *Isotopes in Hydrology*. International Atomic Energy Agency, Vienna, Austria, pp. 93–113.

Dincer, T., T. Florkowski, and S. Salamba. (1983). Discharge measurements of the River Rufiji (Tanzania) using artificial tritium. In: *Isotope Hydrology*. International Atomic Energy Agency, Vienna, Austria, pp. 681–693.

Dockins, K. O., A. E. Bainbridge, J. C. Houtermans, and H. E. Suess. (1967). Tritium in the mixed layer of the North Pacific Ocean. In: *Radioactive Dating and Methods of Low-Level Counting*. International Atomic Energy Agency, Vienna, Austria, pp. 129–142.

Eikenberg, J., A. Tricca, G. Vezza, P. Stille, S. Bajo, and M. Ruethi. (2001). $^{228}Ra/^{226}Ra/^{224}Ra$ and $^{87}Sr/^{86}Sr$ isotope relationships for determining interactions between ground water and river water in the upper Rhine Valley. *Journal Environmental Radioactivity*, 54, 133–162.

Ellins, K. K., A. Roman-Mas, and R. Lee. (1990). Using ^{222}Rn to examine groundwater/surface discharge interaction in the Rio Grande De Manati, Puerto Rico. *Journal of Hydrology*, 115, 319–341.

Elsinger, R. J., and W. S. Moore. (1983). Gas exchange in the Pee Dee River based on ^{222}Rn evasion. *Geophysical Research Letters*, 10, 443–446.

Engesgaard, P., K. H. Jenses, J. Molson, E. O. Frind, and H. Olsen. (1996). Large-scale dispersion in a sandy aquifer: Simulation of subsurface transport of environmental tritium. *Water Resources Research*, 32, 3253–3266.

Eriksson, E. (1963). Atmospheric tritium as a tool for the study of certain hydrologic aspects of river basins. *Tellus*, 15, 303–308.

Foster, S. S. D., and A. Smith-Carrington. (1980). The interpretation of tritium in the Chalk unsaturated zone. *Journal of Hydrology*, 46, 343–364.

Gaye, C. B., and W. M. Edmunds. (1996). Groundwater recharge estimation using chloride, stable isotopes and tritium profiles in the sands of northwestern Senegal. *Environmental Geology*, 27, 246–251.

Genereux, D. P., and H. F. Hemond. (1990). Naturally occurring radon-222 as a tracer for streamflow generation: Steady state methodology and field example. *Water Resources Research*, 26, 3065–3075.

Genereux, D. P., H. F. Hemond, and P. J. Mulholland. (1993). Use of ^{222}Rn and calcium as tracers in a three-end-member mixing model for streamflow generation on the West Fork of Walker Branch watershed. *Journal of Hydrology*, 142, 167–211.

Gibson, J. J., P. Aggarwal, J. Hogan, C. Kendall, L. A. Martinelli, W. Stichler, D. Rank, I. Goni, M. Choudhry, J. Gat, S. Bhattacharya, A. Sugimoto, B. Fekete, A. Pietroniro,

T. Bruhn, H. Panarello, D. Stone, P. Seyler, L. Maurice-Bourgoin, and A. Herzeg. (2002). Isotope studies in large river basins: A new global research focus. *EOS*, 83, 613–617.

Goni, M. A., M. B. Yunker, R. W. Macdonald, and T. I. Eglinton. (2005). The supply and preservation of ancient and modern components of organic carbon in the Canadian Beaufort Shelf of the Arctic Ocean. *Marine Chemistry*, 93, 53–73.

Guerin, M. (2001). Tritium and ^{36}Cl as constraints on fast fracture flow and percolation in the unsaturated zone at Yucca Mountain. *Journal of Contaminant Hydrology*, 51, 257–288.

Haas, H., D. W. Fisher, D. C. Thorstenson, and E. P. Weeks. (1983). $^{13}CO_2$ and $^{14}CO_2$ measurements on soil atmosphere sampled in the sub-surface unsaturated zone. *Radiocarbon*, 25, 301–314.

Hainsworth, L. J., A. C. Mignerey, G. R. Helz, P. Sharma, and P. W. Kubik. (1994). Modern chlorine-36 deposition in southern Maryland, USA. *Nuclear Instruments and Methods in Physics Research Section B*, 92, 345–349, (Elsevier Science, Holland).

Hawley, N., J. A. Robbins, and B. J. Eadie. (1986). The partitioning of beryllium-7 in fresh water. *Geochimica et Cosmochimica Acta*, 50, 1127–1131.

Hedges, J. I., J. R. Ertel, P. D. Quay, P. M. Grootes, J. E. Richey, A. H. Devol, G. W. Farwell, F. W. Schmidt, and E. Salati. (1986). Organic carbon-14 in the Amazon River system. *Science*, 231, 1129–1131.

Houle, D., R. Carignan, and R. Ouinet. (2001). Soil organic sulfur dynamics in a coniferous forest. *Biogeochemistry*, 53, 105–124.

Houle, D., R. Carignan, and J. Roberge. (2004). The transit of $^{35}SO_4^{-2}$ and $^{3}H_2O$ added in situ to soil in a boreal coniferous forest. *Water, Air, and Soil Pollution: Focus*, 4, 501–516.

Izbicki, J. A., J. Radyk, and R. L. Michel. (2000). Water movement through a thick unsaturated zone underlying an intermittent stream in the western Mojave Desert, Southern California, USA. *Journal of Hydrology*, 238, 194–217.

Jia, G., M. Belli, U. Sansone, S. Rosamilia, and S. Gaudino. (2006). Concentration and characteristics of depleted uranium in biological and water samples collected in Bosnia and Herzegovina. *Journal of Environmental Radioactivity*, 89, 172–187.

Johnson, C. A., M. A. Mast, and L. Kestern. (2001). Use of $^{17}O/^{16}O$ to trace atmospherically-deposited sulfate in surface waters: A case study in alpine watersheds in the Rocky Mountains. *Geophysical Research Letters*, 28, 4483–4486.

Jonnson, K., H. Johansson, and A. Worman. (2003). Hyporheic exchange of reactive and conservative solutes in streams – tracer methodology and model interpretation. *Journal of Hydrology*, 278, 153–171.

Jull, A. J. T., G. S. Burr, J. W. Beck, D. J. Donahue, D. Biddulph, A. L. Hatheway, T. E. Lange, and L. R. McHargue. (2003). Accelerator mass spectrometry at Arizona; geochronology of the climate record and connections with the ocean. *Journal of Environmental Radioactivity*, 69, 3–19.

Kao, S. J., and K. K. Liu. (1996). Particulate organic carbon export from a subtropical mountainous river (Lanyang Hsi) in Taiwan. *Limnology and Oceanography*, 41, 1749–1757.

Katz, B. G., J. K. Bohlke, and H. D. Hornsby. (2001). Timescales for nitrate contamination of spring water. *Chemical Geology*, 179, 167–186.

Kennedy, V. C., C. Kendall, W. G. Zellweger, T. A. Wyerman, and R. J. Avanzino. (1986). Determination of the components of streamflow using water chemistry and environmental isotopes, Mattole River Basin, California. *Journal of Hydrology*, 84, 107–140.

Kester, C. L., J. L. Baron, and JT. Turk. (2003). Isotopic study of sulfate sources and residence times in a subalpine watershed. *Environmental Geology*, 43, 606–613.

Koide, M., R. Michel, E. D. Goldberg, M. M. Herron, and C. C. Langway. (1982). Characterization of radioactive fallout from pre and post moratorium tests to polar glaciers. *Nature*, 296, 544–547.

Kraemer, T. F., M. W. Doughten, and T. D. Bullen. (2002). Use of ICP/MS with ultrasonic nebulizer for routine determination of uranium activity ratios in natural water. *Environmental Science and Technology*, 36, 4899–4904.

Lal, D., and B. Peters. (1967). Cosmic ray produced radioactivity on the earth. In: *Encyclopedia of Physics* (Ed. S. Flugge). Springer, New York, pp. 407–434.

Lal, D., T. Rama, and P. K. Zutshi. (1960). Radioisotopes ^{32}P, ^{7}Be, and ^{35}S in the atmosphere. *Journal of Geophysical Research*, 65, 669–674.

Lam, D. C. L., and R. W. Durham. (1984). Finite element analysis of a radioactive tritium patch and a waste heat plume observed near the Pickering Nuclear Power Generation Station, Lake Ontario. *Journal of Great Lakes Research*, 10, 59–67.

Lee, C. W. L., and M. H. Thiemens. (2001). The $\delta^{17}O$ and $\delta^{18}O$ measurements of atmospheric sulfate from a coastal and high alpine region: A mass-independent isotopic anomaly. *Journal of Geophysical Research*, 106, 17353–17359.

Lee, R., and E. F. Hollyday. (1987). Radon measurement in streams to determine location and magnitude of ground water seepage. In: *Radon, Radium, and Other Radioactivity in Ground Water* (Ed. B. Graves). Lewis Publishers, Chelsea, MI, pp. 241–249.

Lin, R., and K. Wei. (2006). Tritium profiles of pore water in the Chinese loess unsaturated zone: Implications for estimation of groundwater recharge. *Journal of Hydrology*, 328, 192–199.

Maloszewski, P., W. Rauert, W. Stichler, and A. Herrmann. (1983). Application of flow models in an Alpine catchment area using tritium and deuterium. *Journal of Hydrology*, 66, 319–330.

Mangini, A., C. Sonntag, G. Bertsch, and E. Müller. (1979). Evidence for a higher natural uranium content in world rivers. *Nature*, 278, 337–339.

Martell, E. A. (1959). Atmospheric aspects of strontium-90 fallout. *Science*, 129, 1197–1206.

Masiello, C. A., and E. R. M. Druffel. (2001). Carbon isotope geochemistry of the Santa Clara River. *Global Biogeochemical Cycles*, 15, 407–416.

McGlynn, B., J. Mc-Donnell, M. Stewart, and J. Seibert. (2003). On the relationships between catchment scale and streamwater mean residence time. *Hydrologic Processes*, 17, 175–181.

Merlivat, L., J. Jouzel, J. Robert, and C. Lorius. (1977). Distribution of artificial tritium in firn samples from East Antarctica. *International Association of Science and Hydrological Publication*, 118, 138–145.

Michel, R. L. (1989). Tritium deposition in the continental United States, 1953–1983. *US Geological Survey Water Resources Investigation Report 89-4072*, 45pp.

Michel, R. L. (1992). Residence times in river basins as determined by analysis of long-term tritium records. *Journal of Hydrology*, 130, 367–378.

Michel, R. L. (2004). Tritium hydrology of the Mississippi River basin. *Hydrologic Processes*, 18, 1255–1269.

Michel, R. L., D. Campbell, D. Clow, and J. T. Turk. (2000). Timescales for migration of atmospherically derived sulphate through and alpine/subalpine watershed, Loch Vale, Colorado. *Water Resources Research*, 36, 27–36.

Michel, R. L., and D. L. Naftz. (1995). Use of sulphur-35 and tritium to study runoff from an alpine glacier, Wind River Range, Wyoming. In: *Biogeochemistry of Seasonally Snow-Covered Catchments* (Eds K. A. Tonnessen, W. W. Williams, and M. Tranter). International Association of Hydrologic Sciences, Oxforshire, UK. (IAHS Publication 228), pp. 441–444.

Miller, M. M., J. S. Leventhal, and W. F. Libby. (1965). Tritium in Mount Everest ice: Annual glacier accumulation and climatology at great equatorial altitudes. *Journal of Geophysical Research*, 70, 3885–3888.

Milliman, J. D., and J. P. M. Syvitski. (1992). Geomorphic-tectonic control of sediment discharge to the ocean: The importance of small mountainous rivers. *Journal of Geology*, 100, 525–544.

Milton, G. M., J. C. D. Milton, S. Schiff, P. Cook, T. G. Kotzer, and L. D. Cecil. (2003). Evidence for chlorine recycling – hydrosphere, biosphere, atmosphere – in a forested wet zone on the Canadian Shield. *Applied Geochemistry*, 18, 1027–1042.

Miskel, J. A. (1973). Production of tritium by nuclear weapons. In: *Tritium* (Ed. A. Moghissi). Messenger Graphics, Las Vegas, NV, pp. 79–85.

Morgenstern, U., L. K. Fifield, and A. Zondervan. (2000). New frontiers in glacier ice dating: Measurement of natural ^{32}Si by AMS. *Nuclear Instruments and Methods in Physics Research B*, 172, 605–609.

Moysey, S., S. N. Davis, M. Zreda, and L. D. Cecil. (2003). The distribution of meteoric $^{36}Cl/Cl$ in the United States: A comparison of models. *Hydrogeology Journal*, 11, 615–627.

Muller, R. A. (1977). Radioisotope dating with a cyclotron. *Science*, 196, 489–494.

Murphy, C. E. (1993). Tritium transport and cycling in the environment. *Health Physics*, 65, 683–697.

Naftz, D. L., R. W. Klusman, R. L. Michel, P. F. Schuster, M. M. Reddy, H. E. Taylor, T. M. Yanosky, and E. A. McConnaughey. (1996). Little Ice Age evidence from s south-central North American ice core, USA. *Arctic and Alpine Researches*, 28, 35–41.

Naftz, D. L., R. L. Michel, and K. A. Miller. (1993). Isotopic indicators of climate change in ice cores, Wind River Range, Wyoming. In: *Climate Change in Continental Isotopic Records, Geophysical Monograph 78* (Eds P. K. Swarat, K. C. Lohmann, J. McKenzie, and S. Savin). American Geophysical Union, Washington, DC, pp. 55–66.

Novak, M., R. L. Michel, E. Prechova, and M. Stepanova. (2004). The missing flux in a ^{35}S budget for the soils of a small polluted catchment. *Water, Air, and Soil Pollution: Focus*, 4, 517–529.

Phillips, F. M. (2000). Chlorine-36. In: *Envrionmental Tracers in Subsurface Hydrology* (Eds P. G. Cook and A. L. Herczeg). Kluwer Academic Publishers, Boston, MA, pp. 2199–2348.

Phillips, F. M., J. L. Mattick, and T. A. Duval. (1988). Chlorine-36 and tritium from nuclear weapons fallout as tracers for long-term liquid and vapor movement in desert soils. *Water Resources Research*, 24, 1877–1891.

Picciotto, E., G. Crozoz, W. Ambach, and H. Eisner. (1967). Lead-210 and strontium-90 in an alpine glacier. *Earth and Planetary Science Letters*, 3, 237–242.

Plummer, L. N., E. C. Prestemon, and D. L. Parkhurst. (1994). An Interactive Code (NETPATH) for Modeling NET Geochemical Reactions along a Flow PATH – Version 2.0. US Geological Survey Water-Resources Investigations Report 94-4169, 130p.

Rank, D., L. Alder, L. A. Arauguas, K. Froehlich, K. Rozanski, and W. Stichler. (1998). Hydrological parameters and climatic signals derived from long-term tritium and stable isotopes time series of the River Danube. In: *Isotopic Techniques in the Study of Environmental Change*. International Atomic Energy Agency, Vienna, Austria, pp. 191–205.

Rank, D., W. Papesch, V. Rajner, and G. Riehl-Herwirsch. (1992). Environmental isotopes study at the Breitenau experimental landfill (Lower Austria). In: *Tracer Hydrology* (Eds H. Hötzl and A. Werner). Balkema, Brookfield, 464pp.

Rat'ko, A. I., V. P. Samodurov, and V. P. Kol'nenkov. (2001). Tritium sorption by modified natural aluminosilicates. *Radiochemistry*, 43, 519–522.

Raymond, P. A., and J. E. Bauer. (2001a). Use of ^{14}C and ^{13}C natural abundances for evaluating riverine, estuarine, and coastal DOC and POC sources and cycling: A review and synthesis. *Organic Geochemistry*, 32, 469–486.

Raymond, P. A., and J. E. Bauer. (2001b). Riverine export of aged terrestrial organic matter to the North Atlantic Ocean. *Nature*, 409, 497–500.

Revelle, R., and H. E. Suess. (1957). Carbon dioxide exchange between atmosphere and ocean and the question of an increase of atmospheric CO_2 during the past decades. *Tellus*, 9, 18–27.

Riotte, J., and F. Chabaux. (1999). $^{234}U/^{238}U$ activity ratios in freshwater as tracers of hydrological processes: The strenbach watershed (Vosges, France). *Geochimica et Cosmochimica Acta*, 63, 1263–1275.

Robinson, H. D., and J. R. Gronow. (2007). Tritium Levels in Leachates and Condensates from Domestic Wastes in Landfill Sites. *Water and Environment Journal*, 10(6), 391–398.

Rogers, A. S. (1958). *Physical Behavior and Geologic Control of Radon in Mountain Streams*. US Geological Survey Bulletin 1052E, pp. 187–212.

Rose, S. (1993). Environmental tritium systematics of baseflow in Piedmont Province watersheds, Georgia, USA. *Journal of Hydrology*, 143, 191–216.

Rose, T. P., M. L. Davisson, and R. E. Criss. (1996). Isotope hydrology of voluminous cold springs in fractured rock from an active volcanic region, northeastern California. *Journal of Hydrology*, 179, 207–236.

Rosen, A. A., and M. Rubin. (1964). Natural carbon-14 activity of organic substances in streams. *Science*, 143, 1163–1164.

Schaeffer, O. A., S. O. Thompson, and N. L. Lark. (1960). Chlorine-36 radioactivity in rain. *Journal of Geophysical Research*, 65, 4013–4016.

Schell, W. R., G. Sauzay, and B. R. Payne. (1970). Tritium injection and concentration distribution in the atmosphere. *Journal of Geophysical Research*, 75, 2251–2266.

Schuster, P. F., D. Krabbenhoft, D. L. Naftz, L. D. Cecil, M. L. Olson, J. F. Dewild, D. D. Susong, J. R. Green, and M. L. Abbott. (2002). Atmospheric mercury deposition during the last 270 years: A glacial ice core of natural and anthropogenic sources. *Environmental Science and Technology*, 36, 2303–2310.

Shiklomanov, I. A. (1998). *World Water Resources – A New Appraisal and Assessment for the 21st Century*. UNESCO Technical Document, UNESCO, Paris.

Snow, D. D., and R. F. Spaulding. (1994). Uranium isotopes in the Platte River drainage basin of the North American High Plains region. *Applied Geochemistry*, 9, 271–278.

Spieker, E. C., and M. Rubin. (1975). Petroleum pollutants in surface and groundwater as indicated by the carbon-14 activity of dissolved organic carbon. *Science*, 187, 61–63.

Steinmann, P., T. Billen, J. L. Loizeau, and J. Dominik. (1999). Beryllium-7 as a tracer to study mechanisms and rates of metal scavenging from lake surface waters. *Geochimica et Cosmochimica Acta*, 63, 1621–1633.

Stewart, G. L. (1966). Experiences using tritium in Scientific Hydrology. In: *Proceedings of the Sixth International Conference on Radiocarbon and Tritium Dating*. US Atomic Energy Agency Report, pp. 643–658.

Stewart, M. K., and C. B. Taylor. (1981). Environmental isotopes in New Zealand hydrology. 1 Introduction: The role of oxygen-18, deuterium and tritium in hydrology. *New Zealand Journal of Science*, 24, 295–311.

Sueker, J. K., J. T. Turk, and R. L. Michel. (1999). Use of cosmogenic ^{35}S for comparing ages of water from three alpine-subalpine basins in the Colorado Front Range. *Geomorphology*, 27, 61–74.

Suess, H. E. (1969). Tritium geophysics as an international research project. *Science*, 163, 1405–1410.

Suess, H. E. (1986). Secular variations of cosmogenic ^{14}C on earthTheir discovery and interpretation. *Radiocarbon*, 28, 259–265.

Tate, C. M., R. E. Broshears, and D. M. McKnight. (1995). Phosphate dynamics in an acidic mountain stream: Interactions involving algal uptake, sorption by iron oxide, and photoreduction. *Limnology and Oceanography*, 40, 938–946.

Taylor, C. B. (1966). Tritium in Southern Hemisphere precipitation. *Tellus*, 18, 105–131.

Thompson, L. G., M. E. Davis, E. Mosley-Thompson, T. A. Sowers, K. A. Henderson, V. S. Zagorodnov, P.-N. Lin, V. N. Mikhalenko, R. K. Campen, J. F. Bolzan, J. Cole-Dai, and B. Francou. (1998). A 25,000-year tropical climate history from Bolivian ice cores. *Science*, 282, 1858–1864.

Thorstenson, D. C., E. P. Weeks, H. Hass, and D. W. Fisher. (1983). Distribution of gaseous $^{12}CO_2$, $^{13}CO_2$, and $^{14}CO_2$ in the sub-soil unsaturated zone of the western US Great Plains. *Radiocarbon*, 25, 315–346.

Turner, J. V., and C. J. Barnes. (1998). Modeling of isotopes and hydrogeochemical responses in catchment hydrology. In: *Isotope Tracers in Catchment Hydrology* (Eds C. Kendall and J. J. McDonnell). Elsevier Press, Amsterdam, The Netherlands, pp. 723–760.

Tuzova, T. V. (1986). Study of the runoff distribution and estimation of water resources of the Sarydzhaz River basin from uranium isotopic composition data. *Water Resources*, 12, 99–105, (Translation of *Vodnye Resursy*).

Tuzova, T. V., and K. S. Filkin. (1991). Estimation of the contribution of the subsurface component of the runoff of small mountain rivers by the uranium isotopic method. *Water Resources*, 17, 162–165, (Translation of *Vodnye Resursy*).

Tuzova, T. V., and V. N. Novikov. (1991). Uranium isotope-related features of streamflow formation for Pyandzh River. *Water Resources Research*, 18, 59–65, (Translation of *Vodnye Resursy*).

Wannakhof, R., P. J. Millholland, and J. W. Elwood. (1990). Gas exchange rates for a first-order stream determined with deliberate and natural tracers. *Water Resources Research*, 26, 1621–1630.

Watwood, M. E., and J. W. Fitzgerald. (1988). S transformation in forest litter and soil: Results of laboratory and field incubations. *Soil Science Society of America Journal*, 52, 1478–1483.

Weiss, W., and W. Roether. (1975). Discharge of tritium in the Rhine, 1961–1973. *Deutsche Gewässerkundliche Mitteilungen*, 19, 1–5.

Wietersen, R. C., T. C. Daniel, K. J. Fermanich, B. D. Girard, K. McSweeney, and B. Lowery. (1993). Atrazine, alachlor and metolachlor mobility through two sandy Wisconsin soils. *Journal of Environmental Quality*, 22, 811–818.

Williams, P. M., and E. R. M. Druffel. (1987). Radiocarbon in dissolved organic matter in the central North Pacific Ocean. *Nature*, 330, 246–248.

Wood, W. W., and W. E. Sanford. (1995). Chemical and isotopic methods for quantifying ground-water recharge in a regional, semiarid environment. *Ground Water*, 33, 458–468.

Yang, I. C. (2002). Percolation flux and transport velocity in the unsaturated zone, Yucca Mountain, Nevada. *Applied Geochemistry*, 17, 807–817.

Yurtsever, Y. (1999). Use of environmental tritium to study catchment dynamics: case study from the Danube River basin. In: *Integrated Methods in Catchment Hydrology – Tracer, Remote Sensing and New Hydrometric Techniques* (Eds C. Leibundgut, J. McDonnell, and G. Schultz). pp. 167–174 (IAHS Publication No. 258).

Zielinski, R. A., S. Asher-Bolinder, A. L. Meier, C. A. Johnson, and B. J. Szabo. (1997). Natural or fertilizer derived uranium in irrigation drainage: A case study in southeastern Colorado, USA. *Applied Geochemistry*, 12, 9–21.

Zielinski, R. A., K. R. Simmons, and W. H. Orem. (2000). Use of ^{234}U and ^{238}U isotopes to identify fertilizer-derived uranium in the Florida Everglades. *Applied Geochemistry*, 15, 369–383.

Zimmermann, U., K. O. Munnich, W. Roether, W. Kreutz, K. Schubach, and O. Siegel. (1966). Tracers determine movement of soil moisture and evapotranspiration. *Science*, 152, 346–347.

5.2. Radionuclides and Transient Gas Tracers in Studies of Lakes and Inland Seas

Robert L. Michel[1], Thomas Kraemer[2] *and* L. DeWayne Cecil[3]

[1]*US Geological Survey, MS 434, Menlo Park, CA, USA*
[2]*US Geological Survey, 430 National Center, Reston, VA, USA*
[3]*US Geological Survey, INL, Idaho Falls, ID, USA*

5.2.1. Introduction

Radionuclides have served a variety of purposes in the study of hydrologic and chemical processes occurring in lake systems. Most studies have been oriented toward obtaining timescales for physical processes occurring within the lake, such as residence times of lake waters, and the internal dynamics of mixing in the lakes. Radionuclides have been used to follow pathways of chemical cycling within the lakes, and interactions of the lakes with the local groundwater systems. They have also provided rates for sedimentation within the lakes and have been used to study issues such as climatic variations and changes in pollutant loading to the lake system. The radionuclides used in lake studies are essentially the same species as shown in Table 1.1 of the surface waters section of this chapter.

Lake waters are frequently used for irrigation or potable water. The abstracted lake water can impart a distinctive radioisotopic signature that can be used to study the movement of these waters through the local hydrologic systems. This is especially true where lake water is used as the source of water in agricultural systems. It can also be of value when tracing leakage of water from lakes contained by dams. Thus, knowledge of the isotopic history of the lake is important in studying the local hydrology.

5.2.2. Tritium

Tritium, as a part of the water molecule, has been widely used in the study of lake hydrology. Tritium provides information both on timescales for water retention in lakes and on the dynamic processes occurring within lakes. Figure 2.1 shows vertical tritium profiles that might be encountered in a lake or inland sea. Figure 2.1A is the type of tritium depth distribution that would be found in a lake that is mixed to the bottom. The mixing results in a constant profile with tritium concentrations being the same at all depths. Frequently, small lakes are constantly mixed to the bottom and this type of tritium profile is present at all times. The lack of stratification can also be present in other conservative tracers like temperature and salinity. However, short seasonal effects may create gradients in many properties but

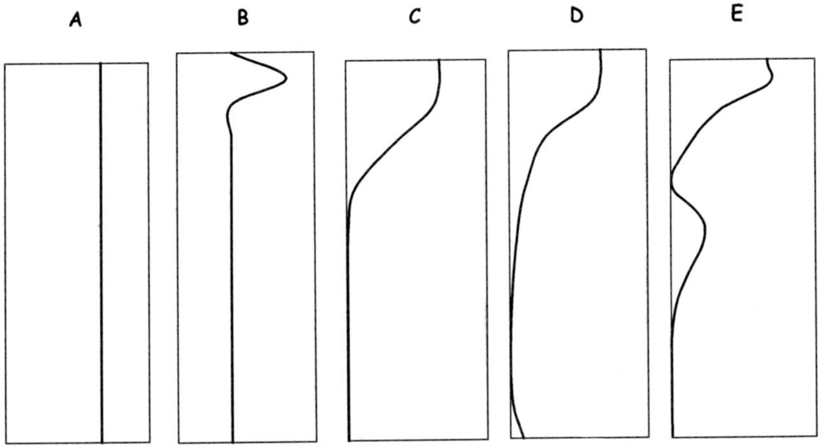

Figure 2.1 Possible tritium distributions in lakes: (A) Lake well mixed to the bottom on a frequent basis. (B) An example of a lake where a seasonal influx of water with higher or lower tritium concentrations occurs during the spring and/or summer. (C) Lake where a permanent pycnocline inhibits mixing to the bottom. (D) Similar to (C) except that dense water is formed, probably near a shelf break, and sinks under the lighter water to the bottom. (E) Similar to (C) except groundwater formed during the tritium transient intrudes into the lake below the pycnocline.

not in tritium. Many lakes will be stratified for part of the year but overturn on a seasonal basis, usually in the winter when the surface water cools and the density increases. Samples collected during the part of the year when overturn occurs will also exhibit a tritium pattern similar to that shown in Figure 2.1A. When a pycnocline develops, generally during the summer and early fall, it is the result of the warming of the surface water, producing a less-dense water that overlies the hypolimnion. The distribution of other physical and chemical properties can be impacted during this period, depending on the physicochemical processes that affect them. Stable isotopes in the upper layer frequently exhibit a shift to higher values during the summer due to evaporation (Gat, 1981). Nutrients and other biologically active chemicals can be removed from the surface layer and transmitted to deeper lake waters or the underlying sediments, resulting in a change from the vertical profile. Tritium is less sensitive to these processes, resulting in fewer and smaller changes to the tritium profile. Surface concentrations of tritium may be impacted by processes such as inflow of surface water from the adjoining watershed or rivers with different tritium concentrations. The changes brought about by inflow depend on the differences in tritium concentrations between the lake water and inflowing water. It is also possible that a lake will mix in some years but not others, depending on climatic effects. Lake Tahoe is one such lake, which appears not to mix to the bottom every year. In the winter of 1973, nitrate data suggested that the lake had overturned to the bottom (Paerl et al., 1975). Measurements made on tritium later that year supported the nitrate data and showed a uniform tritium concentration with depth (Imboden et al., 1977; Imboden, 1979).

An example of the most likely seasonal change in the tritium profile of a lake with a seasonal pycnocline is exhibited in Figure 2.1B. As it is over 40 years since the last major injection of tritium into the hydrosphere, tritium concentrations are continuing to decline and lakes with a water retention time of longer than a few years will usually have a tritium concentration higher than incoming water from streams (Michel and Kraemer, 1995). If a strong stratification develops in spring, the incoming water will dilute the tritium concentration in the upper layer of the lake, resulting in the pattern shown in Figure 2.1B. When the seasonal overturn occurs again in the winter, the pattern of a constant tritium concentration with depth will reappear although the concentration will be lower than that found in previous years.

Figure 2.1C is a profile typical of lakes that do not overturn, but have remained stratified throughout the period of the bomb transient. These lakes can be, but are not necessarily, deep lakes. A typical lake of this type would be Lake Tanganyika, a lake in the rift zone of Eastern Africa, with a depth of about 900 m. The surface tritium concentration was 6–7 TU in 1973, but concentrations decreased sharply in the top 150 m to less than 1 TU (SIO, 1975). The tritium data indicated that the lake had not mixed to the bottom since at least the 1950s, and had probably been stratified for a longer time. In these systems, mixing of surface water into the deeper layers below the pycnocline can generally only occur via diffusive mixing across isopycnals, which is a slow process (decadal or longer). Changes in the physical properties of waters flowing into lakes, or changes in the lake itself, such as lowering or increasing of the lake level, can result in a change from one regime to the other. The Dead Sea had a strong stratification in both tritium and chemical properties during the 1960s and 1970s with tritium profiles similar to Figure 2.1C (Steinhorn, 1985). In February 1979, the Dead Sea mixed to the bottom resulting in a profile similar to that in Figure 2.1A.

In oceanic systems, it is possible for tritium to mix below the local pycnocline through a variety of processes (Ostlund and Fine, 1979). The most common method of mixing tritium into deeper layers is by the outcropping of dense water at a given location, either through wind-stirring or cooling. The dense water picks up tritium from the atmosphere and then flows along density surfaces which sink under lighter water layers in other locations (McPhaden and Fine, 1988). In the oceans, the dense water typically outcrops at high latitudes where temperatures are low, and then flows toward lower latitudes, sinking under warmer and less-dense water. Thus, bomb tritium is able to penetrate well below the pycnocline and, depending on the tritium concentration where the water outcrops, can produce profiles similar to Figure 2.1D. This type of profile may be found in some large lakes and inland seas but is rarely found in smaller lakes.

Another method of injecting recent tritium into the water beneath the pycnocline is by the creation of dense water, usually in shallow near-shore areas, that can then sink below the lighter waters (Deacon, 1937; Foster et al., 1987). Under certain temperature or salinity conditions, water will be created that is dense enough to sink to the bottom. It is also possible to produce water of a density that is between those of the surface and bottom waters, resulting in an injection of surface water at an intermediate depth (Carmack and Killworth, 1978; Foster, 1995). Thus, it is possible to produce profiles similar to that of Figure 2.1E where tritiated water is seen to occur both above and below tritium-free water. Another possibility for introducing water with a different tritium concentration into deep layers of lakes is inflow of groundwater. Groundwater is known to be a major component of the water balance of some lakes and is frequently introduced below the pycnocline. The tritium concentration of groundwater varies from zero to concentrations well above ambient levels. If the lake maintains a density gradient in the deeper layers, the entering groundwater will remain in a small density range, producing a structure similar to Figure 2.1E.

Tritium has also proven to be useful to determine source waters for lakes and wetlands (Cheng, 1996). The inflowing waters can carry solutes and pollutants that can have a negative impact on water quality. Many times, especially for management issues, it is important to know the age of groundwater feeding into lakes, and tritium and tritium/helium-3 can provide age estimates. Age estimates can give resource managers information on how rapidly management decisions can effect changes in pollutant loads (Bohlke and Denton, 1995).

Tritium has frequently been used to study parameters involved in water balances and residence times of water in lakes. Tritium fluxes in lakes involve the same processes that control the flux of deuterium and ^{18}O in lakes (Gat, 1981). Frequently, these types of studies can be carried out using non-isotopic methods, but occasionally isotopes can provide information on a process not easily available from other methods (Krabbenhoft et al. 1990; Kumar et al., 2001). The tritium balance of a lake can be given by:

$$\frac{dC_l}{dt} = -\lambda^* C_l + \mathrm{P}^* C_p + \mathrm{R}^* C_r + \mathrm{G}_i^* C_g - \mathrm{E} - \mathrm{G}_o^* C_l - \mathrm{O}^* C_l \qquad (2.1)$$

where λ is the decay constant of tritium (0.056 yr^{-1}), C_l the tritium concentration in the lake, C_p the tritium concentration in precipitation, C_g the tritium concentration in inflowing groundwater, C_r the tritium concentration in inflowing surface water, E a term representing the change in concentration due to evaporation and molecular exchange, P, R, and G_i

are the fraction of the lake that comes in by precipitation, runoff, and groundwater inflow, and O and G_o are the fractions of the lake that leave via outflow or as groundwater. The evaporation term is a function of evaporation rate, relative humidity, and the tritium concentration in the lake water and in the atmospheric water vapor over the lake. For stable isotopes in a steady state, the left side of the equation would be zero, but as tritium is a transient tracer, the concentration in the lake will exhibit a yearly change. The tritium concentration in the lake at the time of measurement is a function of the sum of these changes during the tritium transient.

Some of the terms in Equation (2.1) may be relatively easy to obtain such as surface water inflow and outflow, and precipitation and evaporation rates. Groundwater fluxes are generally unknown. Tritium concentrations for lake water, groundwater, inflowing water, and precipitation can be measured, but frequently for inflowing water and precipitation there are seasonal fluctuations that require sampling over a full year. Isotopic concentrations in water vapor are difficult to measure and provide one of the most difficult problems for using isotopes in lake balance studies. However, tritium concentrations in precipitation are generally close to those in water vapor and can be substituted in many cases. The above approach has been applied to Lake Naini in India (Kumar et al., 2001), the Finger Lakes in New York (Michel and Kraemer, 1995) and the Neusiedlersee in Austria (Rank et al., 1983), and estimates of residence times for all systems were obtained. Using vertical distributions of tritium and other constituents in Lake Naini, Kumar et al. (2001) were also able to show that residence times of water in all lake strata are short.

What is desired for tritium balance studies in lakes is time series of measurements during a significant part of the tritium transient. Such time series are generally not available. Tritium concentrations in precipitation are the easiest time series to obtain as data from a large number of long-term monitoring sites are available from the International Atomic Energy Agency (IAEA, 2001). For periods when data are not available, it is possible to use correlations from other stations to bridge gaps (IAEA, 1992). It has also proved possible at some locations to obtain tritium concentrations in precipitation by analysis of vintage wines when they are available (Roether, 1967). There are several long-term data series for tritium concentrations in rivers (Weiss and Roether, 1975; Michel, 1992), but very few series are available for lakes and inland seas (Michel and Suess, 1978; Rank et al., 1983; Herczeg and Imboden, 1988). The data from the Neusiedlersee (Rank et al., 1983) represent one of the longest continuous data set published in any lake, starting in the mid-1960s. The data were used to try and determine the importance of the interaction of lake water and local groundwater in recharge. Occasionally, tritium concentrations are measured in different research projects over an extended length of

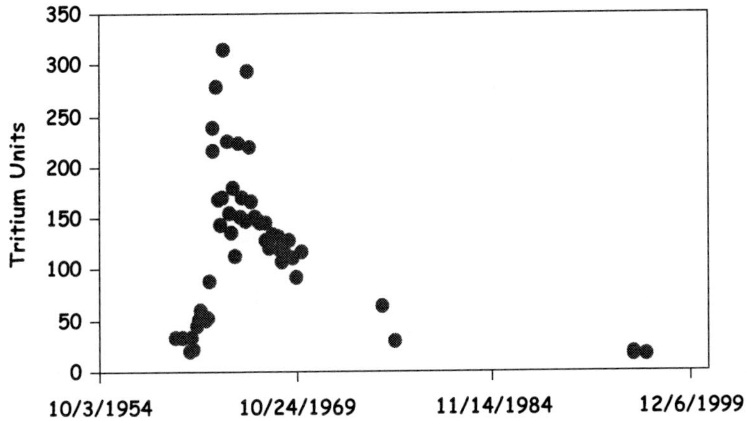

Figure 2.2 Long-term database of tritium concentrations in the Caspian Sea compiled from Michel and Suess (1978), Romanov (1974), and Peeters et al. (2000).

time. Tritium was measured in the Caspian Sea (Figure 2.2) from 1960 to 1970 by the La Jolla Tritium Laboratory and further measurements were obtained by other researchers over the decades since that period (Romanov, 1974; Froehlich et al., 1998; Peeters et al., 2000). This probably represents the longest tritium data base from any lake or inland sea. Crater Lake in Oregon was sampled in 1967 and 1981 to observe changes in tritium, ^{90}Sr and ^{137}Cs in the water column. The 1967 sampling program was carried out to determine if there was enhanced fallout of nuclear debris over water surfaces. Crater Lake is a deep lake that mixes to the bottom and has a very small watershed, making it ideal for this type of experiment. The results for all three nuclides indicated that the total fallout was similar to that found on land (Volchok et al., 1970). The follow-up sampling in 1981 found that tritium and ^{90}Sr were in the water column in the predicted amounts (Volchok et al., 1983). Plutonium and ^{137}Cs had been preferentially removed from the water column.

5.2.3. Dating of subsurface lake waters with transient gas tracers

Transient gas tracers have been used to estimate the times that the subsurface waters reside in lakes. Among the most important of these tracers for lake studies are the tritium/helium-3 pair and chlorofluorocarbons (CFCs). Although CFCs are nonradioactive environmental tracers, they are included in this chapter because they are of transient nature and often used in combination with tritium. Due to its continuing increase in the environment, sulfur hexafluoride (SF_6) is becoming more important in these types of studies (Maiss and Brenninkmeijer, 1998; Hofer et al., 2002). Also available, although utilized less at the present time, is ^{85}Kr which has continued to increase with time due to its release by the nuclear industry (Rozanski, 1979). All these gases have the advantage over many other tracers in that they can give an "apparent age" for a parcel of water that sinks below the surface layer and is removed from interactions with the atmosphere.

The CFCs have a relatively well-known history of changes in global concentrations in the atmosphere (Gamlen et al., 1986). While the concentrations tend to be global, with a slight hemispheric difference, there are areas where CFC concentrations are well above the average and use of CFCs is restricted in these areas. Most large cities have excess CFC in the air and areas such as Western Europe and the Eastern United States tend to have high and changing concentrations at all times. Thus, CFC techniques are frequently unreliable in these areas. However, in most other areas, the CFC concentrations in the atmosphere will reflect the global mean and can be use as an effective tracer for dating water.

The CFCs in the atmosphere equilibrate with surface water, with Henry's Law constants that are dependent on gas properties, temperatures, and salinities (Warner and Weiss, 1985). CFCs are not reactive in most waters, so when the gas is removed from contact with the atmosphere, they will behave conservatively and retain the concentrations initially reached at the surface. Measuring this concentration and knowing the temperature and salinity of the water, one can use Henry's Law to determine the atmospheric concentration with which the water last equilibrated (Plummer and Busenberg, 2000). This concentration can then be related to the atmospheric record of CFC concentrations and the year of equilibration can be calculated. This method works best for the period prior to the late 1990s when these tracers increased at a well-known rate. Presently, concentrations of most CFCs in the atmosphere are decreasing due to international protocols. Due to the present decrease of most CFCs, the tracer of choice for very recent waters has become SF_6 whose concentration is still increasing rapidly in the atmosphere as no restrictions have been placed on its use. The main problem with SF_6 is that it can have a natural source. However, this generally does not impact concentrations in lakes.

Dynamic mixing may also influence the concentration of these gases in the near-surface waters of the lake. Weiss et al. (1991) found that CFC-12 in surface water at Lake Baikal was generally not in equilibrium with the atmosphere, being about 18% lower than expected. They attribute this to occasional upwelling of deep, CFC-12 poor water from below 250 m and occasional seasonal stratification near the surface. Thus, this under-saturation of CFCs must be taken into account when determining ages of deep water.

The other major gas tracer used to study subsurface lake water is ^{3}He, the daughter product of tritium decay:

$$^{3}H \rightarrow {}^{3}He^{+} + \beta^{-} + \nu$$

The ^{3}He technique that was originally developed for studying subsurface waters in oceanography (Jenkins et al., 1972), provides another method of determining how long ago the water sampled has been removed from the surface. Helium equilibrates very rapidly with the atmosphere, but when the water parcel is removed from the surface, the ^{3}He produced by beta decay will be retained. As helium is a rare gas, there are no chemical reactions and the ^{3}He will accumulate over time. The $^{3}He/^{4}He$ ratio will increase, and the ^{3}He produced by the decay of ^{3}H can be used to calculate the "age" of the water parcel, meaning the time elapsed since it last equilibrated with the atmosphere. This ^{3}He-age τ can be calculated by:

$$\tau = \frac{1}{\lambda} \ln\left(1 + \frac{^{3}He_{tri}}{^{3}H}\right) \tag{2.2}$$

where λ is the decay rate of tritium, $^{3}He_{tri}$ is ^{3}He produced by tritium decay (tritiogenic helium), and ^{3}H the measured tritium concentration. In groundwater, corrections to this simple approach have to be made to account for excess ^{4}He produced by alpha decay, and excess ^{3}He that may be released from the mantle. Problems with excess air being entrained at the water table also have to be considered. Some of these issues may be a factor for subsurface lake waters (Kipfer et al., 1994; Clark and Hudson, 2001), though they are generally less important in these systems. In most lakes, there will be mixing of waters, either through diffusion or advection, and these processes will affect the ages calculated. Generally, it can be assumed for ^{3}He that the waters with higher tritium concentrations will have a dominant effect on determining the age (Kipfer et al., 2002). In ideal circumstances, ^{3}He, CFCs, and SF_6 can all be collected at the same time to see if the ages determined are concordant.

One of the first works on the use of gas tracers for age-dating subsurface water in lakes was the work by Torgersen et al. (1977) on some of the Great Lakes in North America. Figure 2.3 comes from data in Table 3 of that publication showing temperature, tritium concentration, and age as determined by $^{3}H/^{3}He$ with depth for a station in Lake Huron in August 1974. A strong thermocline is present at this site at 18–20 m depth. Tritium concentrations are essentially uniform with depth, indicating that surface water has mixed into deeper layers during the recent past. Models can be developed which will place limits on the age of the deep water to within a few years. The use of the $^{3}H/^{3}He$ data constrains the possible ages to a more discreet range, in this case giving ages on the order of 3 months since the deep water had been exposed to the surface. The helium data indicated

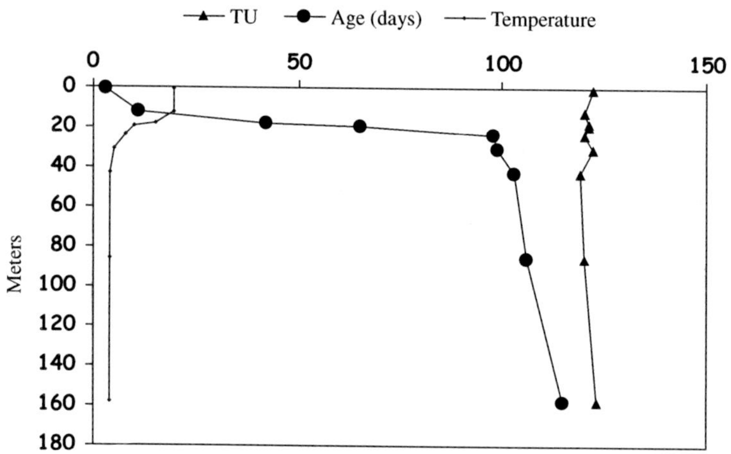

Figure 2.3 Tritium concentrations and ages derived for Lake Huron (Torgersen et al., 1977).

that mixing in these lakes was seasonal with a complex pattern resulting in interleaving of water with different ages throughout the water column. It showed that this method was a powerful technique that could be used to elucidate mixing processes and lake dynamics.

The ages obtained by the $^3H/^3He$ method can also be used to study biogeochemical cycling in subsurface lake waters. Fischer et al. (1985) used ^3He with tritium to estimate ages and gas exchange rates for deep waters in the Königsee in Germany. They used the gas exchange rates to estimate oxygen utilization in the subsurface and concluded that the lake would be sensitive to increased inputs of nutrients. A more extensive study was carried out on Green Lake in New York by Torgersen et al. (1981), studying both isotopic and chemical distributions. There are very sharp chemical gradients in this lake with dissolved oxygen decreasing to zero below 15 m while other constituents such as Ca, CH_4, and S^{2-} increase with depth. Carbon-14 ages increase from about 4,000 years at the surface to about 7–8,000 years with depth. Tritium concentrations decrease rapidly below the thermocline (about 15 m) to the bottom. Helium-3 shows three distinct layers, a young surface layer, an intermediate layer with an age of 3–4 years, and the deeper layer with an age of about 6 years. Part of the chemical differences in the lake can be attributed to age of the water and part to inflow of groundwater.

There have been several studies of large lakes utilizing tritium/helium-3 to date the renewal of subsurface waters. Lake Baikal, the deepest freshwater lake on earth, has received extensive study using gas tracers. Initial tritium measurements carried out in 1964–1965 found much higher tritium concentrations in the southern and central basins of the lake than in the northern basin (Verbolov and Shimaraev, 1973). This discrepancy was attributed to the higher inflow of surface water in the south. Profiles of CFC-12 collected in the three basins by Weiss et al. (1991) in 1988 showed that deep water age varies, with values up to 16 years. The profiles showed younger bottom water was present below older middepth water at all three sites. They suggested that mixing of surface water into deeper layers was not a simple overturn event but a complex process involving formation of dense water under certain conditions. In a series of cruises from 1992 to 1995, extensive sampling of $^3H/^3He$ was conducted to determine what the timescales were for renewal of deep water in Lake Baikal (Hohmann et al., 1998). Ages of deep water were found to be greatest in the central and southern basin and much lower in the northern basin. Furthermore, average deep water ages in the southern basin increased from 1992 to 1995, suggesting that deep water renewal had slowed during that period.

As noted above, the Caspian Sea has one of the longest tritium records of any major inland body of water (Figure 2.2). Early work (Michel and Suess, 1978; Romanov, 1981) calculated residence times for the sea using only surface data and assuming it to be well mixed. Two cruises were carried out in the Caspian Sea in 1995 and 1996 to study the hydrologic and tracer properties of the sea with depth (Froehlich et al., 1999; Peeters et al., 2000). Their results indicated that the age of the deep waters was on the order of 15–20 years. The physical data suggested that overturn to the bottom required inputs of cold, saline water. The tracer data were also consistent with a scenario where no overturn has occurred since the mid-1970s.

5.2.4. Carbon-14

Carbon-14 can be used to study the flux and cycling of both DIC and organic compounds in lakes. The DIC in lake waters can have a very complex origin, and control of the ^{14}C content is determined by the size of the lake and the possible sources of DIC. One control on the ^{14}C content of DIC in lakes is the rate of exchange between atmospheric CO_2 and dissolved CO_2 in lake waters. The rate of exchange of atmospheric CO_2 with surface water is relatively slow compared to the residence time of waters in most lakes (Revelle and Suess, 1957). Only lakes where waters have long residence times would generally have the ^{14}C in DIC controlled by atmospheric exchange. Another control is the dissolution of minerals by the waters that enter the lake (Broecker and Walton, 1959). A significant fraction of the water entering some lakes, whether in streamflow or directly as seepage, can be derived from shallow groundwater. These waters contain DIC from the shallow groundwater aquifers. For many systems, this is the main factor controlling the ^{14}C of DIC in lake water.

For some lakes with sharp permanent pycnoclines, it may be possible to use ^{14}C in the DIC to estimate ages for subsurface waters. However, there should be no possibility that DIC is being introduced by water flowing in below the pycnocline. One such system where this is possible is Lake Vanda in Wright Valley, Antarctica (Jones and Ostlund, 1971). This is a system where there is no groundwater input, and the inflowing melt water is much less saline and dense than the deep water, precluding it from sinking into the deeper layers. A sharp chemocline and pycnocline are present at about 50 m. Carbon-14 measurements showed that the DIC in the deep brine layer had an age of over 2000 years, whereas the upper layers were clearly younger and even contaminated with ^{14}C from productivity studies. These results were in accordance with other studies that suggested that the deep brine layer was the result of climatic changes within the past two centuries (Wilson, 1964). A similar study, using ^{14}C and other isotopes, was conducted on two meromictic lakes in northern Canada by Page et al. (1984). Both lakes had deep brine layers that resulted from a complex process involving brine rejection and inflow of brines from the permafrost. The brine layer at Lake Garrow had a ^{14}C age of over 2,000 years, indicating that this layer had probably been cut off from exchange with the atmosphere and surface waters since the formation of the lake. The DIC in the brine layer in Lake Sophia contained bomb ^{14}C which they interpreted as seawater infiltrating through the permafrost.

Deevey and Stuiver (1964) used ^{14}C balances to calculate the rate of groundwater inflow into the deep layers of lakes in New England. The residence time of water in their lake was very short, and the depletion of ^{14}C found in the deep layer during winter when the lake was ice-covered was related to inflowing groundwater. It has also been found that the ^{14}C of plant matter in lakes is dependent on the source of water the plant uses and the source of carbonate in that water (Deevey et al., 1954; Deevey and Stuiver, 1964). Plants growing in sections of the lake where pore water was primarily derived from inflowing groundwater with low ^{14}C reflected this fact in the ^{14}C of their tissue.

5.2.5. Chlorine-36

The application of $^{36}Cl/Cl$ ratios to determine the source of Cl^- to ice-covered closed-basin lakes in Antarctica was reported by Lyons and others (1998). The three lakes studied were Lake Bonney, Lake Fryxell, and Lake Hoare in the McMurdo Dry Valley, southern Victoria Land, Antarctica. The authors speculated that the source of solutes in the hypolimnia of these lakes could be from: (1) volcanic sources, (2) relict seawater, and/or (3) evapotranspiration of glacier melt during a different climatic regime than presently exists in Antarctica.

The three lakes lie within the same geologic domain and climate regime but their geochemistry is quite different, with the hypolimnions having varying salinities and major-ion ratios. The researchers found that the bottom waters of Lake Bonney had low $^{36}Cl/Cl$ ratios ($<6 \times 10^{-15}$) that were probably derived from a mixture of seawater (3×10^{-15}) and a nearby saline seep (9×10^{-15}). For the bottom and surface water of Lake Hoare, the least saline of the three lakes, the $^{36}Cl/Cl$ ratios were similar to those found in chloride salts in the local soils and meteoric sources (220×10^{-15} and 440×10^{-15}), respectively. The water from Lake Fryxell had an intermediate ratio between that of the other two lakes that suggest a mixture of meteoric and "older" chloride as an input.

The authors found that these three lakes, within 15 km of each other and with the same general watershed geology and climatology, could have such different $^{36}Cl/Cl$ profiles. They concluded that the past and present sources of chloride to Lakes Bonney and Fryxell are different. For Lake Bonney, the major source of chloride is from a subsurface reservoir, probably marine salts, while the source for chloride to Lake Fryxell was diffusion from the sediments on the lake's bottom or from groundwater seepage. In contrast, the chloride source for Lake Hoare appeared to be solely from surface stream runoff and glacial melt.

5.2.6. U and Th series radionuclides: physical and chemical tracers of water in lakes and inland seas

Uranium and Thorium series radionuclides have been used to estimate processes occurring in lakes and inland seas, such as mixing, groundwater inflow, and diffusion. The Dead Sea is a saline, density-stratified meromictic lake that has overturned on an intermittent basis recently and is fed by the Jordan River and saline springs. Water export is by evaporation only. Radium-226 has been used by several groups to study the dynamics of the water body. Stiller and Chung (1984) calculated a meromixic stability of approximately 250 years based on the differences of radium activity in the upper, intermediate, and lower water bodies, assuming the radium inventory of the lake is at steady state and the existence of a uniform radium profile top to bottom before onset of meromixis, and that radium enters only into the mixolimnion (upper layer) while the monimolimnion (deep layer) is isolated from any radium addition or removal other than by radioactive decay. Chan and Chung (1987) refined the model of the Dead Sea by adding a factor for particle scavenging of radium, which had not been included in Stiller and Chungs's model. With the model adjusted for scavenging, the age of the meromixic period is reduced to approximately 100 years.

Somayajulu and Rengarajan (1987) used the radium distribution in the Dead Sea to calculate vertical eddy diffusion coefficients at two profiles within the lake. Their approach was to examine ^{228}Ra and ^{226}Ra activities in both the upper and lower layers of the stratified lake at the two locations. Below the pycnocline they found an increasing gradient downward of ^{228}Ra along with an increasing $^{228}Ra/^{226}Ra$ activity ratio. They reasoned that the gradient could be attributed to vertical eddy diffusion from bottom sediments and upward diffusion in the water column. They estimated the vertical eddy diffusion of the two locations to be 0.4 and 2.0 $cm^2\,s^{-1}$.

Chung and Craig (2004) refined the stratification model of the Dead Sea further by developing a transient box model that assumed a closed system for the lower water body with varying scavenging rates for Ra (and Ba) in order to evaluate the isolation time (time since previous overturn) of the lower water body. They also constructed an open system model (mixing of the lower water body with the upper water body) that included particle scavenging. The transient/closed system model showed the lower water body isolation time to be a maximum of 317 years. The open system model at steady-state conditions gave a maximum isolation time of 339 years.

The Black Sea is another water body that has received attention in terms of radium studies. Falkner et al. (1991), O'Neill et al. (1992), Moore and Falkner (1999) and Moore and O'Neill (1991) have noted a decrease in ^{226}Ra activity in the surface water of the Black Sea between 1957 and 1967 as measured by Grashchenko et al. (1960) and Moore (1969). Between 1967 and 1988, the ^{226}Ra activity had decreased markedly as compared to the earlier activity (Falkner et al., 1991; O'Neill et al., 1992) from about 21 dpm 100 L^{-1} to about 6 dpm 100 L^{-1} (Falkner et al., 1991). Moore and O'Neill (1991) calculated that this decrease could be accounted for by the enhanced radium activity in the recently settled "fluff layer" (FL) of high porosity sediment found at the top of the sediment column, implying recent (within 5 years) removal of the missing water-column radium to the solid phase, based on similar $^{228}Ra/^{226}Ra$ activity ratios of the sediment FL and overlying water. Moore and O'Neill (1991) previously proposed that, based on the $^{228}Ra/^{226}Ra$ activity ratio in the FL, this layer was on the order of 5 years old or less, in that this ratio was similar to the activity ratio in the overlying surface waters. Using a suite of nuclides (^{226}Ra, ^{228}Ra, ^{210}Pb, ^{134}Cs, ^{137}Cs) and the occurrence of the Chernobyl nuclear reactor accident in 1986, they inferred that the FL formed between May 1986 and July 1988, the result of an unusual biological productivity event or events in the water column. Moore and Falkner (1999) reported that since about 1987 the activity of radium in Black Sea waters has been rising slightly, although still well below 1957 levels, from a low of about 6 dpm 100 L^{-1} in 1987 to a high in 1995 of about 10 dpm 100 L^{-1}. The decrease in radium activity from 1967 to 1987 was hypothesized as due mostly to enhanced biological primary productivity with associated formation of ^{226}Ra-bearing barite as a result of recent anthropogenic increases of nutrients to the surface waters draining into the Black Sea (Moore and Falkner, 1999).

In a freshwater environment, Krest and Harvey (2003) used radium isotopes to quantify groundwater exchange with the overlying Everglades (FL, USA). They developed a model to account for water flow between an underlying aquifer, through a peat layer, to the overlying surface water environment that made use of gradients in the dissolved ^{223}Ra and ^{224}Ra isotopes found in the intervening peat layer and the aquifer below. By examining the shape of the ^{223}Ra and ^{224}Ra activity curves in the peat layer and knowing the activities in the water above and below the layer (the aquifer had abundant ^{223}Ra and ^{224}Ra activity due to the production by alpha particle recoil from the clastic material) and values of K_d (distribution coefficient), they modeled vertical advective velocities of water through the peat layer to be on the order of 1 cm d^{-1}.

Kraemer (2005) examined radium isotopes (^{226}Ra, ^{228}Ra, ^{223}Ra, ^{224}Ra) in freshwater Cayuga Lake in New York, USA (one of the eleven "Finger Lakes" of Central New York), and found them to be useful in identifying areas of surface and groundwater flow into the lake during and after a spring inflow event. The ^{228}Ra/^{226}Ra activity ratio was used to infer a groundwater input at the southern end of the lake during and just after the spring inflow event caused by snow melt and precipitation (Figure 2.4). The bulk (old) lake water had a ^{228}Ra/^{226}Ra activity ratio of approximately

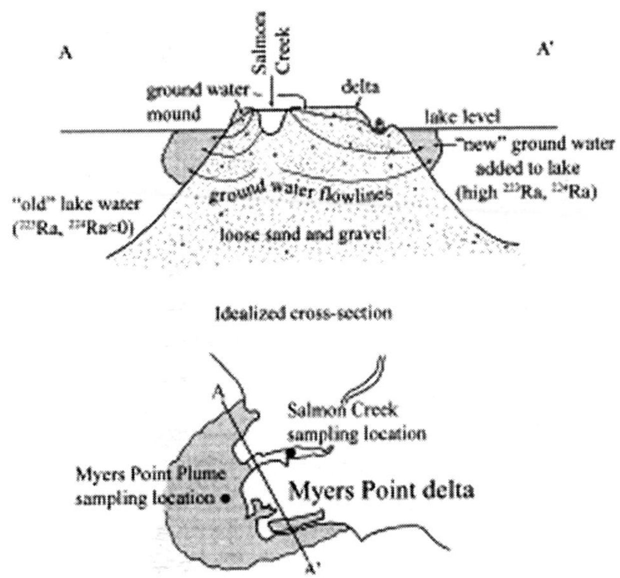

Figure 2.4 Cross-section of the Meyers Point delta showing the mounding of groundwater during the high waters of a vertical inflow. Groundwater and flow from the channel of Salmon Creek infiltrates through the sand and gravel of the delta, gaining short-lived ^{223}Ra and ^{224}Ra from the solid phase, which then flows into the lake below the surface as groundwater (from Kraemer, 2005).

0.7–0.8, while the "anomalous" water mass had a ratio nearer to 0.5. Because there was no significant surface water inflow with this lower ratio, it was concluded that the source was groundwater entering the lake. The $^{223}Ra/^{226}Ra$ and $^{224}Ra/^{228}Ra$ activity ratios of water in this and in a second recently introduced water mass were higher than in the bulk lake water, whose "age" was greater than the half-lives of the short-lived isotopes. These data were used to determine that the masses were behaving as discrete pods within the lake over a period of several weeks, although some mixing of bulk lake water with at least one of the water masses did occur. Additionally, suspended sediment was found to be capable of supporting the small amount of ^{223}Ra and ^{224}Ra present in the bulk lake water. Diffusion from the bed sediments was thought to be required to explain the ^{228}Ra levels in the bulk lake water, considering that decay of ^{228}Ra (half-life of 5.7 years) should have resulted in much lower $^{228}Ra/^{226}Ra$ activity ratios than those measured.

Brennan et al. (2004) examined the behavior of ^{226}Ra entering Florida lakes with groundwater and concluded that the radium is largely absorbed onto particulate matter, which then settles to the bottom. This has significant implications for ^{210}Pb dating (not covered in this chapter) in that a basic assumption in the method is that any radium determined in a sediment sample is assumed to be in equilibrium with its daughter ^{210}Pb and an equal activity of the ^{226}Ra is subtracted from the total ^{210}Pb activity found for the sample. The remaining ^{210}Pb activity after this subtraction is termed "excess" ^{210}Pb and is used to determine sedimentation rates. However, Brenner et al. (2004) contend that the radium entering the lake with groundwater and adsorbing onto particulate matter does not have enough time to generate an equilibrium amount of ^{210}Pb while it is in the upper layers of the sediment. Hence, in cases where a lake is receiving a large amount of radium from groundwater sources, the correction to calculate "excess" ^{210}Pb will not produce valid ^{210}Pb excess curves for the sediment.

Uranium isotopes have been used to infer the source of water within lakes and determine when water entered the lake. Blake et al. (1998) used uranium in the Hawes Water area of north-west England to infer that the water supply to the lake must come from surface runoff and shallow groundwater (springs) rather than deep groundwater, the former two sources having a $^{234}U/^{238}U$ activity ratio greater than 1.00, similar to water in the lake, while deep groundwater, as revealed from borehole samples, had a significantly lower ratio. Even though there was a small seasonal variability to the $^{234}U/^{238}U$ activity ratio in the surface water, the ratio in the lake remained largely constant due to the residence time of the lake (0.4 years) damping or integrating the variation in the lake water body. They also noted that the $^{234}U/^{238}U$ activity ratio of bottom sediments preserves a reasonable record of lake disequilibrium which may prove useful as monitors of long-term changes in catchment hydrology.

Similar findings have been reported for the much larger Lake Baikal. A series of papers has been published focusing on the unusual nature of the uranium isotopic activity ratios of the lake and the Selenga River, the major tributary, and its application as a climate change indicator (Pampura et al., 1996; Edgington et al., 1996; Chebykin et al., 2004). Waters draining the local Baikal drainage area have $^{234}U/^{238}U$ activity ratios of 1.6 or below, for the most part, but the Selenga River, which rises in Khazakstan and Baikal, and the lake itself presently have UAR values of about 2.0. This unusually high ratio is also reflected in the uranium of the surficial bottom sediments. Deeper in the sediment column the UAR values decrease from 2.0 to about 1.0 in correlation with other paleoclimate change indicators such as diatoms, and the pattern is strikingly similar to the deep-sea oxygen isotope record. The uranium isotopic pattern mirrored the timescales of the deep-sea record when estimates of sedimentation and sediment disposition age rate were made using the uranium and thorium isotopes of the sediments.

Chalov et al. (1964) determined the $^{234}U/^{238}U$ activity ratio (UAR) of several rivers and lakes in the Issyk-Kul basin for the purpose of determining the age of water in the lakes of this closed-basin region. By measuring the UAR values of rivers flowing into the lakes (Issyk-Kul and Chatyr-Kul) and measuring the values of the uranium in the lake waters themselves, they estimated the age of the lake (actually the number of years the lakes had been receiving water) using the equation:

$$\left(\frac{^{234}U}{^{238}U}\right)_t - 1 = \left[\left(\frac{^{234}U}{^{238}U}\right)_o - 1\right]\left(\frac{1 - e^{-\lambda t}}{\lambda t}\right) \tag{2.3}$$

For Issyk-Kul, the calculated ages were 110–300 × 10^3 years, which was in agreement with geological evidence. The variation represented uncertainty as to how long a major river had been feeding the lake. For Chatyr-Kul, where large differences in UAR existed between present water entering the lake and the lake itself, the age of the lake was estimated at 1.7×10^6 years, again in agreement with other evidence. These calculations implicitly assume constant addition of ^{234}U and ^{238}U over time and no removal of ^{234}U (or ^{238}U) from water except by radioactive decay.

5.2.7. Radionuclides tracing chemical fluxes and contaminants in lakes

Many radionuclides can be used to trace the flux of chemicals through the biogeochemical system of a lake. These tracers can originate as cosmogenic nuclides ($^{32,33}P$, ^{35}S, ^{7}Be) or be derived from lithogenic sources such as thorium. These nuclides follow the pathway of the stable element or similar chemical elements in lake systems. They can be particularly important in the cycling of nutrients and the scavenging by biogeochemical processes. The radioisotopes of phosphorus (^{32}P = 14.3d, ^{33}P = 25.3d) have been used in oceanography to determine phosphorus cycling in the surface waters of the ocean (Benitez-Nelson and Karl, 2002). Because of the different half-lives, the activity ratio of $^{33}P/^{32}P$ will increase with increasing age. Thus, the relative ages of phosphorus within any given pool is given by the ratio of the two isotopes. Benitez-Nelson and Karl (2002) found that the age of total dissolved phosphate increased as primary productivity increased. They attributed this to the fact that, in areas of high productivity, biologically active phosphorus was removed, leaving a pool of older and less-reactive phosphate in the water column. Though these types of studies have only been carried out in the ocean, they clearly have applications to phosphate dynamics in lake systems. Natural ^{7}Be was used as a surrogate for metals to study scavenging processes at Lake Lugano in Switzerland (Steinmann et al., 1999). They determined that colloid sorption could be an effective mechanism for removal of metals from lake waters.

Sulfur-35 is a radioisotope of sulfur ($t_{1/2}$ = 87 days) that has been found to be a useful isotope in the study of the hydrology and sulfur cycling in small lakes. Michel et al. (2002) used ^{35}S to study sulfur fluxes in lakes in the Flattops Wilderness of Colorado. The main hydrologic event for these lakes is snowmelt, which usually occurs by June. During snowmelt, a significant fraction of the water in the lakes is probably replaced, and with it, the sulfate. Assuming that, initially, very little sulfate is lost to biogeochemical reactions and that the ^{35}S concentration in melting snow is known, the ^{35}S in the lakes in early summer represent the fraction of water that is replaced in the lake that year. Even if some sulfate is lost via biological uptake, the ^{35}S can give a minimum value for the fraction of the lake water replaced each year. Using ^{35}S as a surrogate for water, Michel et al. (2002) calculated that more than half of the lake water was replaced by meltwater each year.

Measurements throughout the summer for 3 years at these lakes showed that ^{35}S concentrations decreased more rapidly than would be expected if only decay affected concentrations (Figure 2.5). In the fall of each year, ^{35}S concentrations then suddenly increased in both lakes. Sulfate concentrations and sulfur isotope ratios changed from year to year, but were unchanged throughout the summer–fall sampling period of any given year. Dilution by groundwater inflow was ruled out due to the lack of change in other chemical constituents (Campbell et al., 1991). The decrease of ^{35}S in summer was presumed to be due to exchange of sulfur between the sulfate in the water column and the biological sulfur pool. The exchange, which would not effect the sulfate concentrations or the sulfur stable isotope ratios, would result in the loss of radioactive sulfur and its replacement with "dead" sulfur. The cause of the concentration increase in fall was possibly caused by the

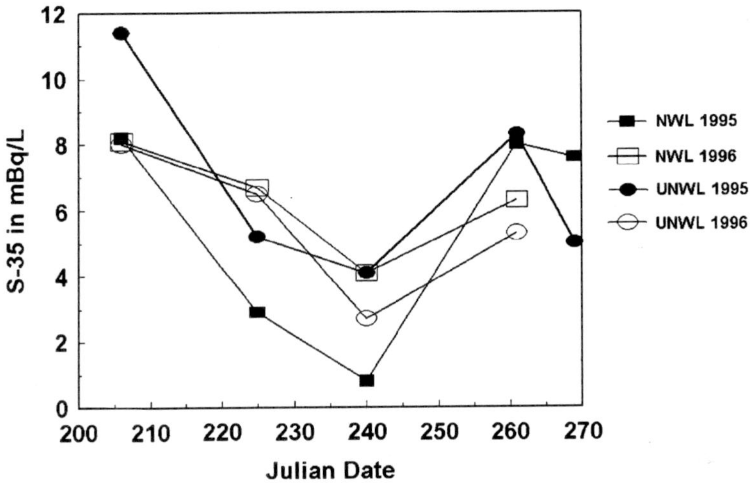

Figure 2.5 Ratio of measured versus expected ^{35}S concentrations for two small lakes in the Flattops Wilderness Area (from Michel et al., 2002). If decay were the only component influencing ^{35}S concentrations, all ratios would be one throughout the year. In most years, there is a decrease in the ratio during the summer which is likely due to exchange of sulfate in water with sulfate in the biological component, which will be mostly devoid of ^{35}S. The cycling rate is calculated to be about 0.2–0.3 eq L d^{-1} of sulfate. Increases in the fall are likely due to increased precipitation washing ^{35}S from the watershed into the lake.

occurrence of fall rains that washed ^{35}S that had deposited in the watershed as dryfall during the summer into the lakes. In larger lakes where timescales for water exchange are relatively large, ^{35}S concentrations are generally very low. Thus, this isotope appears to be best suited for small hydrologic systems where sulfate cycling is on the same timescale as the hydrologic processes (Cooper et al., 1991).

Most studies of contaminant concentrations of radioactivity in lakes focus on potential health hazards and the loads of radioactivity released to lakes. Studies can also focus on the partitioning of the radiochemicals in the system and what implications that can have for the biosphere (Joshi, 1991). However, it is also possible to use the introduction of contaminant levels of radioactivity into a lake to study lake dynamics. Lam and Durham (1984) applied a finite element approach to study the mixing dynamics of a contaminant released from a nuclear power station on the Lake Ontario. The cooling water from the plant resulted in a constant temperature plume. Tritium is occasionally released resulting in a temporary spike. They were able to model these two tracers to determine physical properties affecting contaminant dispersion into the lake waters.

Spiking lake water with tritium or other radionuclides is rare due to both legal and technical problems. Most studies using artificial radioactivity are carried out in microcosms and not within the lake itself. However, Quay et al. (1980) did inject tritium into two small experimental lakes in Canada to determine horizontal and vertical mixing rates. They determined that the upper layers mixed very rapidly, but very little mixing occurred below the thermocline. In another study in Canada, Hesslein et al. (1982) studied the cycling of organic carbon by releasing radiocarbon in a small lake.

REFERENCES

Benitez-Nelson, C. R., and D-M. Karl. (2002). Phosphorus cycling in the North Pacific Subtropical Gyre using cosmogenic ^{32}P and ^{33}P. *Limnology and Oceanography*, 47, 762–770.

Blake, W. H., A. S. Plater, and J. F. Boyle. (1998). Seasonal trends in the uranium-series isotopic signatures of lake waters and sediments: Hawes Water, north-west England. *Journal of Paleolimnology*, 20, 1–14.

Bohlke, J. K., and J. L. Denton. (1995). Combined use of groundwater dating, chemical, and isotopic analyses to resolve the history and fate of nitrate contamination in two agricultural watersheds, Atlantic coastal plain, Maryland. *Water Resources Research*, 31, 2319–2339.

Brenner, M., C. L. Schelske, and W. F. Henney. (2004). Inputs of dissolved and particulate ^{226}Ra to lakes and implications for ^{210}Pb dating recent sediments. *Journal of Paleolimnology*, 32, 53–66.

Broecker, W. S., and A. F. Walton. (1959). The geochemistry of ^{14}C in freshwater systems. *Geochimica et Cosmochimica Acta*, 16, 15–38.

Campbell, D. C., J. T. Turk, and N. W. Spahr. (1991). Response of Ned Wilson Lake Watershed, Colorado, to changes in atmospheric deposition of sulfate. *Water Resources Research*, 27, 2047–2060.

Carmack, E. C., and P. D. Killworth. (1978). Formation and interleaving of abyssal water masses off Wilkes Land, Antarctica. *Deep Sea Research*, 25, 357–369.

Chan, L. H., and Y. Chung. (1987). Barium and radium in the Dead Sea. *Earth and Planetary Science Letters*, 85, 41–53.

Chalov, P. I., T. V. Tuzova, Ya. A. Musin. (1964). The isotope ratio $^{234}U/^{238}U$ in natural water and its use in nuclear geochronology. Bulletin, Academy of Sciences, USSR, Geophysics Series No. 10 (English Translation) Izv. Geophysics Series, 10, pp. 1552–1561.

Chebykin, E. P., D. N. Edgington, E. L. Goldberg, M. A. Phedorin, N. S. Kulikova, T. O. Zheleznyakova, S. S. Vorobyova, O. M. Khylstov, O. V. Levina, G. A. Ziborova, and M. A. Grachev. (2004). Uranium-series isotopes as proxies of late Pleistocene climate and geochronometers in bottom sediments of Lake Baikal. *Russian Geology and Geophysics*, 45, 539–556.

Cheng, S. (1996). Isotopic tracing of the source water for Cedar Bog in west-central Ohio, USA. *Journal of Hydrology*, 186, 31–42.

Chung, Y., and H. Craig. (2004). Radium redux in the Dead Sea: Profiles and transient Ra/Ba models. *Earth and Planetary Science Letters*, 218, 291–299.

Clark, J. F., and G. B. Hudson. (2001). Quantifying the flux of hydrothermal fluids into Mono Lake by use of helium isotopes. *Limnology and Oceanography*, 46, 189–196.

Cooper, L. W., C. R. Olsen, D. K. Solomon, L. I. Larsen, R. B. Cook, and J. M. Grebmeier. (1991). Stable isotopes of oxygen and natural and fallout radionuclides used for tracing runoff during snowmelt in an Arctic watershed. *Water Resources Research*, 27, 2171–2180.

Deacon, G. E. R. (1937). The hydrology of the Southern Ocean. *Discovery Reports*, 15, 124.

Deevey, E. S., M. S. Gross, E. Hutchinson, and H. L. Kraybill. (1954). The natural ^{14}C contents of materials from hard-water lakes. *Proceedings of the National Academy of Sciences*, 40, 285–288.

Deevey, E. S., and M. Stuiver. (1964). Distribution of natural isotopes of carbon in Lindley Pond and other New England lakes. *Limnology and Oceanography*, 9, 1–11.

Edgington, D. N., J. A. Robbins, S. M. Coleman, K. A. Orlandini, and M. P. Gustin. (1996). Uranium-series disequilibrium, sedimentation, diatom frustules, and paleoclimate change in Lake Baikal. *Earth and Planetary Science Letters*, 142, 29–42.

Falkner, K. K., D. J. O'Neill, J. F. Todd, W. S. Moore, and J. M. Edmond. (1991). Depletion of barium and radium-226 in Black Sea surface waters over the past thirty years. *Nature*, 350, 491–494.

Fischer, K. W., W. Weiss, O. Siebeck, and R. Muller. (1985). Implications of slow deep water renewal of Lake Konigsee derived from combined tritium-helium-3 measurements. *Archiv fuer Hydrobiologie*, 102, 409–414.

Foster, T. D. (1995). Abyssal water mass formation off the eastern Wilkes Land coast of Antarctica. *Deep Sea Research*, 42, 501–522.

Foster, T. D., A. Foldvik, and J. H. Middleton. (1987). Mixing and bottom water formation in the shelf break region of the southern Weddell Sea. *Deep Sea Research*, 34, 1771–1794.

Froehlich, K., D. Imboden, R. Kipfer, and K. Rozanski. (1998). Dynamics of the Caspian Sea – Preliminary results of isotope studies. In *Isotope Techniques in the Study of Environmental Change*. Proceeding Symposium on 14–18 April 1997, IAEA, Vienna, pp. 249–263.

Froehlich, K., K. Rozanski, P. Povinec, B. Oregioni, and J. Gastaud. (1999). Isotope studies in the Caspian Sea. *Science of the Total Environment*, 238, 419–427.

Gamlen, P. H., B. C. Lane, P. M. Midgley, and J. M. Steed. (1986). The production and release of CCl_3F and CCL_2F_2 (Chlorofluorocarbons CFC11 and CFC12). *Atmospheric Environment*, 20, 1077–1085.

Gat, J. R. (1981). Lakes. In: *Stable Isotope Hydrology: Deuterium and Oxygen-18 in the Water Cycle*. Technical Report Series No. 210, International Atomic Energy Agency, Vienna, Austria, pp. 203–221.

Grashchenko, S. M., D. S. Nikolaev, L. B. Kolyadin, Y. V. Kuznetsov, and K. F. Lazerev. (1960). *Doklady Akademii Nauk SSSR*, 132, 1171–1172.

Herczeg, A. L., and D. M. Imboden. (1988). Tritium hydrologic studies in four closed-basin lakes in the Great Basins, USA. *Limnology and Oceanography*, 33, 157–173.

Hesslein, R. H., W. S. Broecker, P. D. Quay, and D. W. Schindler. (1982). Whole lake radiocarbon experiment in an ogilotrophic lake at the Experimental Lakes Area, northwestern Ontario. *Canadian Journal of Fisheries and Aquatic Sciences*, 37, 454–463.

Hofer, M., F. Peeters, W. Aeschbach-Hertig, M. Brennwald, J. Holocher, D. M. Livingstone, V. Romanovski, and R. Kipfer. (2002). Rapid deep-water renewal in Lake Issyk-Kul (Kyrgyzstan) indicated by transient tracers. *Limnology and Oceanography*, 47, 1210–1216.

Hohmann, R., M. Hofer, R. Kipfer, F. Peeters, D. M. Imboden, H. Baur, and M. N. Shimaraev. (1998). Distribution of helium and tritium in Lake Baikal. *Journal of Geophysical Research*, 103, 12823–12838.

Imboden, D. M. (1979). Complete mixing in Lake Tahoe, California-Nevada traced by tritium. In: *Applications of Nuclear Techniques to the Study of Lake Dynamics.* International Atomic Energy Agency, Vienna, Austria, pp. 209–212.

Imboden, D. M., R. F. Weiss, H. Craig, R. L. Michel, and C. R. Goldman. (1977). Lake Tahoe geochemical study: 1. Lake chemistry and tritium mixing study. *Limnology and Oceanography*, 22, 1039–1051.

International Atomic Energy Agency. (1992). *Statistical Treatment of Data on Environmental Isotopes in Precipitation.* Technical Report Series No. 331, International Atomic Energy Agency, Vienna, Austria, 781p.

International Atomic Energy Agency. (2001). Isotope hydrology information system: The ISOHIS database. Available at: http://isohis.iaea.org

Jenkins, W. F., M. A. Beg, W. B. Clarke, P. J. Wangersky, and H. Craig. (1972). Excess ^{3}He in the Atlantic Ocean. *Earth and Planetary Science Letters*, 16, 122–126.

Jones, L. M., and H. G. Ostlund. (1971). Carbon-14 age and tritium content of Lake Vanda, Wright Valley. *Antarctic Journal of the United States*, 6, 200–201.

Joshi, S. R. (1991). Radioactivity in the Great Lakes. *Science of the Total Environment*, 100, 61–104.

Kipfer, R., W. Aeschbach-Hertig, H. Baur, D. M. Imboden, and P. Signer. (1994). Injection of mantle type helium into Lake Van (Turkey): The clue for quantifying deep water renewal. *Earth and Planetary Science Letters*, 125, 357–370.

Kipfer, R., W. Aeschbach-Hertig, F. Peeters, and M. Stute. (2002). Noble gases in lakes and groundwaters. In: *Noble Gases in Geochemistry and Cosmochemistry* (Eds D. Porcelli, C. J. Ballentine, and R. Wieler), *Reviews in Mineralogy and Geochemistry*, 47, 615–700.

Krabbenhoft, D. P., C. J. Bowser, M. P. Anderson, and J. W. Valley. (1990). Estimating groundwater exchange with lakes. 1. The stable isotope mass balance method. *Water Resources Research*, 26, 2445–2453.

Kraemer, T. (2005). Radium isotopes in Cayuga Lake, New York: Indicators of inflow and mixing processes. *Limnology and Oceanography*, 50, 158–168.

Krest, J., and J. Harvey. (2003). Using natural distributions of short-lived radium isotopes to quantify groundwater discharge and recharge. *Limnology and Oceanography*, 48, 290–298.

Kumar, U. S., N. Jacob, S. V. Navada, S. M. Rao, R. P. Nachiappan, B. Kumar, and J. S. R. Murthy. (2001). Environmental isotope study on hydrodynamics of Lake Naini, Uttar Pradesh, India. *Hydrological Processes*, 15, 425–439.

Lal, D., and B. Peters. (1967). Cosmic ray produced radioactivity on the earth. In: *Encyclopedia of Physics* (Ed. S. Flügge). Springer, New York, pp. 407–434.

Lam, D. C. L., and R. W. Durham. (1984). Finite element analysis of a radioactive tritium patch and a waste heat plume observed near the Pickering Nuclear Power Generating Station, Lake Ontario. *Journal of Great Lakes Research*, 10, 59–67.

Lyons, W. B., K. A. Welch, and P. Sharma. (1998). Chlorine-36 in the waters of the McMurdo Dry Valley lakes, southern Victoria Land, Antarctica: Revisited. *Geochimica et Cosmochimica Acta*, 62, 185–192.

Maiss, M., and C. A. M. Brenninkmeijer. (1998). Atmospheric SF_6: Trends, sources and prospects. *Environmental Science and Technology*, 32, 3077–3086.

McPhaden, M. J., and R. A. Fine. (1988). A dynamical interpretation of the tritium maximum in the Central Equatorial Pacific. *Journal of Physical Oceanography*, 18, 1454–1457.

Michel, R. L. (1992). Residence times in river basins as determined by analysis of long-term tritium records. *Journal of Hydrology*, 130, 367–378.

Michel, R. L., and T. F. Kraemer. (1995). Use of isotopic data to estimate water residence times of the Finger Lakes, New York. *Journal of Hydrology*, 164, 1–18.

Michel, R. L., T. F. Kraemer, and L. D. Cecil. (2009). Use of radioisotopes in the study of timescales and processes for surface waters and the unsaturated zone (this chapter).

Michel, R. L., and H. E. Suess. (1978). Tritium in the Caspian Sea. *Earth and Planetary Science Letters*, 39, 309–312.

Michel, R. L., J. T. Turk, D. H. Campbell, and M. A. Mast. (2002). Use of natural ^{35}S to trace sulphate cycling in small lakes, Flattops Wilderness Area, Colorado, USA. *Water, Air, and Soil Pollution: Focus*, 2, 5–18.

Moore, W. S. (1969). Oceanic concentration of ^{228}Ra. *Earth and Planetary Science Letters*, 6, 437–446.

Moore, W. S., and K. K. Falkner. (1999). Cycling of radium and barium in the Black Sea. *Journal of Environmental Radioactivity*, 43, 247–254.

Moore, W. S., and D. J. O'Neill. (1991). Radionuclide distributions in recent Black Sea sediments. In: *Black Sea Oceanography* (Eds E. Izdar and J. Murray). Kluwer, London, pp. 257–270.

O'Neill, D. J., J. F. Todd, and W. S. Moore. (1992). ^{226}Ra in the Black Sea and the Sea of Marmara. *Earth and Planetary Science Letters*, 110, 7–21.

Ostlund, H. G., and R. A. Fine. (1979). Oceanic mixing and transport of tritium. In: *Behaviour of Tritium in the Environment*. International Atomic Energy Agency, Vienna, Austria, pp. 303–314.

Paerl, H. W., R. C. Richards, R. L. Leonard, and C. R. Goldman. (1975). Seasonal nitrate cycling as evidence for complete vertical mixing in Lake Tahoe, California-Nevada. *Limnology and Oceanography*, 20, 1–8.

Page, P., M. Ouellet, C. Hillaire-Marcel, and M. Dickman. (1984). Isotopic analyses (^{18}O, ^{13}C, ^{14}C) of two meromictic lakes in the Canadian Arctic Archipelago. *Limnology and Oceanography*, 29, 564–573.

Pampura, V. D., I. V. Sandimirov, I. S. Lomonosov, A. Ye. Gapon, Ye. I. Ulybusheva, A. G. Arsenl'yeva, and G. I. Popovskaya. (1996). Geochemistry and formation of the isotopic composition of U and Th in Lake Baykal basin surface waters. *Geochemistry International*, 33, 90–103.

Peeters, F., R. Kipfer, D. Achermann, M. Hofer, W. Aeschbach-Hertig, U. Beyerle, D. M. Imboden, K. Rozanski, and K. Froehlich. (2000). Analysis of deep-water exchange in the Caspian Sea based on environmental tracers. *Deep Sea Research*, 47, 621–654.

Plummer, L. N., and E. Busenberg. (2000). Chlorofluorcarbons: Tools for dating and tracing young groundwater. In: *Subsurface Hydrology* (Eds P. Cook and A. Herczeg). Kluwer Academic Publishers, Boston, MA, pp. 441–478.

Quay, P. D., W. S. Broecker, R. H. Hesslein, and D. W. Schindler. (1980). Vertical diffusion rated determined by tritium tracer experiments in the thermocline and hypolimnion of two lakes. *Limnology and Oceanography*, 25, 201–218.

Rank, D., V. Ranjner, W. Papesch, J. Dreher, and J. Reitiger. (1983). Study of the interrelationships between groundwater and lake water at Neusiedlersee, Austria. In: *Isotope Hydrology*. International Atomic Energy Agency, Vienna, Austria, pp. 67–81.

Revelle, R., and H. E. Suess. (1957). Carbon dioxide exchange between atmosphere and ocean and the question of an increase of atmospheric CO_2 during the past decades. *Tellus*, 9, 18–27.

Roether, W. (1967). Estimating the tritium to ground water from wine samples: Groundwater and direct runoff contribution to central European surface waters. In: *Isotope Techniques in Groundwater Hydrology*. International Atomic Energy Agency, Vienna, Austria, pp. 73–91.

Romanov, V. V. (1974). Use of natural tritium for studying mixing of sea and river water. *Water Resources*, 1, 85–98.

Rozanski, K. (1979). Krypton-85 in the atmosphere 1950-1977, a data review. *Environment International*, 2, 139–143.

Scripps Institution of Oceanography. (1975). *Lake Tanganyika Geochemical and Hydrographic Study: 1973 Expedition*. SIO Reference Series 75-5, Scripps Institution of Oceanography, La Jolla, CA, 83p.
Somayajulu, B. L. K., and R. Rengarajan. (1987). ^{228}Ra in the Dead Sea. *Earth and Planetary Science Letters*, 85, 54–58.
Steinhorn, I. (1985). The disappearance of the long term meromictic stratification of the Dead Sea. *Limnology and Oceanography*, 30, 451–472.
Steinmann, P., T. Billen, J. L. Loizeau, and J. Dominik. (1999). Beryllium-7 as a tracer to study mechanisms and rates of metal scavenging from lake surface waters. *Geochimica et Cosmochimica Acta*, 63, 1621–1633.
Stiller, M., and Y. C. Chung. (1984). Radium in the Dead Sea: A possible tracer for the duration of meromixis. *Limnology and Oceanography*, 29, 574–586.
Torgersen, T., D. E. Hammond, W. B. Clarke, and T. H. Peng. (1981). Fayetteville, Green Lake, New York: ^{3}H-^{3}He water mass ages and secondary chemical structure. *Limnology and Oceanography*, 26, 110–122.
Torgersen, T., Z. Top, W. B. Clarke, W. J. Jenkins, and W. S. Broecker. (1977). A new method for physical limnology – Tritium-helium-3 ages – results for Lakes Erie, Huron and Ontario. *Limnology and Oceanography*, 22, 181–193.
Verbolov, V. I., and M. N. Shimaraev. (1973). Water exchange in Lake Baikal. *Hydrology of Lakes*, 109, 65–71, (IAHS-AISH Publication).
Volchok, H. L., V. T. Bowen, W. R. Clark, and L. R. Ball. (1983). Crater Lake revisited: concentration changes in water column radionuclides, 1967–1981. *Journal of Geophysical Research*, 88, 4531–4533.
Volchok, H. L., M. Feiner, H. J. Simpson, W. S. Broecker, V. E. Noshkin, V. T. Bowen, and E. Willis. (1970). Ocean fallout: The Crater Lake experiment. *Journal of Geophysical Research*, 75, 1084–1091.
Warner, M. J., and R. F. Weiss. (1985). Solubilities of chlorofluorocarbons 11 and 12 in water and seawater. *Deep Sea Research*, 32, 1485–1497.
Weiss, R. F., E. C. Carmack, and V. M. Korpalov. (1991). Deep-water renewal and biological production in Lake Baikal. *Nature*, 349, 665–669.
Weiss, W., and W. Roether. (1975). Discharge of tritium in the Rhine, 1961–1973. *Deutsche Gewässerkundliche Mitteilungen*, 19, 1–5.
Wilson, A. T. (1964). Evidence from chemical diffusion of a climatic change in the McMurdo Dry Valleys 1200 years ago. *Nature*, 201, 176–177.

5.3. Dynamics and Pollution of Groundwater

Klaus Froehlich

Viktor-Wittner-Gasse 36/7, 1220 Vienna, Austria

5.3.1. Introduction

Groundwater is known to be subsurface water stored in permeable geological units (aquifers) and moved by gravity. More than 1.5 billion urban dwellers use groundwater as the only source of their water supply.

The quantity and quality of groundwater resources are threatened by accelerating human intervention and the impact of global warming. A sustainable groundwater resources development without such adverse effects requires a comprehensive understanding of groundwater systems.

Environmental isotope techniques are proven tools that can provide relevant information on the replenishment and flow regime of groundwater systems and their vulnerability against pollution (Aggarwal et al., 2007). Moreover, for the assessment of groundwater resources in semiarid and arid zones, the application of isotope techniques is cost-effective and indispensable for identification of nonrenewable resources. These techniques take advantage of natural variations in the isotopic composition of water and its solutes and are able to (1) identify sources of groundwater; (2) study surface–groundwater interactions and interconnections between different aquifers; (3) determine groundwater age/residence time distribution and, thus, derive quantitative information on the groundwater recharge and flow regime; (4) identify the sources and behavior of contaminants in groundwater.

The most frequently used stable and radioactive environmental nuclides include those of the elements of the water molecule, hydrogen (^{2}H-deuterium and ^{3}H-tritium) and oxygen (^{18}O), as well as of the elements carbon (^{13}C and ^{14}C), nitrogen (^{15}N), sulfur (^{34}S), chlorine (^{36}Cl, ^{37}Cl) and others, for example, uranium (^{234}U, ^{238}U), radium (^{223}Ra, ^{224}Ra, ^{226}Ra, ^{228}Ra), that occur in water as constituents of dissolved inorganic or organic compounds. Noble gases, such as the radiogenic helium isotopes (^{3}He, ^{4}He) and the radioactive noble gases radon (^{222}Rn) and krypton (^{81}Kr, ^{85}Kr) are also important groundwater tracers. Many of the environmental isotopes are measured as the atomic ratio between the given isotope and the most abundant isotope of the element in question, for example, ^{2}H/^{1}H, ^{3}H/^{1}H, ^{18}O/^{16}O, and ^{14}C/^{12}C, ^{13}C/^{12}C, ^{15}N/^{14}N, ^{234}U/^{238}U.

The following is an overview of the specific potential of selected environmental isotopes in groundwater studies and practical applications in assessing the quantity and quality of groundwater resources. As most practical studies of groundwater systems require the combined use of stable and radioactive isotopes, a brief summary of the fundamentals of stable isotopes will be included, although, in accordance with the scope of this book, the main emphasis is on radioactive isotopes. Comprehensive reviews on isotopes in groundwater studies are given, *inter alia*, by Cook and Herczeg, (1999) and Geyh (2000).

5.3.2. Stable isotopes

The slight mass difference between stable isotopes of chemical elements causes isotope fractionations, that is, changes of the isotopic composition of reactants containing the corresponding element. Isotope fractionations

Table 3.1 Stable environmental isotopes used in groundwater studies.

Isotope	Ratio	Average natural abundance	Commonly measured phases
^{2}H	$^{2}H/^{1}H$	1.55×10^{-4}	H_2O, CH_2O, CH_4, H_2, OH^- minerals
^{3}He	$^{3}He/^{4}He$	1.38×10^{-6}	He in water or gas, crustal fluids, rocks
^{6}Li	$^{6}Li/^{7}Li$	7.5×10^{-2}	Saline waters, rocks
^{11}B	$^{11}B/^{10}B$	0.80	Saline waters, sewage effluents clays, borate, rocks
^{13}C	$^{13}C/^{12}C$	1.11×10^{-2}	CO_2, carbonate, DIC, CH_4, organics
^{15}N	$^{15}N/^{14}N$	3.66×10^{-3}	N_2, NH_4^+, NO_3^-, N-organics
^{18}O	$^{18}O/^{16}O$	2.04×10^{-3}	H_2O, CH_2O, CO_2, sulfates, NO_3^-, carbonates, silicates, OH^- minerals
^{34}S	$^{34}S/^{32}S$	4.22×10^{-2}	Sulfates, sulfides, H_2S, S-organics
^{37}Cl	$^{37}Cl/^{35}Cl$	0.242	Saline waters, evaporites, solvents, rocks
^{81}Br	$^{81}Br/^{79}Br$	0.493	Saline waters, sewage effluents
^{87}Sr	$^{87}Sr/^{86}Sr$	0.709939	Water, carbonates, sulfates, feldspar

occur in natural processes such as phase changes (evaporation, condensation) and chemical/biological reactions (e.g., mass exchange, assimilation, denitrification). These processes result in an isotopic labeling of water and other compounds in the hydrological cycle (Mook, 2000). Among the stable isotopes used for hydrological studies (Table 3.1), the stable hydrogen and oxygen isotopes are of particular interest because they are isotopes of the elements of the water molecule (often called "water isotopes").

5.3.2.1. Stable hydrogen and oxygen isotopes

The hydrogen and oxygen isotope ratios ($^{2}H/^{1}H$, $^{18}O/^{16}O$) in precipitation are slightly higher than in the water vapor from which precipitation has been formed by condensation. The ratio $\alpha = R_W/R_V$ ($R = {}^{2}H/^{1}H$ or $^{18}O/^{16}O$) called the fractionation factor is a function of the condensation temperature (α increases with decreasing condensation temperature (Majoube, 1971)). The isotope measurements are commonly expressed as the relative deviation (δ) of the isotope ratio R_S measured in a given sample from the corresponding ratio in a standard R_{St}:

$$\delta = \frac{R_S - R_{St}}{R_{St}} = \frac{R_S}{R_{St}} - 1$$

The standard of the water isotopes is Vienna Standard Mean Ocean Water (VSMOW) with $(^{2}H/^{1}H)_{St} = (158 \pm 2) \times 10^{-6}$ and $(^{18}O/^{16}O)_{St} = (1993.4 \pm 2.5) \times 10^{-6}$. Using the δ notation, the isotope fractionation is expressed in terms of the isotope enrichment factor ε, which is defined by:

$$\varepsilon = \alpha - 1$$

At a condensation temperature of 18°C, the enrichment factor is $^{18}\varepsilon = 0.001(= 10‰)$ for oxygen-18 and $^{2}\varepsilon = 0.087(= 87‰)$ for deuterium. If water and vapor are in isotopic equilibrium, the relationship between their isotopic composition can be expressed by the isotopic ratios $R_w/R_V = \alpha$ or by the δ values $\delta_W - \delta_V = \varepsilon$. At the temperature of 18°C, the latter difference is 9.7‰ for oxygen-18 and 82‰ for deuterium. Evaporation of water is characterized by a slightly higher enrichment factor because of an additional effect (kinetic fractionation) caused by diffusive and turbulent transport of the evaporating vapor from the water surface to the free atmosphere (e.g., Froehlich, 2000).

Figure 3.1 illustrates the change of the oxygen-18 values due to rainout of atmospheric water vapor that was formed by evaporation from the ocean surface. The isotope fractionation between ocean water ($\delta^{18}O_{ocean} = 0‰$) and vapor is slightly higher (in this example a value of 14‰ is assumed) than that between condensing precipitation and remaining vapor in the clouds (about 10‰ at the assumed temperature of 18°C). Every rainout process gives rise to an incremental decrease of the $\delta^{18}O$ values in the remaining cloud, which is described by the Rayleigh relationship

$$\frac{d\,\delta^{18}O_V}{1 + \delta^{18}O_V} = {}^{18}\varepsilon_e \frac{dm_V}{m_V} \tag{3.1}$$

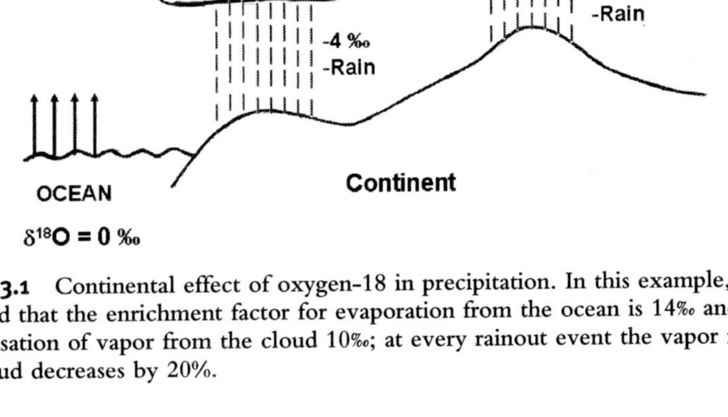

Figure 3.1 Continental effect of oxygen-18 in precipitation. In this example, it is assumed that the enrichment factor for evaporation from the ocean is 14‰ and for condensation of vapor from the cloud 10‰; at every rainout event the vapor mass of the cloud decreases by 20%.

where $^{18}\varepsilon_e$ is the (equilibrium) enrichment factor for condensation and m_V is the amount (moles) of the (remaining) vapor. In Figure 3.1, it is assumed that $dm_V/m_V = 0.2$ for every rainout event, that is, $d\delta^{18}O_V = 2‰$. The rain is in isotopic equilibrium with the cloud, that is, $\delta^{18}O_P = \delta^{18}O_V + {}^{18}\varepsilon_e$ (in this example $^{18}\varepsilon_e = 10‰$). Consequently, both water vapor and precipitation become gradually depleted in ^{18}O by rainout of the vapor mass during the movement from the ocean to the continent (continental effect). Since the deuterium isotopic composition also follows the Rayleigh relationship, the change of the deuterium values is proportional to the change of oxygen-18. The isotopic composition of precipitation has been measured in many stations distributed over the globe within the framework of the IAEA/WMO Global Network of Isotopes in Precipitation (GNIP) (IAEA/WMO, 2006).

The degree of condensation of a vapor mass depends on the temperature. This explains the relation existing between isotope composition of precipitation and the temperature of its formation (Dansgaard, 1964) – the δ-values of precipitation decrease with decreasing condensation temperature. This temperature effect is manifested in (1) seasonal variations of the isotope composition of precipitation (winter precipitation is depleted in heavy isotopes with respect to summer precipitation), and (2) a decrease of the isotope values with increasing latitude (latitude effect) and increasing altitude (altitude effect). The altitude effect allows delineation of recharge areas in mountainous regions (e.g., Ambach et al., 1968; Ramspacher et al., 1992) and the temperature effect can be used to identify palaeowater, that is, groundwater formed under past climatic conditions. Palaeowater is important for water supply in arid and semiarid regions.

Given the linearity of the oxygen and hydrogen isotope fractionation, the δ^2H and $\delta^{18}O$ values of precipitation fit straight lines with a slope close to 8 (Figure 3.2). The intercept with the δ^2H axis depends on the origin of the vapor: an intercept at +10‰ represents vapor that has been formed in oceanic regions with humidity of about 85% (Merlivat and Jouzel, 1979), higher intercepts indicating vapor sources from areas with lower humidity (e.g., eastern Mediterranean); smaller intercepts have been found for palaeowater, suggesting higher humidity under past climatic conditions (Sonntag et al., 1979). The isotope values of water bodies that are subject to evaporation (lakes, reservoirs) fit straight lines with slopes lower than 8. Lines connecting freshwater values with that of sea water represent mixing lines between these two components.

The applications of stable hydrogen and oxygen isotopes in groundwater studies make use of the variations of the isotopic composition in atmospheric precipitation, that is, in the input to the groundwater system. Mixing in the unsaturated zone and bias of seasonal recharge attenuate the seasonal isotopic variation (O'Driscoll et al., 2005), and the degree of

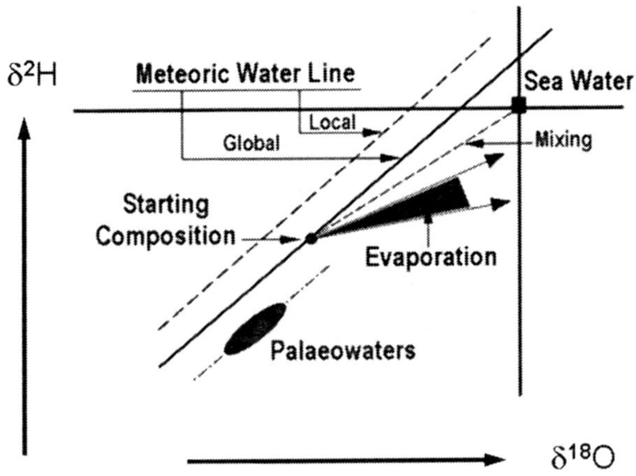

Figure 3.2 Schematic representation of δ^2H and δ^{18}O values of natural waters (Adapted from Aggarwal et al., 2007 with permission from Eolss).

attenuation can be used as an indicator of residence time of young groundwater. In the saturated zone, there is usually no further change in the isotopic composition of the groundwater. (Only in high-temperature geothermal systems, can exchange of oxygen between water of rocks modify the oxygen-18 composition of geothermal water.) The stable hydrogen and oxygen isotopic composition of groundwater is thus related to that of precipitation at the time of the recharge. As the isotopic composition of precipitation replenishing the groundwater depends on the climatic conditions prevailing during recharge, palaeowater serves as a climate archive (Edmunds, 2007). Such palaeoclimatic studies require the combined use of stable and radioactive isotopes; the latter provide the timescale for palaeoclimatic records (Figure 3.3).

Groundwater adjacent to surface water bodies (rivers or lakes) often represents a mixture between local precipitation and surface water (adjacent river or lake water). If the isotopic composition between local precipitation and the surface water is sufficiently different, the proportions of the two recharge components can be derived from the measured stable isotope values of the surface water, precipitation, and groundwater (Figure 3.4).

Until recently, it has been assumed that the abundance of the stable isotope ^{17}O provides no additional information to that of ^{18}O. But Angert et al. (2003) demonstrated that the relationship between ^{17}O and ^{18}O varies with the type of fractionation to which the water was exposed. Thus ^{17}O is expected to become a unique tracer that complements ^{18}O and deuterium in atmospheric (precipitation) and climatic (e.g., polar ice) studies. Consequently, groundwater studies may also benefit from ^{17}O as a tracer, for example, in identification of groundwater sources (palaeowater).

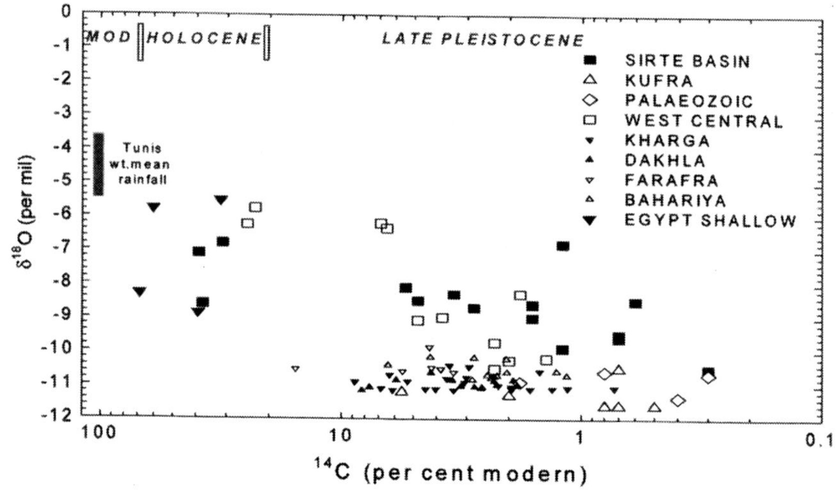

Figure 3.3 Plot of $\delta^{18}O$ against ^{14}C of groundwater from Libya and Egypt. The ^{14}C values indicate an age range from late Pleistocene (low $\delta^{18}O$ values and ^{14}C near 1% modern carbon) to Holocene (higher $\delta^{18}O$ values, ^{14}C above about 25% modern carbon) (Adapted from Edmunds, 2007 with permission from IAEA).

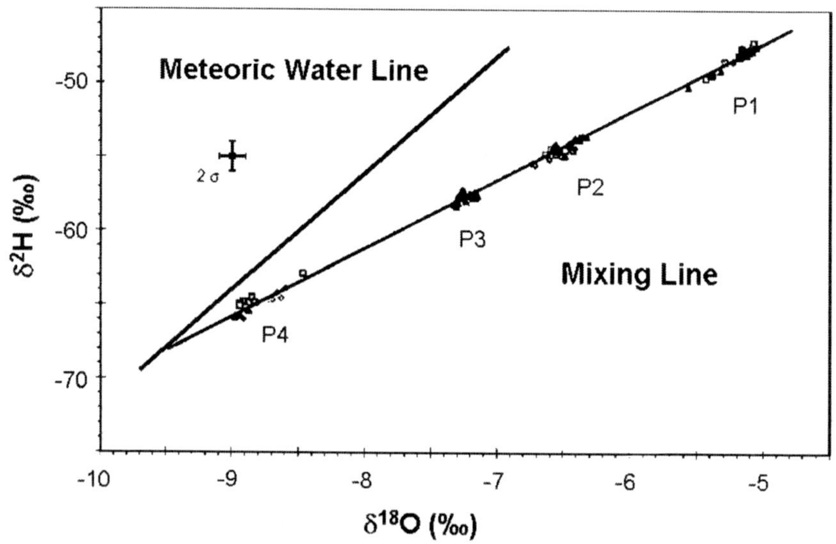

Figure 3.4 Plot of δ^2H against $\delta^{18}O$ of groundwater down-stream of a dredging lake. The cluster at groundwater sampling station P1 represents lake water because this station is located directly on the shore of the lake. The deviation of $\delta^{18}O$ from the Meteoric Water Line is due to evaporation. The other sampling stations are further away from the lake; the respective $\delta^{18}O$ clusters indicate various degree of mixing between lake water and "pure" groundwater (replenished exclusively by local precipitation). The regression line represents the mixing line between these two "end-members" (Adapted from Bertleff et al., 2001 with permission from Landesamt für Geologie, Rohstoffe und Bergbau, Baden Württemberg).

5.3.2.2. Other stable isotopes used in groundwater studies

The isotope ratios of nitrogen ($^{15}N/^{14}N$) and sulfate ($^{34}S/^{32}S$) have been used widely in groundwater pollution studies to (1) identify sources of nitrate and sulfate contaminants and (2) study contaminant processes such as bacterial sulfate and nitrate reduction.

Figure 3.5 shows the normal ranges of $\delta^{15}N$ and NO_3–$\delta^{18}O$ for dominant sources of nitrate and indicates that there are sections where the $\delta^{15}N$ values of different sources overlap. In many situations, it is therefore problematic to determine the contributions from different nitrate sources in groundwater on the basis of nitrogen isotope values alone (Kendall, 1998). The dual-isotope measurement of $\delta^{15}N$ and $\delta^{18}O$ in nitrate has improved the identification of nitrate sources in groundwater (Böttcher et al., 1990). A similar approach is used to identify sulfate sources with $\delta^{34}S$ and SO_4–$\delta^{18}O$ (Mayer, 2007).

Denitrification is a biogeochemical process that is mediated by the bacterium *Thiobacillus denitrificans* and leads to a reduction of dissolved nitrate to gaseous nitrogen in groundwater according to the equation

$$14NO_3^- + 5FeS_2 + 4H^+ \rightarrow 7N_2 + 10SO_4^{2-} + 5Fe^{2+} + 2H_2O \qquad (3.2)$$

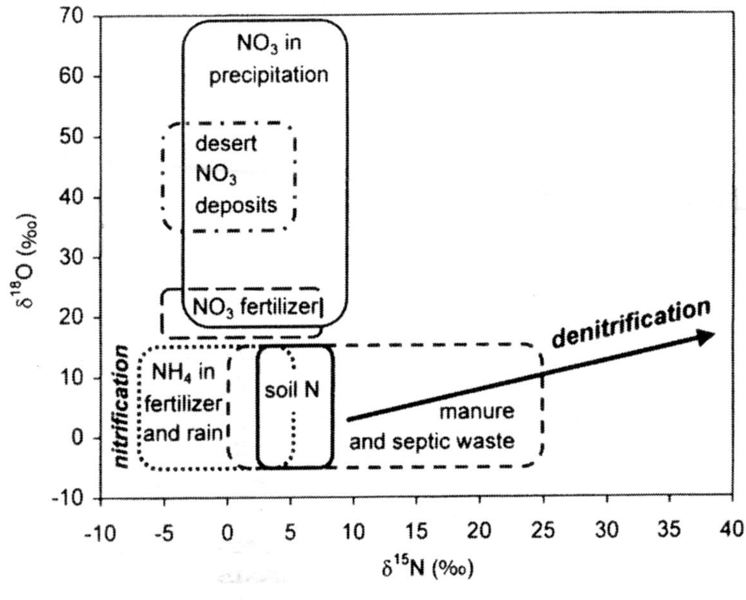

Figure 3.5 Typical ranges of NO_3–$\delta^{18}O$ and $\delta^{15}N$ values of nitrate from several sources: atmospheric deposition (NO_3 in precipitation), synthetic nitrate-containing fertilizers, nitrification in soils, and manure and sewage. The arrow indicates the trend of $\delta^{18}O$ and $\delta^{15}N$ of the remaining nitrate during denitrification (Adapted from Kendall, 1998 with permission from Elsevier).

The microbial denitrification discriminates against the heavy isotopes and causes an enrichment of ^{15}N and ^{18}O in the residual nitrate. This enrichment can be approximated by the Rayleigh equation (Mariotti et al., 1988) where the $\delta^{15}N$ and $\delta^{18}O$ values change proportionally to the logarithm of the remaining nitrate. The increase of the $\delta^{15}N$ values is about twice that of the $\delta^{18}O$ values (Böttcher et al., 1990). Hence, these elevated isotope values are unique for nitrate that has undergone denitrification.

Table 3.2 compiles the stable environmental isotopes that are of special relevance to groundwater studies. In particular, this table includes chemical processes in which isotopes are involved and thus points to the need to complement isotopic studies with chemical investigations and thereby to correlate isotopic with chemical parameters. A typical example is the identification of the origin of salinity in groundwater. The correlation between $\delta^{18}O$ isotope values and salinity of the groundwater allows distinction of whether the measured salinity is due to dissolution of salt deposits in the aquifer or mixing between fresh and saline groundwater, for example, sea water intrusion (Jones et al., 1999).

5.3.3. Radioactive environmental nuclides

The radioactive decay of environmental nuclides (Chapters 1 and 2) makes them a unique tool for determining groundwater residence time, often called groundwater age, defined as the length of time a given water parcel has been isolated from the atmosphere (Davis and Bentley, 1982). The age of groundwater is a powerful parameter for calibrating groundwater flow and solute (contaminant) transport models (Zuber and Maloszewski, 2000). In particular, it is used to constrain hydrogeological parameters such as groundwater recharge rate, groundwater flow velocity, and solute (contaminant) transport velocity. The following two simple examples demonstrate the general approach in groundwater dating.

1. The vertical age profile in an unconfined aquifer of constant thickness, H, is described by

$$\tau = \frac{\mathrm{pH}}{R}\ln\left(\frac{H}{H-z}\right) \tag{3.3}$$

where τ is the travel time (age), z the depth below the water table, p the porosity, and R the recharge rate (Vogel, 1967; Jordan and Froehlich, 1990). If the radionuclide concentration at the time of recharge, $C(t_0)$, is known and the concentration at depth z, $C(t_0 + \tau)$, has been measured, the travel time can be derived from the radioactive decay law

$$C(t_0 + \tau) = C(t_0)e^{-\lambda\tau} \tag{3.4}$$

Table 3.2 Application of stable isotopes in groundwater studies.

Isotopes	Chemical constituent	Chemical processes	Hydrological processes/application
$^{2}H/^{1}H$	Cl^-, Br^-	Mixing of freshwater with seawater	Mechanism of salinization
$^{18}O/^{16}O$	SO_4^{2-}	Evaporative enrichment	Recycling of irrigation water
	TDS, EC	Leaching of salt deposits	Origin of salinity
$^{11}B/^{10}B$	$B(OH_4)^-$	Mixing between fresh and contaminated water	Identification of pollution from sewage effluent
$^{14}C/^{12}C$	CO_2, H_2CO_3, HCO_3^-, CO_3^{2-}	Dissolution/precipitation of calcite, dolomite	^{14}C groundwater dating
		$CO_2 + H_2O \rightarrow H_2CO_3 \rightarrow HCO_3^- + H^+ \rightarrow 2H^+ + CO_3^{2-}$	^{14}C Initial concentration
$^{13}C/^{12}C$		$CO_2 + H_2O + CaCO_3 \rightarrow Ca^{2+} + 2HCO_3^-$	Validation of groundwater models
$^{2}H/^{1}H$	CH_4	$2CH_2O + SO_4^{2-} \rightarrow (bact.)CO_2 + CH_4$	Methanogenesis (e.g., in landfills)
$^{15}N/^{14}N$	NO_3^-, NH_4^+, N_2	$14NO_3^- + 5FeS_2 \rightarrow 7N_2 + 10SO_4^{2-} + 5Fe^{2+}$	Nitrate pollution sources
$18O/^{16}O$			Microbial denitrification
$^{34}S/^{32}S$	SO_4^{2-}, H_2S, HS^-	Dissolution of gypsum/anhydrite	Tracing sulfate origin
$^{18}O/^{16}O$		Oxidation of sulfide	Sulfate contamination
		$FeS_2 + 3.5O_2 + H_2O \rightarrow Fe^{2+} + 2SO_4^{2-} + 2H^+$	
		Reduction of sulfate	Sources of H_2S
		$2CH_2O + SO_4^{2-} \rightarrow 2HCO_3^- + H_2S$	
$^{37}Cl/^{35}Cl$	Cl^-, chlorinated hydrocarbons	Mixing between fresh and contaminated water	Sources of salinity Identification of chlorinated hydrocarbons

The commonly measured phases and the chemical processes relevant to the various applications are also listed.

where λ is the decay rate of the given radioisotope. Combining this equation with Equation (3.3), the recharge rate can be estimated from the measured radioisotope concentrations.

2. If groundwater moves like a piston (without dispersion and mixing) in a confined aquifer with a constant velocity v, the travel time $\tau \equiv (t_2 - t_1)$ between two groundwater sampling sites at locations x_1 and x_2, is calculated by

$$\tau = \frac{x_2 - x_1}{v} \tag{3.5}$$

The concentration of a given radioisotope decreases from x_1 to x_2 according to the radioactive decay law

$$C(x_2) = C(x_1)e^{-\lambda\tau} \tag{3.6}$$

Thus, the combination of Equation (3.5) with Equation (3.6) shows that the flow velocity can be determined by measuring the concentration of a given radioisotope along the groundwater flow path.

In both examples, the parameter $\lambda\tau = \ln(2)\tau/T_{1/2}$ ($T_{1/2}$ = half-life) controls the change of the radioisotope concentration along the groundwater flow (according to Equation (3.4) and Equation (3.6)). Thus, the half-life defines the dating range that can be covered by the given radioisotope. The available radioactive isotopes have half-lives ranging from days to hundreds of thousands of years and thus allow dating of groundwater over a wide age range. Commonly, the following classification is used: (1) young groundwater – age < 60 years (dating limit of CFCs); (2) old groundwater – age between about 100 years (dating limit of ^{3}H) and 50,000 years (dating limit of ^{14}C); (3) very old groundwater – age beyond dating limit of ^{14}C. Table 3.3 lists the most frequently used cosmogenic and anthropogenic environmental radionuclides.

5.3.3.1. Dating of young groundwater

Young groundwater is dominant in shallow unconfined aquifers and it is used for drinking water supply (e.g., springs in mountainous regions) and agricultural purposes (e.g., irrigation). In unconfined aquifers with modern recharge, age, recharge rate, and porosity are well correlated, as shown by Equation (3.3). Thus, groundwater dating by radioisotopes is able to constrain recharge rates in unconfined aquifers (Portniaguine and Solomon, 1998). The age of young groundwater typically ranges from a few years to some hundreds of years, in dry areas up to about thousand years. Suitable radioactive nuclides for dating of young groundwater are ^{3}H, ^{85}Kr, and nuclear-weapon produced ^{36}Cl. Also the parent–daughter pair tritium–helium-3 and some nonradioactive atmospheric gases of anthropogenic

Table 3.3 Cosmogenic and anthropogenic radionuclides used in groundwater dating.

Radioisotope	Half-life in years	Atmospheric production		*In situ* production	
		Reaction	Initial content[a] in 1 m^3 of groundwater	Reaction	Equilibrium content[a] in 1 m^3 of groundwater
^{85}Kr	10.76	Anthropogenic	2.5×10^{-3} Bq	non	0
^{3}H	12.43	$^{14}N(n, t)$	10^3 Bq	$^{6}Li(n, \alpha)$	$2-10^{-2}$ Bq
^{32}Si	140	$^{40}Ar(n, 4p5n)$	10^{-2} Bq	non	0
^{39}Ar	269	$^{40}Ar(n, 2n)$	8.5×10^6 atoms	$^{39}K(n, p)$	$0.2–2 \times 10^6$ atoms
^{14}C	5730	$^{14}N(n, p)$	3 Bq	$^{17}O(n, \alpha)$	3×10^{-2} Bq
^{81}Kr	2.1×10^5	$^{80}Kr(n, \gamma)$	10^6 atoms	Non	0
^{36}CI	3.01×10^5	$^{35}CI(n, \gamma)$	$10^8–10^{11}$ atoms	$^{35}CI(n, \gamma)$	5×10^{10} atoms
^{129}I	16×10^6	Spontaneous fission	5×10^7 atoms	Spontaneous fission	3×10^8 atoms

[a]If the "content" is given in Bq, conventional decay counting (e.g., with liquid scintillation spectrometry) can be used for radionuclide measurements; in the other cases ("atoms") a high measuring sensitivity is required which can only be achieved by AMS (see also Chapter 9).

origin (CFCs and SF_6) have shown potential to complement and/or substitute for ^3H as a dating tool. IAEA (2006) gives a comprehensive review on CFCs and other complementary dating tools such as SF_6, $^3H/^3He$, and ^{85}Kr. The following focuses on the use of ^3H (including $^3H/^3He$) and ^{85}Kr. A review of all dating methods for young groundwater has recently been given by Plummer (2005).

5.3.3.1.1. Tritium (^3H). The temporal variation of the tritium concentration in groundwater is controlled by the variation of tritium input and the radioactive decay during the residence time in the groundwater system. Nir (1964) developed a model allowing for the first time to calculate the groundwater residence time on the basis of tritium measurements. Subsequently numerous studies were carried out making use of transient bomb tritium to determine recharge rates of young groundwater (Robertson and Cherry, 1989). However, at the present time (and in future) the fraction of bomb tritium in groundwater is negligible and much higher analytical sensitivity is required to measure the generally low tritium concentrations (Chapter 9, this book).

Alternatively young groundwater can be dated by the $^3H/^3He$ method, by which the tritium concentration at the time of recharge (initial concentration) is reconstructed from the ^3H and ^3He concentration measured in the given groundwater sample (see Section 5.2.3, Equation (2.2)). This method has first been suggested by Tolstikhin and Kamensky (1969). Subsequently, the method was used in numerous studies, often in combination with other approaches to date young groundwater. Figure 3.6 shows two practical examples where the $^3H/^3He$ method was used to date vertical profiles for determining recharge rates in unconfined aquifers. The $^3H/^3He$ method and complementary dating techniques (chlorofluorocarbons and ^{85}Kr) have been reviewed by Cook and Solomon (1997).

5.3.3.1.2. Krypton-85 (^{85}Kr). The major sources for atmospheric ^{85}Kr, a radioactive noble gas, are reprocessing facilities for nuclear fuel (Winger et al., 2005). It is dissolved in precipitation and enters with the recharge of the groundwater region. The primary advantage of ^{85}Kr is that it is chemically inert in groundwater; it is only dependent on radioactive decay and hydraulic properties of the aquifer. The measured $^{85}Kr/Kr$ ratio yields directly the specific ^{85}Kr activity (Bq per cm^3 of krypton) and thus is not affected by recharge temperature and krypton solubility in water. Its half-life of 10.76 years makes ^{85}Kr very suitable for dating shallow (young) groundwater (Rozanski and Florkowski, 1979). A typical example of ^{85}Kr groundwater dating is shown in Figure 3.7. In several multitracer applications (^3H, $^3H/^3He$, CFCs, SF_6, and ^{85}Kr) for dating young (shallow) groundwater, ^{85}Kr has been proven to be a useful complementary dating tool (Smethie et al., 1992; Ekwurzel et al., 1994; Loosli, 1992; Bauer et al., 2001). It is expected

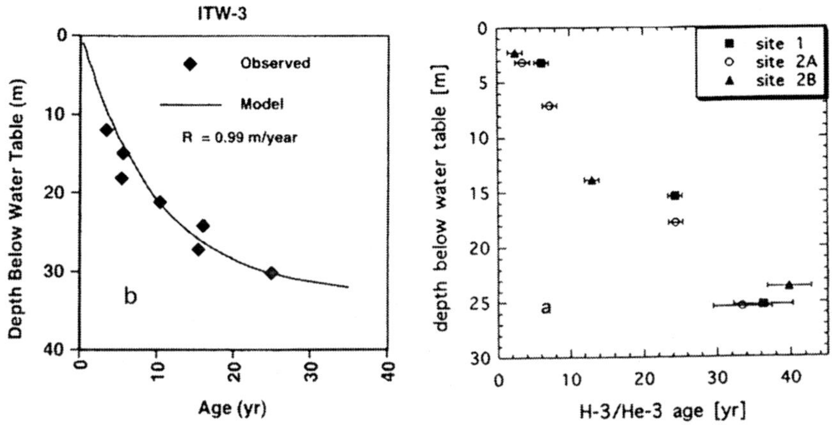

Figure 3.6 (a) Apparent $^3H/^4He$ age derived from measured $^3H/^4He$ values as a function of depth in an unconfined aquifer in southern New Jersey, USA (Adapted from Szabo et al., 1996 with permission from AGU); (b) observed and simulated groundwater ages as a function of depth. The recharge rate was adjusted to fit the simulated age distribution (Equation 3.3) to the measured values (Adapted from Solomon et al., 1995 with permission from Blackwell).

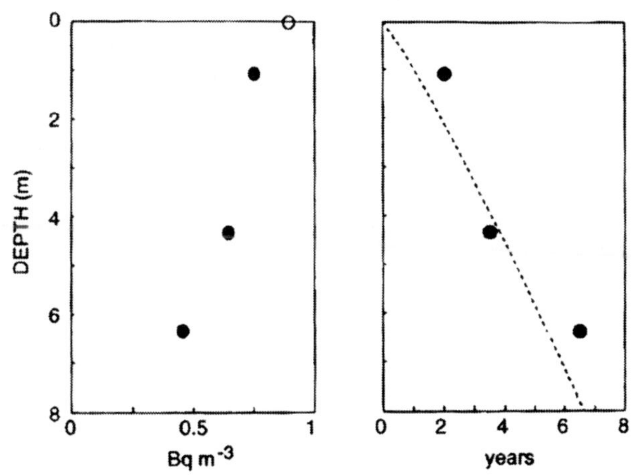

Figure 3.7 Specific activity of ^{85}Kr (Bq m^{-3}) and ^{85}Kr age (years) in a vertical profile in the Bordon aquifer, ON, Canada (Smethie et al., 1992). The open circle in the left diagram indicates the ^{85}Kr value in the soil gas and the atmosphere. The broken line in the left diagram depicts the vertical groundwater age profile determined by a two-dimensional flow model (Adapted from Cook and Solomon, 1997 with permission from Elsevier).

that the atmospheric concentration of ^{85}Kr will continue to increase in the foreseeable future and that this will allow extensive hydrological applications of this radioisotope during the coming decades (Plummer, 2005).

5.3.3.2. Dating of old groundwater

Old groundwater can be found in greater depth of unconfined and in confined aquifers with lateral extensions up to some tens of kilometers. It is preferentially used for drinking water supply, and it is less vulnerable to pollution than shallow groundwater.

5.3.3.2.1. ^{39}Ar and ^{32}Si. Given their half-lives, the cosmogenic radioactive nuclides ^{32}Si (ca. 140 years) and ^{39}Ar (269 years) appear to be suited to date groundwater of ages in the range from about 100 to 1,000 years. However, the practical application of these nuclides for groundwater dating is still in an exploratory stage.

^{39}Ar samples are difficult to collect and to analyse (Forster et al., 1992a); moreover, it is difficult to disentangle cosmogenic from *in situ* produced ^{39}Ar of a given groundwater sample (Forster et al., 1992b). The first studies on the use of ^{39}Ar for groundwater dating were published by Loosli and Oeschger (1969, 1980).

^{32}Si was first measured in groundwater by Nijampurkar et al. (1966). Attempts have been made to use ^{32}Si as a groundwater dating tool. Froehlich et al. (1987a) measured ^{32}Si and tritium in groundwater taken from aquifers of several rock types. Under some favorable conditions, the ^{32}Si concentration was found to be in agreement with the measured tritium values. In particular, in a phreatic sandy aquifer, the ^{32}Si concentration decreased with depth, which corresponded with the increase in groundwater age. Similar results were found by Morgenstern et al. (1995) who measured ^{32}Si and ^{14}C in limestone aquifers in Estonia, Germany and Poland. Yet, to arrive at decisive conclusions on the potential of ^{32}Si as an age indicator for young groundwater, further studies of ^{32}Si in groundwater and its geochemical behavior are still needed.

5.3.3.2.2. Carbon-14 (^{14}C). The most suitable radioactive nuclide for dating of old groundwater is ^{14}C (Geyh, 2005). With its half-life of 5,730 years, an age range for groundwater dating from about 1,000 to 30,000 years can be covered. Münnich (1957) published the first results on ^{14}C in inorganic carbon dissolved in groundwater (DIC) and, with his team at Heidelberg University, Germany, developed the ^{14}C groundwater dating method (Brinkmann et al., 1959). Subsequently, it was found that, in addition to the ^{14}C radioactive decay, a number of biogeochemical processes can affect the ^{14}C concentration in groundwater (Münnich et al., 1967). These processes include: (1) ^{14}C dilution by isotopic exchange between dissolved HCO_3^- and $CaCO_3$ of the aquifer matrix, (2) isotopic fractionation accompanying the soil CO_2 dissolution, (3) sulfate reduction, and (4) methanogenesis. A comprehensive theoretical description of the evolution of carbon isotopic ratios ($^{14}C/^{12}C$ and $^{13}C/^{12}C$) in groundwater has been given by Wigley et al. (1978). To facilitate routine use of the ^{14}C

dating method in groundwater studies, computer codes have been developed by which the geochemical and isotopic evolution of DIC in groundwater is traced. The interactive geochemical model NETPATH has been used widely for evaluating ^{14}C measurements in terms of groundwater ^{14}C ages; it takes into account carbon exchange reactions and their effect on the evolution of the $^{14}C/^{12}C$ ratio along a groundwater flow path (Plummer et al., 1994).

A typical example of this approach has been published by Kennedy and Genereux (2007) who measured ^{14}C together with relevant geochemical parameters in groundwater of confined aquifers in North Carolina, USA, and used NETPATH to evaluate the data in terms of groundwater age (Figure 3.8). The ^{14}C concentration decreased in the flow direction indicating increasing groundwater age. The ^{14}C ages were "adjusted for geochemical and diffusion effects" and used to estimate the long-term average groundwater flow velocity. For the upper aquifer, the estimated value was ca. 2.5 m yr^{-1}; a slightly higher value resulted for the lower aquifer.

Apart from ^{14}C in DIC, ^{14}C in the DOC fraction can also be used for groundwater dating (Murphy et al., 1989). The concentrations of DOC in groundwater are between about 0.1 and 1 mg/L; they are too low for conventional ^{14}C counting. Hence, ^{14}C in DOC requires the use of accelerator mass spectrometry (AMS) (see also Chapter 9). Additional measurements of ^{14}C in DOC facilitate a better assessment of the biogeochemical reactions for the evaluation of the ^{14}C measurements. It was found that fulvic acid from the high-molecular fraction in DOC is most suitable for ^{14}C dating (Geyer et al., 1993; Artinger et al., 1996).

Figure 3.8 Left: Cross-section showing the two studied aquifers, the Black Creek and the Upper Cape Fear aquifer, in the Coastal Plain of North Carolina, USA. The aquitards between the aquifers are shown in black. The wells used for collecting samples from the Black Greek and Upper Cape Fear along the main flow direction are also shown (Adapted from Kennedy and Genereux, 2007 with permission from Radiocarbon). Right: Measured (uncorrected) ^{14}C values as a function of the distance from the recharge area of the corresponding aquifer.

5.3.3.3. Dating of very old groundwater

Groundwater with ages beyond the dating limit of ^{14}C occurs in deep sedimentary basins and large aquifer systems, most of them in arid regions such as the Nubian Sandstone Aquifer System in North Africa (Thorweihe and Heinl, 2002) and the Great Artesian Basin in Australia (Torgersen et al., 1991). These groundwater resources are often the only source of drinking water supply for the local population. They are unlikely to be replenished under present climatic conditions, thus their exploitation is synonymous with groundwater mining. The rational management of the water resources in these systems requires improved understanding of their replenishment and flow regime. Groundwater dating is a very powerful tool in this regard.

The dating methods for very old groundwater are based on the (1) radioactive decay of long-lived radionuclides (^{81}Kr, ^{36}Cl, ^{129}I) produced by cosmic rays in the atmosphere, (2) accumulation of ^{4}He generated in the natural decay series of uranium and thorium in rocks of the aquifer and adjacent strata, and (3) disequilibrium between uranium isotopes (^{234}U, ^{238}U) due to the combined effect of water–rock interaction and radioactive decay.

5.3.3.3.1. ^{81}Kr, ^{36}Cl, ^{129}I. Of these three radionuclides, ^{81}Kr (half-life 229 kyr) has the greatest potential for dating of very old groundwater because it is an inert gas (no chemical processes affecting its concentration), and its *in situ* production in rocks or groundwater is negligible. Therefore, the conversion of the measured $^{81}Kr/Kr$ ratios by the decay law into groundwater ages is rather straightforward. However, the extremely low ^{81}Kr concentrations ($<1{,}000$ atoms per liter) of water make the sampling and analysis of ^{81}Kr very difficult (Chapter 9). Using cyclotron AMS, the first ^{81}Kr groundwater ages were determined for the Great Artesian Basin (Collon et al., 2000) and the Nubian Sandstone Aquifer System (Sturchio et al., 2004). The volume of groundwater sample required for ^{81}Kr determination was about 16 m^3, from which about 0.5 cm^3 (STP) krypton gas was extracted. Despite this progress in measuring ^{81}Kr, analytical facilities for routine applications for dating very old groundwater are not yet available (Lehmann et al., 2003), but there are good prospects of lowering the required sample amount to only a few liters.

In contrast to ^{81}Kr, ^{36}Cl has already been used for more than 20 years for dating very old groundwater. According to the ^{36}Cl half-life of 301 kyr, the dating range is from <100 kyr to nearly 1,000 kyr. The ^{36}Cl studies of the Great Artesian Basin, Australia (Bentley et al., 1986; Torgersen et al., 1991), the Milk River aquifer, Canada (Phillips et al., 1986), and the *Continental Intercalaire* aquifer of the northern Sahara (Fontes et al., 1985; Guendouz and Michelot, 2006) showed that, apart from radioactive decay of cosmogenic ^{36}Cl, other process (e.g., *in situ* production of ^{36}Cl and dissolution/diffusion of Cl^- from adjacent aquitards) can also affect the evolution of the ^{36}Cl along the groundwater flow path. These processes may cause ambiguities in the interpretation of ^{36}Cl groundwater ages (Andrews and Fontes, 1992). More recent papers (Love et al., 2000; Patterson et al., 2005) demonstrated that, through combined use of complementary dating tools, this ambiguity can be resolved.

Although inappropriate for dating of fresh groundwater, ^{129}I (half-life 1.6×10^7 yr) has been used to date old brines (Fehn et al., 1990) and to study the evolution of halides in groundwater, especially in connection with ^{36}Cl dating (Fabryka-Martin et al., 1987).

5.3.3.3.2. Helium-4 (^4He). Of the noble gases occurring in groundwater, ^4He has shown special potential to date old groundwater. The *in situ* ^4He production is largely due to neutralization of alpha particles produced in rocks by alpha decay in the natural decay series of ^{238}U, ^{232}Th and, to a lesser extent, ^{235}U (Chapter 1). Part of this ^4He diffuses into groundwater where the concentration of ^4He increases linearly with time. A second source to be taken into account in estimating the ^4He increase is diffusion of crustal helium into the aquifer (Torgersen and Clarke, 1985). The ^4He accumulation rate is highly variable between different aquifers. Furthermore, small admixtures of very old groundwater with high ^4He concentrations can drastically change the ^4He concentration along a groundwater flow path. Therefore, ^4He groundwater dating is considered as a semiquantitative rather than a quantitative dating method (Lehmann et al., 2003).

5.3.3.3.3. Uranium isotope disequilibrium. In groundwater, the activity concentration of ^{238}U is commonly different from that of ^{234}U ($^{234}U/^{238}U \neq 1$), that is, the isotopes are in radioactive disequilibrium. The activity concentration ratio $^{234}U/^{238}U$ is mainly controlled by preferential leaching of ^{234}U and alpha recoil transfer of its progenitor ^{234}Th from the aquifer rock, radioactive decay of ^{234}U and sorption-desorption processes between water and rock. Assuming that the change of the $^{234}U/^{238}U$ in the groundwater flow direction is only controlled by radioactive decay (^{234}U half-life $= 2.45 \times 10^5$ yr), the dating range would be from about 100 to 800 kyr. However, the $^{234}U/^{238}U$ ratio can also change due to chemical reactions in the aquifer (Kigoshi, 1973). The dating model suggested by Froehlich and Gellermann (1987b) considered reversible sorption, irreversible dissolution, and precipitation that may affect the evolution of the uranium isotopic composition along the groundwater flow path. It was shown that precipitation of uranium may even exceed the effect of the ^{234}U decay (Ivanovich et al., 1991b). A short review of the use of uranium disequilibrium in groundwater studies has been given by Davis and Bentley (1982). The various attempts to use the uranium isotope disequilibrium as a dating tool have been reviewed by Froehlich and Gellermann (1987b).

The study of the age structure and dynamics of systems with very old groundwater requires multiisotope approaches including extensive geochemical analyses. A typical example in this regard is the complex investigation of the Milk River Aquifer system (Ivanovich et al., 1991a). Among other isotope techniques, uranium isotopes and ^{36}Cl have been measured along the two flow paths identified in this system to estimate the groundwater flow velocity. The measured concentrations of ^{234}U and ^{36}Cl have been evaluated with models taking water–rock interactions including *in situ* production of ^{36}Cl into account. The flow velocities estimated by ^{36}Cl ranged from 0.04 to 0.14 m yr^{-1} (Nolte et al., 1991), while the values derived from the uranium isotope measurements fell in the range from 0.13 to 0.65 m yr^{-1} (Ivanovich et al., 1991b).

In most applications carried out so far, the uranium isotope disequilibrium appeared to be a more suitable indicator of origin and mixing of groundwater than of groundwater age. Recently, the uranium isotope disequilibrium has been employed to identify sources and mixing of groundwater in the Nubian Sandstone aquifer beneath several oases in the Western Desert of Egypt (Dabous and Osmond, 2001). Correlating the $^{234}U/^{238}U$ activity ratio with the inverse of the uranium concentration, it has been found that sources from deep aquifers predominate. Contributions of groundwater replenished during the last pluvial period ranged from 5 to about 26%.

5.3.4. Geothermal systems

Under favorable geological conditions, rainwater penetrates deep into the earth's crust and becomes trapped in vast underground reservoirs where it is heated by the magma to very high temperatures. Sometimes geothermal water comes out at the surface and forms hot springs or geysers. Very hot water can also be extracted by drilling into the geothermal reservoir to generate steam for electric power production. Geothermal energy is used in about 20 countries. Although the USA is the largest producer of geothermal energy (nearly 3 GWe), the share of total energy production is rather low. However, in some countries including the Philippines, Mexico, Nicaragua, and El Salvador, geothermal energy satisfies about 25–40% of the national energy demand.

In most geothermal development projects, isotope methods are considered a well-established tool that can be used to

- identify the origin of the geothermal fluid;
- indicate the host rock of the geothermal water;
- delineate fluid flow directions and its pathways in the area;
- determine rates of fluid flow in and out of the reservoir;
- select potential drilling sites;
- identify suitable sites for production of steam and reinjection of waste fluids.

5.3.4.1. Stable isotopes

Amongst the stable environmental isotopes, those of the elements hydrogen, oxygen, carbon and sulfur are of main interest because these elements are important constituents of geothermal fluids and rocks. The most frequently used stable isotopes are those of oxygen and hydrogen on which the following is focused. A more comprehensive presentation of isotopes in geothermal fluids is given by Clark and Fritz (1997) and Geyh (2000).

Compared to meteoric water, rocks, and magmatic water are generally enriched in ^{18}O ($\delta^{18}O > 0$) but depleted in ^{2}H ($\delta^{2}H < 0$). Hot magma in contact with water quickly exchanges ^{18}O and ^{2}H with the newly formed minerals so that no isotope fractionation can occur. Given the high oxygen concentration in the rock compared to that in the fluid reservoir (which has a much smaller mass than the rocks), the fluid takes on the ^{18}O isotopic composition of the rock. In contrast, hydrogen represents only a minor component of the minerals, so that the hydrogen isotope composition of the rock evolves during crystallization along with that of the fluid. Therefore, in the $\delta^{2}H$–$\delta^{18}O$ diagram, the isotopic composition of geothermal fluids fits a horizontal line (no increase in $\delta^{2}H$). The initial isotopic composition of the various geothermal fluids corresponds with the isotopic composition of local meteoric water. Consequently, the oxygen and hydrogen isotopic composition of the geothermal fluids is a unique indicator of the origin and evolution of geothermal fluids.

Giggenbach (1992) found that horizontal lines are not always characteristic for geothermal fluids. Waters in geothermal systems associated with circum-Pacific andesitic volcanism are enriched both in ^{18}O and ^{2}H, not only due to water–rock interaction but mainly because of mixing between meteoric and magmatic waters, so called andesitic water (Figure 3.9). These acidic fluids are suggested to be formed by (i) condensation of H_2S-rich steam into groundwater to form acid–sulfate waters, (ii) direct input of sulfur dioxide and other volcanic gases, and (iii) remobilization of sulfur deposits at depth. Acidic fluids are corrosive and cause fast deterioration of wells and pipelines. Therefore, the understanding of their actual origin and occurrence enables better development strategies; specifically, drilling targets can be better defined. Correlation of $\delta^{34}S$ with $\delta^{18}O$ in $SO_4^{=}$ as well as with other geochemical parameters can help identify the origin of acidic fluids and study processes in which they are involved (Robinson et al., 1987).

5.3.4.2. Radioactive isotopes

Water–rock interaction at high temperature and the kinetics of chemical reactions strongly affect the behavior of cosmogenic radioisotopes that enter the geothermal reservoirs with infiltrating meteoric water. The effect of the

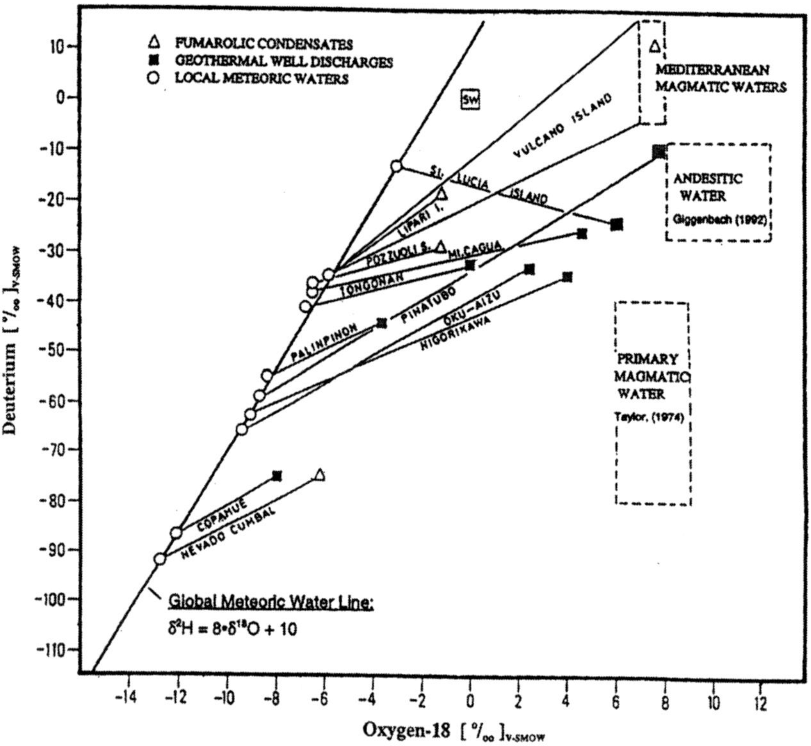

Figure 3.9 Isotope characteristics of major geothermal systems of the world (Adapted from Froehlich et al., 1998 with permission from IRD).

radioactive decay is masked by sorption and precipitation processes, which prevent their use as timers in geothermal reservoirs. Only tritium as a constituent of the water molecule is in practice not influenced by these processes and thus proves to be a suitable tracer and timer of geothermal processes, especially in thermal springs to determine the fraction of young meteoric water (e.g., Ahmad et al., 2002; Casey, 2007).

However, radioisotopes of the natural decay series have proven to be very useful in geothermal studies. The radioactive disequilibrium observed in fluid (brine) and solid (rock) phases allows estimation of rates of brine – rock chemical interaction and parameters of geological relevance such as brine flow rates and residence time in geothermal systems. For example, the residence time of brines in the Salton Sea Geothermal field, CA, USA, was found to be in the range from 100 to 1,000 years, with a best estimate of 350 years (Hammond et al., 1988). Measuring the four radium isotopes (^{223}Ra, ^{224}Ra, ^{228}Ra, ^{226}Ra) and ^{222}Rn in geothermal fluids as well as ^{226}Ra and ^{228}Ra in rocks of the Central Italy Geothermal Area and evaluating the data on the basis of a conceptual model, Froehlich et al. (1995) were able to estimate parameters of the water–rock interaction including recoil ejection, chemical leaching, and residence time of the fluid in the geothermal system. Radium isotopes and ^{222}Rn were also measured in fluids and solids in the Coso Geothermal System, South-eastern California (Leslie and Hammond, 2007). Residence times of fluids and fluid velocities in the geothermal system could be estimated on the basis of the measured activity ratios $^{224}Ra/^{226}Ra$ and $^{228}Ra/^{226}Ra$. Radon-222 was found to be an effective tracer for steam velocities within the geothermal system. Evaluation of carbon dioxide and ^{222}Rn data indicated that the residence time of steam (time since separation from the liquid) is in the order of a few days only.

In conclusion, measurements of radium isotopes and ^{222}Rn as well as gaseous phase constituents in geothermal fluids and rocks can help quantify and constrain the processes and characteristics (e.g., surface area/volume, residence times) of geothermal systems.

REFERENCES

Aggarwal, P. K., K. Froehlich, and K. M. Kulkarni. (2007). Environmental isotopes in groundwater studies. In: *Groundwater – Encyclopedia of Live Support Systems* (Eds L. Silveira and E. J. Usunoff). EOLSS Publishers/UNESCO, Paris, pp. 69–92.

Ahmad, M., W. Akram, N. Ahmad, M. A. Tasneem, M. Rafiq, and Z. Latif. (2002). Assessment of reservoir temperatures of thermal springs of the northern areas of Pakistan by chemical and isotope geothermometry. *Geothermics*, 31(5), 613–631.

Ambach, W., W. Dansgaard, H. Eisner, and J. Moeller. (1968). The altitude effect on the isotopic composition of precipitation and glacier ice in the Alps. *Tellus*, 20(4), 595–600.

Andrews, J. N., and J.-Ch. Fontes. (1992). Importance of in situ production of chlorine-36, argon-36 and carbon-14 in hydrology and hydrogeochemistry. In: *Isotope Techniques in Water Resources Development 1991*. IAEA, Vienna, Austria, pp. 245–269.

Angert, A., C. D. Cappa, and D. J. Depaolo. (2003). ^{17}O as a new tracer of the hydrological cycle. *American Geophysical Union, Fall Meeting 2003*. Abstract. 2003AGUFM. A42F.05A.

Artinger, R., G. Buckau, J. I. Kim, S. Geyer, and M. Wolf. (1996). Influence of sedimentary organic matter on dissolved fulvic acids in groundwater. – Significance for groundwater dating with ^{14}C dissolved organic matter. In: *Isotopes in Water Resources Management*. IAEA, Vienna, Austria, pp. 57–72.

Bauer, S., C. Fulda, and W. Schäfer. (2001). A multi-tracer study in a shallow aquifer using age dating tracers ^{3}H, ^{85}Kr, CFC-113 and SF_6 – Indication for retarded transport of CFC-113. *Journal of Hydrology*, 248, 14–34.

Bentley, H. W., F. M. Phillips, S. N. Davis, M. A. Habermehl, P. L. Airey, G. E. Calf, D. Elmore, H. E. Gove, and T. Torgensen. (1986). Chlorine-36 dating of very old groundwater. 1. The Great Artesian Basin. *Water Resources Research*, 22, 1991–2001.

Bertleff, B., H. Plum, J. Schuff, W., Stichler, D. H. Storch, and C. Trapp. (2001). Wechselwirkung zwischen Baggerseen und Grundwasser. Landesamt für Geologie, Rohstoffe und Bergbau, Baden-Württemberg, 64p.

Böttcher, J., O. Strebel, S. Voerkelius, and H. L. Schmidt. (1990). Using isotope fractionation of nitrate-nitrogen and nitrate-oxygen for evaluation of microbial denitrification in a sandy aquifer. *Journal of Hydrology*, 114, 413–424.

Brinkmann, R., K. O. Münnich, and J. C. Vogel. (1959). ^{14}C-Alterbestimmung von Grundwasser. *Naturwissenschaften*, 46, 10–12.

Casey, C. A. (2007). An assessment and comparison of the geochemical signatures of thermal springs in the Animas River Valley, South-western Colorado. *Geological Society of America Abstracts with Programs*, 39(6), 324.

Clark, I., and P. Fritz. (1997). *Environmental Isotopes in Hydrogeology*. Lewis Publishers, New York, 328 p.

Collon, P., W. Kutschera, H. H. Loosli, R. E. Lehmann, R. Purtschert, A. Love, L. Sampson, D. Anthony, D. Cole, B. Davids, D. J. Morrissey, B. M. Sherrill, M. Steiner, R. C. Pardo, and M. Paul. (2000). ^{81}Kr in the great Artesian Basin, Australia: A new method for dating very old groundwater. *Earth and Planetary Science Letters*, 182, 103–113.

Cook, P. and A. L. Herczeg. (Eds) (1999). *Environmental Tracers in Subsurface Hydrology*. Kluwer Academic Publishers, Boston, Dordrecht, London, 552p.

Cook, P. G., and D. K. Solomon. (1997). Recent advances in dating young groundwater: Chlorofluorocarbons, ^{3}H/^{3}He and ^{85}Kr. *Journal of Hydrology*, 191, 245–265.

Dabous, A. A., and J. K. Osmond. (2001). Uranium isotopic study of artesian and pluvial contributions to the Nubian Aquifer, Western Desert, Egypt. *Journal of Hydrology*, 243, 242–253.

Dansgaard, W. (1964). Stable isotopes in precipitation. *Tellus*, 16, 436–468.

Davis, N. S., and H. W. Bentley. (1982). Dating groundwater – A short review. In: *Nuclear and Chemical Dating Techniques – Interpreting the Environmental Record* (Ed L. A. Curie). *American Chemical Society Symposium Series* 176, pp. 187–222.

Edmunds, W. M. (2007). Groundwater as an archive of climatic and environmental change. In: *Isotopes in the Water Cycle* (Eds P. Aggarwal, J. Gat, and K. Froehlich). Springer, Dordrecht, pp. 341–352.

Ekwurzel, B., P. Schlosser, W. M. J. Smethie, L. N. Plummer, E. Busenberg, R. L. Michel, R. Weppernig, and M. Stute. (1994). Dating of shallow groundwater: Comparison of the transient tracers ^{3}H/^{3}He, chlorofluorocarbons, and ^{85}Kr. *Water Resources Research*, 30(6), 1693–1708.

Fabryka-Martin, J., D. O. Whittemore, S. N. Davis, P. K. Kubik, and P. Sharma. (1987). Geochemistry of halogens in the Milk River aquifer, Alberta, Canada. *Applied Geochemistry*, 6, 447–464.

Fehn, U., S. Tullai-Fitzpatrick, R. T. D. Teng, H. E. Gove, P. W. Kubik, P. Sharma, and D. Elmore. (1990). Dating of oil field brines using ^{129}I. *Nuclear Instruments and Methods B*, 52, 446–450.

Fontes, J.-Ch., A. Guendouz, and J. L. Michelot. (1985). *Etude par le Chlore-36 des Eaux Souterraines du Sahara Septentrional*. Report RAF/8/007, IAEA, Vienna, Austria.

Forster, M., P. Maier, and H. H. Loosli. (1992a). Current techniques for measuring the activity of ^{37}Ar and ^{39}Ar in the environment. In: *Isotopes of Noble Gases as Tracers in*

Environmental Studies (Proceedings of Consultants Meeting). IAEA, Vienna, Austria, pp. 63–72.

Forster, M., K. Ramm, and P. Maier. (1992b). Argon-39 dating of groundwater and its limiting conditions. In: *Isotope Techniques in Water Resources Development 1991*. IAEA, Vienna, Austria, pp. 203–214.

Froehlich, K. (2000). Evaluating the water balance of inland seas using isotopic tracers: The Caspian Sea experience. *Hydrological Processes*, 14, 1371–1383.

Froehlich, K., A. Battaglia, A. Ceccarelli, A. Ridolfi, and C. Panichi. (1995). Radium isotopes contribution to geothermal exploration in Central Italy. In: *Proceedings of the World Geothermal Congress*, Vol. II, pp. 1059–1064.

Froehlich, K., T. Franke, R. Gellermann, D. Hebert, and P. Jordan. (1987a). Silicon-32 in different aquifer types and implications for groundwater dating. In: *Isotope Techniques in Water Resources Development*. IAEA, Vienna, Austria, pp. 149–163.

Froehlich, K., and R. Gellermann. (1987b). On the potential use of uranium isotopes for groundwater dating. *Chemical Geology (Isotope Geoscience Section)*, 65, 67–77.

Froehlich, K., K. Rozanski, and L. Araguas-Araguas. (1998). Isotope hydrology: Applied discipline in Earth sciences. In: *Hydology and Isotope Geochemistry* (Eds C. Gausse and F. Gasse). pp. 55–72. Orstom Editions, 305p.

Geyer, S., M. Wolf, L. I. Wassenaar, P. Fritz, G. Buckau, J. I. Kim. (1993). Isotope investigations on fractions of dissolved organic carbon for ^{14}C groundwater dating. In: *Isotope Techniques in the Study of Past and Current Environmental Changes in the Hydrosphere and Atmosphere*. IAEA, Vienna, Austria, pp. 359–380.

Geyh, M. A. (2000). Groundwater – Saturated and unsaturated zone. In: *Environmental Isotopes in the Hydrological Cycle – Principles and Applications* (Ed. W. G. Mook). UNESCO/IAEA, Vol. IV, 205p.

Geyh, M. A. (2005). Dating of old groundwater. History, potential, limits and future. In: *Isotopes in the Water Cycle* (Eds P. K. Aggarwal, J. R. Gat, and K. Froehlich). Springer, Dordrecht, The Netherlands, pp. 221–241.

Giggenbach, W. F. (1992). Isotopic shifts in waters from geothermal and volcanic systems along convergent plate boundaries and their origin. *Earth and Planetary Science Letters*, 113, 495–510.

Guendouz, A., and J. L. Michelot. (2006). Chlorine-36 dating of deep groundwater from northern Sahara. *Journal of Hydrology*, 328, 572–580.

Hammond, D. E., J. G. Zukin, and T-L. Ku. (1988). The kinetics of radioisotope exchange between brine and rock in a geothermal system. *Journal of Geophysical Research*, 93(B11), 13175–13186.

IAEA. (2006). *Use of Chlorofluorocarbon in Hydrology – A Guidebook*. IAEA, 277p.

IAEA/WMO. (2006). *Global Network of Isotopes in Precipitation. The GNIP Database*. Accessible at: http://isohis.iaea.org

Ivanovich, M., K. Froehlich, and M. J. Hendry (Eds.) (1991a). Dating of very old groundwater, Milk River Aquifer, Alberta, Canada. *Applied Geochemistry* 6(4), 367–475 (Special Issue).

Ivanovich, M., K. Froehlich, and M. J. Hendry. (1991b). Uranium-series radionuclides in fluids and solids, Milk River aquifer, Alberta, Canada. *Applied Geochemistry*, 6, 405–418.

Jones, B. F., A. Vengosh, E. Rosenthal, Y. Yecheli. (1999). Geochemical investigations. In: *Seawater Intrusions in Coastal Aquifers*. Kluwer Publishers, Dordrecht, pp. 51–74.

Jordan, H., and K. Froehlich. (1990). Groundwater dating in monitoring groundwater resources. In: *Groundwater Monitoring and Management*, pp. 139–151 (IAHS Publication 173).

Kendall, C. (1998). Tracing nitrogen sources and cycling in catchments. In: *Isotope Tracers in Catchment Hydrology* (Eds C. Kendall and J. J. McDonnel). Elsevier, The Netherlands, pp. 519–576.

Kennedy, C. D., and D. P. Genereux. (2007). ^{14}C groundwater age and the importance of chemical fluxes across aquifer boundaries in confined Cretaceous aquifers of North Carolina, USA. *Radiocarbon*, 49(3), 1181–1203.

Kigoshi, K. (1973). Uranium 234/238 disequilibrium and age of underground water. Presented at IAEA Panel on Interpretation of Uranium Isotope Disequilibrium as a Hydrological Tracer. IAEA, Vienna, Austria.

Lehmann, B. E., A. Love, R. Purtschert, P. Collon, H. H. Loosli, W. Kutschera, U. Beyerle, W. Aeschbach-Hertig, R. Kipfer, S. K. Frape, A. Herczeg, J. Moran, I. N. Tostikhin, and M. Gröning. (2003). A comparison of groundwater dating with ^{81}Kr, ^{36}Cl and ^{4}He in four wells of the Great Artesian Basin, Australia. *Earth and Planetary Science Letters*, 211, 237–250.

Leslie, B. W., and D. Hammond. (2007). Use of U and Th decay-series disequilibrium to characterize geothermal systems: An example from the Coso Geothermal System. *American Geophysical Union* 2007AGUFM.V54C..07L.

Loosli, H. H. (1992). Application of ^{37}Ar, ^{39}Ar, and ^{85}Kr in hydrology, oceanography and atmospheric studies. Current state of the Art. In: *Isotopes of Noble Gases as Tracers in Environmental Studies* (Proceedings of Consultants Meeting) IAEA, Vienna, Austria, pp. 73–85.

Loosli, H. H., and H. Oeschger. (1969). ^{37}Ar and ^{81}Kr in the atmosphere. *Earth and Planetary Science Letters*, 7, 67–71.

Loosli, H. H., and H. Oeschger. (1980). Use of ^{39}Ar and ^{14}C for groundwater dating. *Radiocarbon*, 22(3), 863–870.

Love, A. J., A. L. Herczeg, L. Sampson, R. G. Cresswell, and L. K. Fifield. (2000). Sources of chloride and implications for ^{36}Cl dating of old groundwater, southwestern Great Artesian Basin, Australia. *Water Resources Research*, 36, 1561–1574.

Majoube, M. (1971). Fractionnement en oxygène-18 et en deutérium entre l'eau et sa vapeur. *Journal of Chemical Physics*, 197, 1423–1436.

Mariotti, A., A. Landreau, and B. Simon. (1988). ^{15}N isotope biogeochemistry and natural denitrification process in groundwater: Application to the chalk aquifer of northern France. *Geochimica et Cosmochimica Acta*, 52, 1869–1878.

Mayer, B. (2007). Assessing sources and transformations of sulphate and nitrate in the hydrosphere using isotope techniques. In: *Isotopes in the Water Cycle* (Eds P. Aggarwal, J. Gat, and K. Froehlich). Springer, Dordrecht, The Netherlands, pp. 67–88.

Merlivat, L., and J. Jouzel. (1979). Global climate interpretation of the deuterium-oxygen 18 relationship for precipitation. *Journal of Geophysical Research*, 84, 5029–5033.

Mook, W. G. (2000). Environmental Isotopes in the Hydrological Cycle: Principles and applications, Volume 1: Introduction: Theorie, Methods Review. International Hydrological Programme IHP-V. Technical Documents in Hydrology, No. 39, Vol 1, UNESCO, Paris.

Morgenstern, U., R. Gellermann, D. Hebert, I. Boerner, W. Stolz, R. Vaikmae, R. Rajamae, and H. Putnik. (1995). ^{32}Si in limestone aquifers. *Chemical Geology*, 120, 127–134.

Münnich, K. O. (1957). Messung des ^{14}Gehaltes von hartem Grundwasser. *Naturwissenschaften*, 44, 32–33.

Münnich, K. O., W. Roether, and L. Thilo. (1967). ^{14}C in fractions of dissolved organic carbon in groundwater. *Nature*, 337, 153–155.

Murphy, E. M., S. N. Davis, A. Long, D. Donahue, and A. J. T. Jull. (1989). Characterization and isotopic composition of organic and inorganic carbon in the Milk River Aquifer. *Water Resources Research*, 25(8), 1893.

Nijampurkar, V. N., B. S. Amin, D. P. Kharkar, and D. Lal. (1966). Dating of groundwaters of ages younger than 1000–1500 years using natural Silicon-32. *Nature*, 210, 478–480.

Nir, A. (1964). On the interpretation of tritium 'age' measurements of groundwater. *Journal of Geophysical Research*, 69, 2589–2595.

Nolte, E., P. Krauthan, G. Korschinek, P. Maloszewski, P. Fritz, and M. Wolf. (1991). Measurements and interpretations of ^{36}Cl in groundwater, Milk River aquifer, Alberta, Canada. *Applied Geochemistry*, 6, 435–445.

O'Driscoll, M. A., D. R. Dewalle, K. J. McGuire, and W. J. Gburek. (2005). Seasonal ^{18}O variations and groundwater recharge for three landscape types in central Pennsylvania, USA. *Journal of Hydrology*, 303(1-4), 108–124.

Patterson, L. J., N. C. Sturchio, B. M. Kennedy, M. C. van Soest, M. Sultan, Z. T. Lu, B. Lehmann, R. Purtschert, Z. El Alfy, B. El Kaliouby, Y. Dawood, and A. Abdallah. (2005). Cosmogenic, radiogenic, and stable isotopic constraints on groundwater residence time in the Nubian Aquifer, Western Desert of Egypt. *Geochemistry Geophysics Geosystems*, 6, Q01005, doi:10.1029/2004GC000779.

Phillips, F. M., H. W. Bentley, S. N. Davis, D. Elmore, and G. B. Swanick. (1986). Chlorine-36 dating of very old groundwater. 2. Milk river aquifer, Alberta, Canada. *Water Resources Research*, 22, 2003–2016.

Plummer, L. N. (2005). Dating of young groundwater. In: *Isotopes in the Water Cycle* (Eds P. K. Aggarwal, J. R. Gat, and K. Froehlich). Springer, Dordrecht, pp. 193–218.

Plummer, L.N., Prestemon, E.C., Parkhurst, D.L. (1994). *An Interactive Code (NETPATH) for Modelling net Geochemical Reactions along a Flow Path, Version 2.0.* US Geological Survey Water Resources Investigation Report 94-4169, 130p.

Portniaguine, O., and D. K. Solomon. (1998). Parameter estimation using groundwater age and head data, Cape Cod, Massachusetts. *Water Resources Research*, 34(4), 637–645.

Ramspacher, P., H. Zojer, K. Froehlich, and W. Stichler. (1992). The recharge of large springs from a carbonate aquifer near Innsbruck applying environmental tracers. In: *Tracer Hydrology* (Eds L. Hötzl and A. Werner). A.A. Balkema, Rotterdam/Brookfield, pp. 251–257.

Robertson, W. D., and J. A. Cherry. (1989). Tritium as an indicator of recharge and dispersion in a groundwater system in Central Ontario. *Water Resources Research*, 25(6), 1097–1109.

Robinson, B., L. Villasenor, and V. Clemente. (1987). Preliminary stable isotope investigations of acidic fluids in geothermal systems of the Philippines. In: *Proceedings of 9th New Zealand Geothermal Workshop*, pp. 73–78.

Rozanski, K., and T. Florkowski. (1979). Krypton-85 dating of groundwater. In: *Isotope Hydrology 1978*, IAEA, Vienna, Austria, pp. 949–961.

Smethie, W. M., D. K. Solomon, S. L. Schiff, and G. Mathieu. (1992). Tracing groundwater flow in the Bordon aquifer using ^{85}Kr. *Journal of Hydrology*, 130, 279–297.

Solomon, D. K., R. J. Poreda, P. G. Cook, and A. Hunt. (1995). Site characterization using $^{3}H/^{3}He$ groundwater ages, Cape Cod, MA. *Ground Water*, 33(6), 988–996.

Sonntag, C., E. Klitzsch, E. P. Löhnert, E. M. El-Shazly, K. O. Münnich, Ch. Junghans, U. Thorweihe, K. Weistroffer, and F. M. Swailem. (1979). Palaeoclimatic information from deuterium and oxygen-18 in carbon-14-dated North Saharan Groundwaters. In: *Isotope Hydrology 1978*. Vol. II, IAEA, Vienna, Austria, pp. 569–581.

Sturchio, N. C., X. Du, R. Purtschert, B. E. Lehmann, M. Sultan, L. J. Patterson, Z.-T. Lu, P. Müller, T. Bigler, K. Bailey, T. P. O'Cooner, L. Young, R. Lorenzo, R. Becker, Z. El Alfy, B. El Kaliouby, Y. Dawood, and A. M. A. Abdallah. (2004). One million year old groundwater in the Sahara revealed by krypton-81 and chlorine-36. *Geophysical Research Letters*, 31, L05503.

Szabo, Z., D. E. Rice, L. N. Plummer, E. Busenberg, S. Drenkard, and P. Schlosser. (1996). Age dating of shallow groundwater with chlorofluorocarbons, tritium/helium 3, and flow path analysis, southern New Jersey, coastal plain. *Water Resources Research*, 32(4), 1023–1038.

Thorweihe, U., and M. Heinl. (2002). Groundwater Resources of the Nubian Aquifer System NE-Africa – Synthesis. Observatoire du Sahara et du Sahel (OSS, Paris).

Tolstikhin, I. N., and I. L. Kamensky. (1969). Determination of groundwater age by the T-^{3}He method. *Geochemistry International*, 6, 810–811.

Torgersen, T., and W. B. Clarke. (1985). Helium accumulation in groundwater, I: An evaluation of sources and the continental flux of crustal ^{4}He in the Great Artesian Basin, Australia. *Geochimica et Cosmochimica Acta*, 49, 1211–1218.

Torgersen, T., M. A. Habermehl, F. M. Phillips, D. Elmore, P. Kubik, B. G. Jones, T. Hemmick, and H. E. Gove. (1991). Chlorine-36 dating of very old groundwater III. Further studies in the Great Artesian Basin, Australia. *Water Resources Research*, 27(12), 3201–3213.

Vogel, J. C. (1967). Investigation of groundwater flow with radiocarbon. In: *Isotopes in Hydrology. Proceedings of International Conference, Vienna, 1966.* IAEA, Vienna, Austria, pp. 355–369.

Wigley, T. M. L., L. N. Plummer, and F. J. Pearson, Jr. (1978). Mass transfer and carbon isotope evolution in natural water systems. *Geochimica et Cosmochimica Acta*, 42, 1117–1139.

Winger, K., J. Feichter, M. B. Kalinowski, H. Sartorius, and P. Schlosser. (2005). A new compilation of the atmospheric 85krypton inventories from 1945 to 2000 and its evaluation in a global transport model. *Journal of Environmental Radioactivity*, 80, 183–215.

Zuber, A., and P. Maloszewski. (2000). Lumped parameter models. In: *Environmental Isotopes in the Hydrological Cycle – Principles and Applications* (Ed. W. G. Mook). UNESCO/IAEA, Vol. VI, pp. 5–31.

CHAPTER 6

Examining Processes and Rates of Landscape Change with Cosmogenic Radionuclides

Susan Ivy-Ochs[1,2,*] *and* Mirjam Schaller[3]

Contents

*Corresponding author. Tel: +41-44-633-3886; Fax: +41-44-633-1067
E-mail address: ivy@phys.ethz.ch

[1] Institute of Particle Physics, ETH Zurich, CH-8093 Zurich, Switzerland
[2] Department of Geography, University of Zurich, CH-8057 Zurich, Switzerland
[3] Department of Geological Sciences, University of Michigan, Ann Arbor, MI 48109.1005, USA

Radioactivity in the Environment, Volume 16
ISSN 1569-4860, DOI 10.1016/S1569-4860(09)01606-4

1. Introduction

The rate at which the surface of the Earth is modified varies from imperceptible (slow weathering of bare bedrock surfaces) to catastrophic (landslides or volcanic eruptions). Sites of rapid fluvial incision contrast with hillslope interfluves where slow bedrock weathering dominates. As cosmogenic nuclides measure how long mineral grains have been within the upper few meters of the Earth's surface, the timing of landform construction and rates of landscape modification can be determined. Cosmogenic nuclides build up predictably with time within the lattices of minerals as a result of reactions induced by cosmic rays.[1] The ability to measure the concentrations of cosmogenic nuclides in rock and sediment has revolutionized geomorphology and played a central role in catapulting it from a largely descriptive to a quantitative science (Summerfield, 2005). The unique power of cosmogenic nuclides is that erosional surfaces and depositional units that were themselves formed by rivers or glaciers can now be dated directly. Prior to the advent of cosmogenic nuclide methods landform ages were estimated or dated indirectly through correlation; or with bracketing ages from underlying and overlying sediments.

In 1955, the possibility of using ^{36}Cl produced *in situ* in terrestrial samples was discussed and tested by Davis and Schaeffer (1955). Because of low production rates at the Earth's surface (in contrast to higher ones in space) combined with measurement limitations, application to terrestrial studies was not possible. Real advancement in the field awaited the development of highly sensitive mass spectrometers, especially accelerator mass spectrometers, in the early 1980s (Klein et al., 1982; Elmore and Phillips, 1987; Raisbeck et al., 1987). Subsequently, successful exposure-age determinations in terrestrial samples using ^{3}He (Kurz, 1986), ^{10}Be and ^{26}Al (Klein et al., 1986; Nishiizumi et al., 1986), ^{21}Ne (Marti and Craig, 1987), and ^{36}Cl (Kubik et al., 1984; Phillips et al., 1986) were reported. The measurement of ^{14}C in terrestrial rock samples followed quickly (Jull et al., 1992).

[1] See Chapter 1.

In 1991, Lal presented the framework for using cosmogenic nuclides to determine either exposure ages or erosion rates. A detailed review of all aspects of cosmogenic nuclides in Earth sciences including the appropriate equations can be found in Gosse and Phillips (2001). Additional insights may be found in Nishiizumi et al. (1993), Cerling and Craig (1994b), Zreda and Phillips (2000), Bierman et al. (2002), Morris et al. (2002), Niedermann (2002), Bierman and Nichols (2004), Cockburn and Summerfield (2004), Granger (2006), Nichols et al. (2006), and von Blanckenburg (2006).

Cosmogenic nuclides of interest for geomorphological studies, their half-lives and the suitable minerals are given in Table 1. In this chapter, we discuss the radionuclides ^{10}Be, ^{14}C, ^{26}Al, and ^{36}Cl in detail; where appropriate, information on ^{3}He and ^{21}Ne is included.

2. Principles of Cosmogenic Nuclide Research

2.1. Systematics of *in situ*-produced cosmogenic nuclides

Cosmic rays entering the atmosphere consist primarily of protons.[2] Due to interactions with target nuclei in the atmosphere, a cascade of secondary particles is produced. At the surface of the Earth, this secondary cosmic ray flux is dominated by neutrons and muons (Lal and Peters, 1967; Lal, 1991). The main reactions for production of ^{10}Be, ^{14}C, ^{26}Al, and ^{36}Cl in rocks and sediment are listed in Table 2. Most cosmogenic nuclides are produced through spallation of the target element. In this reaction, the target element is broken apart by interaction with a neutron or proton of sufficient energy. Although the newly formed cosmogenic nuclide is not bound within the crystal lattice, apart from weathering or diffusion processes, it generally remains trapped. Cosmogenic nuclides are also produced through interactions of muons with the target element (capture of slow muons and stopping of fast muons). Finally, cosmogenic nuclides can be produced through the capture of epithermal and thermal (low-energy) neutrons (Gosse and Phillips, 2001, and references therein). The total production rate is given by the sum of all production mechanisms for that nuclide from the target elements present in that mineral (Table 2):

$$P_{(0)} = P_{(0)\mathrm{nuc}} + P_{(0)\mu\mathrm{stopped}} + P_{(0)\mu\mathrm{fast}} + P_{(0)\mathrm{thermal}} + P_{(0)\mathrm{epithermal}} \quad (1)$$

where $P_{(0)\mathrm{nuc}}$, $P_{(0)\mu\mathrm{stopped}}$, $P_{(0)\mu\mathrm{fast}}$ $P_{(0)\mathrm{thermal}}$, and $P_{(0)\mathrm{epithermal}}$ [atoms g^{-1} yr^{-1}] are the production rates at the Earth's surface for spallation (nucleonic), stopped muonic, fast muonic production, capture of thermal and epithermal neutrons, respectively, at the Earth's surface. The low-energy neutron capture mechanisms are relevant only for ^{36}Cl, thus for ^{10}Be and ^{26}Al, the

[2] See Chapter 1.

Table 1 Mineral–nuclide overview.

Nuclide	Half-life	Other isotopes	Measuring method	Target elements	Production rate (atoms g^{-1} yr^{-1})	Advantages/minerals used	Disadvantages
^{10}Be	1.51 Myr	^{9}Be	AMS	O	5	Quartz resistant and ubiquitous	Low production rate, ^{10}B interference in AMS
				Si			Generally restricted to quartz (no meteoric ^{10}Be)
^{26}Al	716 kyr	^{27}Al	AMS	Si	31	High production rate	Restricted to quartz (low Al)
						Quartz resistant and ubiquitous	Accurate determination of ^{27}Al required
^{36}Cl	301 kyr	^{35}Cl, ^{37}Cl	AMS	Ca	Composition dependent	Low detection limit (low AMS background)	Complicated production
				K		Any rock type, silicates and carbonates	^{36}S interference in AMS
				^{35}Cl	10 granite		Accurate determination of total Cl required
					20 limestone		Determination of rock composition required

^{14}C	5.73 kyr	^{12}C, ^{13}C	AMS	O	16	Useful for short timescales	Short half-lfe
						Quartz resistant and ubiquitous	Atmospheric ^{14}C contamination
^{3}He	Stable	^{4}He	Static mass spectrometry	Many	120	High production rate	Diffuses out of quartz or volcanic groundmass
						Useful for long timescales	Radiogenic/ nucleogenic/ magmatic correction
						Pyroxene, olivine	Beware pre-exposure
^{21}Ne	Stable	^{20}Ne, ^{22}Ne	Static mass spectrometry	Mg	20	Useful for long timescales, >50 ka	Nucleogenic/magmatic correction
				Si		Quartz, olivine, pyroxene	High air background possible
							Beware pre-exposure

Production rates are at sea level and high latitude (for production rate compilation see Gosse and Phillips, 2001).

Table 2 Main reactions for production of cosmogenic nuclides ^{10}Be, ^{14}C, ^{26}Al, and ^{36}Cl in rocks at the surface of the Earth.

Target element	Spallation	Low-energy neutron capture	Negative muon capture
O	$^{16}O(n, 2pn)\ ^{14}C$		$^{16}O(\mu^-, pn)\ ^{14}C$
	$^{16}O(n, 4p3n)\ ^{10}Be$		$^{16}O(\mu^-, 3p3n)\ ^{10}Be$
Si	$^{28}Si(n, x)\ ^{10}Be$		$^{28}Si(\mu^-, x)\ ^{10}Be$
	$^{28}Si(n, p2n)\ ^{26}Al$		$^{28}Si(\mu^-, 2n)\ ^{26}Al$
Cl		$^{35}Cl(n, \gamma)\ ^{36}Cl$	
K	$^{39}K(n, 2p2n)\ ^{36}Cl$	$^{39}K(n, \alpha)\ ^{36}Cl$	$^{39}K(\mu^-, p2n)\ ^{36}Cl$
Ca	$^{40}Ca(n, 3p2n)\ ^{36}Cl$		$^{40}Ca(\mu^-, \alpha)\ ^{36}Cl$
	$^{40}Ca(n, \alpha p)\ ^{36}Cl$		

For production reactions for ^{3}He and ^{21}Ne see Niedermann (2002).

last two terms are not required. The noble gases, ^{3}He and ^{21}Ne, are produced only through spallation reactions ($P_{(0)nuc}$).

Production rates are altitude and latitude dependent (Figure 1) (Gosse and Phillips, 2001). The Earth's magnetic field impedes particles with lesser energies at the equator and deflects particles toward the poles. Hence production rates are higher at higher latitude. In traversing the atmosphere, the flux of cosmic ray secondary particles decreases. As a consequence production rates are lower at lower altitude. As production due to muon reactions decreases less with decreasing altitude than production due to spallation, the two types of reactions must be scaled separately. Several physical models have been presented for the scaling of production rates from their sea level and high latitude values (Table 1) to the altitude and latitude of the sampling site (Lal, 1991; Dunai, 2000, 2001; Stone, 2000; Desilets and Zreda, 2001; Pigati and Lifton, 2004; Muzikar, 2005). In any case, it is crucial when calculating an exposure age to use the same scaling formalities as were used in the original production rate determinations.

Production rates of cosmogenic nuclides have been determined: (a) by measurement of nuclide concentrations in rocks from the Earth's surface that were exposed for a "known" amount of time (Nishiizumi et al., 1989; Zreda et al., 1991; Cerling and Craig, 1994a; Niedermann et al., 1994; Phillips et al., 1996b; Stone et al., 1996, 1998; Kubik et al., 1998; Licciardi et al., 1999, 2006; Evans, 2001); (b) through experiments involving exposure of artificial targets to cosmic rays over several years (Nishiizumi et al., 1996; Brown et al., 2000; Graham et al., 2000); and (c) through calculations based on measured or estimated cross sections for given production reactions (Masarik and Reedy, 1995; Masarik, 2002).

Variations of nuclide production over time are related to variations in the primary galactic cosmic ray flux, to changes in solar activity, and to changes in the shielding (reduction of cosmic ray flux) by the Earth's

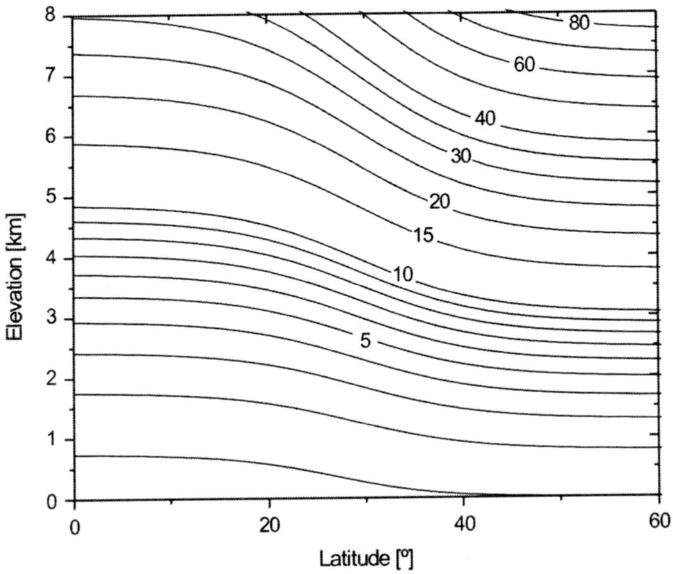

Figure 1 Production of cosmogenic nuclides [atoms g^{-1} yr^{-1}] as a function of geomagnetic latitude and altitude (based on Lal and Peters, 1967; Lal, 1991; Kubik et al., 1998). The production rates have been normalized to sea level and high latitude. At low latitude, production rates are lower than at high latitude. Production rates increase exponentially with increasing altitude.

magnetic field (Lal and Peters, 1967; Lal, 1991). Since most applications span over a certain time interval, the moderating effect of variable production rates over time is smoothed out (Lal, 1991). Several different models for correcting for the effect of past magnetic field changes on production rates have been proposed (Dunai, 2001; Masarik et al., 2001; Pigati and Lifton, 2004; Desilets et al., 2006b). In Figure 2, the variations of integrated production rates over time for different latitudes due to changes in paleomagnetic intensity are shown for the last 150 kyr (based on the method of Masarik et al., 2001).

Cosmic rays are attenuated as they pass through rock. Therefore, production of cosmogenic nuclides by spallation and muon-induced reactions decreases with depth (Figure 3). ^{10}Be, ^{14}C, and ^{26}Al are produced predominantly by spallation at the very surface. As muons are less apt to react, they penetrate much deeper into the Earth's surface. Thus, production due to muons becomes more important than production due to spallation below depths of about 2 m in a rock of density 2.7 g cm^{-3}. The depth-dependent production rates for ^{10}Be, ^{14}C, and ^{26}Al are given as follows:

$$P_{(x)} = P_{(0)\text{nuc}} e^{-(x\rho/\Lambda_{\text{Nuc}})} + \mathbf{P}_{(0)\mu\text{stopped}} e^{-(x\rho/\Lambda_{\mu\text{stopped}})} + P_{(0)\mu\text{fast}} e^{-(x\rho/\Lambda_{\mu\text{fast}})} \quad (2)$$

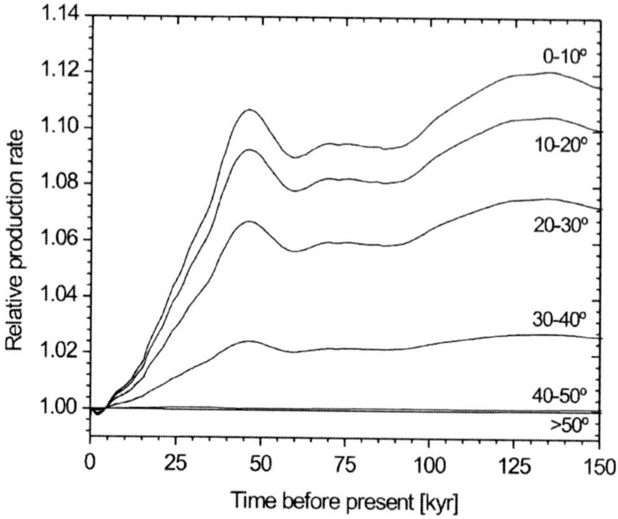

Figure 2 Production rate change over time due to paleomagentic field variations (after Masarik et al., 2001). The relative mean production rates of long-lived cosmogenic radionuclides are shown for samples exposed between the time given on the abcissa and the present day. The different lines show the production rate changes for various latitude bins (figure courtesy of J. Masarik).

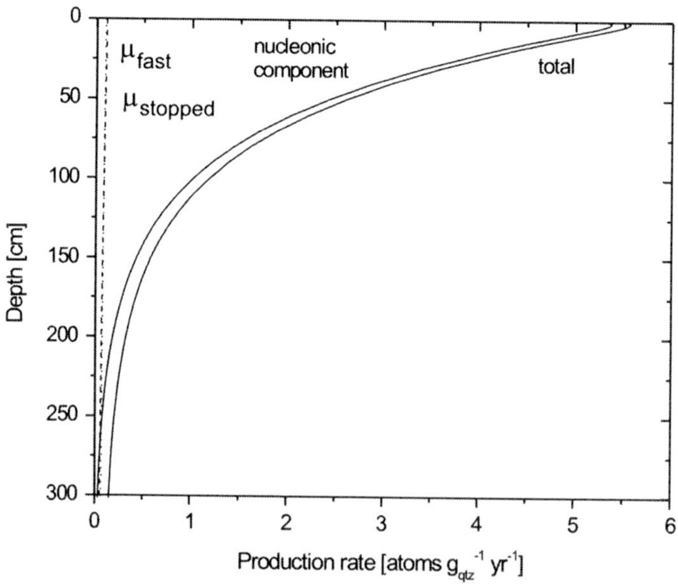

Figure 3 Production rate of ^{10}Be in quartz as a function of depth. The production rates are for sea level and high latitude. The total production is a composite of the production by neutron spallation, stopped muons, and fast muons.

where x (cm) is the depth below the surface, ρ (g cm^{-3}) the density of the overburden, and Λ_{Nuc}, $\Lambda_{\mu stopped}$, and $\Lambda_{\mu fast}$ (g cm^{-2}) the attenuation length for nucleons, stopped, and fast muons, respectively. Suggested values for Λ_{Nuc}, $\Lambda_{\mu stopped}$, and $\Lambda_{\mu fast}$ are 150, 1,300, and 5,000 g cm^{-2}, respectively (Braucher et al., 2003) (cf. Gosse and Phillips, 2001). Several authors have suggested that the depth profiles for the muon-induced production mechanisms may be better approximated by a combination of several exponential terms (Granger and Smith, 2000; Schaller et al., 2001, 2002). It is important to note that production is similarly reduced if snow, sediment or vegetation covers sampled surfaces during exposure (Benson et al., 2004; Schildgen et al., 2005).

^{36}Cl is produced not only through spallation and muon reactions but also through low-energy (thermal and epithermal) neutron capture (Table 2). Low-energy neutrons can diffuse back out of a rock surface. As a result of this neutron leakage, production of ^{36}Cl by neutron capture peaks about 15–20 cm down into the rock (Figure 4), depending on the rock

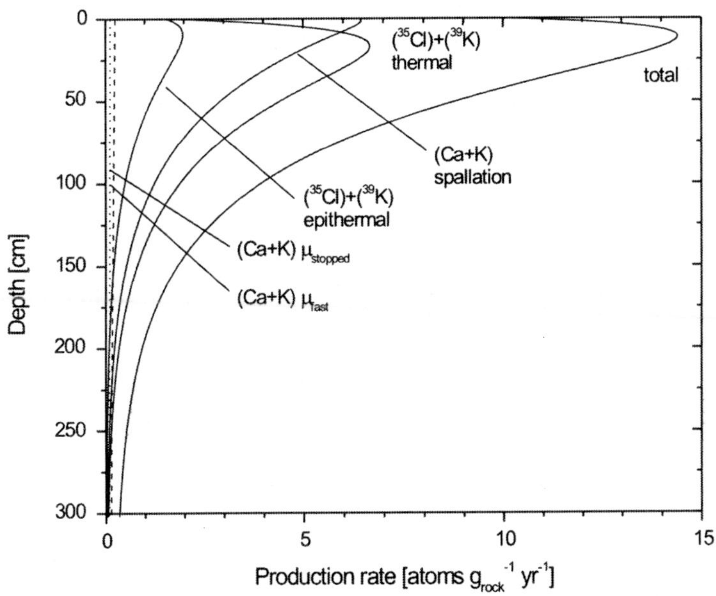

Figure 4 Production rate of ^{36}Cl in granite with 200 ppm of Cl as a function of depth. The production rates are for sea level and high latitude. The total production is a composite of production by neutron spallation, thermal and epithermal neutrons, stopped and fast muons. The maximum production by thermal neutrons is highest at about 20 cm depth in a rock with a given density of 2.7 g cm^{-3}. Diffusion of thermal neutrons back out of the rock in the uppermost few centimeters causes this reduction in production rate. Variation of production rates with depth calculated after Liu et al. (1994), Stone et al. (1996, 1998), Evans (2001), and Phillips et al. (2001).

composition (proportion of target elements) and density (Fabryka-Martin, 1988; Liu et al., 1994; Phillips et al., 2001). The depth-dependent production rate for ^{36}Cl is similar to Equation (2), but terms for the depth dependence of thermal and epithermal neutron capture reactions must be included (see Gosse and Phillips, 2001).

Production at a specific sampling site is reduced if the sampled surface was dipping or was shielded by surrounding topography. Detailed equations for making this correction are given in Lal (1991), Dunne et al. (1999), and Dunne and Elmore (2003). The quantification of this production rate reduction not only is important in the case of exposure dating, but also must be taken into account in the determination of catchment-wide erosion rates (Section 4.9) especially in drainage basins with steep slopes.

Following the above discussion on production systematics, we may now state the basic assumptions that are implicit in the use of *in situ*-produced cosmogenic nuclides to address problems of landscape evolution:

- the half-life is known,
- production pathways and production rates including their variation in space and time are known,
- the initial nuclide concentration (inheritance) is zero or can be determined,
- the mineral has remained a closed system, that is, there has been no gain or loss of the nuclide except due to production or decay (or through erosion).

2.2. Determination of age and/or erosion rate

2.2.1. Exposure ages and erosion rates

The concentration of cosmogenic nuclides in a mineral is mainly a function of production rate, decay rate of the nuclide, exposure time of the surface and erosion rate. The concentration of cosmogenic nuclides (atoms g^{-1}) in a surface exposed to secondary cosmic rays is governed by the following general equation:

$$C_{(t)} = \frac{P_{(0)}}{\lambda + \rho\varepsilon/\Lambda}\left(1 - e^{-(\lambda + (\rho\varepsilon/\Lambda)t)}\right) + C_{\text{in}} e^{-\lambda t} \qquad (3)$$

where $P_{(0)}$ (atoms g^{-1} yr^{-1}) is the production rate at the Earth's surface, t (yr) the exposure age of the surface, λ (yr^{-1}) the decay constant, ρ (g cm^{-3}) the density of the irradiated material, ε (cm yr^{-1}) the erosion rate, Λ (g cm^{-2}) the attenuation length, and C_{in} (atoms g^{-1}) the nuclide concentration already present at the beginning of exposure (inherited). Inheritance (or pre-exposure) is that portion of the nuclide inventory that is already in the rock before the beginning of the final exposure period. In the first instance, exposure ages are calculated assuming zero inheritance (the second term in Equation (4) drops out). For calculating an age assuming no erosion of the rock surface, Equation (3) reduces to:

$$C_{(t)} = \frac{P_{(0)}}{\lambda}(1 - e^{-\lambda t}) \tag{4}$$

An exposure age is calculated from the measured nuclide concentrations by solving Equation (4) for t. The production rate is scaled for site latitude and longitude, altitude as well as thickness or topographic shielding.

Radionuclides buildup with time but they are also decaying. After about three to four half-lives, production of radionuclides equals loss due to decay (Figure 5). This balance in production and decay is called saturation, secular equilibrium, or nuclide steady state. Thus, when there is no erosion, the time ranges useful for exposure dating are about 4–5 Myr for ^{10}Be, 2–3 Myr for ^{26}Al, 1 Myr for ^{36}Cl, and 20 kyr for ^{14}C. With more rapid erosion secular equilibrium is reached earlier and at a lower nuclide concentration (Figure 6). Note that, because of the production depth profile (Figure 4), this rule does not strictly hold for ^{36}Cl (cf. Gosse and Phillips, 2001). As shown in Figure 7, in stark contrast to the behavior shown by other nuclides, a higher concentration of ^{36}Cl can buildup when the rock surface

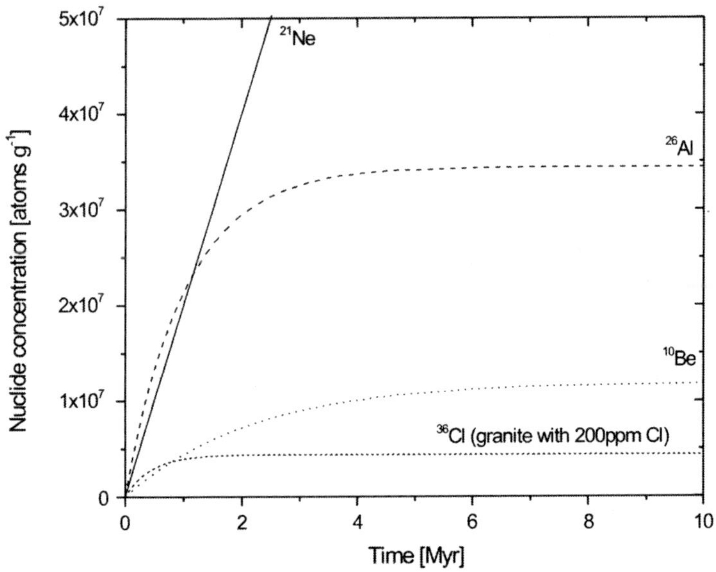

Figure 5 Increase in concentration of the radionuclides ^{10}Be, ^{26}Al, ^{36}Cl, and the stable nuclide ^{21}Ne with time on a flat, non-eroding surface. The production of ^{10}Be, ^{26}Al, and ^{21}Ne is in quartz. A granite with 200 ppm Cl has been assumed for the production of ^{36}Cl. All production values are for sea level and high latitude. Secular equilibrium, where production of radionuclides equals radioactive decay, is approached after 4–5 half-lives. Secular equilibrium sets the limit of the maximum exposure age that can be determined with a given radionuclide. The determined age may nevertheless be a minimum age for the landform.

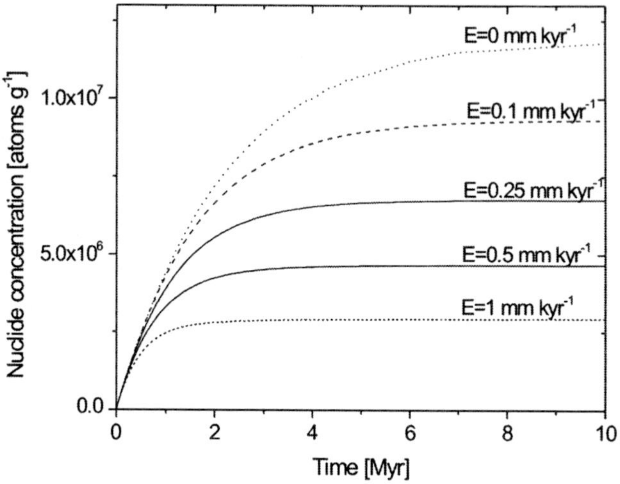

Figure 6 Increase of the ^{10}Be concentration with time taking into account different steady-state erosion rates of the exposed surface. Secular equilibrium, where radionuclide gain due to production equals loss due to radioactive decay and erosion, is approached earlier for more rapid erosion rates.

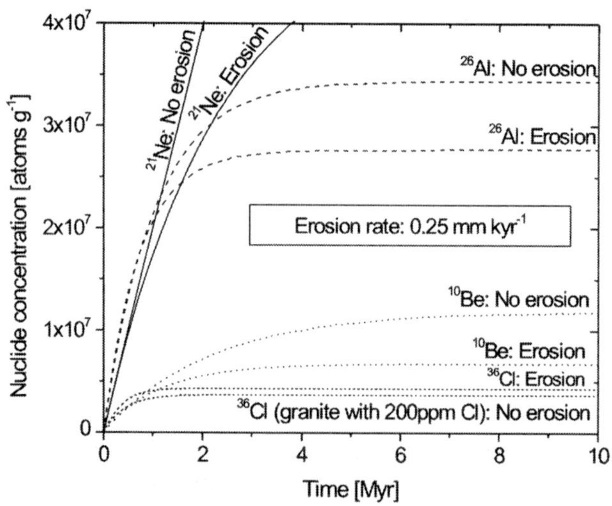

Figure 7 Effect of erosion on growth of nuclide concentrations for ^{21}Ne, ^{26}Al, ^{10}Be, and ^{36}Cl. Note that, for this rock composition (granite with 200 ppm Cl), the ^{36}Cl concentration reaches higher concentrations when the rock surface is eroding. For all other nuclides, the concentrations in eroding surfaces are lower than for non-eroding surfaces after an equivalent amount of time.

is eroding in comparison to when there is no erosion. For the rock composition and erosion rate in the example shown in Figure 7, the ^{36}Cl exposure age of an eroding rock surface will be older than the true age.

Cosmogenic nuclide concentrations can be used to determine erosion rates. The critical assumption is that the nuclide concentration is at steady state (erosion and decay balance production). When this condition is satisfied the term, *t*, goes to infinity and Equation (3) reduces to:

$$C_{(t)} = \frac{P_{(0)}}{\lambda + ((\rho\varepsilon)/\Lambda)} \tag{5}$$

Assuming that enough time has passed so that the nuclide concentrations have reached steady state, measuring the nuclide concentrations allows calculation of the maximum erosion rate. If the steady-state condition has not been satisfied, the true erosion rate can be significantly lower than the calculated rate.

2.2.2. Paired analysis: Exposure age, erosion rate, and burial age determinations

Both exposure age and erosion rate can be determined when the concentrations of two nuclides are measured in the same rock surface or sediment sample. Two nuclide data are portrayed in erosion-island plots (Klein et al., 1986; Lal, 1991; Nishiizumi et al., 1991). As both Be and Al can be extracted from a single quartz separate (Section 3.2), the pair of radionuclides ^{10}Be and ^{26}Al have often been used to determine exposure ages, erosion rates, as well as burial histories. In addition to the $^{26}Al/^{10}Be$ pair, other pairs can be used (e.g., $^{14}C/^{10}Be$, $^{21}Ne/^{10}Be$, and $^{36}Cl/^{10}Be$). In principle, the shorter the half-life, the better the nuclide is suited to study faster erosion rates on shorter timescales. A greater difference in half-life increases the effective time range that can be examined. Here we discuss the $^{26}Al/^{10}Be$ versus ^{10}Be plot (Figure 8). Four broad scenarios as defined by the different areas in the diagram are commonly considered: (i) constant exposure with no erosion, (ii) steady-state erosion, (iii) forbidden zone, and (iv) complex exposure scenario with burial and exhumation.

In the constant exposure scenario, as time of exposure increases, the data point follows a trajectory forming the constant exposure curve (Figure 8a). Because the production rates of ^{26}Al and ^{10}Be are about 31 and 5, respectively (Table 1), the $^{26}Al/^{10}Be$ ratio in short exposures (where decay is not a factor) is about 6. As the exposure time increases (^{10}Be and ^{26}Al concentrations increase) and as ^{26}Al and ^{10}Be decay with different rates (but ^{26}Al faster), the $^{26}Al/^{10}Be$ ratio decreases (Figure 8a). Where ^{26}Al and ^{10}Be production and decay reach the point of secular equilibrium (the concentrations of both no longer increase because production and decay

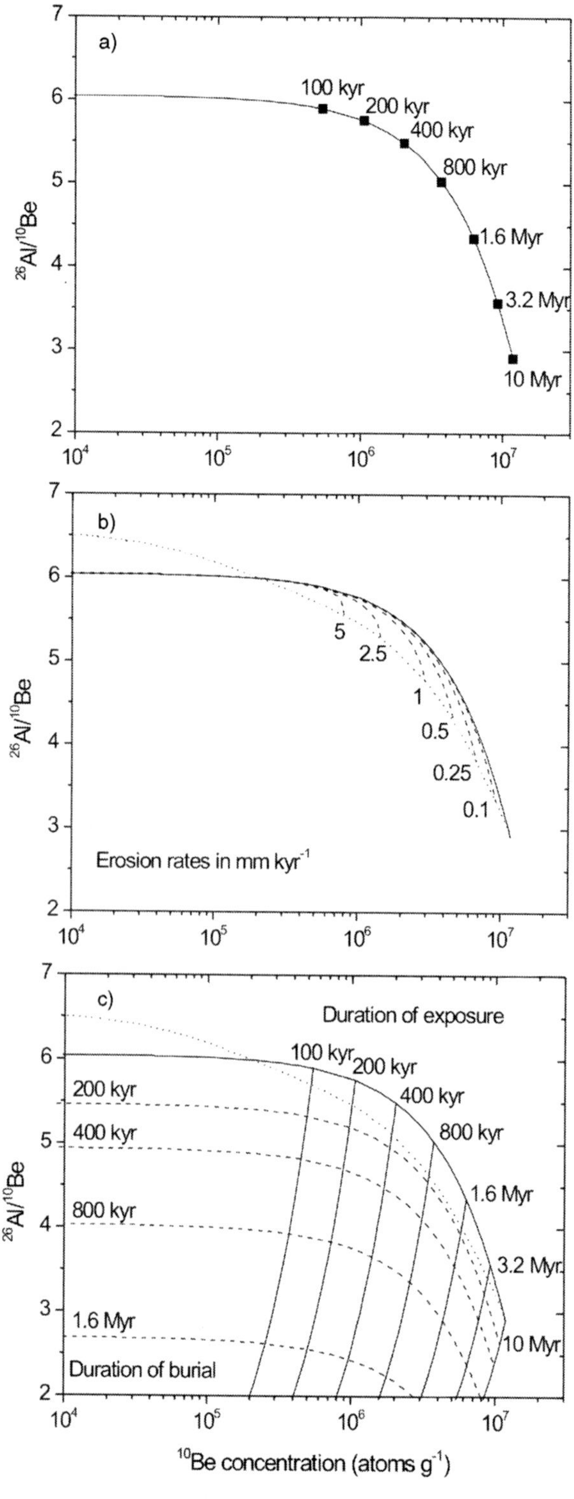
a)
100 kyr
200 kyr
400 kyr
800 kyr
1.6 Myr
3.2 Myr
10 Myr
b)
5
2.5
1
0.5
0.25
0.1
Erosion rates in mm kyr⁻¹
c)
Duration of exposure
100 kyr
200 kyr
400 kyr
800 kyr
1.6 Myr
3.2 Myr
10 Myr
200 kyr
400 kyr
800 kyr
1.6 Myr
Duration of burial
²⁶Al/¹⁰Be
¹⁰Be concentration (atoms g⁻¹)

are balanced; Figure 5), the $^{26}Al/^{10}Be$ ratio also reaches secular equilibrium at a $^{26}Al/^{10}Be$ ratio of about 3.

The thinner lines which fan down from the constant exposure curve are trajectories for rocks continuously exposed, but undergoing continuous steady-state erosion (see Figure 8b). The different erosion rates are indicated in mm kyr^{-1} (m Myr^{-1}). For each erosion trajectory, the ratio of the endpoint (final ratio of ^{26}Al to ^{10}Be) is higher than the endpoint ratio reached assuming no erosion. The area defined by the curve connecting the endpoints of the erosion trajectories and the constant no-erosion exposure curve is called the erosion island.

The region above the continuous exposure curve is called the forbidden zone (Figure 8b). No simple geologic process leads to data points that plot in this zone. Samples that were exposed at higher elevation may plot to the right of the erosion island (Nishiizumi et al., 1991). Problems with accurate ^{27}Al determinations may lead to data points that plot in the field above the erosion island.

The field below the erosion island is the field of complex exposure (Figure 8c). A period of exposure is followed by complete (or partial) shielding due to burial by snow, ice, or sediment. As no new nuclides are produced, and ^{26}Al is decaying faster than ^{10}Be, the $^{26}Al/^{10}Be$ ratio and the ^{10}Be concentration both decrease. The data point is shifted down and to the left on the diagram. Note that cases of partial shielding versus complete shielding lead to different scenarios (Bierman et al., 1999). Sample data points may also plot below the erosion island if thick (tens of centimeter) slabs of rock have spalled off. After spalling, the $^{26}Al/^{10}Be$ ratio in the new rock surface is the same as before the spalling. However, the concentration of each nuclide is lower as the new surface was located deeper within the rock (Small et al., 1997; Bierman et al., 1999). Therefore, the data point shifts straight to the left on the diagram.

As a result of the fact that ^{26}Al and ^{10}Be decay at different rates, the age of deeply buried sediments (e.g., in caves) can be determined in sediments completely shielded from further nuclide production

---

Figure 8 Plots of $^{26}Al/^{10}Be$ ratios versus ^{10}Be concentration showing the evolution of the $^{26}Al/^{10}Be$ ratios with time under different conditions. (a) Continuously exposed, non-eroding surfaces evolve along the black line. (b) Continuously exposed surfaces eroding with steady-state erosion follow the trajectories (dotted lines) that splay downward from the no-erosion line. The prescribed area is called the steady-state erosion island. (c) Samples that plot below the steady-state erosion island experienced a more complex exposure that may involve periods of burial. The dashed line represents burial isochrons of samples subjected to different exposure/erosion histories, but the same burial time. However, a given $^{26}Al/^{10}Be$ ratio and ^{10}Be concentration may be the result of several combinations of burial and exposure/erosion phases.

(e.g., Granger et al., 1997, 2001; Granger and Muzikar, 2001; Granger, 2006). Burial dating is based on the following assumptions: (i) the inherited nuclide concentration was acquired during steady erosion in the bedrock outcrop, (ii) transport and intermediate storage were short, and (iii) burial was rapid and there was only one burial event with complete shielding. As the sample is completely shielded from further irradiation (no production if muon-related production is neglected), Equation (3) reduces to:

$$C_{(t)} = C_{\mathrm{in}} e^{-\lambda t} \tag{6}$$

The ratio of two nuclides is then given by:

$$\frac{C_1}{C_{2(t)}} = \left(\frac{C_1}{C_2}\right)_{\mathrm{in}} e^{-(\lambda_1 - \lambda_2)t} \tag{7}$$

As $(C_1/C_2)_{\mathrm{in}}$ is given by the production ratio at the Earth's surface ($\sim$6) and $(C_1/C_2)_{(t)}$ can be determined through measurement, Equation (7) can be solved for t. In this way, not only the total burial time can be calculated, but also the steady-state (paleo) erosion rate of the sediment source region (Figure 8c).

In Figure 9, we show for comparison the $^{36}Cl/^{10}Be$ erosion-island plot. It differs from the $^{26}Al/^{10}Be$ erosion-island plot (Figure 8) in two respects (Figure 9). First, the half-life of ^{36}Cl is significantly shorter than the half-life

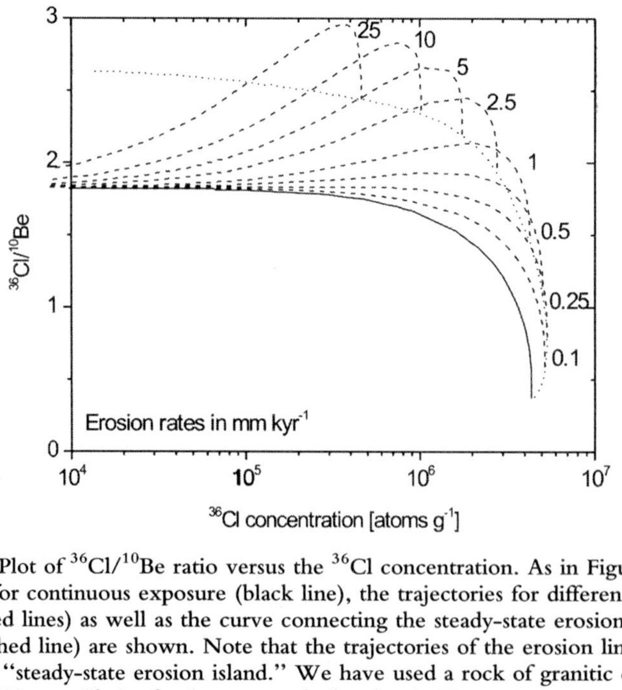

Figure 9 Plot of $^{36}Cl/^{10}Be$ ratio versus the ^{36}Cl concentration. As in Figure 8b the trajectory for continuous exposure (black line), the trajectories for different erosion rates (dotted lines) as well as the curve connecting the steady-state erosion end points (dashed line) are shown. Note that the trajectories of the erosion lines lie outside ot the "steady-state erosion island." We have used a rock of granitic composition with 200 ppm Cl. Production rates calculated as in Figure 4.

of ^{10}Be (301 kyr in comparison to 1.51 Myr). This difference in half-lives makes the ^{36}Cl/^{10}Be pair useful for geologic events with much shorter exposure histories and/or more rapid erosion rates. Second, as described above, the depth profile for ^{36}Cl is rock composition dependent and the nuclide concentration does not necessarily decrease consistently with depth as in the case for ^{10}Be (Figures 3 vs. 4). This has the effect of opening up the erosion island. But note that the trajectories of the erosion lines pass outside of the erosion island on their way to the erosion saturation points. This may lead to non-unique solutions for specific age and erosion rate combinations.

2.2.3. The depth profile

Measurements of nuclide concentrations from different depths can be used to determine exposure age, burial age, erosion (or aggradation) rate of the surface, and nuclide inheritance in sedimentary deposits (Figure 10). The cosmogenic nuclide inventory in most deposits consists not only of the nuclide concentration produced by post-depositional irradiation, but also of inherited nuclides produced before deposition, for example, during hillslope erosion and sediment transport. The total nuclide concentration, $C_{\text{total}}(z)$ at any depth in a deposit, is given by:

$$C_{\text{total}}(z) = C_{\text{in}} \times e^{-(\lambda t)} + C_{\text{dep}}(z) \quad (8)$$

where t (yr) is the deposition age of the sediment, λ (yr^{-1}) the decay constant, C_{in} (atoms g^{-1}) the inherited nuclide concentration, and $C_{\text{dep}}(z)$

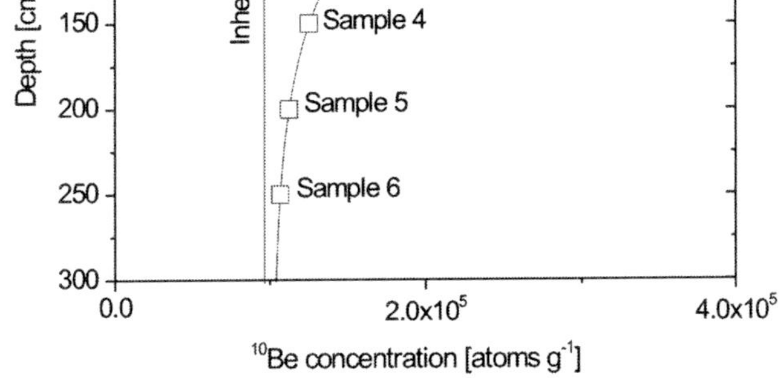

Figure 10 Principles of age determination in sedimentary deposits. The nuclide concentration in a sample from the surface is a composite of the inherited and the post-depositional nuclide concentration. The inherited concentration can be approximated by measuring several samples from different depths.

(atoms g^{-1}) the concentration of nuclides generated by post-depositional irradiation at a given depth. C_{in} is a function of the erosion rate in the sediment source area (see Equation (5)). $C_{dep}(z)$ is a function of the age of the deposit, the depth below the surface at which samples were taken, the density of the sediment deposit and subsequent aggradation or degradation of the top of the deposit. $C_{dep}(z)$ can be calculated using the following equation:

$$C_{dep(z)} = \frac{P_{(0)}}{\lambda + \rho\varepsilon/\Lambda} e^{-(x\rho/\Lambda)}(1 - e^{-(\lambda + (\rho\varepsilon/\Lambda)t)}) \quad (9)$$

Equation (9) is valid for nuclides produced by spallation reactions only (^{3}He and ^{21}Ne). In principle, the total of the nuclide inventory is comprised of all of the different production mechanisms. This means that a term similar to the one above must be provided for each production mechanism (Equation (1) with the terms appropriate for the nuclide of interest) (see Gosse and Phillips, 2001 for complete equations).

The depth profile technique is especially useful for dating fluvial terraces or alluvial fans. When no erosion (or aggradation) of the top surface can be assumed, the age and the inheritance can be obtained by simple linear regression of the depth profile data (Figure 10). Where the rate of the upper surface erosion cannot be clearly constrained by field data or morphostratigraphic relationships, the measured depth profile may be difficult to interpret. In Figure 11, we show several single-nuclide depth profiles into bedrock or a single rapidly deposited sedimentary unit where no nuclide inheritance is

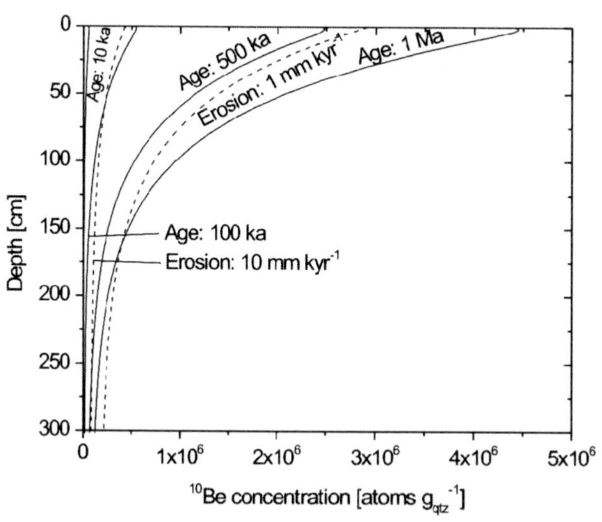

Figure 11 The concentration of ^{10}Be with depth into a rock surface. The solid lines represent the nuclide concentration in a rock surface that is continuously exposed for the time periods given but not eroding. The dashed lines show concentrations in continuously exposed and eroding rock surfaces that have reached secular equilibrium (see Figure 6).

present. The shape of the curves is dependent only on exposure time and erosion rate. With longer exposure periods the curve reaches farther and farther to the right. Note the similarity in the shapes of the curves produced in non-eroding surfaces after exposure for a certain period of time and the curves for rapidly eroding and infinitely exposed surfaces (nuclide concentrations at steady state). For many depth profile data sets an infinite array of combinations of age and erosion rate are possible.

3. Mineral–Nuclide Pairs

The radionuclides, ^{10}Be, ^{14}C, ^{26}Al, and ^{36}Cl, and the stable noble gases, ^{3}He and ^{21}Ne, are routinely measured today, the former using accelerator mass spectrometry (AMS) and the latter using static noble gas magnetic sector mass spectrometry. Depending on site and age, sample sizes for the radionuclides are in the tens of grams range, while those for the noble gases are in the hundreds of milligrams range or less. The noble gases differ from the radionuclides in that interferences from non-cosmogenic components are more common (Niedermann, 2002). In addition to the six commonly measured nuclides, below we discuss briefly ^{53}Mn, ^{41}Ca, and ^{38}Ar. We also note that, because of its short half-life (269 years), ^{39}Ar has been suggested as a useful nuclide (Froehlich and Lubbert, 1973). Recently, ^{15}N was measured successfully in terrestrial samples (Marty et al., 2006).

3.1. Radionuclides

3.1.1. ^{10}Be

Quartz is used in nearly all ^{10}Be studies (half-life 1.51 Myr). This is because it is a ubiquitous, resistant mineral that can be consistently cleaned of meteoric ^{10}Be. Meteoric ^{10}Be is produced in the atmosphere at a rate much higher than it is produced in terrestrial minerals. Inclusion of even minor amounts of this contaminating non-*in situ* ^{10}Be would lead to spurious results. Meteoric ^{10}Be may adsorb to mineral outer surfaces or may be incorporated in secondary minerals that form during weathering. Several studies have shown that meteoric ^{10}Be can be effectively removed from the outer surface of quartz grains with several steps with very dilute HF (Brown et al., 1991; Kohl and Nishiizumi, 1992). Accordingly, ^{10}Be is not measured in whole rock samples (Klein et al., 1997). Landscapes underlain by quartz-rich crystalline rocks (e.g., granites, granodiorites, tonalites; gneiss, schist especially with quartz veins), by silicic volcanic rocks (rhyolites, dacites, etc.) and by resistant sandstones and quartzites are suitable. A pure quartz mineral separate is obtained by etching crushed whole rock or sediment samples with dilute HF (Kohl and Nishiizumi, 1992). Further purification with heavy liquids or magnetic separation may be necessary. Extraction of Be and Al from quartz is well established (Kohl and Nishiizumi, 1992; Ivy-Ochs, 1996; von Blanckenburg et al., 1996; Ochs and Ivy-Ochs, 1997; Child et al., 2000; Bierman et al., 2002). Be and Al fractions are separated and purified using cation and anion exchange columns and selective pH precipitations. Ratios of ^{10}Be/^{9}Be are measured with AMS.

Beryllium has also been successfully extracted from chert (flint, microcrystalline quartz) (Boaretto et al., 2000; Ivy-Ochs et al., 2001; Verri et al., 2004, 2005) (Section 5.1).

^{10}Be concentrations have been measured in olivine (Nishiizumi et al., 1990; Seidl et al., 1997; Shimaoka et al., 2002, 2004) and pyroxene (Ivy-Ochs et al., 1998b; Braucher et al., 2006), although problems in removing meteoric ^{10}Be contamination (Seidl, 1993; Seidl et al., 1997; Ivy-Ochs et al., 1998b; Shimaoka et al., 2002) that is likely found in clays formed during mineral weathering (Ivy-Ochs et al., 1998b) may limit their utilization.

Kober et al. (2005) reported success using phenocryts of the alkali feldspar sanidine for ^{10}Be studies of 20 Myr old rhyolite ignimbrite bedrock from Chile. Good agreement with ^{10}Be concentrations measured in co-existing quartz and sanidine phenocrysts verifies that no meteoric ^{10}Be had invaded the sanidine grains. The sanidine phenocrysts in that study are fresh and unaltered with sharp grain boundaries. Production rates of ^{10}Be in sanidine are of the order of 90% of that in quartz (Kober et al., 2005; Ivy-Ochs et al., 2007b).

Braucher et al. (2005) compared ^{10}Be and ^{36}Cl concentrations in limestone with ^{10}Be concentrations in interbedded chert in samples from a depth profile. They reported that rinsing with water removes meteoric ^{10}Be from calcite. In contrast, Granger et al. (2005) reported ^{10}Be concentrations higher than expected in samples from a limestone quarry of Roman age. It is possible that carbonate rocks from different regions or even from the same site may exhibit no or random degrees of incorporation of meteoric ^{10}Be, depending on the degree of recrystallization in the rock. Merchel et al. (2008) found that the affinity of ^{10}Be for clay minerals poses a significant obstacle to the use of ^{10}Be in carbonate rocks.

3.1.2. ^{14}C

Because of its ubiquity, resistance to weathering and simple chemical composition, quartz is commonly used for *in situ* ^{14}C (half-life 5,730 yr) studies (Jull et al., 1992; Lifton et al., 2001). The most important concern is the removal of atmospheric ^{14}C contamination from the surface and crevices of the quartz before final melting (Lifton et al., 2001). *In situ* radiocarbon is extracted from about 5 g of quartz by flux ($LiBO_2$) melting at 1,100°C (Lifton et al., 2001). Resulting CO_2 is converted to graphite using standard procedures, and carbon ratios are measured with AMS.

3.1.3. ^{26}Al

The use of ^{26}Al (half-life 716 kyr) is restricted to minerals with low ^{27}Al content. Too much ^{27}Al would yield a ^{26}Al/^{27}Al ratio below the detection limit of AMS. Conveniently, quartz satisfies the requirements for both ^{10}Be and ^{26}Al so that both can be extracted from a single quartz dissolution (see ^{10}Be above). Meteoric ^{27}Al is less of a problem as suitable target elements do not occur in abundance in the atmosphere. The main disadvantage of ^{26}Al is that a separate accurate measurement (with its own inherent uncertainties) is required to determine the ^{27}Al content of the quartz. Sanidine is not used because of the abundance of inherent ^{27}Al. Nishiizumi et al. (1990) presented consistent ^{26}Al and ^{10}Be results from olivines extracted from Hawaiian lava flows.

3.1.4. ^{36}Cl

In contrast to Be and Al, Cl is hydrophilic. In most cases, contaminating meteoric ^{36}Cl (or ^{37}Cl or ^{35}Cl) can be removed with rinsing procedures, so that any rock type or mineral separate can be used for ^{36}Cl determinations. As a consequence of the numerous production pathways (Tables 1 and 2), production rates for ^{36}Cl must be calculated individually for each rock analyzed. In addition to determination of major element oxides, concentrations of B, Gd, and Sm are also required. These elements are strong neutron absorbers and influence the proportion of low-energy neutrons that are available for neutron capture reactions on ^{35}Cl and ^{39}K. U and Th concentrations are needed to correct for background (non-cosmogenic) ^{36}Cl production (Fabryka-Martin, 1988). Quartz is used infrequently for ^{36}Cl studies because of the lack of a target element (production only through low-energy neutron capture on ^{35}Cl found in fluid inclusions). The recent example of the measurement of ^{10}Be, ^{21}Ne, and ^{36}Cl in the mineral sanidine opens up numerous possibilities for multi-nuclide studies (Ivy-Ochs et al., 2007b).

Sample preparation procedures for ^{36}Cl are given in Zreda (1994), Ivy-Ochs (1996), Stone et al. (1996, 1998), Ivy-Ochs et al. (2004), and Desilets et al. (2006a). Carbonates are dissolved with HNO_3 and silicates with HF. A carrier of known isotopic composition (pure ^{35}Cl, ^{37}Cl, or a mixture of both) is added. Sulfur (^{36}S interferes with AMS measurement of ^{36}Cl) is removed by addition of $Ba(NO_3)_2$ and precipitation of $BaSO_4$. The possibility to use isotope dilution to measure both the total rock Cl concentration and ^{36}Cl in a single target using an AMS setup (Elmore et al., 1997; Ivy-Ochs et al., 2004) has led to marked improvements in both precision and accuracy in ^{36}Cl exposure ages (cf. Desilets et al., 2006a).

3.1.5. ^{41}Ca

The half-life of ^{41}Ca, 105 kyr, makes it a nuclide that may provide an important bridge between the long-lived (and stable) nuclides to radiocarbon, which has in comparison a very short half-life. The primary target element is ^{40}Ca, but depending on mineral chemistry Fe can also be important. Carbonate samples are prepared by dissolution with HNO_3 and precipitation of CaF_2 or production of CaH_2 (Finkel and Suter, 1993). $^{41}Ca/^{40}Ca$ ratios measured with AMS in tufa (calcite), dolomite, and aragonite samples are low because of high ^{40}Ca contents (Middleton et al., 1989; Fink et al., 1990; Klein et al., 1991). Even the ratios for very old surfaces that ought to show near-saturation values were still very close to blank values.

3.1.6. ^{53}Mn

The long half-life (3.7 Myr) of ^{53}Mn makes it an interesting nuclide for the study of old slowly eroding landscapes (Gladkis et al., 2007; Schaefer et al., 2006b) especially where mafic volcanic and intrusive rocks are present, for example in Antarctica and regions of southern Africa. The target element is ^{54}Fe, thus mafic minerals (olivine, pyroxene) are suitable. Schaefer et al. (2006b) found good agreement between pyroxene and whole rock ^{53}Mn results. Sample preparation involves dissolution after addition of carrier. ^{53}Cr (an interfering isobar in ^{53}Mn AMS measurements) is separated with chromatography or organic extractions (Lahiri et al., 2006). Information about changes in erosion rates with time over many millions of years ago may be gleaned by combining ^{53}Mn results with ^{10}Be, ^{26}Al, ^{3}He, and ^{21}Ne data from co-existing minerals.

3.2. Noble gases

3.2.1. ^{3}He

^{3}He is produced by spallation of nearly all elements. Quartz is not used because ^{3}He is not retained by the crystal structure and diffuses out of it (Trull et al., 1991; Brook and Kurz, 1993). Because of their low diffusion coefficients, olivine and pyroxene phenocrysts and microphenocrysts are most often used for ^{3}He (Kurz et al., 1990; Anthony and Poths, 1992; Cerling and Craig, 1994a; Bruno et al., 1997; Licciardi et al., 1999, 2006; Fenton et al., 2001, 2002, 2004; Oberholzer et al., 2003; Marchetti et al., 2005; Duffield et al., 2006; Williams et al., 2005). ^{3}He is especially useful for landscape change studies in regions underlain by mafic volcanic (basalts) and intrusive (dolerites) rocks or for the dating of landforms comprised of volcanic rocks (moraines, flood deposits, etc.). Mineral separates are pre-concentrated using heavy liquids and/or magnetic separation methods, then final sample material is briefly treated with dilute acids (Bruno et al., 1997) before it is handpicked (or cleaned) under a binocular microscope. The presence of nucleogenic ^{3}He produced from Li may limit study to rocks with young crystallization ages (cf. Williams et al., 2005). The contribution due to trapped (magmatic) helium may be revealed by crushing *in vacuo* prior to final melting (Kurz and Brook, 1994; Fenton et al., 2001; Niedermann, 2002; Marchetti and Cerling, 2005; Marchetti et al., 2005). ^{3}He has been measured in garnets (Dunai and Roselieb, 1996; Gayer et al., 2004) but radiogenic He is often present and may complicate calculation of young exposure ages (Niedermann, 2002). By comparing with ^{21}Ne measured in co-existing quartz, Kober et al. (2005) found that ^{3}He can be measured in Fe–Ti oxides (magnetite, titanomagnetite, and ilmenite) (cf. Margerison et al., 2005) from Chilean ignimbrites. The systematics of distinguishing cosmogenic and nucleogenic ^{3}He in apatite, titanite (sphene) and zircon have been examined recently by Farley et al. (2006).

3.2.2. ^{21}Ne

^{21}Ne is produced by spallation of Si, Al, Mg, and Na. An important advantage of ^{21}Ne is that, in contrast to ^{3}He, it can be measured in quartz (Graf et al., 1991; Staudacher and Allégre, 1991). Thus three cosmogenic nuclides, ^{10}Be, ^{26}Al, and ^{21}Ne can be determined on aliquots of a single quartz mineral separate (Hetzel et al., 2002a, 2002b; Kober et al., 2007). ^{21}Ne is also measured in pyroxene and olivine (Marti and Craig, 1987; Staudacher and Allégre, 1991; Poreda and Cerling, 1992; Bruno et al., 1997; Schäfer et al., 1999). Kober et al. (2005) found recently that the alkali feldspar sanidine is retentive for ^{21}Ne, opening up to study regions underlain by silicic to intermediate volcanic rocks that lack quartz or pyroxene (e.g., trachytes). Interferences due to trapped (Hetzel et al., 2002a) and nucleogenic (both non-cosmogenic) neon are deconvoluted using stepwise heating and may be identified by plotting data on the neon three isotope (^{20}Ne, ^{21}Ne, ^{22}Ne) plots (Graf et al., 1991; Niedermann, 2002). The measurement of exposure ages less than about 50 kyr with ^{21}Ne may be complicated by the high air background and/or nucleogenic neon especially in rocks with old crystallization ages (depending on site latitude and elevation).

3.2.3. ^{38}Ar

Cosmogenic ^{38}Ar has been measured in apatite, fluorite (Renne et al., 2001) and pyroxene (Niedermann et al., 2007), and garnet (Knight et al., 2003), where the main target element is ^{40}Ca, although it is also produced from ^{39}K (Niedermann, 2002). At present, focus is on constraining ^{38}Ar production rates, including production of ^{36}Ar due to decay of cosmogenic ^{36}Cl (Renne et al., 2001; Knight et al., 2003; Niedermann et al., 2007).

4. Determining Rates of Landscape Evolution with Cosmogenic Nuclides

4.1. Sampling considerations

The choice of which nuclide to use for a given geological/geomorphological question is limited by two considerations: (i) the half-life (or stability) of the nuclide and (ii) the bedrock geology (mineralogy) of the study area. As described in detail above (Section 3), certain nuclides can only be measured in certain minerals. ^{36}Cl exposure ages can be determined in whole rock samples or mineral separates from all lithologies. ^{10}Be (with or without ^{26}Al) is used in quartz-rich terrains and ^{3}He in regions of mafic terrains (where olivine or pyroxene is present). The lower age limit for ^{10}Be, ^{26}Al, and ^{36}Cl is limited predominantly by measurement capabilities. Upper age limits are constrained by nuclide half-life and rock weathering rate (Section 2.2). The noble gases are especially useful for studies in regions where landforms are tens of millions of years old because they do not decay (Schäfer et al., 1999; Dunai et al., 2005). The longer-lived and stable nuclides have longer memories and are suited for determining denudation rates integrated over hundreds of thousands to millions of years, while shorter-lived nuclides, especially ^{14}C, are useful for revealing rapid rock weathering rates over the last 15,000 years.

Sampling procedures and field considerations for different geomorphic applications of cosmogenic nuclides have certain elements in common. To obtain a date that reflects the time of formation of a landform (e.g., moraine or landslide), the sampled rock surface must have undergone single-stage (no pre-exposure), continuous (not covered) exposure in the same position (not shifted), and have undergone only minimal surface weathering (not spalled) or erosion. Where boulders are present, the largest ($>$1.5 m high), broadest, and flattest boulders located in optimal positions (Figure 12a) (moraine crest; flat tread of a terrace) are sampled. This is because they have a greater probability of having remained stable and of not having been covered during exposure. Similarly, bedrock samples should be taken from areas where field relationships indicate no coverage by sediment, ice, or snow during exposure. This may require sampling a steeply dipping surface (Nishiizumi et al., 1989; Kelly et al., 2006) and making the appropriate shielding corrections. Samples for bedrock erosion rates are taken from bedrock interfluves in regions where regolith cover has been minimal over time (Small et al., 1997; Kober et al., 2007). Flat surfaces at least 20–30 cm away from an edge are sampled to avoid geometry effects (Masarik et al., 2000; Masarik and Wieler, 2003; Lal and Chen, 2005). Several hundred grams of the upper few centimeters of the rock surface are chipped off with

Figure 12 Boulder on glacially polished bedrock located just north of Grimsel Pass. Samples were taken both from the top surface of the boulder (a) and from the glacially scoured bedrock (b) to determine the timing of deglaciation and rate of downwasting of the Aare Glacier at the end of the Last Glacial Maximum.

a hammer and chisel (Figure 12b). Sample size depends on location, exposure age, and mineralogy of the sampled surface.

An abandoned sedimentary unit (such as fluvial terraces, alluvial fans, debris-flow deposits, and landslide deposits) can be dated with cosmogenic nuclides. Such units can be dated by analyzing samples from the tops of

Figure 13 View of debris-flow fan leves on Shephard Creek fan, Owen's Valley, California. Note person in center left of photo for scale. The tops of the largest boulders were sampled to determine the timing of lobe formation (Dühnforth et al., 2007).

large boulders as described above (Figure 13). Where no boulders are present (landforms older than about 100 kyr) amalgamated clast samples are analyzed. Repka et al. (1997) showed that the random effects of some clasts being affected by post-depostional processes (i.e., "too young") and some being affected by inheritance (i.e., "too old") can be smoothed out by taking a large number of clasts. More than 50 clasts of similar size (several centimeters in diameter) and lithology are collected from the flat part of a depositional surface well away from modifying channels (Repka et al., 1997). In order to assess and correct for the inherited component, several samples (amalgamated clast samples as above) are taken at tens of centimeters intervals from a depth profile (Figure 10, Section 2.2) (Anderson et al., 1996). An important caveat with respect to depth-profiling methods is that the depositional unit must be either a single rapidly emplaced package or individual units must have been deposited over a short period of time (Hancock et al., 1999). Another important advantage of analyzing samples from different depths is that one can check for possible erosion and/or mixing of the uppermost few tens of centimeters of sediment (Phillips et al., 1998; Perg et al., 2001; Lal and Chen, 2005, 2006).

For determination of catchment-wide erosion rates, about one-half kilogram of sand-sized sediment is collected from sand bars in the active river channel during periods of low flow. Depending on the expected erosion rate, the quartz content in the sediment, the average production rate and the shielding in the catchment, more or less sediment of a given grain size must be collected. In order to determine paleoerosion rates, samples are collected from depths of more than several meters in sedimentary deposits (e.g., marine terraces, river terraces, cave sediments) from which the age is known independently or can be determined with cosmogenic nuclides (Section 4.9).

4.2. Establishing sequences of events in volcanic terrains

The sequence of events in a volcanic field can be established with cosmogenic nuclides. To determine the eruptive age of the lava flow and develop a chronology for the various flows, original eruptive surfaces must be unequivocally identified. This may be verified by the presence of primary eruptive features (Figure 14) (spatter, frothy glassy texture of the cooling rinds on flows or bombs, tumuli ridges) (Cerling and Craig, 1994b; Fenton et al., 2001; Phillips, 2003). The original flow top of Pahoehoe flows often exhibits a distinctive ropey texture. As pointed out by Cerling and Craig (1994a), single flows may have layer upon layer of onlapping pahoehoe features. Lavas of intermediate to silicic composition are chaotic and blocky making identification of flow tops, and thus

Figure 14 Sampled ropey surface of the the Bar Ten pahoehoe basalt flow (Arizona); approximate exposure age 88 ± 6 kyr based on ^{3}He measured in olivine phenocrysts (Fenton et al., 2001).

sampling, difficult. ^{3}He and/or ^{21}Ne are used for mafic (basaltic) flows (Section 3.2). ^{21}Ne and ^{10}Be in quartz and/or sanidine phenocrysts have been used to date ignimbrites (welded tuffs) (Albrecht et al., 1993; Libarkin et al., 2002; Kober et al., 2005). Determination of ^{36}Cl concentrations from whole rock samples allows the dating of aphyric (phenocryst-free) lava flows (Zreda et al., 1993; Phillips, 2003). Lava flows that are difficult to date with argon methods as they are either too young, barren of K-rich phases (mafic flows) and/or when excess argon may be a problem can now be dated with cosmogenic nuclides (Fenton et al., 2001; Phillips, 2003; Duffield et al., 2006). In addition, cosmogenic nuclide results can be compared with K–Ar or ^{40}Ar/^{39}Ar dates from the same flow providing opportunities for calibration of production rates (Braucher et al., 2006) (Section 5.4).

4.3. Glacial landscapes: moraines and bedrock

Exposure dating of both boulders and bedrock surfaces has been used to constrain the timing of past ice margin fluctuations (for a recent compilation see Reuther et al., 2006b). Several examples are listed here:

Europe: Alps (Ivy-Ochs et al., 1999, 2006; Kerschner et al., 2006), Scotland (Everest and Kubik, 2006; Everest et al., 2006), Ireland (Ballantyne et al., 2006), Iceland (Principato et al., 2006), Finland (Tschudi et al., 2000; Rinterknecht et al., 2004), Poland (Rinterknecht et al., 2005), north-western Russia (Linge et al., 2006b).
Asia: Turkey (Akçar et al., 2007, 2008b), eastern Russia (Gualtieri et al., 2000), Wrangel Island (Gualtieri et al., 2005), central Asia and Himalaya (Schäfer et al., 2002; Finkel et al., 2003; Owen et al., 2003a; Zech et al., 2003, 2005).
North America: California (Phillips et al., 1996a; James et al., 2002; Owen et al., 2003b), Rocky Mountains (Gosse et al., 1995a, 1995b; Licciardi et al., 2001, 2004), western Canada (Jackson et al., 1997, 1999), Alaska (Briner et al., 2005b), eastern United States (Balco et al., 2002; Balco and Schaefer, 2006), and eastern Canada (Briner et al., 2005a; Miller et al., 2005).
South America: Argentina (Kaplan et al., 2004, 2005), Peru (Smith et al., 2005b), and Chile (Zech et al., 2006).
Australia, Tasmania and New Zealand: (Ivy-Ochs et al., 1999; Barrows et al., 2001, 2004; Schaefer et al., 2006a).
Antarctica: (Brown et al., 1991; Brook et al., 1995b, Bruno et al., 1997; Oberholzer et al., 2003; Ackert and Kurz, 2004; Bentley et al., 2006; Staiger et al., 2006).

In addition, exposure dating has been used to decipher the timing of deposition of glacial outwash (Phillips et al., 1997) and boulders on ice-dammed lake margins (Davis et al., 2006) as well as ice-dam related catastrophic outburst flood deposits (Cerling et al., 1994; Reuther et al., 2006a).

4.3.1. Dating of moraines

Moraines record the location of the margins of a glacier in the past. Hence, moraines provide a chronological structure to glacier fluctuations (Gosse, 2005). The changes in glacier volume and length themselves reflect changes in temperature and precipitation patterns in a region with time (Kerschner, 2005). Concentrations of cosmogenic nuclides are measured in boulders from the crests of moraines (Figure 15). The exposure age represents the length of time since the boulder stabilized or emerged from the finer grained sediment (Hallet and Putkonen, 1994; Zreda et al., 1994; Zreda and Phillips, 1995; Putkonen and Swanson, 2003; Putkonen and O'Neal, 2006; Reuther et al., 2006b; Ivy-Ochs et al., 2007a). Remarkably consistent data sets have been obtained when only large (>1.5 m) boulders in stable positions on glacial landforms are sampled. A boulder must be large enough to have not toppled or shifted and high enough to have protruded above the matrix since moraine deposition. Limiting factors are the weathering of the boulder surface and degradation of the moraine itself. Old moraines contain few boulders and the spread in boulder ages is greater, making interpretation of the deposition age difficult (Kaplan et al., 2005; Smith et al., 2005a).

Boulders deposited in a moraine originate from rock fall onto the glacier surface in steep-walled valleys and/or as a result of plucking from the glacier bed. In either case, it is possible that the rock surface sampled was exposed prior to final deposition in the moraine. At a site on Baffin Island, Davis et al. (1999) found no evidence for inherited ^{26}Al and ^{10}Be concentrations in boulders recently deposited and bedrock recently uncovered by the receding Tumbling Glacier. Consistent exposure ages from cirque glacier moraines at several sites in the European Alps indicate negligible inherited nuclide concentrations (Ivy-Ochs et al., 2006). In contrast, Briner et al. (2005b) found inheritance in several boulders on moraines located close to the cirque walls at sites in Alaska. Overall, only a few percent of analyzed moraine boulders deposited by valley glaciers were identified to contain pre-exposure (Putkonen and Swanson, 2003). In some cases, the inheritance may be due to reworking (moving of a boulder into a new landform) rather than to inheritance acquired in the bedrock setting (Ivy-Ochs et al., 2007a).

4.3.2. Dating of buried till deposits

Balco et al. (2005) used $^{26}Al/^{10}Be$ ratios in quartz to determine the age of deeply buried paleosols and underlying till units. By assuming a burial history based on regional correlations and independent age data for the overlying units, they were able to determine the age of the buried soil and

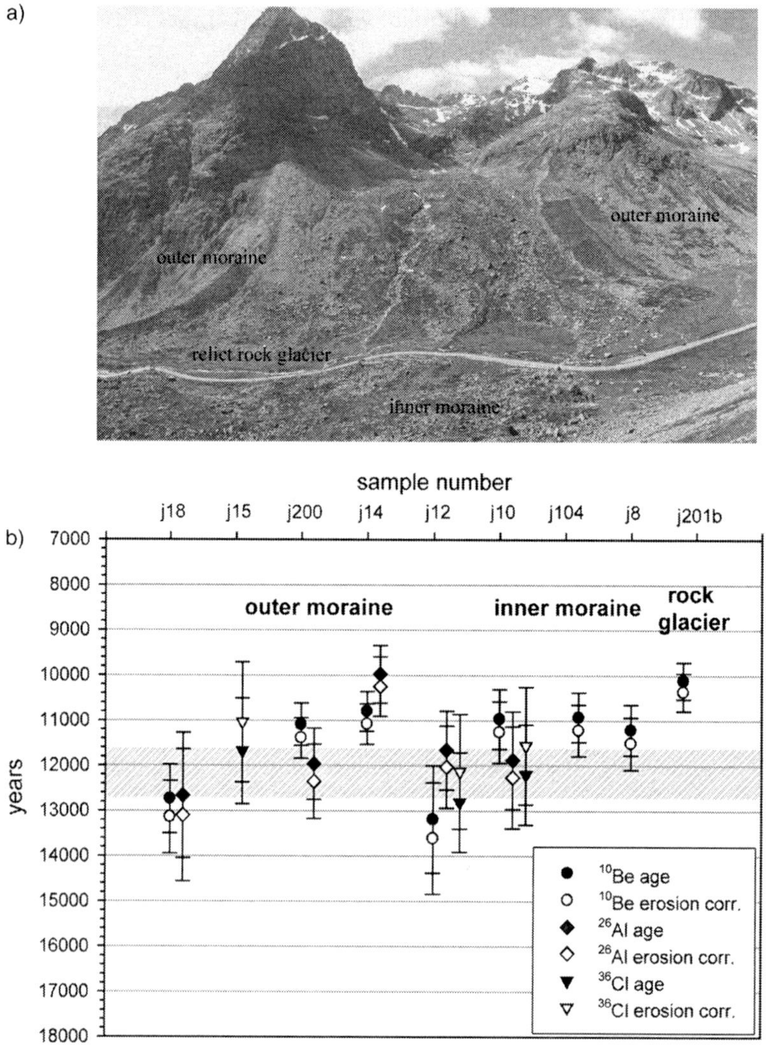

Figure 15 Exposure age of formation time of the moraine complex at Julier Pass. (a) Photo looking up into the Lagrev catchment; outer moraine, inner moraine, rock glacier correspond to the data shown in (b). ^{10}Be, ^{26}Al, and ^{36}Cl data verify that the moraine complex formed during the Younger Dryas cold interval $11{,}600 \pm 1{,}000$ yr ago. Rock glacier activity at that site lasted until 10,400 yr ago (figures modified from Ivy-Ochs et al., 2006, 2009).

till. This application is a variation of burial dating as there has been more than one period of burial. The age they determined for the lowermost loess unit and underlying till deposit is 0.65 ± 0.14 Myr. This age agrees well with the age of a correlative till unit that underlies an ash layer that has been independently dated to 600 kyr.

4.3.3. Glacially modified bedrock surfaces and rates of sub-glacial erosion

Initial deglaciation of the valley bottom and rates of glacier downwasting can be determined by analyzing cosmogenic nuclides in glacially scoured bedrock surfaces (e.g., roche moutonnées). However, results from bedrock surfaces should be viewed with caution. If 3 m or more of bedrock have not been removed by sub-glacial erosion during the final glaciation, then the rock surface may contain inherited nuclide concentrations. This may occur where the glacier is not highly erosive (e.g., high up on the trough walls or at the base where the glacier is frozen to its bed). In the case where the timing of deglaciation is independently known, the nuclide concentration measured in glacially scoured bedrock can be used to determine sub-glacial erosion rates. Determined rates are on the order of 0.1 to 1 mm yr^{-1} (Briner and Swanson, 1998; Colgan et al., 2002; Fabel et al., 2004).

Cosmogenic nuclide data is one of the bases for a new understanding of the extent and thermal regime of past ice sheets. Detailed studies have been done in eastern Canada (Miller et al., 2006; Marquette et al., 2004; Staiger et al., 2005; Briner et al., 2006; Davis et al., 2006b), Sweden (Fabel et al., 2002; Stroeven et al., 2002), Norway (Linge et al., 2006a), Scotland (Stone and Ballantyne, 2006; Phillips et al., 2006), and Antarctica (Sugden et al., 2005). Exposure ages and nuclide ratios document spatial patterns of erosive warm-based ice versus less-erosive cold-based ice that is frozen to its bed. Past patterns of ice streams are mapped (Staiger et al., 2005; Briner et al., 2006). The addition of ^{26}Al and the use of the erosion-island plot (see Section 2.2) led to recognition of periods of burial after exposure (Bierman et al., 1999). The $^{26}Al/^{10}Be$ data points plot in the complex exposure field below the erosion island (Figure 8c). Assuming one event of exposure followed by complete shielding, the minimum duration of exposure and shielding (during which the nuclides decay) can be determined. However, numerous periods of exposure and burial cannot be excluded. Marquette et al. (2004) state "Tors and felsenmeer do survive one or more glacial periods with minimal erosional disturbance by ice sheets." These studies unequivocally proved that cold-based ice that is frozen to its bed leaves underlying bedrock largely unmodified. The measurement of two nuclides is crucial. Single-nuclide data may lead to an interpretation of misleadingly young exposure ages.

4.4. Bare bedrock erosion rates

Erosion (weathering or *in situ* down-wearing) rates of bare bedrock surfaces can be determined by measuring the concentrations of cosmogenic nuclides. Sampled sites comprise bedrock hillslope interfluves or tors that have never been covered by regolith and are outside areas affected by flowing ice or water. The sampled bedrock surfaces must satisfy the steady-state assumption, otherwise the actual erosion rates are lower than the calculated erosion rates.

Small et al. (1997) determined average bare granitic bedrock erosion rates of 8 mm kyr^{-1} or less in unglaciated high-elevation sites in the Rocky Mountains and the Sierra Nevada (USA). Results of their study support the hypothesis that summit flats are formed by cryoplanation processes. Small et al. (1997) were able to identify recently spalled surfaces using the $^{26}Al/^{10}Be$ ratio. Immediately after a spalling event the surface does not satisfy the steady-state assumption. This means that analysis of several samples as well as measurement of two nuclides is necessary to determine the average bedrock erosion rate. On the summit flats investigated by Small et al. (1997), the rate of bedrock weathering was an order of magnitude lower than the rate of incision (glacial and fluvial) in the adjacent valleys, contributing to an increase in relief with time (Small and Anderson, 1998).

Cosmogenic nuclides have shown that in regions with hyperarid climate extremely low denudation rates have dominated the landscape for millions of years (Cockburn and Summerfield, 2004). ^{10}Be in combination with ^{26}Al data showed that many surfaces in Antarctica have experienced single-stage continuous exposure with very low erosion rates (less than several mm kyr^{-1}) for at least several million years (Nishiizumi et al., 1991; Brook et al., 1995a; Ivy-Ochs et al., 1995; Summerfield et al., 1999; Matsuoka et al., 2006). Similarly, very low erosion rates were measured in bedrock in Namibia (Cockburn et al., 1999, 2000; Fleming et al., 1999; Bierman and Caffee, 2001), south-central Australia (Bierman and Turner, 1995; Bierman and Caffee, 2002; Belton et al., 2004), and in the northern Atacama desert in Chile (Dunai et al., 2005; Nishiizumi et al., 2005; Kober et al., 2007).

Nuclide steady state is a fundamental assumption for calculation of outcrop scale or catchment-wide denudation rates (Section 2.2). Often this is difficult to prove without circular reasoning. Steady state with respect to the landscape is not necessarily the same as, and need not operate on the same timescales as, steady state with respect to the nuclides. For example, in the study of Kober et al. (2007) in northern Chile, the landscape itself is likely at steady state. Nevertheless, $^{21}Ne/^{10}Be$ data points do not fall on the erosion saturation line, as they fall inside of the erosion island. Kober et al. (2007) could show that this was likely due to infrequent spalling of flakes a few decimeters in thickness.

4.5. Soil production and creep rates

The quantification of the generally slow soil production and creep rates has proven difficult in the past. Cosmogenic nuclide data measured in soil sections and at soil/bedrock interfaces allow the determination of these slow but important processes for landform evolution. Insights into soil creeping rates can be gained from measuring the nuclide concentration in sediment transported downhill (Nichols et al., 2002). The soil production rate is quantified by measuring the *in situ*-produced cosmogenic nuclide concentration in quartz in the bedrock close to the soil/bedrock interface (Heimsath et al., 1997, 1999, 2000, 2001; Small et al., 1999).

The soil production function can be used to explore the role of geology and climate in landscape evolution. Applications on soils of different depth in California (USA), Oregon (USA), and Australia reveal an exponential decline of soil production rates with increasing soil depth (Heimsath et al., 1997, 1999, 2000, 2001). The observed decline suggests that these landscapes are not in morphologic equilibrium (Heimsath et al., 1997, 2000). The inverse linear relationship between topographic curvature and soil depth implies that simple creep of the soil does not adequately characterize hillslope processes (Heimsath et al., 2000; Braun et al., 2001). Simple and depth-dependent soil creep combined with overland flow better predict soil thicknesses and the evolution of soil distribution. The combination of the soil production technique with thermoluminescence dating allows a quantitative characterization of grain-scale transport processes within creeping soil (Heimsath et al., 2002). Results show that grains move independently in the soil column, repeatedly visiting the surface before being transported out of the system.

Small et al. (1999) determined regolith production rates in alpine settings by combining *in situ*-produced cosmogenic nuclide measurements with mass balance models. In contrast to the general findings of Heimsath et al. (1997, 1999, 2000, 2001), Small et al. (1999) found higher regolith production rates beneath a regolith cover than on bare rock surfaces. The higher regolith production may be attributed to faster weathering of bedrock due to the presence of water in regolith covering the bedrock.

4.6. Dating of large mass movements: bedrock and boulders

Landsliding is an important process for the modification of valley slopes and landscape denudation in general. The effect of high-magnitude/low-frequency events on landscape change can be evaluated when time series of mass movements for a given region are constructed (Cockburn and Summerfield, 2004). By determining the timing of recurrence intervals, we can begin to understand the mechanisms of large-scale downslope movement of rock, and the interplay between various possible causes and triggers. The latter to these include inherited tectonic structures, debuttressing following retreat of glaciers, climate (precipitation, temperature, permafrost extent etc.), and/or seismic events.

Several catastrophic landslides have been dated using cosmogenic nuclides: the Koefels landslide in Austria (Ivy-Ochs et al., 1998a; Kubik et al., 1998), the Flims landslide in Switzerland (Ivy-Ochs et al., 2009), the

Fernpass landslide in Austria (Prager et al., 2009), the Beinn Alligin rock avalanche in Scotland (Ballantyne and Stone, 2004), the Blackhawk landslide in California, USA (Nichols et al., 2006), the Roaring River slide in the Rocky Mountains, USA (Gosse in Morris et al., 2002), slides in Argentina (Hermanns et al., 2001, 2004), and northern India (Barnard et al., 2001), as well as both deep-seated and shallow landslides in Hong Kong (Sewell et al., 2006). Bigot-Cormier and coworkers (2005) dated several gravitational scarps of an active landslide at La Clapière in France.

Although new rock surfaces are created during mass movements, exposure dating of landslides can have mixed results. Each possible sampling site (head scarps, sliding planes, large boulders) has its own advantages and disadvantages. In the breakaway zone, recent rock fall can be difficult to identify. Sliding planes with low dip angles may yield too young ages due to coverage by snow and/or vegetation (Ivy-Ochs et al., 2009). Because of the movement mechanisms of landslides (e.g., Erismann and Abele, 2001), rocks from the outer surface of the pre-slide bedrock frequently end up as boulders on top of the deposit. As a result, "too old" exposure ages with respect to independent age controls are not uncommon in landslide boulders (Nichols et al., 2006; Sewell et al., 2006; Ivy-Ochs et al., 2009).

4.7. Fluvial incision rates

Fluvial incision rates can be determined by dating (i) bedrock such as strath terraces or fluvially sculpted channel walls, (ii) fill-cut terraces (fluvial and/or debris-flow deposits) as well as (iii) cave sediments. The first method makes use of surface-exposure dating (generally single nuclide), the second method uses dating based on single nuclides measured in a depth profile (Section 2.2), whereas the third method employs the technique of burial dating (two nuclides required).

Based on the height and age of abandoned strath terraces located along a given reach of a river, fluvial incision rates can be calculated (Burbank et al., 1996; Leland et al., 1998; Pratt et al., 2002; Pratt-Sitaula et al., 2004). Sampled surfaces that bear scars of fluvial erosion have not weathered or eroded appreciably. However, it is important that the sampled surface were not covered by sediment at any time during exposure. As straths are planar remnants of old bedrock channel floors, they are separated by erosional steps. Thus, fluvially sculpted channel walls (Figure 16) may contain a more continuous record of river incision rates (Schaller et al., 2005).

When the age of a fluvial terrace can be determined successfully, then the height above the active river channel can be used to reconstruct incision rates over time (Schildgen et al., 2002; Ward et al., 2005). Boulders or amalgamated clast samples on debris flows on ancient straths are dated in order to determine incision rates (Fenton et al., 2001, 2004; Marchetti and

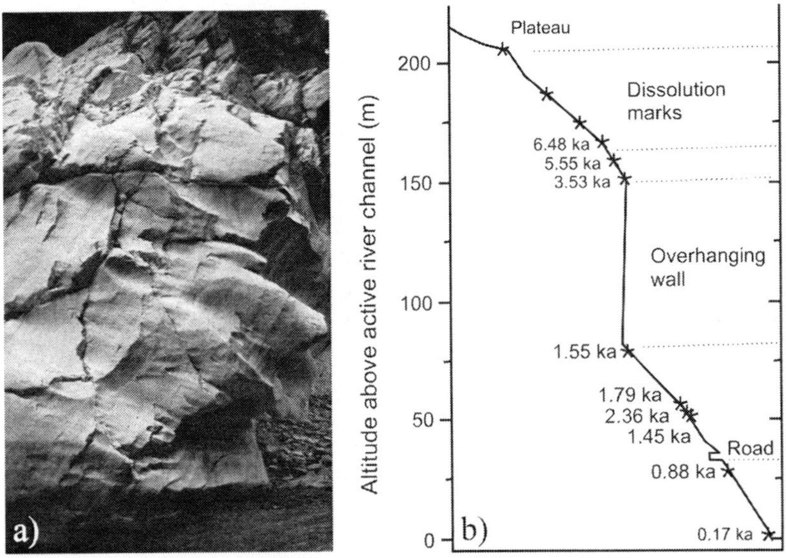

Figure 16 Incision rates determined with surface exposure dating of fluvially sculpted channel walls in the Taroko gorge, Taiwan. (a) Well-developed flutes in marble were sampled for ^{36}Cl analysis. (b) Diagram giving an overview of age constraints from fluvially sculpted surfaces at Taroko gorge. Exposure ages generally increase with increasing height above the active river channel. An average incision rate of $26 \pm 3\ mm\ yr^{-1}$ was determined (Schaller et al., 2005).

Cerling, 2005; Marchetti et al., 2005a). Two considerations are crucial: (i) the amount of inheritance in the clasts must be estimated and (ii) the degradation or aggradation rate of the top surface must be constrained independently or estimated. High inheritance in the clasts of a river terrace may result from scavenging of material from older terraces at higher elevation (Ward et al., 2005). In addition, long travel times or slow erosion in the source area of the sediment may lead to inheritance in clasts or boulders. Wolkowinsky and Granger (2004) used the $^{26}Al/^{10}Be$ pair to burial date sediment in terraces along the San Juan River, UT (USA). The calculated age of 1.36 Ma based on a deep depth profile (greater than ca. 10 m) is twice the ages obtained from amalgamated clast samples collected at the surface, yielding significantly lower incision rates. In addition, their data allowed calculation of erosion rates of about $15\ mm\ kyr^{-1}$ for the terrace top surface.

Granger and co-workers (Granger et al., 1997; Granger and Muzikar, 2001) showed that sediment deeply buried in caves can be explicitly dated using the ratios of two cosmogenic nuclides (Section 2.2). Combining burial ages of cave sediments with the height of the cave above the active river channel, allows calculating past incision rates and their variations with time (Granger et al., 1997; Granger and Muzikar, 2001; Stock et al., 2004).

4.8. Rates of tectonic activity

Cosmogenic nuclides can be used to determine rates of tectonic activity in two ways: (i) by dating offset features along faults and (ii) by dating bedrock fault surfaces directly.

Slip rates on strike-slip faults or rates of uplift on normal faults have been determined by taking the offset of the dated features and their age (Bierman et al., 1995; Ritz et al., 1995, 2003; Siame et al., 1997, 2002; Brown et al., 1998a; van der Woerd et al., 1998, 2006; Zehfuss et al., 2001; Hetzel et al., 2002a, 2002b, 2004a, 2004b; Mériaux et al., 2004, 2005; Matmon et al., 2005; Ryerson et al., 2006). If amalgamated clast samples are analyzed, an estimate of the inherited component is needed. This is best determined by analyzing samples from a depth profile. As in the case of the dating of terraces for determining fluvial incision rates, a well-constrained estimate of the erosion rate of the dated surface is needed. Finally, an important issue is the interpretation of the distribution of exposure ages from landforms that are built up over a non-instantaneous time interval. Does the spread in ages approximate the time interval of deposition or does it reflect natural scatter in age data largely due to post-depositional effects and inheritance? Similarly, cautious interpretation of field evidence for offset distances (piercing point, offset terrace risers vs. treads; offset stream channels; Mériaux et al., 2004) is required.

^{36}Cl has been used to date bedrock fault surfaces directly (Zreda and Noller, 1998; Mitchell et al., 2001; Benedetti et al., 2002, 2003; Phillips et al., 2003; Palumbo et al., 2004). To be suitable for dating with cosmogenic nuclides, the exposed fault surface must be fresh, uneroded, several meters high and extend continuously across the landscape for kilometers. Primary structures related to the fault movement itself, both fine-scale (slickensides) and medium-scale features (broad undulating surfaces) are preserved well in carbonate rocks and serve to verify low erosion. Samples are taken at centimeters intervals along the entire exposed fault surface. On a single fault face, the recurrence interval of earthquakes along the fault segment can be reconstructed based on ^{36}Cl concentrations. Earthquakes that occurred as recently as only a thousand years ago can be dated.

4.9. Catchment-wide erosion rates from river sediment

Over the past 10 years, the technique of cosmogenic nuclides has been extended to determine erosion rates on catchment scale. This approach makes use of the fact that river sediment represents the erosion products of an entire landscape (Lal, 1991). For a recent summary of the technique see von Blanckenburg (2006). Erosion is not uniform in space. However, sediments become well mixed through hillslope and fluvial transport processes. The nuclide concentration measured in river sediment can be

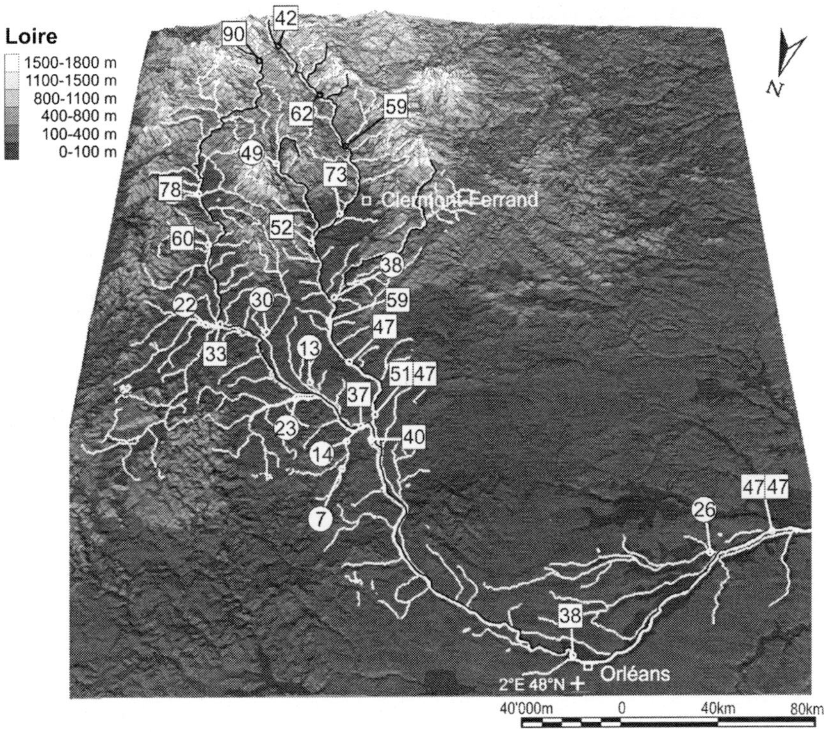

Figure 17 Catchment-wide erosion rates (mm kyr^{-1}) derived from *in situ*-produced cosmogenic nuclides measured in river sediment from the Loire and Allier rivers, France (Schaller et al., 2001). River sediment was collected from sand bars and the 0.5–1.0 mm grain size fraction were analyzed for cosmogenic nuclide concentrations. Erosion rates from the main streams (squares) and tributaries (circles) generally decrease downstream.

used to determine catchment-wide erosion rates (Figure 17) (Brown et al., 1995; Bierman and Steig, 1996; Granger et al., 1996). The nuclide concentration is inversely proportional to the catchment-wide erosion rate. We need to be aware of the assumptions the technique is based on and the potential violation of these assumptions by nature. The assumption of steady state is an essential prerequisite of the technique. The production of cosmogenic nuclides in the landscape equals the transport of cosmogenic nuclides out of the landscape (Figure 18). This condition is only fulfilled in the case of constant erosion rates. When the erosion rate is changing over time, then the influx of cosmogenic nuclide concentration into the system is in disequilibrium with the flux out of the system. The system needs some time to reach equilibrium again. The cosmogenic nuclide-derived erosion rate lags behind the true erosion rate in a catchment, where erosion rates have changed over time (Schaller and Ehlers, 2006). Due to this time lag,

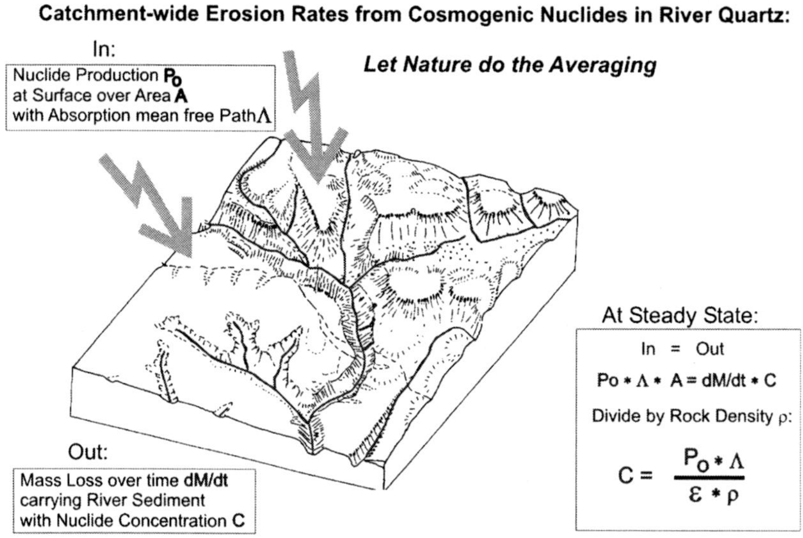

Figure 18 Principles for the calculation of catchment-wide erosion rates using cosmogenic nuclides measured in river sediments.

changes in erosion rates are smoothed and an average erosion rate is reported. The average erosion rate might be higher or lower than the actual erosion rate. The memory of past erosion rates integrated in the average erosion rate is removed more rapidly in fast-eroding landscapes than in slowly eroding settings. As pointed out by Bierman and Steig (1996), this technique is not applicable to recently glaciated catchments. Depending on how long ago and how deeply the bedrock was eroded by the ice (Section 4.3), the landscape may not have had enough time to reach steady state since deglaciation. Another assumption of the technique is that the analyzed mineral is homogeneously distributed in the catchment. The selective dissolution of regolith, and the consequent relative enrichment of quartz, should be addressed in basin-averaged erosion rate studies (Small et al., 1999; Riebe et al., 2001a). As the analyzed catchments often contain a great variety of lithologies and regolith thicknesses, it is difficult to determine a generally applicable correction factor.

The technique has been validated in small (1–10 km^2), non-glaciated catchments on the eastern side of the Sierra Nevada, CA (Granger et al., 1996). Accumulation rates of alluvial fans were compared to erosion rates derived from the cosmogenic nuclide contents of sediments supplied by the corresponding hillside catchments. Their comparison revealed excellent agreement between the two methods integrating erosion rates over the same time scales. Intercomparison of cosmogenic nuclide-derived erosion rates from river sediments allows the study of the influence of changing climatic (Riebe et al., 2001b), tectonic (Riebe et al., 2000; Schaller et al., 2001), and lithological conditions (Clapp et al., 2000; Morel et al., 2003) on erosion rates. The analysis of cosmogenic nuclides in bedrock outcrops, soils, mass wasting sites, and river sediment can be used to characterize sediment generation and source areas (Brown et al., 1995, 1998b; Clapp et al., 2000, 2001, 2002; Bierman and Caffee, 2001; Nichols et al., 2005). Further insight into these processes may be obtained by the analysis of different grain-size fractions in river sediments (Brown et al., 1995; Brown et al., 1998b; Matmon et al., 2003a).

Cosmogenic nuclide-derived erosion rates have been compared with erosion rate estimates from techniques ranging over different timescales. This comparison allows study of the changes of erosion over time due to tectonic, climatic, and anthropogenic influences. Erosion rates calculated from river load gauging (timescales <100 yr) and rates derived from thermochronometers as well as metamorphic assemblages (timescale >1 Myr) are frequently used. Cosmogenic nuclide-derived erosion rates agree with rates derived from river load gauging (Clapp et al., 2001; Matmon et al., 2003b; Vance et al., 2003). This agreement may be interpreted to show that the landscape is in equilibrium. In other studies, cosmogenic nuclide-derived erosion rates were higher than rates from river load gauging (Kirchner et al., 2001; Schaller et al., 2001). The difference has been explained by systematic underestimation of modern river loads due to the stochastic nature of erosion events as well as the difficulty of measuring river load accurately. Another explanation is that the cosmogenic erosion rate is influenced by an inherited erosion signal from the past reflecting different tectonic and climatic conditions. In some catchments, however, the cosmogenic nuclide-derived erosion rate is lower than rates based on river load gauging (Brown et al., 1995, 1998b; Clapp et al., 2000; Hewawasam et al., 2003). One explanation for this observation is that rates based on river load gauging are affected by human impact, whereas cosmogenic nuclide-derived erosion rates reflect the natural background erosion rate. This comparison offers the possibility to quantify directly a decrease or increase of soil erosion due to agricultural land use (Brown et al., 1995, 1998b; Hewawasam et al., 2003).

4.9.1. Paleoerosion rates derived from sedimentary deposits

One of the unique advantages of the method to determine catchment-wide erosion rates is that it is not only applicable on modern river bedload, but also on sedimentary deposits on which the age can be determined. The measured cosmogenic nuclide concentration in these deposits can be corrected for post-depositional irradiation and for decay if the age is independently known or can be determined by cosmogenic nuclide pairs (burial dating). The remaining nuclide concentration can be attributed to irradiation during erosion and transport of the sediment (Schaller et al., 2002, 2004). If fluvial transport is rapid and erosion in the catchment is in steady state, then the inherited nuclide concentration is solely a product of irradiation during erosion and hence a function of the erosion rate. The older the deposit, the higher is the correction of the post-depositional concentration on the total nuclide concentration and the lower the inherited concentration. In order to have a meaningful signal resulting from the inherited concentration in samples from old terraces, the samples need to be collected from great depth, that is, >10 m (Granger and Muzikar, 2001). The time corrections on the measured nuclide concentration can only be done if the age of the sediment deposit is independently known. If the deposition age is not known, the measured nuclide concentration can still be used as a face value of the erosion rate in the source area at the time of sediment deposition or the transport time of the sediment in the river system (Anderson et al., 1996; Repka et al., 1997; Hancock et al., 1999).

The technique applied on a sequence of river terraces with known age allows study of the change of catchment-wide erosion rates over time. Terraces deposited during glacial/interglacial transitions (e.g., the late Pleistocene to Holocene transition) reveal higher catchment-wide erosion rate during glacial than interglacial epochs in Middle Europe (Schaller et al., 2002). However, changes in paleoerosion rates determined with cosmogenic nuclides cannot always be attributed unequivocally to climate due to limitations of the cosmogenic nuclide technique to reflect climatic changes (Schaller and Ehlers, 2006). The technique applied in a 1.3 Ma terrace sequence of the Meuse River (Schaller et al., 2004) revealed an increase in paleoerosion rate from 25 to 80 $mm\,kyr^{-1}$. This increase is attributed to tectonic and/or climatic changes over time. Similar observations were made by Balco and Stone (2005) in the western USA.

4.9.2. Chemical weathering and physical erosion rates

The combination of cosmogenic nuclides to determine catchment-wide erosion rates with geochemical mass balance methods allows determination of long-term chemical weathering and physical erosion rates (Riebe et al., 2001c, 2003, 2004; Green et al., 2006). Long-term chemical weathering and physical erosion rates are interrelated processes. In order to better understand landscape evolution, it is important to quantify both processes.

Long-term chemical weathering rates can be determined by quantifying the changes in composition when a parent material is converted to a weathered soil. The enrichment of elements which are immobile during weathering processes (e.g., zirconium) can be used to determine chemical weathering losses (April et al., 1986; Brimhall and Dietrich, 1987). The greater the enrichment of an immobile element in the weathered material, the greater the mass loss in the transformation of parent material to a weathered soil. This mass loss can be attributed solely to chemical weathering if physical erosion is negligible. In order to calculate a chemical weathering rate, the soil age must be known independently. Unfortunately, non-eroding soils of known age are rare and the determination of long-term weathering rates sparse. In physically eroding settings, the soil age cannot be determined as soil is continuously removed and fresh rock incorporated in the soil. In such settings, cosmogenic nuclides measured in the erosion products reveal information about the total erosion rate (sum of chemical weathering and physical erosion rate) as well as the soil residence time. For the immobile element analysis, samples of soil and parent material are collected from several locations in the study area. The weathering rates derived from these different locations should reflect the average weathering rate of the landscape rather than reflecting an anomalous rate of a single point. Small catchments underlain with a uniform and quartz-rich lithology are the most suitable.

The approach to determine long-term weathering and physical erosion rates with the cosmogenic nuclide and geochemical mass balance method has been tested in a catchment where weathering rates from independent studies were available (Riebe et al., 2001b, 2003). The study showed that chemical weathering is tightly coupled with the physical erosion rate. Therefore, the coupling between climate and chemical weathering may be obscured by the rate at which parent material is supplied to the soil by physical erosion. In their study areas, no clear correlation between chemical weathering rate and mean annual temperature or average precipitation is observed. However, when comparison of the chemical depletion fraction (chemical weathering rate normalized to the total erosion rate) is made with precipitation and temperature, a coupling with climate does become visible (Riebe et al., 2001b, 2003).

5. New Directions and Outlook

5.1. Archeology

Cosmogenic nuclides can be used to exposure date fossils, lithic artefacts, or monuments directly, to exposure date rocks or sediment associated with fossils or stone tools, or to determine nuclide concentrations to assess tool procurement strategies. Cosmogenic nuclide methods have an as yet largely unrealized potential in archeology (Stuart, 2001; Akçar et al., 2008a).

Cerling and Craig suggested in 1994 that the pyramids, sphinx, Stonehenge, or even the statues on Easter Island can be directly exposure dated. Initial results from the analysis of ^{36}Cl in the bluestones from Stonehenge is controversial (Williams-Thorpe et al., 1995). In every case, the same requirements as for exposure dating a landform must be satisfied: (i) the initial nuclide concentration is zero (no inheritance); (ii) the exposures were continuous (no burial); (iii) the sampled surface must have remained in the same orientation (not reconstructed); and (iv) the sampled rock surfaces are unweathered (no blocks missing).

Farley et al. (2001) successfully measured ^{3}He in the fluorapatite of fossil teeth, opening up the possibility to date fossil remains at archeological sites. Farley et al. (2001) successfully measured ^{3}He in the fluorapatite of fossil teeth from the Turkana Basin, Kenya, opening up the possibility to exposure date fossil remains at archeological sites. However, their preliminary results suggest that some of the fossils they studied comprise lag deposits and are much older than the associated sediment.

Flint (chert, microcrystalline SiO_2) artifacts have been analyzed by Boaretto et al. (2000), Ivy-Ochs et al. (2001) and Verri et al. (2004, 2005). Ivy-Ochs et al. (2001) determined ^{10}Be surface-exposure ages for chert tools collected from flat-lying limestone benches near Luxor, Egypt. The exposure ages of several hundred thousand years are in good agreement with age estimates based on tool typology, although the amount of inheritance is difficult to gauge. Verri et al. (2004, 2005) used ^{10}Be concentrations in tools buried in caves in Israel to determine if raw materials for tool production (flint nodules interbedded in limestone strata) were simply gathered from the surface or were mined from pits. Their results showed that unworked chert nodules collected from the desert surface contain random and variable concentrations of ^{10}Be. Tools from the lower layer in the Tabun Cave (Israel), corresponding to 400–200 ka, contained consistently low concentrations. This implies those tools were made from chert nodules collected from pits more than several meters deep. Surface exposure dating with cosmogenic nuclides has an unparalleled advantage over other archeological dating methods in that the tool itself is dated directly.

The determination of the age of cave sediments is not only important for landscape evolution (incision rates) but also in archeology. Hominid fossils are frequently found buried in cave sediments. The age of the sedimentary deposit can be determined by U–Th speleothem dating, paleomagnetic stratigraphy, or cosmogenic nuclide burial dating. Using burial dating, Granger and co-workers determined ages of 4 Myr for sand lenses enclosing fossils of early hominids in caves in South Africa (Clark et al., 2003; Partridge et al., 2003; Granger, 2006).

5.2. Multi-nuclide studies

The combination of radionuclides with contrasting half-lives and radionuclides with stable noble gases provides more information about rates of landscape change than using a single nuclide. A crucial use of two nuclide data is distinguishing single-stage continuous exposure from complex exposure. This is especially critical when only a stable noble gas is measured from a given site. Noble gases do not decay; therefore they record all periods of exposure, even those that are millions of years apart. Multi-nuclide studies are not only needed to identify complex exposures but changes in erosion rates over time can also be uncovered.

For multi-nuclide studies to be possible, the lithology (mineralogy) of the field area must be suitable (Section 3). Presently, the most practical combinations are ^{10}Be, ^{26}Al, and ^{21}Ne in quartz (Nishiizumi et al., 2005; Ivy-Ochs et al., 2006; Kober et al., 2007; Graf et al., 2007), ^{10}Be, ^{21}Ne, and ^{36}Cl in sanidine (Ivy-Ochs et al., 2007b), and noble gases (^{3}He and ^{21}Ne) in olivine or pyroxene phenocrysts combined with ^{36}Cl in the whole rock (mafic volcanic rocks). As mineralogical considerations may limit the possible combinations, new nuclide/mineral pairs are searched for to expand the potential of the method, extend the potential geologic settings, cross-calibrate results from different nuclides, and unravel complex scenarios. In principle, the measurement of many nuclides measured in different co-existing minerals can help to establish production rate ratios (Kober et al., 2005).

The use of burial dating (measuring concentrations of two nuclides from samples taken at depths of more than 10 m) is showing much promise. By using burial dating in terrace sediment, Wolkowinsky and Granger (2004) showed that the assumption of no erosion of the terrace surfaces must be used with caution. Burial dating may provide more accurate ages than those obtained from surface samples. Balco and Stone (2005) used ^{26}Al and ^{10}Be ratios in quartz to determine the age of deeply buried paleosols and underlying till units (Section 4.3). This method has great potential, especially in the dating of old, buried glacial deposits.

5.3. Combining cosmogenic nuclides with other methods

Erosion rates vary over time. But there are few methods that can be used to constrain past erosion rates. Thermobarometry in metamorphic belts (Philpotts, 1990) and fission track analysis (Gleadow and Brown, 2000) aid the determination of long-term exhumation rates, but they do not allow the detailed reconstruction of erosion histories on timescales that are the same as climate and sea level change (10^3–10^6 yr). Sedimentary records of natural closed basins (Trustrum et al., 1999; Hinderer, 2001) or the dissection of datable morphological surfaces (Burbank et al., 1996; Abbott et al., 1997) may provide such detailed information. Unfortunately, such paleoerosion archives are sparse and in many cases temporally restricted. Cosmogenic nuclides measured in sedimentary deposits (e.g., fluvial or marine terraces) provide additional information.

The combining of techniques that integrate erosion rates over different timescales will allow the study of changes of erosion rates over time (Kirchner et al., 2001). Comparing cosmogenic nuclide-derived rates with rates based on thermochronology data will allow the study of rates over 1–100 kyr and 1 Myr timescales, respectively. Such comparison may show that erosion rates over short-term periods are higher in contrast to long-term rates. Climatic influences on erosion rates may overprint long-term rates set by tectonic activity (Pratt et al., 2002; Pratt-Sitaula et al., 2004; Schaller and Ehlers, 2006). Cosmogenic nuclide-derived erosion rates (timescale of 1–100 kyr) may be used as a benchmark of natural erosion rates. For example, in Sri Lanka, cosmogenic nuclide data combined with river gauging data (timescale of 100 yr) shows an increase of erosion rate since the onset of anthropogenic perturbation (Hewawasam et al., 2003).

Cosmogenic nuclides measured in river sediments are widely used to determine catchment-wide erosion rates. The assumption made for the calculation of catchment-wide erosion rates is that the sediment collected has been eroded equally from the entire drainage basin. This assumption might not hold true and sediment might be liberated preferentially in restricted zones (e.g., zones of landslides (Niemi et al., 2005)). Recently, it has been suggested that thermochronology data of single mineral grains collected in river sediment can be used to pinpoint from where in the catchment the sediment has been eroded (Stock et al., 2006; Vermeesch, 2007). The combination of this technique with cosmogenic nuclide methods will prove a powerful tool.

5.4. What are the current issues?

Concern about methodological uncertainties, such as those associated with the production rates, the site latitude, and altitude scaling factors as well as the effect of past changes in the Earth's magnetic field, has led to the establishment of an international consortium made up of CRONUS-EARTH (www.physics.purdue.edu/cronus) and CRONUS-EU (www.cronus-eu.net). Studies aimed at determining nuclide production rates are being completed on both natural sites with independent age constraints and on artificial targets. Scaling factors are being assessed with neutron monitors and other techniques. Numerical modeling is being used to determine production rates and scaling factors both now and in the past. The half-lives of radioactive nuclides must be accurately known. In the case of ^{10}Be, two different half-lives have been determined, 1.51 and 1.34 Myr (Granger, 2006). Experiments to measure the half-life of ^{10}Be are in progress as part of the CRONUS agenda. When these factors are better constrained the errors of the final ages will be closer to the range of the AMS and noble gas mass spectrometry measurement uncertainties (of the order of 1–4%).

5.5. Outlook

The ability to use cosmogenic nuclides to determine how long minerals have been exposed at the surface of the Earth provides an unrivaled tool for determining ages of landforms and rates of geomorphic processes. Because of this unique capability, the variety of applications of cosmogenic nuclides will continue to grow. As the knowledge of production rate and scaling formalities improves, the precision of obtained ages improves. But the accuracy of the ages remains a question of geological uncertainties. Samples taken from different parts of a landscape element may have a large geological variance. Similarly, with larger data sets and smaller errors for individual ages, focus shifts to problems in interpretation. Obtained exposure ages must be evaluated individually for conformity with field relationships, including local terrace or moraine stratigraphy and regional morphostratigraphic relationships; as well as with independent age constraints for the same or correlative features. It is now becoming possible to date events that are closely spaced in time (hundreds of years), such as deposition of individual lobes of a landslide or a fan (Benn et al., 2006; Dühnforth et al., 2007), or buildup of nested or cross-cutting moraines (Ivy-Ochs et al., 2006, 2009). There is an immense potential for revealing long-term variations in denudation rates when cosmogenic nuclide methods are combined with thermochronometric methods. Such studies will lead to characterization of the steadiness of denudational processes and may yield information on perturbation and response times of catchments to climatic or tectonic forces (Schaller and Ehlers, 2006). Cosmogenic nuclides provide a powerful and multifaceted tool whose potential has yet to be fully realized. But this power is tempered with the need for careful sampling in light of detailed field mapping.

ACKNOWLEDGMENTS

We sincerely thank F. Kober for help at many stages; C. Fenton, F. Kober, and R. Wieler for insightful discussions, critical reading and innumerable suggestions that greatly improved this manuscript; and P. W. Kubik, H. -A. Synal, and M. Suter at the PSI/ETH tandem accelerator facility for their unwavering support.

REFERENCES

Abbott, L. D., E. A. Silver, R. S. Anderson, R. Smith, J. C. Ingle, K. A. King, D. W. Haig, E. E. Small, J. Galewsky, and W. V. Sliter. (1997). Measurement of tectonic surface uplift rate in a young collisional mountain belt. *Nature*, 385, 501–507.

Ackert, R. P., and M. D. Kurz. (2004). Age and uplift rates of Sirius Group sediments in the Dominion Range, Antarctica, from surface exposure dating and geomorphology. *Global and Planetary Change*, 42, 207–225.

Akçar, N., S. Ivy-Ochs, and C. Schlüchter. (2008a). Application of in-situ produced terrestrial cosmogenic nuclides to archaeology: A schematic review. *Quaternary Science Journal*, 57, 226–238.

Akçar, N., V. Yavuz, S. Ivy-Ochs, P. W. Kubik, M. Vardar, and C. Schlüchter. (2007). Paleoglacial records from Kavron Valley, NE Turkey: Field and cosmogenic exposure dating evidence. *Quaternary International*, 164–165, 170–183.

Akçar, N., V. Yavuz, S. Ivy-Ochs, P. W. Kubik, M. Vardar, and C. Schlüchter. (2008b). A Case for a down wasting mountain glacier during Termination-I, Verçenik Valley, NE Turkey. *Journal of Quaternary Science*, 23, 273–285.

Albrecht, A., G. F. Herzog, J. Klein, B. Dezfouly-Arjomandy, and F. Goff. (1993). Quaternary erosion and cosmic-ray-exposure history derived from ^{10}Be and ^{26}Al produced in situ – An example from Pajarito plateau, Valles caldera region. *Geology*, 21, 551–554.

Anderson, R. S., J. L. Repka, and G. S. Dick. (1996). Explicit treatment of inheritance in dating depositional surfaces using ^{10}Be and ^{26}Al. *Geology*, 24, 47–51.

Anthony, E. Y., and J. Poths. (1992). He-3 surface exposure dating and its implications for magma evolution in the Potrillo volcanic field, Rio Grande Rift, New Mexico, USA. *Geochimica et Cosmochimica Acta*, 56, 4105–4108.

April, R., R. Newton, and L. T. Coles. (1986). Chemical weathering in two Adirondack watersheds-past and present-day rates. *Geological Society of America Bulletin*, 97, 1232–1238.

Balco, G., and J. Schaefer. (2006). Cosmogenic-nuclide and varve chronologies for the deglaciation of southern New England. *Quaternary Geochronology*, 1, 15–28.

Balco, G., and J. O. H. Stone. (2005). Measuring middle Pleistocene erosion rates with cosmic-ray-produced nuclides in buried alluvial sediment, Fisher Valley, southeastern Utah. *Earth Surface Processes and Landforms*, 30, 1051–1067.

Balco, G., J. O. H. Stone, and C. Jennings. (2005). Dating Plio-Pleistocene glacial sediments using the cosmic-ray-produced radionuclides Be-10 and Al-26. *American Journal of Science*, 305, 1–41.

Balco, G., J. O. H. Stone, S. C. Porter, and M. W. Caffee. (2002). Cosmogenic-nuclide ages for New England coastal moraines, Martha's Vineyard and Cape Cod, Massachusetts, USA. *Quaternary Science Reviews*, 21, 2127–2135.

Ballantyne, C. K., D. McCarroll, and J. O. Stone. (2006). Vertical dimensions and age of the Wicklow Mountains ice dome, Eastern Ireland, and implications for the extent of the last Irish Ice Sheet. *Quaternary Science Reviews*, 25, 2048–2058.

Ballantyne, C. K., and J. O. Stone. (2004). The Beinn Alligin rock avalanche, NW Scotland: cosmogenic Be-10 dating, interpretation and significance. *The Holocene*, 14, 448–453.

Barnard, P. L., L. A. Owen, M. C. Sharma, and R. C. Finkel. (2001). Natural and human-induced landsliding in the Garhwal Himalaya of northern India. *Geomorphology*, 40, 21–35.

Barrows, T. T., J. O. Stone, and L. K. Fifield. (2004). Exposure ages for Pleistocene periglacial deposits in Australia. *Quaternary Science Reviews*, 23, 697–708.

Barrows, T. T., J. O. Stone, L. K. Fifield, and R. G. Cresswell. (2001). Late Pleistocene Glaciation of the Kosciuszko Massif, Snowy Mountains, Australia. *Quaternary Research*, 55, 179–189.

Belton, D. X., R. W. Brown, B. P. Kohn, D. Fink, and K. A. Farley. (2004). Quantitative resolution of the debate over antiquity of the central Australian landscape: Implications for the tectonic and geomorphic stability of cratonic interiors. *Earth and Planetary Science Letters*, 219, 21–34.

Benedetti, L., R. Finkel, G. King, R. Armijo, D. Papanastassiou, R. J. Ryerson, F. Flerit, D. L. Farber, and G. Stavrakakis. (2003). Motion on the Kaparelli fault (Greece) prior to the 1981 earthquake sequence determined from ^{36}Cl cosmogenic dating. *Terra Nova*, 15, 118–124.

Benedetti, L., R. Finkel, D. Papanastassiou, G. King, R. Armijo, F. Ryerson, D. Farber, and F. Flerit. (2002). Post-glacial slip history of the Sparta fault (Greece) determined by

Cl-36 cosmogenic dating: Evidence for non-periodic earthquakes. *Geophysical Research Letters*, 29, 87.1–87.4.

Benn, D. I., L. A. Owen, R. C. Finkel, and S. Clemmens. (2006). Pleistocene lake outburst floods and fan formation along the eastern Sierra Nevada, California: Implications for the interpretation of intermontane lacustrine records. *Quaternary Science Reviews*, 25, 2729–2748.

Benson, L., R. Madole, W. M. Phillips, G. P. Landis, T. Thomas, and P. W. Kubik. (2004). The probable importance of snow and sediment shielding on cosmogenic ages of north-central Colorado Pinedale and pre-Pinedale moraines. *Quaternary Science Reviews*, 23, 193–206.

Bentley, M. J., C. J. Fogwill, P. W. Kubik, and D. E. Sugden. (2006). Geomorphological evidence and cosmogenic ^{10}Be/^{26}A1 exposure ages for the last glacial maximum and deglaciation of the Antarctic Peninsula Ice Sheet. *Geological Society of America Bulletin*, 118, 1149–1159.

Bierman, P., and M. Caffee. (2001). Slow rates of rock surface erosion and sediment production across the Namib Desert and Escarpment, Southern Africa. *American Journal of Science*, 301, 326–358.

Bierman, P., and M. Caffee. (2002). Cosmogenic exposure and erosion history of Australian bedrock landforms. *GSA Bulletin*, 114, 787–803.

Bierman, P. R., M. W. Caffee, P. T. Davis, K. Marsella, M. Pavich, P. Colgan, D. Mickelson, and J. Larsen. (2002). Rates and timing of earth surface processes from in-situ produced cosmogenic ^{10}Be. In: *Beryllium: Mineralogy, Petrology, and Geochemistry* (Ed E. Grew). Reviews in Mineralogy and Geochemistry, pp. 147–204.

Bierman, P. R., A. R. Gillespie, and M. W. Caffee. (1995). Cosmogenic ages for earthquake recurrence intervals and debris flow fan deposition, Owens Valley, California. *Science*, 270, 447–450.

Bierman, P. R., K. A. Marsella, C. Patterson, P. T. Davis, and M. Caffee. (1999). Mid-Pleistocene cosmogenic minimum-age limits for pre-Wisconsian glacial surfaces in southwestern Minnesota and southern Baffin Island: A multiple nuclide approach. *Geomorphology*, 27, 25–39.

Bierman, P., and K. K. Nichols. (2004). Rock to sediment-slope to sea with ^{10}Be-rates of landscape change. *Annual Review of Earth and Planetary Sciences*, 32, 215–255.

Bierman, P., and E. J. Steig. (1996). Estimating rates of denudation using cosmogenic isotope abundances in sediment. *Earth Surface Processes*, 21, 125–139.

Bierman, P. R., and J. Turner. (1995). ^{10}Be and ^{26}Al evidence for exceptionally low rates of Australian bedrock erosion and the likely existence of pre-Pleistocene landscapes. *Quaternary Research*, 44, 378–382.

Bigot-Cormier, F., R. Braucher, D. Bourlès, Y. Guglielmi, M. Dubar, and J.-F. Stèphan. (2005). Chronological constraints on processes leading to large active landslides. *Earth and Planetary Science Letters*, 235, 141–150.

Boaretto, E., D. Berkovits, M. Hass, S. K. Hui, A. Kaufmann, M. Paul, and S. Weiner. (2000). Dating of prehistoric caves sediments and flints using ^{10}Be and ^{26}Al in quartz from Tabun Cave (Israel): Progress report. *Nuclear Instruments and Methods in Physics Research*, B172, 767–771.

Braucher, R., P.-H. Blard, L. Benedetti, and D. L. Bourlès. (2006). Extending ^{10}Be applications to carbonate-rich and mafic environments. In: *In Situ Cosmogenic Nuclides and their Applications in Earth Sciences* (Eds L. L. Siame, D. L. Bourlès, and E. T. Brown). Geological Society of America, Special Paper 415, pp. 17–28.

Braucher, R., L. Benedetti, D. Bourlès, E. T. Brown, and D. Chardon. (2005). Use of in situ-produced ^{10}Be in carbonate-rich environments: A first attempt. *Geochimica et Cosmochimica Acta*, 69, 1473–1478.

Braucher, R., E. T. Brown, D. L. Bourlès, and F. Colin. (2003). In situ produced ^{10}Be measurements at great depths: Implications for production rates by fast muons. *Earth and Planetary Science Letters*, 211, 251–258.

Braun, J., A. M. Heimsath, and J. Chappell. (2001). Sediment transport mechanisms on soil-mantled hillslopes. *Geology*, 29, 683–686.

Brimhall, G. H., and W. E. Dietrich. (1987). Constitutive mass balance relations between chemical composition, volume, density, porosity, and strain in metasomatic hydrochemical systems-results on weathering and pedogenesis. *Geochimica et Cosmochimica Acta*, 51, 567–587.

Briner, J. P., G. H. Miller, P. T. Davis, and R. C. Finkel. (2005a). Cosmogenic exposure dating in arctic glacial landscapes: Implications for the glacial history of northeastern Baffin Island, Arctic Canada. *Canadian Journal of Earth Sciences*, 42, 67–84.

Briner, J. P., G. H. Miller, P. T. Davis, and R. C. Finkel. (2006). Cosmogenic radionuclides from fiord landscapes support differential erosion by overriding ice sheets. *Geological Society of America Bulletin*, 118, 406–420.

Briner, J. P., D. S. Kaufman, W. F. Manley, R. C. Finkel, and M. W. Caffee. (2005b). Cosmogenic exposure dating of late Pleistocene moraine stabilization in Alaska. *Geological Society of America Bulletin*, 117, 1108–1120.

Briner, J. P., and T. Swanson. (1998). Using inherited cosmogenic ^{36}Cl to constrain glacial erosion rates of the Cordilleran ice sheet. *Geology*, 26, 3–6.

Brook, E. J., E. T. Brown, M. D. Kurz, R. P. Ackert, G. Raisbeck, and F. Yiou. (1995b). Constraints on age, erosion, and uplift of Neogene glacial deposits in the Transantarctic Mountains determined from in situ cosmogenic ^{10}Be and ^{26}Al. *Geology*, 23, 1063–1066.

Brook, E. J., and M. D. Kurz. (1993). Surface-exposure chronology using in-situ cosmogenic ^{3}He in Antarctic quartz sandstone boulders. *Quaternary Research*, 39, 1–10.

Brook, E. J., M. D. Kurz, R. P. Ackert, G. M. Raisbeck, and F. Yiou. (1995a). Cosmogenic nuclide exposure ages and glacial history of late Quaternary Ross Sea drift in McMurdo sound, Antarctica. *Earth and Planetary Science Letters*, 131, 41–56.

Brown, E. T., D. L. Bourlès, B. C. Burchfiel, D. Qidong, L. Jun, P. Molnar, G. M. Raisbeck, and F. Yiou. (1998a). Estimation of slip rates in the southern Tien Shan using cosmic ray exposure dates of abandoned alluvial fans. *Geological Society of America Bulletin*, 110, 377–386.

Brown, E. T., J. M. Edmond, G. M. Raisbeck, F. Yiou, M. D. Kurz, and E. J. Brook. (1991). Examination of surface exposure ages in Antarctic moraines using in-situ produced ^{10}Be and ^{26}Al. *Geochimica et Cosmochimica Acta*, 55, 2269–2283.

Brown, E. T., R. F. Stallard, M. C. Larsen, D. L. Bourlès, G. Raisbeck, and F. Yiou. (1995). Denudation rates determined from the accumulation of in-situ produced ^{10}Be in the Luquillo Fortes, Puerto Rico. *Earth and Planetary Science Letters*, 129, 695–702.

Brown, E. T., R. F. Stallard, M. C. Larsen, D. L. Bourlès, G. M. Raisbeck, and F. Yiou. (1998b). Determination of predevelopment denudation rates of an agricultural watershed (Cayaguas River, Puerto Rico) using in-situ-produced Be-10 in river-borne quartz. *Earth and Planetary Science Letters*, 160, 723–728.

Brown, E. T., T. W. Trull, P. Jean-Baptiste, G. Raisbeck, D. L. Bourlès, F. Yiou, and B. Marty. (2000). Determination of cosmogenic production rates of ^{10}Be, ^{3}He and ^{3}H in water. *Nuclear Instruments and Methods in Physics Research*, B172, 873–883.

Bruno, L. A., H. Baur, T. Graf, C. Schlüchter, P. Signer, and R. Wieler. (1997). Dating of Sirius Group tillites in the Antarctic Dry Valleys with cosmogenic ^{3}He and ^{21}Ne. *Earth and Planetary Science Letters*, 147, 37–54.

Burbank, D. W., J. Leland, E. Fielding, R. S. Anderson, N. Brozovic, M. R. Reid, and C. Duncan. (1996). Bedrock incision, rock uplift and threshold hillslopes in the northwestern Himalayas. *Nature*, 379, 505–510.

Cerling, T. E., and H. Craig. (1994a). Cosmogenic ^{3}He production rates from 39°N to 46°N latitude, western USA and France. *Geochimica et Cosmochimica Acta*, 58, 249–255.

Cerling, T. E., and H. Craig. (1994b). Geomorphology and in-situ cosmogenic isotopes. *Annual Review of Earth and Planetary Sciences*, 22, 273–317.

Cerling, T. E., R. J. Poreda, and S. L. Rathburn. (1994). Cosmogenic ^{3}He and ^{21}Ne age of the Big Lost River flood, Snake River plain, Idaho. *Geology*, 22, 227–230.

Child, D., G. Elliot, C. Mifsud, A. M. Smith, and D. Fink. (2000). Sample processing for earth science studies at ANTARES. *Nuclear Instruments and Methods in Physics Research*, B172, 856–860.

Clapp, E. M., P. Bierman, and M. Caffee. (2002). Using ^{10}Be and ^{26}Al to determine sediment generation rates and identify sediment source areas in an arid region drainage basin. *Geomorphology*, 45, 89–104.

Clapp, E. M., P. Bierman, A. P. Schick, J. Lekach, Y. Enzel, and M. Caffee. (2000). Sediment yield exceeds sediment production in arid region drainage basins. *Geology*, 28, 995–998.

Clapp, E. M., P. R. Bierman, K. K. Nichols, M. Pavich, and M. Caffee. (2001). Rates of sediment supply to arroyos from upland erosion determined using in situ produced cosmogenic Be-10 and Al-26. *Quaternary Research*, 55, 235–245.

Clark, R. H., T. Partiridge, D. E. Granger, and M. W. Caffee. (2003). Dating the Sterkfontein Fossils. *Science*, 301, 596–597.

Cockburn, H. A. P., R. W. Brown, M. A. Summerfield, and M. A. Seidl. (2000). Quantifying passive margin denudation and landscape development using a combined fission-track thermochronology and cosmogenic isotope analysis approach. *Earth and Planetary Science Letters*, 179, 429–435.

Cockburn, H. A. P., M. A. Seidl, and M. A. Summerfield. (1999). Quantifying denudation rates on inselbergs in the central Namib Desert using in situ produced cosmogenic ^{10}Be and ^{26}Al. *Geology*, 27, 399–402.

Cockburn, H. A. P., and M. A. Summerfield. (2004). Geomorphological applications of cosmogenic isotope analysis. *Progress in Physical Geography*, 28, 1–42.

Colgan, P. M., P. Bierman, D. M. Mickelson, and M. Caffee. (2002). Variation in glacial erosion near the southern margin of the Laurentide ice sheet, south-central Wisconsin, USA; implications for cosmogenic dating of glacial terrains. *Geological Society of America Bulletin*, 114, 1581–1591.

Davis, N. K., W. W. Locke, K. L. Pierce, and R. C. Finkel. (2006a). Glacial Lake Musselshell: Late Wisconsin slackwater on the Laurentide ice margin in central Montana, USA. *Geomorphology*, 75, 330–345.

Davis, P. T., J. P. Briner, R. D. Coulthard, R. C. Finkel, and G. H. Miller. (2006b). Preservation of Arctic landscapes overridden by cold-based ice sheets. *Quaternary Research*, 65, 156–163.

Davis, P. T., P. Bierman, K. A. Marsella, M. W. Caffee, and J. R. Southon. (1999). Cosmogenic analysis of glacial terrains in the eastern Canadian Arctic: a test for inherited nuclides and the effectiveness of glacial erosion. *Annals of Glaciology*, 28, 181–188.

Davis, R., and O. A. Schaeffer. (1955). Chlorine-36 in Nature. *Annals of the New York Academy of Sciences*, 62(5), 107–121.

Desilets, D., and M. Zreda. (2001). On scaling cosmogenic nuclide production rates for altitude and latitude using cosmic-ray measurements. *Earth and Planetary Science Letters*, 193, 213–225.

Desilets, D., M. Zreda, P. F. Almasi, and D. Elmore. (2006b). Determination of cosmogenic ^{36}Cl in rocks by isotope dilution: innovations, validation and error propagation. *Chemical Geology*, 233, 185–195.

Desilets, D., M. Zreda, and T. Pradu. (2006a). Extended scaling factors for in situ cosmogenic nuclides: New measurements at low latitude. *Earth and Planetary Science Letters*, 246, 265–276.

Duffield, W., N. Riggs, D. Kaufman, D. Champion, C. Fenton, S. Forman, W. McIntosh, R. Hereford, J. Plescia, and M. Ort. (2006). Multiple constraints on the age of a Pleistocene lava dam across the Little Colorado River at Grand Falls, Arizona. *Geological Society of America Bulletin*, 118, 421–429.

Dühnforth, M., A. L. Densmore, S. Ivy-Ochs, P. A. Allen, and P. W. Kubik. (2007). Timing and patterns of debris-flow fan deposition on Shepherd and Symmes Creek fans, Owens Valley, California, deduced from cosmogenic ^{10}Be. *Journal of Geophysical Research – Earth Surface*, 112, F03S15, doi:10.1029/2006JF000562.

Dunai, T. J. (2000). Scaling factors for production rates of in situ produced cosmogenic nuclides: a critical re-evaluation. *Earth and Planetary Science Letters*, 176, 157–169.

Dunai, T. J. (2001). Influence of secular variation of the geomagnetic field on production rates of in situ produced cosmogenic nuclides. *Earth and Planetary Science Letters*, 193, 197–212.

Dunai, T. J., G. A. Gonzales Lopez, and J. Juez-Larre. (2005). Oligocene–Miocene age of aridity in the Atacama Desert revealed by exposure dating of erosion-sensitive landforms. *Geology*, 33, 321–324.

Dunai, T. J., and K. Roselieb. (1996). Sorption and diffusion of helium in garnet: Implications for volatile tracing and dating. *Earth and Planetary Science Letters*, 139, 411–421.

Dunne, J., and D. Elmore. (2003). Monte Carlo simulations of low-energy cosmogenic neutron fluxes near the bottom of cliff faces. *Earth and Planetary Science Letters*, 206, 43–49.

Dunne, J., D. Elmore, and P. Muzikar. (1999). Scaling factors for the rates of production of cosmogenic nuclides for geometric shielding and attenuation at depth on sloped surfaces. *Geomorphology*, 27, 3–11.

Elmore, D., X. Ma, T. Miller, K. Mueller, M. Perry, F. Rickey, P. Sharma, and P. Simms. (1997). Status and plans for the PRIME Lab AMS facility. *Nuclear Instruments and Methods in Physics Research*, B123, 69–72.

Elmore, D., and F. M. Phillips. (1987). Accelerator mass spectrometry for measurement of long-lived radioisotopes. *Science*, 236, 543–550.

Erismann, T. H., and G. Abele. (2001). *Dynamics of Rockslides and Rockfalls*. Springer, Berlin, 316 pp.

Evans, J. M. (2001). Calibration of the production rates of cosmogenic chlorine-36 from potassium. PhD Thesis, Australian National University, Canberra, 142 pp.

Everest, J., and P. W. Kubik. (2006). The deglaciation of eastern Scotland: Cosmogenic ^{10}Be evidence for a Lateglacial stillstand. *Journal of Quaternary Science*, 21, 95–104.

Everest, J. D., T. Bradwell, C. J. Forgwill, and P. W. Kubik. (2006). Cosmogenic ^{10}Be age constraints for the Wester Ross readvance moraine: Insights into British Ice Sheet behaviour. *Geografiska Annaler*, 88A, 9–17.

Fabel, D., J. Harbor, D. Dahms, A. James, D. Elmore, L. Horn, K. Daley, and C. Steele. (2004). Spatial patterns of glacial erosion at a valley scale derived from terrestrial cosmogenic Be-10 and Al-26 concentrations in rock. *Annals of the Association of American Geographers*, 94, 241–255.

Fabel, D., A. P. Stroeven, J. Harbor, J. Kleman, D. Elmore, and D. Fink. (2002). Landscape preservation under Fennoscandian ice sheets determined from in situ produced Be-10 and Al-26. *Earth and Planetary Science Letters*, 201, 397–406.

Fabryka-Martin, J. (1988). Production of radionuclides in the Earth and the hydrologic significance, with emphasis on chlorine-36 and iodine-129. PhD Thesis, University of Arizona, Tucson, AZ.

Farley, K. A., T. E. Cerling, and P. G. Fitzgerald. (2001). Cosmogenic ^{3}He in igneous and fossil tooth enamel fluorapatite. *Earth and Planetary Science Letters*, 185, 7–14.
Farley, K. A., J. Libarkin, S. Mukhopadhyay, and W. Amidon. (2006). Cosmogenic and nucleogenic He-3 in apatite, titanite, and zircon. *Earth and Planetary Science Letters*, 248, 451–461.
Fenton, C. R., R. Poreda, B. P. Nash, R. H. Webb, and T. E. Cerling. (2004). Geochemical discrimination of five Pleistocene lava-dam outburst-flood deposits, western Grand Canyon, Arizona. *Journal of Geology*, 112, 91–110.
Fenton, C. R., R. H. Webb, T. E. Cerling, R. J. Poreda, and B. P. Nash. (2002). Cosmogenic ^{3}He ages and geochemical discrimination of lava-dam outburst-flood deposits in western Grand Canyon, Arizona. In: *Ancient Floods and Modern Hazards: Principles and Applications of Paleoflood Hydrology* (Eds P. K House, R. H. Webb, V. R. Baker, and D. R. Levish). Washington, DC, American Geophysical Union, Water Science and Application Series 4, pp. 191–215.
Fenton, C. R., R. H. Webb, P. A. Pearthree, T. E. Cerling, and R. J. Poreda. (2001). Displacement rates on the Toroweap and Hurricane faults: Implications for Quaternary downcutting in the Grand Canyon, Arizona. *Geology*, 29, 1035–1038.
Fink, D., J. Klein, and R. Middleton. (1990). ^{41}Ca: Past, present and future. *Nuclear Instruments and Methods in Physics Research*, B52, 572–582.
Finkel, R. C., L. A. Owen, P. L. Barnard, and M. W. Caffee. (2003). Beryllium-10 dating of Mount Everest moraines indicates a strong monsoon influence and glacial synchroneity throughout the Himalaya. *Geology*, 31, 561–564.
Finkel, R. C., and M. Suter. (1993). AMS in the earth sciences. *Advances in Analytical Geochemistry*, 1, 1–114.
Fleming, A., M. A. Summerfield, J. O. Stone, L. K. Fifield, and R. G. Cresswell. (1999). Denudation rates for the southern Drakensberg escarpment, SE Africa, derived from in-situ-produced cosmogenic Cl-36: Initial results. *Journal of Geological Society*, 156, 209–212.
Froehlich, K., and J. Lubbert. (1973). Use of natural radionuclides ^{41}Ca and ^{39}Ar in measurement of rate of erosion of surface rocks. *Zeitschrift f Angewandte Geologie*, 19, 550–553.
Gayer, E., R. Pik, J. Lave, C. France-Lanord, D. L. Bourlès, and B. Marty. (2004). Cosmogenic ^{3}He in Himalayan garnets indicating an altitude dependence of the ^{3}He/^{10}Be production ratio. *Earth and Planetary Science Letters*, 229, 91–104.
Gladkis, L. G., L. K. Fifield, C. R. Morton, T. T. Barrows, and S. G. Tims. (2007). Manganese-53: Development of the AMS technique for exposure-age dating applications. *Nuclear Instruments and Methods in Physics Research*, B259(1), 236–240.
Gleadow, A. J. W., and R. W. Brown. (2000). Fission-track thermochronology and the long-term denudational response to tectonics. In: *Geomorphology and Global Tectonics* (Ed. M. A. Summerfield). Wiley, pp. 57–75.
Gosse, J. C. (2005). The contributions of cosmogenic nuclides to unraveling alpine paleo-glacier histories. In: *Global Change and Mountain Regions: A State of Knowledge Overview* (Eds U. M. Huber, M. A. Reasoner, and B. Bugmann). Springer, Dordrecht, The Netherlands, pp. 39–50.
Gosse, J. C., E. B. Evenson, J. Klein, B. Lawn, and R. Middleton. (1995a). Precise cosmogenic ^{10}Be measurements in western North America: Support for a global Younger Dryas cooling event. *Geology*, 23, 877–880.
Gosse, J. C., J. Klein, E. B. Evenson, B. Lawn, and R. Middleton. (1995b). Be-10 dating of the duration and retreat of the last Pinedale glacial sequence. *Science*, 268, 1329–1333.
Gosse, J. C., and F. M. Phillips. (2001). Terrestrial in situ cosmogenic nuclides: theory and application. *Quaternary Science Reviews*, 20, 1475–1560.

Graf, A., S. Strasky, S. Ivy-Ochs, P. W. Kubik, M. Burkhard, and C. Schlüchter. (2007). Pre-Last glaciation erratics from the Montoz Area, Jura Mountains, Switzerland. *Quaternary International*, 164–165, 43–52.

Graf, T., C. P. Kohl, K. Marti, and K. Nishiizumi. (1991). Cosmic-ray produced neon in Antarctic rocks. *Geophysical Research Letters*, 18, 203–206.

Graham, I. J., B. J. Barry, R. G. Ditchburn, and N. E. Whitehead. (2000). Validation of cosmogenic nuclide production rate scaling through direct measurement. *Nuclear Instruments and Methods in Physics Research*, B172, 802–805.

Granger, D. E. (2006). A review of burial dating methods using ^{10}Be and ^{26}Al. In: *In Situ Cosmogenic Nuclides and their Applications in Earth Sciences* (Eds L. L. Siame, D. L. Bourlès, and E. T. Brown). Geological Society of America Special Paper 415, pp. 1–16.

Granger, D. E., A. J. Cyr, and E. Rauh. (2005). Beryllium-10 measurement in carbonate rocks: Progress, problems, and potential, AMS-10, Berkeley, CA.

Granger, D. E., D. Fabel, and A. N. Palmer. (2001). Plio-Pleistocene incision of the Green River, Kentucky, from radioactive decay of cosmogenic ^{26}Al and ^{10}Be in Mammoth Cave sediments. *Geological Society of America Bulletin*, 113(7), 825–836.

Granger, D. E., J. W. Kirchner, and R. Finkel. (1996). Spatially averaged long-term erosion rates measured from in-situ-produced cosmogenic nuclides in alluvial sediment. *Journal of Geology*, 104, 249–257.

Granger, D. E., J. W. Kirchner, and R. C. Finkel. (1997). Quaternary downcutting rate of the New River, Virginia, measured from differential decay of cosmogenic ^{26}Al and ^{10}Be in cave-deposited alluvium. *Geology*, 25, 107–110.

Granger, D. E., and P. F. Muzikar. (2001). Dating sediment burial with in situ-produced cosmogenic nuclides: Theory, techniques, and limitations. *Earth and Planetary Science Letters*, 188, 269–281.

Granger, D. E., and A. L. Smith. (2000). Dating buried sediments using radioactive decay and muogenic production of ^{10}Be and ^{26}Al. *Nuclear Instruments and Methods in Physics Research*, B172, 822–826.

Green, E. G., W. E. Dietrich, and J. F. Banfield. (2006). Quantification of chemical weathering rates across an actively eroding hillslope. *Earth and Planetary Science Letters*, 242, 155–169.

Gualtieri, L., J. Brigham-Grette, and O. Glushkova. (2000). Evidence for restricted ice extent during the last glacial maximum in the Koryak Mountains of Chukotka, far eastern Russia. *Geological Society of America Bulletin*, 112, 1106–1118.

Gualtieri, L., S. L. Vartanyan, J. Brigham-Grette, and P. M. Anderson. (2005). Evidence for an ice-free Wrangel Island, northeast Siberia during the Last Glacial Maximum. *Boreas*, 34, 264–273.

Hallet, B., and J. Putkonen. (1994). Surface dating of dynamic landforms-young boulders on ageing moraines. *Science*, 265, 937–940.

Hancock, G. S., R. S. Anderson, O. A. Chadwick, and R. C. Finkel. (1999). Dating of fluvial terraces using ^{10}Be and ^{26}Al profiles, Wind River, Wyoming. *Geomorphology*, 27, 41–60.

Heimsath, A. M., J. Chappell, W. E. Dietrich, K. Nishiizumi, and R. C. Finkel. (2000). Soil production on a retreating escarpment in southeastern Australia. *Geology*, 28, 787–790.

Heimsath, A. M., J. Chappell, N. A. Spooner, and D. G. Questiaux. (2002). Creeping soil. *Geology*, 30, 111–114.

Heimsath, A. M., W. E. Dietrich, K. Nishiizumi, and R. C. Finkel. (1997). The soil production rate function and landscape equilibrium. *Nature*, 388, 358–361.

Heimsath, A. M., W. E. Dietrich, K. Nishiizumi, and R. C. Finkel. (1999). Cosmogenic nuclides, topography, and the spatial variation of soil depth. *Geomorphology*, 27, 151–172.

Heimsath, A. M., W. E. Dietrich, K. Nishiizumi, and R. C. Finkel. (2001). Stochastic processes of soil production and transport: erosion rates, topographic variation and cosmogenic nuclides in the Oregon Coast Range. *Earth Surface Processes and Landforms*, 26, 531–552.

Hermanns, R. L., S. Niedermann, A. V. Garcia, J. S. Gomez, and M. R. Strecker. (2001). Neotectonics and catastrophic failure of mountain fronts in the southern intra-Andean Puna Plateau, Argentina. *Geology*, 29, 619–622.

Hermanns, R. L., S. Niedermann, S. Ivy-Ochs, and P. W. Kubik. (2004). Rock avalanching into a landslide-dammed lake causing multiple dam failure in Las Conchas valley (NW Argentina)-evidence from surface exposure dating and stratigraphic analyses. *Landslides*, 2, 113–122.

Hetzel, R., S. Niedermann, S. Ivy-Ochs, P. W. Kubik, M. Tao, and B. Goa. (2002a). ^{21}Ne versus ^{10}Be and ^{26}Al exposure ages of fluvial terraces: The influence of crustal Ne in quartz. *Earth and Planetary Science Letters*, 201, 575–591.

Hetzel, R., S. Niedermann, M. Tao, P. Kubik, S. Ivy-Ochs, B. Gao, and M. R. Strecker. (2002b). Low slip rates and long-term preservation of geomorphic features in Central Asia. *Nature*, 417, 428–432.

Hetzel, R., M. Tao, S. Niedermann, M. R. Strecker, S. Ivy-Ochs, P. W. Kubik, and B. Gao. (2004a). Implications of the fault scaling law for the growth of topography: mountain ranges in the broken foreland of north-west Tibet. *Terra Nova*, 16, 157–162.

Hetzel, R., M. Tao, S. Stokes, S. Niedermann, S. Ivy-Ochs, B. Gao, M. R. Strecker, and P. W. Kubik. (2004b). Late Pleistocene/Holocene slip rate of the Zhangye thrust (Qilian Shan, China) and implications for the active growth of the northeastern Tibetan Plateau. *Tectonics*, 23, TC6006.

Hewawasam, T., F. von Blanckenburg, M. Schaller, and P. W. Kubik. (2003). Increase of human over natural erosion rates in tropical highlands constrained by cosmogenic nuclides. *Geology*, 3, 597–600.

Hinderer, M. (2001). Late Quaternary denudation of the Alps, valley and lake fillings and modern river loads. *Geodinamica Acta*, 14, 231–263.

Ivy-Ochs, S. (1996). The dating of rock surface using in situ produced ^{10}Be, ^{26}Al and ^{36}Cl, with examples from Antarctica and the Swiss Alps. PhD Thesis No. 11763, ETH, Zürich, 196 pp.

Ivy-Ochs, S., H. Heuberger, P. W. Kubik, H. Kerschner, G. Bonani, M. Frank, and C. Schlüchter. (1998a). The age of the Koefels event. Relative, ^{14}C and cosmogenic isotope dating of an early Holocene landslide in the Central Apls (Tyrol, Austria). *Zeitschrift für Gletscherkunde und Glazialgeologie*, 134, 57–70.

Ivy-Ochs, S., H. Kerschner, A. Reuther, M. Maisch, R. Sailer, J. Schaefer, P. W. Kubik, H.-A. Synal, and C. Schlüchter. (2006). The timing of glacier advances in the northern European Alps based on surface exposure dating with cosmogenic ^{10}Be, ^{26}Al, ^{36}Cl, and ^{21}Ne. In: *In Situ Cosmogenic Nuclides and their Applications in Earth Sciences* (Eds L. L. Siame, D. L. Bourlès, and E. T. Brown). Geological Society of America Special Paper 415, pp. 43–60.

Ivy-Ochs, S., H. Kerschner, and C. Schlüchter. (2007a). Cosmogenic nuclides and the dating of Lateglacial and early Holocene glacier variations: The Alpine perspective. *Quaternary International*, 164–165, 53–63.

Ivy-Ochs, S., F. Kober, V. Alfimov, P. W. Kubik, and H.-A. Synal. (2007b). Cosmogenic ^{10}Be, ^{21}Ne, and ^{36}Cl in sanidine and quartz from Chilean ignimbrites. *Nuclear Instruments and Methods*, B223, 588–594.

Ivy-Ochs, S., H. Kerschner, M. Maisch, M. Christl, P.W. Kubik, C. Schlüchter. (2009). Latest Pleistocene and Holocene glacier variations in the European Alps. *Quaternary Science Reviews*.

Ivy-Ochs, S., P. W. Kubik, J. Masarik, R. Wieler, L. Bruno, and C. Schlüchter. (1998b). Preliminary results on the use of pyroxene for ^{10}Be surface exposure dating. *Schweizerische Mineralogische und Petrographische Mitteilungen*, 78, 375–382.

Ivy-Ochs, S., C. Schlüchter, P. W. Kubik, and G. H. Denton. (1999). Moraine exposure dates imply synchronous Younger Dryas glacier advances in the European Alps and in the Southern Alps of New Zealand. *Geografiska Annaler*, 81A, 313–323.

Ivy-Ochs, S., C. Schlüchter, P. W. Kubik, B. Dittrich-Hannen, and J. Beer. (1995). Minimum ^{10}Be exposure ages of early Pliocene for the Table Mountain plateau and the Sirius Group at Mount Fleming, Dry Valleys, Antarctica. *Geology*, 23, 1007–1010.

Ivy-Ochs, S., H.-A. Synal, C. Roth, and M. Schaller. (2004). Initial results from isotope dilution for Cl and Cl-36 measurements at the PSI/ETH Zurich AMS facility. *Nuclear Instruments and Methods in Physics Research*, B223–24, 623–627.

Ivy-Ochs, S., R. Wüst, P. W. Kubik, H. Müller-Beck, and C. Schlüchter. (2001). Can we use cosmogenic isotopes to date stone artifacts? *Radiocarbon*, 43, 759–764.

Ivy-Ochs, S., A. v. Poschinger, H.-A. Synal, and M. Maisch. (2009). Surface exposure dating of the Flims landslide, Graubünden, Switzerland. *Geomorphology*, 103, 104–112.

Jackson, L. E., F. M. Phillips, and E. C. Little. (1999). Cosmogenic Cl-36 dating of the maximum limit of the Laurentide Ice Sheet in southwestern Alberta. *Canadian Journal of Earth Sciences*, 36, 1347–1356.

Jackson, L. E., F. M. Phillips, K. Shimamura, and E. C. Little. (1997). Cosmogenic Cl-36 dating of the Foothills erratics train, Alberta, Canada. *Geology*, 25, 195–198.

James, L. A., J. Harbor, D. Fabel, D. Dahms, and D. Elmore. (2002). Late Pleistocene glaciations in the northwestern Sierra Nevada, California. *Quaternary Research*, 57, 409–419.

Jull, A. J., A. E. Wilson, G. S. Burr, L. J. Toolin, and D. J. Donahue. (1992). Measurements of cosmogenic ^{14}C produced by spallation in high-altitude rocks. *Radiocarbon*, 34, 737–744.

Kaplan, M. R., R. P. Ackert, B. S. Singer, D. C. Douglass, and M. D. Kurz. (2004). Cosmogenic nuclide chronology of millennial-scale glacial advances during O-isotope stage 2 in Patagonia. *Geological Society of America Bulletin*, 116, 308–321.

Kaplan, M. R., D. C. Douglass, B. S. Singer, and M. W. Caffee. (2005). Cosmogenic nuclide chronology of pre-last glacial maximum moraines at Lago Buenos Aires, 46 degrees S, Argentina. *Quaternary Research*, 63, 301–315.

Kelly, M. A., S. Ivy-Ochs, P. W. Kubik, F. von Blanckenburg, and C. Schlüchter. (2006). Chronology of deglaciation based on ^{10}Be dates of glacial erosional features in the Grimsel Pass region, central Swiss Alps. *Boreas*, 35, 634–643.

Kerschner, H. (2005). Glacier-climate models as palaeoclimatic information sources: Examples from the Alpine Younger Dryas period. In: *Global Change and Mountain Region: A State of Knowledge Overview* (Eds U. M. Huber, H. K. M. Bugmann, and M. A. Reasoner). Springer, Dordrecht, The Netherlands, pp. 73–82.

Kerschner, H., A. Hertl, G. Gross, S. Ivy-Ochs, and P. W. Kubik. (2006). Surface exposure dating of moraines in the Kromer valley (Silvretta Mountains, Austria)-evidence for glacial response to the 8.2 ka event in the Eastern Alps? *The Holocene*, 16, 7–15.

Kirchner, J. W., R. C. Finkel, C. S. Riebe, D. E. Granger, J. L. Clayton, J. G. King, and W. F. Megahan. (2001). Mountain erosion over 10 yr, 10 k.y., and 10 m.y. time scales. *Geology*, 29, 591–594.

Klein, J., D. Fink, R. Middleton, K. Nishiizumi, and J. Arnold. (1991). Determination of the half-life of Ca-41 from measurements of Antarctic meteorites. *Earth and Planetary Science Letters*, 103, 79–83.

Klein, J., R. Giegengack, R. Middleton, P. Sharma, J. Underwood, and R. A. Weeks. (1986). Revealing histories of exposure using in situ produced ^{26}Al and ^{10}Be in Libyan desert glass. *Radiocarbon*, 28, 547–555.

Klein, J., B. Lawn, J. Gosse, and C. Harrington. (1997). Can terrestrial cosmogenic Be-10 be measured in whole rock samples to decipher surface exposure histories? Geological Society of America Abstracts with Programs, pp. 346.

Klein, J., R. Middleton, and H. Q. Tang. (1982). Instrumentation of an FN tandem for the detection of ^{10}Be. *Nuclear Instruments and Methods in Physics Research*, 193, 601–616.

Knight, K. B., P. R. Renne, and K. A. Farley. (2003). Preliminary estimate of production rates for terrestrial cosmogenic ^{38}Ar from calcium, AGU Fall Meeting. EOS Transactions, San Francisco, pp. 84(46).

Kober, F., S. Ivy-Ochs, I. Leya, R. Wieler, H. Baur, and T. Magna. (2005). In situ cosmogenic ^{10}Be and ^{21}Ne in sanidine and in situ cosmogenic ^{3}He in Fe-Ti-oxide minerals. *Earth and Planetary Science Letters*, 236, 404–418.

Kober, F., S. Ivy-Ochs, F. Schlunegger, H. Baur, P. W. Kubik, and R. Wieler. (2007). Denudation rates and a topography-driven precipitation threshold in northern Chile: multiple cosmogenic nuclide data and sediment yield budgets. *Geomorphology*, 83, 97–120.

Kohl, C. P., and K. Nishiizumi. (1992). Chemical isolation of quartz for measurement of in-situ-produced cosmogenic nuclides. *Geochimica et Cosmochimica Acta*, 56, 3583–3587.

Kubik, P. W., S. Ivy-Ochs, J. Masarik, M. Frank., and C. Schlüchter. (1998). ^{10}Be and ^{26}Al production rates deduced from an instantaneous event within the dendro-calibration curve, the landslide of Köfels, Ötz Valley, Austria. *Earth and Planetary Science Letters*, 161, 231–241.

Kubik, P. W., G. Korschinek, and E. Nolte. (1984). Accelerator Mass-Spectrometry with Completely Stripped Cl-36 Ions at the Munich Postaccelerator. *Nuclear Instruments and Methods in Physics Research*, B229, 51–59.

Kurz, M. D. (1986). Cosmogenic helium in a terrestrial rock. *Nature*, 320, 435–439.

Kurz, M. D., and E. J. Brook. (1994). Surface exposure dating with cosmogenic radionuclides. In: *Dating in Exposed and Surface Contexts* (Ed. S. Beck). University of New Mexico Press, Socorro, New Mexico, pp. 139–159.

Kurz, M. D., D. Colodner, T. W. Trull, R. B. Moore, and K. O'Brien. (1990). Cosmic ray exposure dating with in-situ produced cosmogenic ^{3}He: Results from young Hawaiian lava flows. *Earth and Planetary Science Letters*, 97, 177–189.

Lahiri, S., D. Nayak, and G. Korschinek. (2006). Separation of no-carrier-added ^{52}Mn from bulk chromium: A simulation study for accelerator mass spectrometry measurement of ^{53}Mn. *Analytical Chemistry*, 78, 7517–7521.

Lal, D. (1991). Cosmic ray labeling of erosion surfaces: In-situ nuclide production rates and erosion models. *Earth and Planetary Science Letters*, 104, 424–439.

Lal, D., and J. Chen. (2005). Cosmic ray labeling of erosion surfaces II: Special cases of exposure histories of boulders, soils and beach terraces. *Earth and Planetary Science Letters*, 236, 797–813.

Lal, D., and J. Chen. (2006). Erratum to "Cosmic ray labeling of erosion surfaces II: Special cases of exposure histories of boulders, soils and beach terraces". *Earth and Planetary Science Letters*, 241, 360.

Lal, D., and B. Peters. (1967). Cosmic ray produced radioactivity on the Earth. In: *Handbuch der Physik* (Ed. K. Sitte). Vol. 46/2, Springer Verlag, Berlin, pp. 551–612.

Leland, J., M. R. Reid, D. W. Burbank, R. Finkel, and M. Caffee. (1998). Incision and differential bedrock uplift along the Indus River near Nanga Parbat, Pakistan Himalaya, from Be-10 and Al-26 exposure age dating of bedrock straths. *Earth and Planetary Science Letters*, 154, 93–107.

Libarkin, J. C., J. Quade, C. G. Chase, J. Poths, and W. McIntosh. (2002). Measurement of ancient cosmogenic ^{21}Ne in quartz from the 28 Ma Fish Canyon Tuff, Colorado. *Chemical Geology*, 186, 199–213.

Licciardi, J. M., P. U. Clark, E. J. Brook, D. Elmore, and P. Sharma. (2004). Variable responses of western US glaciers during the last deglaciation. *Geology*, 32, 81–84.

Licciardi, J. M., P. U. Clark, E. J. Brook, K. L. Pierce, and M. D. Kurz. (2001). Cosmogenic ^{3}He and ^{10}Be chronologies of the late Pinedale northern Yellowstone ice cap, Montana, US. *Geology*, 29, 1095–1098.

Licciardi, J. M., M. D. Kurz, P. U. Clark, and E. J. Brook. (1999). Calibration of cosmogenic ^{3}He production rates from Holocene lava flows in Oregon, USA, and effects of the Earth's magnetic field. *Earth and Planetary Science Letters*, 172, 261–271.

Licciardi, J. M., M. D. Kurz, and J. M. Curtice. (2006). Cosmogenic ^{3}He production rates from Holocene lava flows in Iceland. *Earth and Planetary Science Letters*, 246, 251–264.

Lifton, N. A., A. J. T. Jull, and J. Quade. (2001). A new extraction technique and production rate estimate for in situ cosmogenic ^{14}C in quartz. *Geochimica et Cosmochimica Acta*, 65, 1953–1969.

Linge, H., E. J. Brook, A. Nesje, G. M. Raisbeck, F. Yiou, and H. Clark. (2006). In situ ^{10}Be exposure ages from southeastern Norway: implications for the geometry of the Weichselian Scandinavian ice sheet. *Quaternary Science Reviews*, 25, 1097–1109.

Linge, H., E. Larsen, K. H. Kjær, I. N. Demidov, E. J. Brook, G. M. Raisbeck, and F. Yiou. (2006). Cosmogenic ^{10}Be exposure age dating across Early to Early to Late Weichselian ice-marginal zones in northwestern Russia. *Boreas*, 35, 576–586.

Liu, B., F. M. Phillips, J. T. Fabryka-Martin, M. M. Fowler, and W. D. Stone. (1994). Cosmogenic ^{36}Cl accumulation in unstable landforms, 1. Effects of the thermal neutron distribution. *Water Resources Research*, 30, 3115–3125.

Marchetti, D. W., and T. E. Cerling. (2005). Cosmogenic ^{3}He exposure ages of Pleistocene debris flows and desert pavements in Capitol Reef National Park, Utah. *Geomorphology*, 67, 423–435.

Marchetti, D. W., T. E. Cerling, and E. W. Lips. (2005). A glacial chronology for the Fish Creek drainage of Boulder Mountain, Utah, USA. *Quaternary Research*, 64, 264–271.

Margerison, H. R., W. M. Phillips, F. M. Stuart, and D. E. Sugden. (2005). Cosmogenic ^{3}He concentrations in ancient flood deposits from the Coombs Hills, northern Dry Valleys, East Antarctica: interpreting exposure ages and erosion rates. *Earth and Planetary Science Letters*, 230, 163–175.

Marquette, G. C., J. T. Gray, J. C. Gosse, F. Courchesne, L. Stockli, G. Macpherson, and R. Finkel. (2004). Felsenmeer persistence under non-erosive ice in the Torngat and Kaumajet mountains, Quebec and Labrador, as determined by soil weathering and cosmogenic nuclide exposure dating. *Canadian Journal of Earth Sciences*, 41, 19–38.

Marti, K., and H. Craig. (1987). Cosmic-ray-produced neon and helium in the summit lavas of Maui. *Nature*, 325, 335–337.

Marty, B., R. Yokochi, J. M. Schaefer, and R. Wieler. (2006). A search for cosmogenic nitrogen in a terrestrial rock. *Geochimica et Cosmochimica Acta*, 70, A396.

Masarik, J. (2002). Numerical simulation of in-situ production of cosmogenic nuclides. *Geochimica et Cosmochimica Acta*, 66(15A), A491.

Masarik, J., M. Frank, J. M. Schaefer, and R. Wieler. (2001). Correction of in-situ cosmogenic nuclide production rates for geomagnetic field intensity variations during the past 800,000 years. *Geochimica et Cosmochimica Acta*, 65, 2995–3003.

Masarik, J., D. Kollar, and S. Vanya. (2000). Numerical simulation of in situ production of cosmogenic nuclides: Effect of irradiation geometry. *Nuclear Instruments and Methods in Physics Research*, B172, 786–789.

Masarik, J., and R. C. Reedy. (1995). Terrestrial cosmogenic-nuclide production systematics calculated from numerical simulations. *Earth and Planetary Science Letters*, 136, 381–395.

Masarik, J., and R. Wieler. (2003). Production rates of cosmogenic nuclides in boulders. *Earth and Planetary Science Letters*, 216, 201–208.

Matmon, A., P. R. Bierman, J. Larsen, S. Southworth, M. Pavich, R. Finkel, and M. Caffee. (2003a). Erosion of an ancient mountain range, the Great Smoky Mountains, North Carolina and Tennessee. *American Journal of Science*, 303, 817–855.

Matmon, A., P. R. Bierman, M. C. Larsen, S. Southworth, M. Pavich, and M. Caffee. (2003b). Temporally and spatially uniform rates of erosion in the southern Appalachian Great Smoky Mountains. *Geology*, 31, 155–158.

Matmon, A., D. P. Schwartz, R. Finkel, S. Clemmens, and T. Hanks. (2005). Dating offset fans along the Mojave section of the San Andreas fault using cosmogenic Al-26 and Be-10. *Geological Society of America Bulletin*, 117, 795–807.

Matsuoka, N., C. E. Thomachot, C. T. Oguchi, T. Hatta, M. Abe, and H. Matsuzaki. (2006). Quaternary bedrock erosion and landscape evolution in the Sør Rondane Mountains, East Antarctica: Reevaluating rates and processes. *Geomorphology*, 81, 408–420.

Merchel, S., R. Braucher, L. Benedetti, O. Grauby, and D. L. Bourlès. (2008). Dating carbonate rocks with in-situ produced cosmogenic ^{10}Be: Why it often fails. *Quaternary Geochronology*, 3, 299–307.

Mériaux, A.-S., F. J. Ryerson, P. Tapponnier, J. Van der Woerd, R. C. Finkel, X. Xu, Z. Xu, and M. W. Caffee. (2004). Rapid slip along the central Altyn Tagh Fault: Morphochronologic evidence from Cherchen He and Sulamu Tagh. *Journal of Geophysical Research.-Solid Earth*, 109(B6), B06401.

Mériaux, A.-S., P. Tapponnier, F. J. Ryerson, X. Xu, G. King, J. Van der Woerd, R. C. Finkel, H. Li, M. W. Caffee, Z. Xu, and W. Chen. (2005). The Aksay segment of the northern Altyn Tagh fault: Tectonic geomorphology, landscape evolution, and Holocene slip rate. *Journal of Geophysical Research – Solid Earth*, 110(B4), B04404.

Middleton, R., D. Fink, J. Klein, and P. Sharma. (1989). ^{41}Ca concentrations in modern bone and their implications for dating. *Radiocarbon*, 31, 305–309.

Miller, G. H., J. P. Briner, N. A. Lifton, and R. C. Finkel. (2006). Limited ice-sheet erosion and complex exposure histories derived from in situ cosmogenic ^{10}Be, ^{26}Al, and ^{14}C on Baffin Island, Arctic Canada. *Quaternary Geochronology*, 1, 74–85.

Miller, G. H., A. P. Wolfe, J. P. Briner, P. E. Sauer, and A. Nesje. (2005). Holocene glaciation and climate evolution of Baffin Island Arctic Canada. *Quaternary Science Reviews*, 24, 1703–1721.

Mitchell, S. G., A. Matmon, P. R. Bierman, Y. Enzel, M. Caffee, and D. Rizzo. (2001). Displacement history of a limestone normal fault scarp, northern Israel, from cosmogenic Cl-36. *Journal of Geophysical Research-Solid Earth*, 106(B3), 4247–4264.

Morel, P., F. von Blanckenburg, M. Schaller, P. W. Kubik, and M. Hinderer. (2003). Lithology, landscape dissection and glaciation controls on catchment erosion as determined by cosmogenic nuclides in river sediment (the Wutach Gorge, Black Forest). *Terra Nova*, 15, 398–404.

Morris, J. D., J. C. Gosse, S. Brachfield, and F. Tera. (2002). Cosmogenic Be-10 and the solid Earth: Studies in geomagnetism, subduction zone processes, and active tectonics. In: *Beryllium: Mineralogy, Petrology, and Geochemistry* (Ed E. Grew). Reviews in Mineralogy and Geochemistry, Vol. 50, Mineralogical Society of America, Washington, DC, pp. 207–270.

Muzikar, P. (2005). Geomagnetic field variations and the accumulation of in-situ cosmogenic nuclides in an eroding landform. *Geochimica et Cosmochimica Acta*, 69, 4127–4131.

Nichols, K. K., P. Bierman, R. L. Hooke, E. M. Clapp, and M. Caffee. (2002). Sediment transport on desert piedmonts using ^{10}Be and ^{26}Al. *Geomorphology*, 45, 105–125.

Nichols, K. K., P. R. Bierman, R. Foniri, A. R. Gillespie, M. Caffee, and R. C. Finkel. (2006). Dates and rates of arid region geomorphic processes. *Geological Society of America Today*, 16, 4–11.

Nichols, K. K., P. R. Bierman, M. Caffee, R. C. Finkel, and J. Larsen. (2005). Cosmogenically enabled sediment budgeting. *Geology*, 33, 133–136.

Niedermann, S. (2002). Cosmic-ray-produced noble gases in terrestrial rocks: Dating tools for surface processes. In: *Noble Gases in Geochemistry and Cosmochemistry* (Eds D. Porcelli, C. J. Ballentine, and R. Wieler). Reviews in Mineralogy and Geochemistry, Vol. 47, Mineralogical Society of America, Washington, DC, pp. 731–784.

Niedermann, S., T. Graf, J. S. Kim, C. P. Kohl, K. Marti, and K. Nishiizumi. (1994). Cosmic-ray-produced ^{21}Ne in terrestrial quartz: The inventory of Sierra Nevada quartz separates. *Earth and Planetary Science Letters*, 125, 341–355.

Niedermann, S., J. M. Schaefer, R. Wieler, and R. Naumann. (2007). The production rate of cosmogenic ^{38}Ar from calcium in terrestrial pyroxene. *Earth and Planetary Science Letters*, 257(3–4), 596–608.

Niemi, N., M. Oskin, D. W. Burbank, A. Heimsath, and E. J. Gabet. (2005). Effects of bedrock landsliding on cosmogenically determined erosion rates. *Earth and Planetary Science Letters*, 237, 480–498.

Nishiizumi, K., M. Caffee, R. C. Finkel, G. Brimhall, and T. Mote. (2005). Remnants of a fossil alluvial fan landscape of Miocene age in the Atacama Desert of northern Chile using cosmogenic nuclide exposure age dating. *Earth and Planetary Science Letters*, 237, 499–507.

Nishiizumi, K., R. C. Finkel, J. Klein, and C. P. Kohl. (1996). Cosmogenic production of ^{7}Be and ^{10}Be in water targets. *Journal of Geophysical Research*, 101, 22225–22232.

Nishiizumi, K., J. Klein, R. Middleton, and H. Craig. (1990). Cosmogenic ^{10}Be, ^{26}Al and ^{3}He in olivine from Maui lavas. *Earth and Planetary Science Letters*, 98, 263–266.

Nishiizumi, K., C. P. Kohl, J. R. Arnold, J. Klein, D. Fink, and R. Middleton. (1991). Cosmic-ray produced ^{10}Be and ^{26}Al in Antarctic rocks-exposure and erosion history. *Earth and Planetary Science Letters*, 104, 440–454.

Nishiizumi, K., C. P. Kohl, J. R. Arnold, J. Klein, R. Dorn, D. Fink, R. Middleton, and D. Lal. (1993). Role of in situ cosmogenic nuclides ^{10}Be and ^{26}Al in the study of diverse geomorphic processes. *Earth Surface Processes*, 18, 407–425.

Nishiizumi, K., D. Lal, J. Klein, R. Middleton, and J. R. Arnold. (1986). Production of ^{10}Be and ^{26}Al by cosmic rays in terrestrial quartz in situ and implications for erosion rates. *Nature*, 319, 134–136.

Nishiizumi, K., E. L. Winterer, C. P. Kohl, J. Klein, R. Middleton, D. Lal, and J. R. Arnold. (1989). Cosmic ray production rates of ^{10}Be and ^{26}Al in quartz from glacially polished rocks. *Journal of Geophysical Research*, 94, 17907–17915.

Oberholzer, P., C. Baroni, J. M. Schaefer, G. Orombelli, S. Ivy-Ochs, P. W. Kubik, H. Baur, and R. Wieler. (2003). Limited Pleistocene glaciation in Deep Freeze Range, Northern Victoria Land, Antarctica, derived from in-situ cosmogenic nuclides. *Antarctic Science*, 15, 493–502.

Ochs, M., and S. Ivy-Ochs. (1997). The chemical behavior of Be, Al, Fe, Ca and Mg during AMS target preparation from terrestrial silicates modeled with chemical speciation calculations. *Nuclear Instruments and Methods in Physics Research*, B123, 235–240.

Owen, L. A., R. C. Finkel, M. Haizhou, J. Q. Spencer, E. Derbyshire, P. L. Barnard, and M. W. Caffee. (2003a). Timing and style of Late Quaternary glaciation in northeastern Tibet. *Geological Society of America Bulletin*, 115, 1356–1364.

Owen, L. A., R. C. Finkel, R. A. Minnich, and A. E. Perez. (2003b). Extreme southwestern margin of late Quaternary glaciation in North America: Timing and controls. *Geology*, 31, 729–732.

Palumbo, L., L. Benedetti, D. Bourlès, A. Cinque, and R. Finkel. (2004). Slip history of the Magnola fault (Apennines, Central Italy) from Cl-36 surface exposure dating: evidence for strong earthquakes over the Holocene. *Earth and Planetary Science Letters*, 225, 163–176.

Partridge, T. C., D. E. Granger, M. W. Caffee, and R. J. Clarke. (2003). Lower Pliocene hominid remains from Sterkfontein. *Science*, 300, 607–612.

Perg, L. A., R. S. Anderson, and R. C. Finkel. (2001). Use of a new ^{10}Be and ^{26}Al inventory method to date marine terraces, Santa Cruz, California, USA. *Geology*, 29, 879–882.

Phillips, F. M. (2003). Cosmogenic ^{36}Cl ages of Quaternary basalt flows in the Mojave Desert, California, USA. *Geomorphology*, 53, 199–208.

Phillips, F. M., J. P. Ayarbe, J. B. J. Harrison, and D. Elmore. (2003). Dating rupture events on alluvial fault scarps using cosmogenic nuclides and scarp morphology. *Earth and Planetary Science Letters*, 215, 203–218.

Phillips, F. M., B. D. Leavy, N. O. Jannik, D. Elmore, and P. W. Kubik. (1986). The accumulation of cosmogenic chlorine-36 in rocks: A method for surface exposure dating. *Science*, 231, 41–43.

Phillips, F. M., W. D. Stone, and J. T. Fabryka-Martin. (2001). An improved approach to calculating low-energy cosmic-ray neutron fluxes near the land/atmosphere interface. *Chemical Geology*, 175, 689–701.

Phillips, F. M., M. G. Zreda, L. V. Benson, M. A. Plummer, D. Elmore, and P. Sharma. (1996b). Chronology for fluctuations in late Pleistocene Sierra Nevada glaciers and lakes. *Science*, 274, 749–751.

Phillips, F. M., M. G. Zreda, M. R. Flinsch, D. Elmore, and P. Sharma. (1996a). A reevaluation of cosmogenic Cl-36 production rates in terrestrial rocks. *Geophys. Res. Lett.*, 23, 949–952.

Phillips, F. M., M. G. Zreda, J. C. Gosse, J. Klein, E. B. Evenson, R. D. Hall, O. A. Chadwick, and P. Sharma. (1997). Cosmogenic ^{36}Cl and ^{10}Be ages of Quaternary glacial and fluvial deposits of the Wind River Range, Wyoming. *Geological Society of America Bulletin*, 109, 1453–1463.

Phillips, W. M., A. M. Hall, R. Mottram, L. K. Fifield, and D. E. Sugden. (2006). Cosmogenic ^{10}Be and ^{26}Al exposure ages of tors and erratics, Cairngorm Mountains, Scotland: Timescales for the development of a classic landscape of selective linear glacial erosion. *Geomorphology*, 73, 222–245.

Phillips, W. M., E. V. McDonald, S. L. Reneau, and J. Poths. (1998). Dating soils and alluvium with cosmogenic ^{21}Ne depth profiles: Case studies from the Pajarito Plateau, New Mexico, USA. *Earth and Planetary Science Letters*, 160, 209–223.

Philpotts, A. R. (1990). *Principles of Igneous and Metamorphic Petrology*. Prentice Hall, Englewood Cliffs, NJ, 498 pp.

Pigati, J. S., and N. A. Lifton. (2004). Geomagnetic effects on time-integrated cosmogenic nuclide production with emphasis on in situ ^{14}C and ^{10}Be. *Earth and Planetary Science Letters*, 226, 193–205.

Poreda, R. J., and T. E. Cerling. (1992). Cosmogenic neon in recent lavas from the Western United States. *Geophysical Research Letters*, 18, 1863–1866.

Prager, C., S. Ivy-Ochs, M. Ostermann, H.-A. Synal, and G. Patzelt. (2009). Geology and dating of the Fernpass rockslide (Tyrol, Austria). *Geomorphology*, 103, 93–103.

Pratt, B., D. W. Burbank, A. Heimsath, and T. Ojha. (2002). Impulsive alluviation during early Holocene strengthened monsoons, central Nepal Himalaya. *Geology*, 30, 911–914.

Pratt-Sitaula, B., D. W. Burbank, A. Heimsath, and T. Ojha. (2004). Landscape disequilibrium on 1000–10,000 year scales Marsyandi River, Nepal, central Himalaya. *Geomorphology*, 58, 233–241.

Principato, S. M., Á. Geirsdóttir, G. E. Jóhannsdóttir, and J. T. Andrews. (2006). Late Quaternary glacial and deglacial history of eastern Vestfirdir, Iceland using cosmogenic isotope (^{36}Cl) exposure ages and marine cores. *Journal of Quaternary Science*, 21, 271–285.

Putkonen, J., and M. O'Neal. (2006). Degradation of unconsolidated Quaternary landforms in western North America. *Geomorphology*, 75, 408–419.

Putkonen, J., and T. Swanson. (2003). Accuracy of cosmogenic ages for moraines. *Quaternary Research*, 59, 255–261.

Raisbeck, G. M., F. Yiou, D. Bourlès, J. Lestringuez, and D. Deboffle. (1987). Measurements of ^{10}Be and ^{26}Al with a Tandetron AMS facility. *Nuclear Instruments and Methods in Physics Research*, B29, 22–26.

Renne, P. R., K. A. Farley, T. A. Becker, and W. D. Sharp. (2001). Terrestrial cosmogenic argon. *Earth and Planetary Science Letters*, 188, 435–440.

Repka, J. L., R. S. Anderson, and R. C. Finkel. (1997). Cosmogenic dating of fluvial terraces, Fremont River, Utah. *Earth and Planetary Science Letters*, 152, 59–73.

Reuther, A. U., J. Herget, S. Ivy-Ochs, P. Borodavco, P. W. Kubik, and K. Heine. (2006b). Constraining the timing of the most recent cataclysmic flood event from ice-dammed lakes in the Russian Altai mountains, Siberia using cosmogenic in-situ ^{10}Be. *Geology*, 34, 913–916.

Reuther, A. U., S. Ivy-Ochs, and K. Heine. (2006a). Application of surface exposure dating in glacial geomorphology and the interpretation of moraine ages. *Zeitschrift fur Geomorphologie Supplementband*, 142, 335–359.

Riebe, C. S., G. Kirchner, and R. C. Finkel. (2003). Long-term rates of chemical weathering and physical erosion from cosmogenic nuclides and geochemical mass balance. *Geochimica et Cosmochimica Acta*, 67, 4411–4427.

Riebe, C. S., J. W. Kirchner, and R. C. Finkel. (2004). Sharp decrease in long-term chemical weathering rates along an altitudinal transect. *Earth and Planetary Science Letters*, 218, 421–434.

Riebe, C. S., J. W. Kirchner, and D. E. Granger. (2001a). Quantifying quartz enrichment and its consequences for cosmogenic measurements of erosion rates from alluvial sediment and regolith. *Geomorphology*, 40, 15–19.

Riebe, C. S., J. W. Kirchner, D. E. Granger, and R. C. Finkel. (2000). Erosional eqilibrium and diseqilibrium in the Sierra Nevada, inferred from cosmogenic ^{26}Al and ^{10}Be in alluvial sediment. *Geology*, 28, 803–806.

Riebe, C. S., J. W. Kirchner, D. E. Granger, and R. C. Finkel. (2001b). Minimal climatic control on erosion rates in the Sierra Nevada, California. *Geology*, 29, 447–450.

Riebe, C. S., J. W. Kirchner, D. E. Granger, and R. C. Finkel. (2001c). Strong tectonic and weak climatic control of long-term chemical weathering rates. *Geology*, 29, 511–514.

Rinterknecht, V. R., P. U. Clark, G. M. Raisbeck, F. Yiou, E. J. Brook, S. Tschudi, and J. P. Lunkka. (2004). Cosmogenic Be-10 dating of the Salpausselka I Moraine in southwestern Finland. *Quaternary Science Reviews*, 23, 2283–2289.

Rinterknecht, V. R., L. Marks, J. A. Piotrowski, G. M. Raisbeck, F. Yiou, E. J. Brook, and P. U. Clark. (2005). Cosmogenic Be-10 ages on the Pomeranian Moraine, Poland. *Boreas*, 34, 186–191.

Ritz, J.-F., D. Bourlès, E. T. Brown, S. Carretier, J. Chéry, B. Enhtuvshin, P. Galsan, R. C. Finkel, T. C. Hanks, K. J. Kendrick, H. Philip, G. M. Raisbeck, A. Schlupp, D. P. Schwartz, and F. Yiou. (2003). Late Pleistocene to Holocene slip rates for the Gurvan Bulag thrust fault (Gobi-Altay, Mongolia) estimated with Be-10 dates. *Journal of Geophysical Research-Solid Earth*, 108(B3), 2162.

Ritz, J. F., E. T. Brown, D. L. Bourlès, H. Philip, A. Schlupp, G. M. Raisbeck, F. Yiou, and B. Enkhtuvshin. (1995). Slip rates along active faults estimated with cosmic-ray-exposure dates: Application to the Bogd fault, Gobi-Altai, Mongolia. *Geology*, 23, 1019–1022.

Ryerson, F. J., P. Tapponnier, R. C. Finkel, A.-S. Mériaux, J. Van der Woerd, C. Laserre, M.-L. Chevalier, X. Xu, H. Li, and G. C. P. King. (2006). Applications of morphochronology to the active tectonics of Tibet. In: *In Situ Cosmogenic Nuclides and their Applications in Earth Sciences* (Eds L. L. Siame, D. L. Bourlès, and E. T. Brown). Geological Society of America Special Paper 415, pp. 61–86.

Schaefer, J. M., G. H. Denton, D. J. A. Barrell, S. Ivy-Ochs, P. W. Kubik, B. G. Andersen, F. M. Phillips, T. V. Lowell, and C. Schlüchter. (2006a). Near-synchronous interhemispheric termination of the last glacial maximum in mid-latitudes. *Science*, 312, 1510–1513.

Schaefer, J. M., T. Faestermann, G. F. Herzog, K. Knie, G. Korschinek, J. Masarik, A. Meier, M. Poutivtsev, G. Rugel, C. Schlüchter, F. Serifiddin, and G. Winckler. (2006b). Terrestrial manganese-53 – A new monitor of Earth surface processes. *Earth and Planetary Science Letters*, 251, 334–345.

Schäfer, J. M., S. Ivy-Ochs, R. Wieler, I. Leya, H. Baur, G. H. Denton, and C. Schlüchter. (1999). Cosmogenic noble gas studies in the oldest landscape on earth: Surface exposure ages of the Dry Valleys, Antarctica. *Earth and Planetary Science Letters*, 167, 215–226.

Schäfer, J. M., S. Tschudi, Z. Zhao, X. Wu, S. Ivy-Ochs, R. Wieler, H. Baur, P. W. Kubik, and C. Schlüchter. (2002). The limited influence of glaciations in Tibet on global climate over the past 170 000 yr. *Earth and Planetary Science Letters*, 194, 287–297.

Schaller, M., and T. A. Ehlers. (2006). Limits to quantifying climate driven changes in denudation rates with cosmogenic radionuclides. *Earth and Planetary Science Letters*, 248, 138–152.

Schaller, M., N. Hovius, S. D. Willet, S. Ivy-Ochs, H.-A. Synal, and H.-C. Chen. (2005). Fluvial bedrock incision in the active mountain belt of Taiwan from in situ-produced cosmogenic nuclides. *Earth Surface Processes and Landforms*, 30, 955–971.

Schaller, M., F. von Blanckenburg, N. Hovius, and P. W. Kubik. (2001). Large-scale erosion rates from in situ-produced cosmogenic nuclides in European river sediments. *Earth and Planetary Science Letters*, 188, 441–458.

Schaller, M., F. von Blanckenburg, N. Hovius, A. Veldkamp, M. W. van den Berg, and P. W. Kubik. (2004). Paleoerosion rates from cosmogenic ^{10}Be in a 1.3 Ma terrace sequence: Response of the River Meuse to changes in climate and rock uplift. *Journal of Geology*, 112, 127–144.

Schaller, M., F. von Blanckenburg, A. Veldkamp, L. A. Tebbens, N. Hovius, and P. W. Kubik. (2002). A 30000 yr record of erosion rates from cosmogenic ^{10}Be in Middle Europe river terraces. *Earth and Planetary Science Letters*, 204, 307–320.

Schildgen, T., D. P. Dethier, P. Bierman, and M. Caffee. (2002). ^{26}Al and ^{10}Be dating of late Pleistocene and Holocene fill terraces: A record of fluvial deposition and incision, Colorado Front Range. *Earth Surface Processes and Landforms*, 27, 773–787.

Schildgen, T. F., W. M. Phillips, and R. S. Purves. (2005). Simulation of snow shielding corrections for cosmogenic nuclide surface exposure studies. *Geomorphology*, 64, 67–85.

Seidl, M. A. (1993). Form and processes in channel incision of bedrock. PhD Thesis, University of California, Berkeley, CA, 163 pp.

Seidl, M. A., R. C. Finkel, M. W. Caffee, G. B. Hudson, and W. E. Dietrich. (1997). Cosmogenic isotope analysis applied to river longitudinal profile evolution: Problems and interpretations. *Earth Surface Processes and Landforms*, 22, 195–209.

Sewell, R. J., T. T. Barrows, S. D. G. Campbell, and L. K. Fifield. (2006). Exposure dating (^{10}Be, ^{26}Al) of natural terrain landslides in Hong Kong, China. In: *In Situ Cosmogenic*

Nuclides and their Applications in Earth Sciences (Eds L. L. Siame, D. L. Bourlès, and E. T. Brown). Geological Society of America Special Paper 415, pp. 131–146.

Shimaoka, A., P. Kong, R. C. Finkel, M. Caffee, and K. Nishiizumi. (2002). The determination of in situ cosmogenic radionuclide in olivine. *Geochimica et Cosmochimica Acta*, 66, A709.

Shimaoka, A., M. Sakamoto, H. Hiyagon, H. Matsuzaki, I. Kaneoka, and M. Imamura. (2004). Meteoric Be-10 in volcanic materials and its behavior during acid-leaching. *Nuclear Instruments and Methods in Physics Research*, B223–24, 591–595.

Siame, L. L., O. Bellier, M. Sébrier, D. L. Bourlès, P. Leturmy, M. Perez, and M. Araujo. (2002). Seismic hazard reappraisal from combined structural geology, geomorphology and cosmic rays exposure dating analyses: The Eastern Precordillera thrust system (NW Argentina). *Geophysical Journal International*, 150, 241–260.

Siame, L. L., D. L. Bourlès, M. Sebrier, O. Bellier, J. C. Castano, M. Araujo, M. Perez, G. M. Raisbeck, and F. Yiou. (1997). Cosmogenic dating ranging from 20 to 700 ka of a series of alluvial fan surfaces affected by the El Tigre fault, Argentina. *Geology*, 25, 975–978.

Small, E. E., and R. S. Anderson. (1998). Pleistocene relief production in Laramide mountain range, western United States. *Geology*, 26, 123–126.

Small, E. E., R. S. Anderson, and G. S. Hancock. (1999). Estimates of regolith production from ^{10}Be and ^{26}Al: Evidence of steady-state alpine hillslopes. *Geomorphology*, 27, 131–150.

Small, E. E., R. S. Anderson, J. L. Repka, and R. Finkel. (1997). Erosion rates of alpine bedrock summit surfaces deduced from in situ ^{10}Be and ^{26}Al. *Earth and Planetary Science Letters*, 150, 413–425.

Smith, J. A., R. C. Finkel, D. L. Farber, D. T. Rodbell, and G. O. Seltzer. (2005a). Moraine preservation and boulder erosion in the tropical Andes: interpreting old surface exposure ages in glaciated valleys. *Journal of Quaternary Science*, 20(7–8), 735–758.

Smith, J. A., G. O. Seltzer, D. L. Farber, D. T. Rodbell, and R. C. Finkel. (2005b). Early local last glacial maximum in the tropical Andes. *Science*, 308, 678–681.

Staiger, J. K. W., J. C. Gosse, J. V. Johnson, J. Fastook, J. T. Gray, D. F. Stockli, L. Stockli, and R. C. Finkel. (2005). Quaternary relief generation by polythermal glacier ice. *Earth Surface Processes and Landforms*, 30, 1145–1159.

Staiger, J. W., D. R. Marchant, J. M. Schaefer, P. Oberholzer, J. V. Johnson, A. R. Lewis, and K. M. Swanger. (2006). Plio-Pleistocene history of Ferrar Glacier, Antarctica: Implications for climate and ice sheet stability. *Earth and Planetary Science Letters*, 243, 489–503.

Staudacher, T., and C. J. Allégre. (1991). Cosmogenic neon in ultramafic nodules from Asia and in quartzite from Antarctica: Earth Planet. *Science Letters*, 106, 87–102.

Stock, G. M., R. S. Anderson, and R. C. Finkel. (2004). Pace of landscape evolution in the Sierra Nevada, California, revealed by cosmogenic dating of cave sediments. *Geology*, 32, 193–196.

Stock, G. M., T. A. Ehlers, and K. A. Farley. (2006). Where does sediment come from? Quantifying catchment erosion with detrital apatite (U-Th)/He thermochronometry. *Geology*, 34, 725–728.

Stone, J. (2000). Air pressure and cosmogenic isotope production. *Journal of Geophysical Research*, 105, 23753–23759.

Stone, J. O., G. L. Allan, L. K. Fifield, and R. G. Cresswell. (1996). Cosmogenic chlorine-36 from calcium spallation. *Geochimica et Cosmochimica Acta*, 60, 555–561.

Stone, J. O., and C. K. Ballantyne. (2006). Dimensions and deglacial chronology of the Outer Hebrides Ice Cap, northwest Scotland: implications of cosmic ray exposure dating. *Journal of Quaternary Science*, 21, 75–84.

Stone, J. O. H., J. M. Evans, L. K. Fifield, G. L. Allan, and R. G. Cresswell. (1998). Cosmogenic chlorine-36 production in calcite by muons. *Geochimica et Cosmochimica Acta*, 62, 433–454.

Stroeven, A. P., D. Fabel, C. Hättestrand, and J. Harbor. (2002). A relict landscape in the centre of Fennoscandian glaciation: Cosmogenic radionuclide evidence of tors preserved through multiple glacial cycles. *Geomorphology*, 44, 145–154.

Stuart, F. M. (2001). In situ cosmogenic isotopes: Principles and potential for Archaeology. In: *Handbook of Archaeological Sciences* (Eds D. R. Brothwell and A. M. Pollard). Wiley, pp. 93–110.

Sugden, D. E., G. Balco, S. G. Cowdery, J. O. Stone, and L. C. Sass. (2005). Selective glacial erosion and weathering zones in the coastal mountains of Marie Byrd Land, Antarctica. *Geomorphology*, 67, 317–334.

Summerfield, M. A. (2005). The changing landscape of geomorphology. *Earth Surface Processes and Landforms*, 30, 779–781.

Summerfield, M. A., F. M. Stuart, H. A. P. Cockburn, D. E. Sugden, G. H. Denton, T. J. Dunai, and D. R. Marchant. (1999). Long-term rates of denudation in the Dry Valleys, Transantarctic Mountains, southern Victoria Land, Antarctica based on in-situ-produced cosmogenic ^{21}Ne. *Geomorphology*, 27, 113–129.

Trull, T. W., M. D. Kurz, and W. J. Jenkins. (1991). Diffusion of cosmogenic ^{3}He in olivine and quartz: Implications for surface exposure dating. *Earth and Planetary Science Letters*, 103, 241–256.

Trustrum, N. A., B. Gomez, L. M. Reid, M. J. Page, and D. M. Hicks. (1999). Sediment production, storage, and output: The relative role of large magnitude events in steepland catchments. *Zeitschrift für Geomorphologie, Suppl. Bd.*, 115, 71–86.

Tschudi, S., S. Ivy-Ochs, C. Schlüchter, P. Kubik, and H. Raino. (2000). ^{10}Be dating of Younger Dryas Salpausselkä I formation in Finland. *Boreas*, 29, 287–293.

Van der Woerd, J., Y. Klinger, K. Sieh, P. Tapponnier, F. J. Ryerson, and A.-S. Meriaux. (2006). Long-term slip rate of the southern San Andreas Fault from ^{10}Be-^{26}Al surface exposure dating of an offset alluvial fan. *Journal of Geophysical Research-Solid Earth*, 111(B4), B04407.

Van der Woerd, J., F. Ryerson, P. Tapponier, Y. Gaudemer, R. C. Finkel, A. Meriaux, M. Caffee, Z. Guogang, and H. Qunlu. (1998). Holocene left-slip rate determined by cosmogenic surface dating on the Xidatan segment of the Kunlun Fault (Qinghai, China). *Geology*, 26, 695–698.

Vance, D., M. Bickle, S. Ivy-Ochs, and P. W. Kubik. (2003). Erosion and exhumation in the Himalaya from cosmogenic isotope inventories of river sediments. *Earth and Planetary Science Letters*, 206, 273–288.

Vermeesch, P. (2007). Quantitative geomorphology of the White Mountains (California), using detrital Apatite Fission Track thermochronology. *Journal of Geophysical Research – Earth Surface*, 112, F03004.10.1029/2006JF000671

Verri, G., R. Barkai, C. Bordeanu, A. Gopher, M. Hass, A. Kaufmann, P. W. Kubik, E. Montanari, M. Paul, A. Ronen, S. Weiner, and E. Boaretto. (2004). Flint mining in prehistory recorded by in situ-produced cosmogenic ^{10}Be. *Proceedings of the National academy of Sciences*, 101, 7880–7884.

Verri, G., R. Barkai, A. Gopher, M. Hass, P. W. Kubik, M. Paul, A. Ronen, S. Weiner, and E. Boaretto. (2005). Flint procurement strategies in the Late Lower Palaeolithic recorded by in situ produced cosmogenic Be-10 in Tabun and Qesem Caves (Israel). *Journal of Archaeological Science*, 32, 207–213.

von Blanckenburg, F. (2006). The control mechanisms of erosion and weathering at basin scale from cosmogenic nuclides in river sediment. *Earth and Planetary Science Letters*, 242, 224–239.

von Blanckenburg, F., N. S. Belshaw, and R. K. O'Nions. (1996). Separation of ^{9}Be and cosmogenic ^{10}Be from environmental materials and SIMS isotope dilution analysis. *Chemical Geology*, 129, 93–99.

Ward, D. J., J. A. Spotila, G. S. Hancock, and J. M. Galbraith. (2005). New constraints on the late Cenozoic incision history of the New River, Virginia. *Geomorphology*, 75, 54–72.

Williams-Thorpe, O., D. G. Jenkins, J. Jenkins, and J. S. Watson. (1995). Chlorine-36 dating and the bluestones of Stonehenge. *Antiquity*, 69(266), 1019–1020.

Williams, A. J., F. M. Stuart, S. J. Day, and W. M. Phillips. (2005). Using pyroxene microphenocrysts to determine cosmogenic ^{3}He concentrations in old volcanic rocks: an example of landscape development in central Gran Canaria. *Quaternary Science Reviews*, 24, 211–222.

Wolkowinsky, A. J., and D. E. Granger. (2004). Early Pleistocene incision of the San Juan River, Utah, dated with ^{26}Al and ^{10}Be. *Geology*, 32, 749–752.

Zech, R., C. Kull, and H. Veit. (2006). Late Quaternary glacial history in the Encierro Valley, northern Chile (29 degrees S), deduced from Be-10 surface exposure dating. *Palaeogeography Palaeoclimatology Palaeoecology*, 234, 277–286.

Zech, R., B. Glaser, P. Sosin, P. W. Kubik, and W. Zech. (2005). Evidence for long-lasting landform surface instability on hummocky moraines in the Pamir Mountains from surface exposure dating. *Earth and Planetary Science Letters*, 237, 453–461.

Zech, W., B. Glaser, U. Abramowski, C. Dittmar, and P. W. Kubik. (2003). Reconstruction of the Late Quaternary Glaciation of the Macha Khola valley (Gorkha Himal, Nepal) using relative and absolute (C-14, Be-10, dendrochronology) dating techniques. *Quaternary Science Reviews*, 22, 2253–2265.

Zehfuss, P. H., P. R. Bierman, A. R. Gillespie, R. M. Burke, and M. W. Caffee. (2001). Slip rates on the Fish Springs Fault, Owens Valley, California, deduced from cosmogenic ^{10}Be and ^{26}Al and soil development on fan surfaces. *Geological Society of America Bulletin*, 113, 241–255.

Zreda, M. (1994). Development and calibration of the ^{36}Cl surface expsoure dating method and its application to the chronology of Late Quaternary glaciations. PhD Thesis, New Mexico Institute of Mining and Technology, Soccorro, 318 pp.

Zreda, M. G., and J. S. Noller. (1998). Ages of prehistoric earthquakes revealed by cosmogenic chlorine-36 in a bedrock fault scarp at Hebgen Lake. *Science*, 282, 1097–1099.

Zreda, M. G., and F. M. Phillips. (1995). Insights into alpine moraine development from cosmogenic Cl-36 buildup dating. *Geomorphology*, 14, 149–156.

Zreda, M. G., and F. M. Phillips. (2000). Cosmogenic nuclide buildup in surficial materials. In: *Quaternary Geochronology: Methods and Applications* (Eds J. S. Noller, J. M. Sowers, and W. R. Lettis). AGU, Washington, DC, pp. 61–76.

Zreda, M. G., F. M. Phillips, and D. Elmore. (1994). Cosmogenic ^{36}Cl accumulation in unstable landforms. 2. simulations and measurements on eroding moraines. *Water Resources Research*, 30, 3127–3136.

Zreda, M. G., F. M. Phillips, P. W. Kubik, P. Sharma, and D. Elmore. (1993). Cosmogenic ^{36}Cl dating of a young basaltic eruption complex, Lathrop Wells, Nevada. *Geology*, 21, 57–60.

Zreda, M. G., F. M. Phillips, D. Elmore, P. W. Kubik, P. Sharma, and R. I. Dorn. (1991). Cosmogenic chlorine-36 production rates in terrestrial rocks. *Earth and Planetary Science Letters*, 105, 94–109.

CHAPTER 7

Soil Erosion and Sedimentation Studies Using Environmental Radionuclides

Felipe Zapata *and* Minh-Long Nguyen*

Contents

1. Introduction

Soil erosion and associated sediment deposition are natural landscape-forming processes. They can be accelerated by human activities (Figure 1)

*Corresponding author. Tel.: +43-1-2600-21648
E-mail address: m.nguyen@iaea.org

Soil and Water Management and Crop Nutrition Section, Joint FAO/IAEA Division of Nuclear Techniques in Food and Agriculture, Department of Nuclear Sciences and Applications, International Atomic Energy Agency – IAEA, Wagramer Strasse 5, A-1400 Vienna, Austria

Radioactivity in the Environment, Volume 16

ISSN 1569-4860, DOI 10.1016/S1569-4860(09)01607-6

Figure 1 Accelerated soil erosion due to bad farming practices. (A) Accelerated soil erosion in cultivated steep hill slopes, Yangjuangou catchment, China. (B) Losses of fertile topsoil (nutrients and organic matter) after crop harvest in the Cerrado region, Brazil.

and may badly affect agricultural production as well as watershed management and conservation of natural resources (Lal, 2000; Walling, 2001). Recent reports highlight the seriousness of soil degradation, in particular, soil erosion at the regional and global levels (UNEP, 2000). Land degradation by water and wind erosion is associated with an irretrievable loss of the basic soil resource and thus is a major threat to global food security. More than three quarters of the surface land area affected by erosion is in developing countries of Africa, Asia and Latin America, with about one-half of the total occurring in Asia. Removal of soil by water is more serious than by wind across all regions.

Soil erosion and deposition cause not only on-site degradation of land resources but also off-site problems such as downstream sediment deposition in fields, floodplains and water bodies. Suspended sediments in water bodies affect water quality and cause pollution because of various agrochemicals adsorbed on the sediments. Sedimentation causes loss of reservoir storage capacity and eutrophication of the water bodies (Clark, 1985). Soil erosion and deposition also have negative ecological and socio-economic impacts. Globally, the current economic costs of the on-site and off-site impacts of erosion of agricultural land have been estimated to amount to some US $400 billion per year (Pimentel et al., 1995). However, it is also important to recognise that effective soil conservation programmes can successfully counter soil erosion losses. For instance, it is estimated that between 1982 and 1997 in the United States, the total erosion on land was reduced by 42%, as a result of stopping agricultural production on cropland that is highly susceptible to erosion, and by improving soil conservation measures (Bernard and Iivari, 2000). It is also reported that the recent reduction of sediment load in the Yellow River in China partly reflects the extensive implementation of soil and water conservation measures within the highly erodible loess region of the middle Yellow River basin (Mou, 1996).

Current concerns about both on-site and off-site adverse effects associated with accelerated soil loss generate an urgent need for obtaining reliable quantitative data on the extent and actual rates of soil erosion worldwide (Ritchie and McHenry, 1990; Walling and Quine, 1995). Such data are required for a more comprehensive assessment of the magnitude of the effects to obtain a better understanding of the main factors involved, to validate new soil erosion/sedimentation prediction models and to provide a basis for developing scientifically sound land-use policies and selecting effective soil conservation measures and land-management strategies, including assessment of their economic and environmental impacts (Walling, 2001; Zapata, 2002). Despite extensive literature on the global, regional and national problems of soil erosion, quantitative and reliable data on the extent and rates of soil erosion are scarce for many regions of the world. Soil erosion research is capital and labour intensive as well as a time-consuming exercise. Well-designed experiments using standardised methodologies should be performed, so that the data obtained are comparable and representative of the study areas (FAO, 1977; Lal, 1988; Zapata et al., 1995). The existing methods to assess soil erosion can be grouped into two main categories: erosion modelling and prediction methods and erosion measurement methods. In all cases, there is a need for direct measurement of soil erosion, which can be done using erosion plots, surveying methods and nuclear techniques. The selection of a particular method basically depends on the objectives of the study and the availability of resources (Elliot et al., 1991). Existing classical techniques such as erosion plots and surveying methods for monitoring soil erosion are capable of meeting some of these requirements, but they have a number of important limitations in terms of the representativeness of the data obtained, their spatial resolution and potential to provide information on long-term rates of soil erosion and associated spatial patterns over extended areas, and the costs involved (Loughran, 1989; Higgitt, 1991). In addition, advances in the use of distributed numerical models and the application of GIS and geostatistics to erosion modelling have highlighted the need for spatially distributed data that represent the spatial variability of soil erosion and deposition rates within the landscape, in response to the local topography and land use/management (Walling, 2001). The quest for alternative techniques of soil erosion assessment to complement existing methods and to meet new requirements has directed attention to the use of radionuclides, in particular to the fallout of anthropogenic ^{137}Cs, as tracers for documenting rates and spatial patterns of soil redistribution within the landscape (Ritchie and McHenry, 1990; Walling and Quine, 1995).

The objectives of this chapter are (a) to describe briefly the use of radionuclides as tracers of soil movement in the landscape, in particular, the developments made in the standardisation of the ^{137}C technique and its application to measure soil erosion/deposition; (b) to overview recent achievements in the use of fallout radionuclides (FRNs) to control soil erosion and (c) to point to future directions on the application of FRNs for sustainable land and water management at the watershed scale. Future studies involving integrated approaches for the application of FRN techniques allied with innovative sediment fingerprinting techniques will allow identification of soil loss and sediment production risk areas. In this way, comprehensive sediment budgets will be established that support an integrated and sustainable land and water management at the watershed scale and thus contribute to achieving the UN Millennium Declaration Goals of food security and environmental sustainability.

2. Radionuclides as Tracers of Soil Movement

In soil erosion research, both the extent and the source of soil loss can be determined by artificially labelling the soil particles with an appropriate radionuclide. Several artificial radionuclides, mainly gamma emitters such as ^{59}Fe, ^{46}Sc, ^{110}Ag, ^{198}Au, ^{134}Cs, ^{51}Cr, etc., have been applied as tracers in field-erosion studies. Dual labelling techniques allowed detection of two particle sizes or the contribution from two sources (Menzel and Smith, 1984). The majority of radionuclide applications are related to environmental radionuclides such as ^{137}Cs, ^{210}Pb, ^{7}Be, ^{240}Pu, ^{14}C, ^{32}Si, ^{26}Al and ^{36}Cl, which showed great potential in assessing soil erosion and deposition (IAEA, 1998). Some of these radionuclides are produced by cosmic rays in the atmosphere (^{7}Be ^{14}C, ^{32}Si, ^{26}Al and ^{36}Cl), others are members of the natural decay series of the primordial radionuclides ^{238}U, ^{235}U and ^{232}Th (e.g. ^{210}Pb). Artificial radionuclides such as ^{134}Cs, ^{137}Cs, $^{239-240}Pu$ and ^{241}Am that have been released into the environment by atmospheric nuclear weapon tests and nuclear facilities have also proved very useful. The radioactive decay of all these radionuclides makes them suited for dating sediments, provided there is an appropriate model of the processes by which they enter the ecosystem and are incorporated in the soil/sediment. The usefulness of a particular radionuclide will therefore depend on (i) its half-life and (ii) the suitability of the model used for the evaluation of the measured radionuclide values. For example, radiocarbon dating based on the decay of ^{14}C (half-life 5,730 years) may provide accurate dates over a period spanning thousands of years, but will be inaccurate when used conventionally for dating sediments younger than about 150 years as a result of anthropogenic disturbances of the natural ^{14}C budget (Crickmore et al., 1990).

When radionuclides such as ^{137}Cs, ^{210}Pb and ^{7}Be reach the soil surface, they are quickly and strongly adsorbed by exchange sites of soil particles and become essentially non-exchangeable in most environments. Biological and chemical processes hardly influence the migration of the adsorbed radionuclides: the movement of radionuclides such as ^{137}Cs, ^{210}Pb and ^{7}Be that are tagged on soil particles is driven by physical processes. Accurately measuring these radionuclides in soil/sediment samples is relatively easy using modern instrumentation (high-purity germanium gamma spectrometry) and standardised protocols for quality assurance control. Measured inventories of the radionuclides ^{7}Be, ^{137}Cs and ^{210}Pb provide information on the short-term (<30 days), medium-term (~ 40 years) and long-term (~ 100 years) average soil redistribution rates and patterns, respectively (Zapata et al., 2002).

2.1. Principles of the use of ^{137}Cs, ^{210}Pb and ^{7}Be in erosion and sedimentation studies

Early work in the 1960s demonstrated the potential of using FRNs such as ^{90}Sr and ^{137}Cs to estimate soil losses by erosion (Menzel, 1960; Tamura, 1964). Menzel (1960) used ^{90}Sr and found a direct relationship between its concentration in soil and the soil loss.

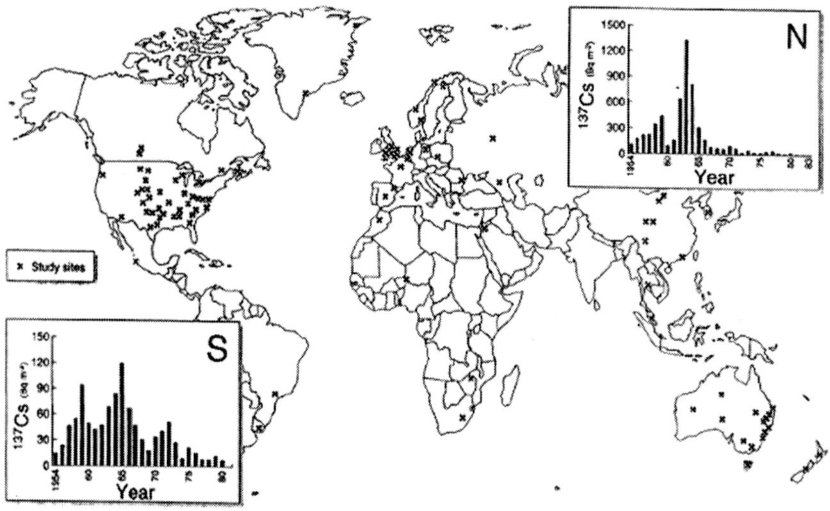

Figure 2 Typical records of bomb-fallout ^{137}Cs for the northern (NY, USA/ Milford Haven, UK) and southern (Adelaide/Brisbane, Australia) hemispheres and locations where ^{137}Cs has been used successfully in soil erosion and related studies (from IAEA, 1998).

2.1.1. ^{137}Cs

The widespread global distribution of ^{137}Cs (half-life 30.2 years) into the environment began with the atmospheric tests of high-yield nuclear weapons in the 1950s and the early 1960s.[1] The global ^{137}Cs fallout reached its peak in about 1963 in the northern and in about 1965 in the southern hemisphere (Figure 2). The atmospheric nuclear bomb explosions after 1963 were of comparatively low influence on the temporal variation of atmospheric ^{137}Cs. The deposition (by precipitation) of fallout ^{137}Cs from the atmosphere to the ground surface, its adsorption on fine soil particles and redistribution by erosion and redeposition are illustrated in Figure 3. Currently, ^{137}Cs is by far the most widely used FRN in soil erosion and sedimentation research, mainly due to its strong adsorption to fine soil particles, its relatively medium-term half-life, ease of measurement by gamma-ray spectrometry and well-defined fallout input patterns (Figure 4). Its redistribution in agro-ecosystems is directly related to erosion, transport and deposition of soil particles occurring during the period extending from the main phase of atmospheric fallout to the time of sampling (Ritchie and McHenry, 1990; Walling and Quine, 1995). The Chernobyl accident in April 1986 resulted in an additional atmospheric input of ^{137}Cs (Chernobyl fallout) in certain areas of Europe and this has also been used to study soil

[1] See Chapters 1 and 3 of this book for more details on the production and distribution of ^{137}Cs.

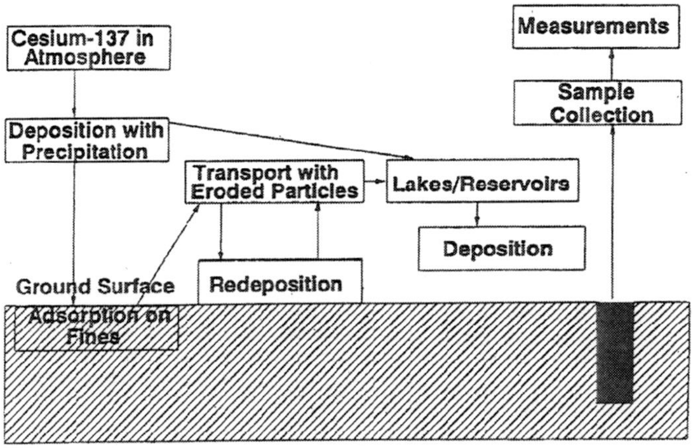

Figure 3 Diagram illustrating processes controlling the distribution of the fallout radionuclide ^{137}Cs (from Walling and Quine, 1993).

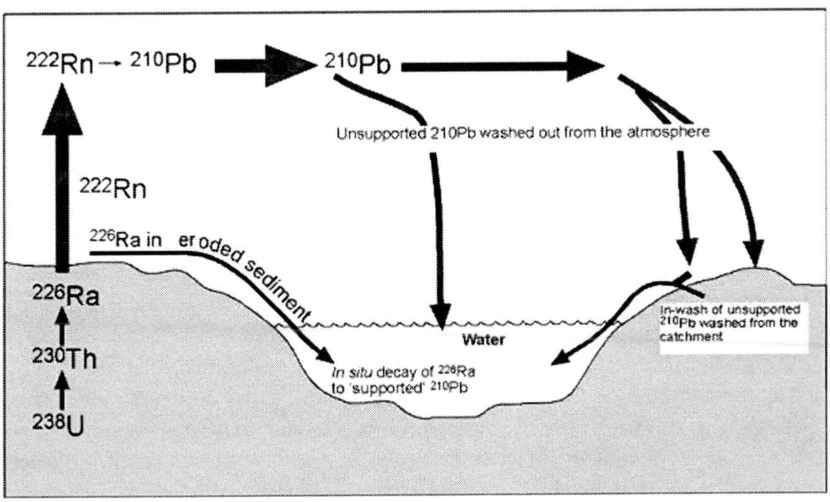

Figure 4 Pathways by which ^{210}Pb reaches the sediment in lakes and reservoirs (from Crickmore et al., 1990).

erosion/redistribution in some of these areas (IAEA, 1991; Golosov, 2002, 2003; Rezzoug et al., 2006).

2.1.2. ^{210}Pb

^{210}Pb is a naturally occurring radionuclide with a half-life of 22.26 years; it is a member of the ^{238}U decay series.[2] The total ^{210}Pb activity

[2] See Chapters 1 and 3.

concentration in soil/sediment columns has two components which are called supported and unsupported (excess) ^{210}Pb. Supported ^{210}Pb is the component that is produced by *in situ* decay of the parent isotope ^{226}Ra present in soil/sediment particles. Unsupported or excess ^{210}Pb derives from the radioactive decay of ^{222}Rn in the air, from where it is deposited to the surface of the soil or the top layer of the sediment. The atmospheric radon itself is formed by decay of ^{226}Ra in the soil, from where it escapes by diffusion into the atmosphere. Figure 4 illustrates the deposition from the atmosphere to the sediment in a lake; the deposition to surface soil follows a similar pattern (Appleby and Oldfield, 1992; Walling, 1998).

^{210}Pb is rapidly and strongly adsorbed by surface soil and behaves in a similar manner as ^{137}Cs. It has been widely used for dating sediment cores, studying soil redistribution and estimating soil erosion within a timescale of 100–150 years with a temporal resolution of 5–10 years (Appleby and Oldfield, 1992). In most situations, it can be assumed that supported ^{210}Pb is in radioactive equilibrium with its progenitor ^{226}Ra and, therefore, the unsupported or excess ^{210}Pb activity at any level is obtained by measuring both total ^{210}Pb and ^{226}Ra and subtracting the supported component. Recent advances in low background, low-energy gamma spectrometry and electronic software for data acquisition have facilitated a routine use of the ^{210}Pb technique (Walling, 2001; Zapata et al., 2002). The atmospheric deposition of ^{210}Pb is a steady process, that is continuous fallout input and supply to the soil surface. The spatial variation of the ^{210}Pb deposition rate ranges from about 30 to 370 $Bq\,m^{-2}\,yr^{-1}$. ^{210}Pb can be used as an alternative to ^{137}Cs in soil erosion studies of areas where ^{137}Cs proves to be inapplicable, for instance where a significant input of Chernobyl fallout was received or where the inventories of bomb-derived ^{137}Cs fallout are too low or too heterogeneous (Walling and He, 1999b).

At undisturbed sites, the unsupported ^{210}Pb concentrations are greatest at the surface and decrease with depth (Figure 5). In contrast, ^{137}Cs concentration in the top layer of soil/sediment is lower than at depth due to the lack of replenishment in the past 30 years. In general, the layer of maximum ^{137}Cs concentration has been displaced a few centimetres below soil surface (Figure 5).

2.1.3. ^{7}Be

Beryllium-7 (^{7}Be) is a naturally occurring radionuclide, which is produced in nuclear spallation reactions of cosmic rays with nitrogen and oxygen atoms in the upper atmosphere.[3] It is washed out from the atmosphere by precipitation and dry deposition together with aerosols (dry and wet fallout). In most environments, ^{7}Be is rapidly adsorbed to the fine soil

[3] See Chapters 1 and 3.

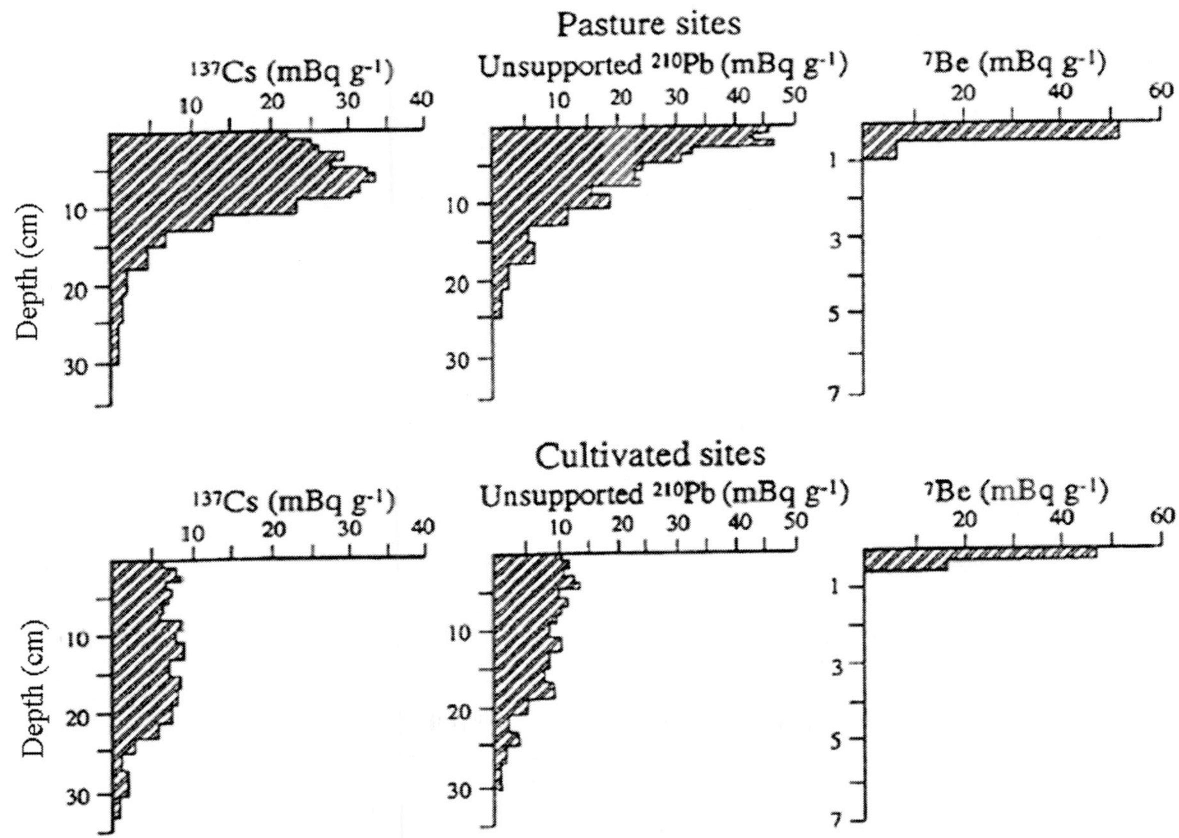

Figure 5 Vertical distributions of ^{137}Cs, ^{210}Pb and ^{7}Be activities within soil cores collected from representative pasture (uncultivated) and cultivated field sites near Exeter, Devon, UK (from IAEA, 1998).

particles after it reaches the soil surface. Daily surface fluxes of ^{7}Be range between 140 and 740 Bq m^{-2}. Gamma-ray spectrometry is used to measure ^{7}Be activities. The comparatively short half-life (53.4 days) enables ^{7}Be to provide information on the dynamics of soil redistribution over short timescales (about 30 days) (Blake et al., 1999; Zapata et al., 2002). Because of its short half-life, ^{7}Be is usually limited to the top layer (about 5–10 mm) of the soil (Figure 5). It can be used to distinguish between soil from the surface and soil derived from depths >10 mm; the latter is free of ^{7}Be (Figure 5). Some authors have used ^{7}Be to distinguish sediment mobilised by sheet erosion from sediment stemming from rill erosion (Burch et al., 1988; Wallbrink and Murray, 1996). Blake et al. (2002) estimated the sediment budget after significant rainfall events using ^{7}Be.

3. Standardisation of the ^{137}Cs Technique to Measure Soil Erosion/Deposition

3.1. Description of the ^{137}Cs technique

Radioactive fallout has been monitored globally since the early 1950s. In 1967, a map of the cumulative deposition of ^{90}Sr, the most dangerous FRN, was published. Based on the available ^{90}Sr data, an assessment of the potential global distribution of ^{137}Cs inputs was made (Garcia Agudo, 1998) and later a database and model were developed to estimate global ^{137}Cs inventories (Walling and He, 2000). It was shown that the ^{137}Cs fallout was much greater in the northern hemisphere than in the southern because more atmospheric nuclear tests were carried out in the northern hemisphere (Figures 6 and 7). Local events, such as the Chernobyl accident, have strong impacts on the fallout patterns in certain regions but are of limited significance for the global fallout patterns and rates (IAEA, 1991).

The value of ^{137}Cs as a tracer lies in its rapid and strong adsorption by fine soil particles, particularly clay minerals and humic materials (Tamura, 1964; Bachhuber et al., 1982). The key assumptions and requirements of the ^{137}Cs technique have been fully described in many publications (Ritchie and McHenry, 1990; Walling and Quine, 1993; Zapata, 2002). The ^{137}Cs technique for determining rates and patterns of soil loss is based on several assumptions; it appears to be attractive in its simplicity. But there are also limitations and uncertainties to be taken into account in any of these applications (Ritchie and McHenry, 1990; Walling and Quine, 1995; Walling, 1998).

The assessment of redistribution is commonly based on a comparison of the ^{137}Cs inventory (total activity per unit area) measured at a given sampling

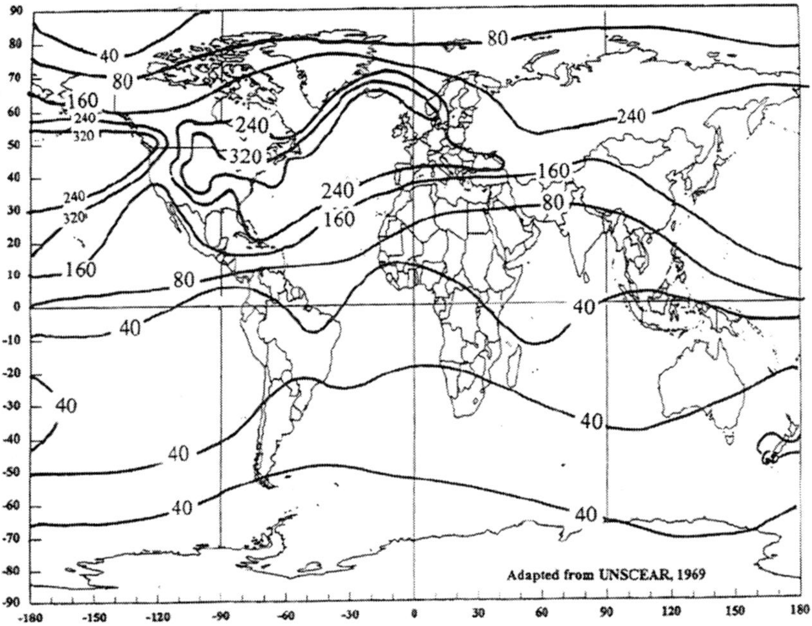

Figure 6 Global inventory of ^{137}Cs from nuclear tests in mBq cm^{-2}, estimated for 1996 (from IAEA, 1998).

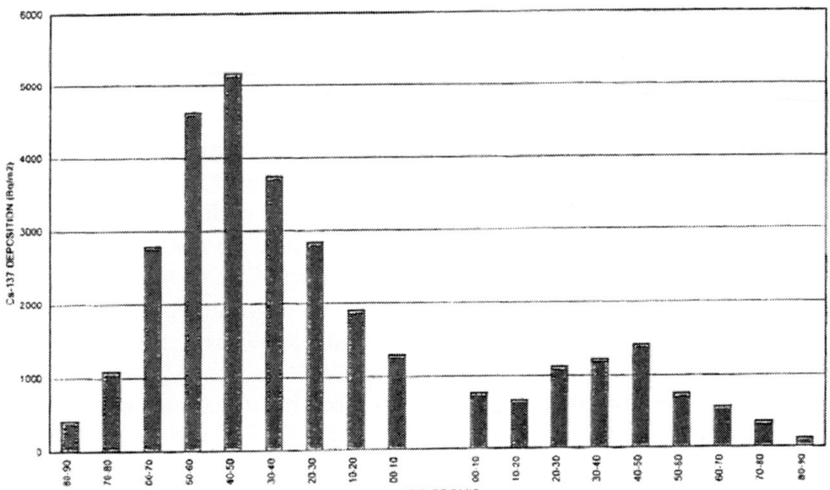

Figure 7 Distribution of the ^{137}Cs deposition along latitude bands (from IAEA, 1998).

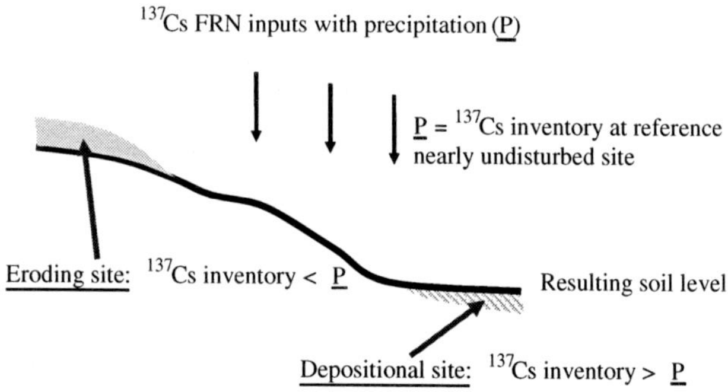

Figure 8 Schematic representation of the ^{137}Cs method to study soil erosion and deposition. It is essential to determine the inventory (P) of the fallout radionuclide (FRN) ^{137}Cs at a reference site (undisturbed by erosion or sedimentation) in the study area. Erosion and deposition sites can be distinguished by comparing the ^{137}Cs inventory of the given site with that at the reference site. Inventories lower than the reference value represent soil losses by erosion and those in excess of the reference level indicate deposition (accumulation) of soil.

site with that in a reference site (inventory representing the cumulative atmospheric fallout input at the site, taking due account of the different behaviour of cultivated and non-cultivated soils). Because direct long-term measurements of atmospheric fallout are rarely available, the cumulative input or reference inventory is usually established by sampling adjacent stable and nearly undisturbed sites, where neither significant erosion nor additional deposition have occurred (Figure 8). The determination of ^{137}Cs inventories requires sampling of soil cores and measuring the ^{137}Cs concentration versus depth in the core.

The magnitude and direction of the measured deviations from the local reference level provide only a qualitative assessment of soil redistribution; this means a general indication of relative erosion/deposition involved and associated patterns (Walling and Quine, 1993). To derive quantitative estimates of soil erosion and deposition rates from ^{137}Cs measurements, it is necessary to establish relationships between the magnitude of the deviation from the reference inventory and the extent of soil loss (erosion) or gain (deposition). A number of calibration procedures are available ranging from simple empirical relationships to proportional models and mass balance models (Ritchie and McHenry, 1990; Walling and Quine, 1993). Many workers have favoured the use of conversion models that are developed on the basis of physical processes influencing the relationship between magnitude of the reduction or increase in the ^{137}Cs inventory and the erosion or deposition rate (Walling and He, 1999a, Walling et al., 2002).

3.2. Advantages and limitations of the ^{137}Cs technique

The choice of a particular method for measuring soil erosion and sedimentation requires a thorough knowledge of its advantages and limitations. The advantages and limitations of the ^{137}Cs technique have been described by Walling and Quine (1995), IAEA (1998), Walling (2001) and Zapata (2002).

The key advantages of the ^{137}Cs technique can be summarised as follows:

- Estimates are based on contemporary sampling and provide a retrospective assessment of medium-term (30–40 years) rates of soil redistribution.
- Estimates can be obtained on the basis of a single site visit.
- Resulting estimates of soil redistribution rates are integrated medium-term, average data and less influenced by extreme events.
- Estimates refer to individual points within the landscape; information on rates and spatial patterns can be assembled.
- Sampling does not require significant disturbance of the landscape or study area.
- The results are compatible with recent developments in physically based distributed modelling and the application of GIS and geostatistics to soil erosion and sediment yield studies.
- The spatial patterns of soil redistribution can be further improved by using geostatistical tools, where appropriate.
- Rates of soil redistribution represent integrated effects of all landscape processes resulting in movement of soil particles under defined land use/management.
- The technique provides information on both erosion and deposition in the same watershed and therefore net rates of sediment export.
- There are no major constraints, for area-wide (watershed scale) application apart from the number of samples to be analysed.
- From data interpretation, the technique permits quantification of processes such as soil tillage redistribution and soil loss and deposition associated with sheet erosion.
- Application in fingerprinting suspended sediment sources and estimating rates of over-bank floodplain accretion.

The following aspects characterise limitations of the ^{137}Cs technique:

- The successful application of the technique requires a multi-disciplinary team. This requirement is difficult to meet in developing countries with limited scientific staff.
- Specialised laboratories with sample preparation and costly gamma-counting equipment are necessary.
- Requirements for quality assurance/control of low-level gamma-spectrometry measurements.

- The technique is effectively limited to documentation of sheet and general surface lowering, but it can be applied to study rill/gully erosion in cultivated areas.
- It is an indirect approach that depends on the link between measured soil redistribution and observed ^{137}Cs redistribution.
- There are uncertainties associated with the selection of appropriate conversion models to estimate erosion and deposition rates from the ^{137}Cs measurements.
- There is an inability to document short-term changes in erosion rates such as those related to changes in cropping systems and management practices.

While some of the limitations are inherent in the technique, several others have been surmounted by coordinated actions within research networks such as the one conducted by the International Atomic Energy Agency (IAEA) that is addressed below. Given the limitations of other techniques, the ^{137}Cs method can also serve as a complementary approach in soil erosion/sedimentation studies.

3.3. Developments in the harmonisation of the ^{137}Cs technique

Early applications of the ^{137}Cs technique in soil erosion research were made by several groups active in Australia (Campbell et al., 1982; Elliott et al., 1990; Loughran et al., 1988; Wallbrink and Murray, 1993); Canada (de Jong et al., 1983; Kachanoski, 1993; Pennock et al., 1995); England (Walling and Bradley, 1988; Walling and Quine, 1991; Walling and He, 1999a, 1999b) and the United States of America (Ritchie and McHenry, 1973, 1990, 1995). These groups provided a solid database on the use of ^{137}Cs to measure soil redistribution. Figure 2 shows the locations where ^{137}Cs has been used successfully in soil erosion and related studies until 1998. One of the lessons learned from these studies is that a worldwide application of this technique requires harmonised protocols.

Recognising that soil erosion and sedimentation are a serious threat to sustainable watershed management, the IAEA examined the potential role of nuclear techniques in soil erosion and siltation studies (IAEA, 1995) and ultimately conducted and coordinated from 1995 to 2001 two research projects[4] to (i) refine, validate and standardise relevant methodologies for documenting soil erosion using the ^{137}Cs technique across a range of environments, which can then be used to test and calibrate existing models of soil erosion, and to (ii) evaluate the effect of specific land-use management on soil erosion for providing data to underpin the selection of soil conservation

[4] The project on 'Assessment of soil erosion through the use of ^{137}Cs and related techniques as a basis for soil conservation, sustainable production and environmental protection' was coordinated by the Soil and Water Management and Crop Nutrition Section, and the other project, entitled 'Sedimentation assessment by environmental radionuclides and their application to soil conservation measures', by the Isotope Hydrology Section of the IAEA (Zapata et al., 1995; Zapata, 2002).

strategies (Pennock and Zapata, 1995; IAEA, 1998). Through the two closely linked projects, standardised methods and protocols for the application of the ^{137}Cs technique were developed and a wealth of information on soil erosion/sedimentation rates was obtained (Zapata, 2001a, 2002, 2003; Queralt et al., 2000; Zapata and Garcia Agudo, 2000). The values determined for the rates of soil loss and sedimentation ranged between zero and more than 100 t ha^{-1} yr^{-1}. It was also found that local land-use practices are of higher influence on these processes than soil type, lithology and climate (Zapata, 2003). The available conversion models for deriving soil erosion/sedimentation rates from ^{137}Cs measurements were tested (Walling and He, 1999a; Zapata, 2003) and specific approaches developed to take due account of the Chernobyl fallout in Russia (Golosov, 2003). Existing conversion models were refined to better suit local conditions (Schuller et al., 2000; Zhang et al., 2000). The potential use of ^{137}Cs measurements for validating soil erosion and sediment delivery models has also been demonstrated (Walling et al., 2003b; Golosov, 2003; Bacchi et al., 2003).

4. Recent Developments of Fallout Radionuclide Techniques

The development, refinement and calibration of the ^{137}Cs technique has provided a universal tool to quantify soil redistribution rates in a range of natural and agro-ecosystems, and has paved the way for a wider application of the technique, particularly in assessing the effectiveness of soil conservation technologies in controlling or mitigating soil erosion and associated degradation processes (Figure 9). Recent studies on soil conservation[5] have emphasised the combined use of FRN such as ^{137}Cs, ^{210}Pb and ^{7}Be, which enables the determination of erosion rates on several timescales (from fast to slow processes) and under different local conditions (from hill slopes to small watersheds). The ^{7}Be technique has been further developed to study short-term soil redistribution for assessing the impacts of tillage systems and forest harvest operations. New and improved conversion models have been developed that provide a high effective spatial and visual representation of soil redistribution data and processes. The model outputs are in good agreement with data derived from field measurements. Furthermore, it was found that FRNs can be very useful in determining parameter/processes such as transit time of water, residence time of sediments, sediment accumulation, sediment sources and bioturbation. These radionuclide techniques have also contributed to better

[5] The studies were conducted from 2002 to 2007 within the framework of the IAEA Coordinated Research Project Assessment of the effectiveness of soil conservation technologies using fallout radionuclides (Turner, 2000; Zapata, 2001b, 2003, 2007).

Figure 9 Sustainable land-use and management strategies for soil conservation and improving soil and water quality. (A) Terrace cultivation to control soil erosion and manage irrigation water. (B) Conservation agriculture systems in the Cerrado region, Brazil – direct planting and use of crop residue cover to protect the soil from erosion.

understanding of redistribution of soil organic matter (SOM) and to assessing the impact of exceptional events (i.e. forest fires, flooding and permafrost melting) on soil redistribution.[6]

[6] A detailed report on recent advances in the field of soil erosion and conservation is given by Zapata and Li (2007). A 'Bibliography of publications of ^{137}Cs studies related to erosion and sediment deposition' that includes 4,038 publications related to ^{137}Cs and other radionuclides was compiled by Ritchie and Ritchie (2008); it can be downloaded under http://www.ars.usda.gov/Main/docs.htm?docid = 16037

5. Applications of Fallout Radionuclides in Soil Erosion/Sedimentation Studies

Much of the initial work focused on the development of refined and standardised procedures and protocols for the application of the ^{137}Cs technique under various environment conditions (Queralt et al., 2000; Zapata, 2003). As the efficacy and value of the ^{137}Cs technique were increasingly recognised, this technique alone or in combination with other FRNs was applied in numerous studies (Zapata and Li, 2007). The unique information derived from estimates of soil redistribution rates and their spatial distribution patterns improved the understanding of the relationships between soil loss/gain and other parameters such as soil quality, soil carbon content, nutrient redistribution and the fate of agrochemicals and other contaminants in the landscape (Table 1).

6. Future Research

Current concerns related to agricultural land-use impacts on the environment have highlighted the important role of sediment (and associated nutrient/chemicals) in degrading water quality and in a range of other environmental problems (Horowitz and Walling, 2005). The loss of topsoil and associated nutrients/chemical inputs as well as the potential degradation of water quality through land erosion enhanced by human activity is a serious threat to food security, sustainable agricultural production and environmental protection in many regions worldwide. These impacts are likely to increase further due to the anticipated intensification of climate variability. Sediment transport is the key to understanding the movement and fate of many nutrients (e.g. soil organic carbon and phosphorus) and contaminants (McDowell et al., 2001; McDowell and Sharpley, 2003), and carbon storage and mobilisation within the watershed (Lal, 2007). These substances drive eutrophication, decrease agricultural productivity and cause environmental degradation. Therefore, there is a clear need to develop a better understanding of sediment mobilisation, transfer and storage in watersheds from source to sink (Walling et al., 1998; Nguyen, 2008). Planning and implementing effective control measures to reduce soil loss/sediment transport and related environmental problems require (i) quantification of sediment loads, (ii) identification and apportioning of sediment sources and (iii) the fate of the sediment-associated pollutants on a watershed scale. The interaction of sediment sources, transfers and sinks across the landscape of a watershed is highly complex.

In future, the FRN techniques will be further developed to enable quantitative determination of the mobilisation, transfer and storage of

Table 1 List of selected applications of fallout radionuclides in soil erosion/sedimentation investigations.

Issue/topic	Location	References
The relationship between soil redistribution and topographic controls	Canada	Pennock (2000, 2003)
	Brazil	Bacchi et al. (2000)
	Argentina	Bujan et al. (2000, 2003)
	Slovakia	Fulajtar (2000, 2003)
	China	Li et al. (2000)
The relationship between soil quality parameters and soil erosion as inferred from ^{137}Cs redistribution	Canada	Pennock (1998, 2000)
	Brazil	Bacchi et al. (2000)
	China	Li et al. (2000)
	Slovakia	Fulajtar (2000)
	Greece	Theocharopoulos et al. (2000)
The influence of tillage in soil redistribution	Canada	Lobb et al. (1995)
	Belgium	Govers et al. (1996)
	China	Zhang et al. (2003), Li and Lindstrom (2001)
	Chile	Schuller et al. (2003, 2004, 2007)
	Canada	Li et al. (2007)
Tracing sources of suspended sediment in river basins	France	Sogon et al. (1999)
	Australia	Wallbrink et al. (1999)
	UK	Walling (2005)
Characterising land-surface erosion from ^{137}Cs profiles in lake and reservoir sediments	China	Zhang and Walling (2005)
Soil erosion/sedimentation control for restoring fish populations in a river basin	Canada	Bernard and Lavardiere (2000)
The relationship between ^{137}Cs-derived soil redistribution and soil organic carbon distribution patterns	Canada	Pennock and Frick (2001)
	USA	McCarty and Ritchie (2002), Ritchie and McCarty (2003), Venteris et al. (2004)
	China	Li et al. (2006, 2007)
The contribution of sediment and sediment-P from erosion processes and land-use types in a mixed land-use catchment	Australia	Wallbrink et al. (2003)
The impact of forest harvesting operations on soil redistribution	Australia	Wallbrink et al. (2002)
	Chile	Schuller et al. (2006)
National reconnaissance survey of soil erosion, National Landcare Program of Australia	Australia	Loughran et al. (1992), Loughran and Elliott (1996), Loughran, personal communication

Table 1. (*Continued*)

Issue/topic	Location	References
Comparison of direct measurement, USLE and ^{137}Cs-based methods for evaluating severe sheet, rill and gully erosion	Russia	Belyaev et al. (2005)
Gully erosion studies	Romania	Ionita and Margineanu (2000)
	Russia	Belyaev et al. (2004)
	China	Li et al. (2003)
Spatial characterisation of soil redistribution data using GIS	Canada	Mabit et al. (2007)
The relationship between soil redistribution and the fate of pesticides in agricultural landscapes	Austria	IAEA (2002)
	Canada	Farenhorst et al. (2008)
	Latin America	Ferris, personal communication
Measuring erosion losses in various environments	Morocco; Japan; Vietnam; Brazil; Turkey; Italy	Nouira et al. (2003), Onda et al. (2003, 2007), Fukuyama et al. (2005), Phan Son Hai et al. (2006), Guimarães et al. (2003), Haciyakupoglu et al. (2006), Porto et al. (2003a, 2003b)
Combined and specific FRNs methodological applications	Australia	Burch et al. (1988)
	UK	Blake et al. (2002)
	Italy	Porto et al. (2006)
	Chile	Sepúlveda et al. (2007)
	UK	Walling et al. (2003a)
	China	Yang et al. (2006), Zhang et al. (2006), Zheng et al. (2007)
Validating distributed erosion/sedimentation models	UK	Walling et al. (2003b)
Refinement of existing models	Chile	Schuller et al. (2000)
	China	Zhang et al. (2000)
Comparison of erosion rates derived from the ^{137}Cs method and erosion model estimates	Brazil	Bacchi et al. (2003)
	Russia	Golosov (2003)
Wind erosion studies	USA	Van Pelt et al. (2007)
Precision conservation studies	USA	Schumacher et al. (2005)
Impacts of severe bush fires on redistribution of soil and organic (nutrient) material	Australia	Wallbrink et al. (2005)

sediment on an area-wide basis over different timescales (Walling, 2005; Walling et al., 2006; Le Cloarec et al., 2007). Furthermore, selected nuclides will be used as fingerprints of sediment source identification. Recent studies using compound specific stable isotope (^{13}C and ^{15}N) techniques (CSI) identified the sources of soils in sediments deposited in an estuary (Gibbs, 2005, 2006). Such applications will help to identify risk areas and thus establish guidelines for the sustainable management of land and water resources (OECD, 2003, 2004).[7]

7. Concluding Remarks

The use of the FRNs techniques affords an effective and valuable means for the assessment of surface erosion and deposition within the landscape over several spatial and temporal scales. The key advantage of this approach is that it can provide integrated rates of erosion/deposition and spatial patterns of soil redistribution, without the need for long-term monitoring programmes. Recently, significant progress has been made in harmonising the ^{137}Cs technique and other FRNs in soil erosion research through cooperation of specialist teams and coordination by the IAEA. The developed methods and protocols provide a standardised framework for the application of these techniques worldwide. This work is an essential pre-requisite to obtaining directly comparable and representative information on soil erosion rates in a wide range of environments and to understanding the influence of the main factors affecting soil loss/sediment production in the landscape. This information is required for a scientifically sound land and water management to mitigate/control soil erosion and associated problems. The efficacy of the techniques has been increasingly recognised in other studies and a range of expanding applications in landscape studies demonstrates its value.

In future, integrated approaches that combine these FRN techniques with sediment fingerprinting techniques will allow identification of soil loss/sediment production risk areas. Techniques and guidelines will be provided to establish comprehensive sediment budgets and foster sustainable land and water management at the watershed scale. In this way, FRN methods contribute to achieving the UN Millennium Declaration Goals of food security and environmental sustainability.

[7] The IAEA envisages a coordinated research project (CRP) entitled 'Tracking sediment sources and transfers at the watershed scale using isotopic and nuclear techniques to ensure sustainable land and water management and environmental protection' (Nguyen, 2008).

REFERENCES

Appleby, P. G., and F. Oldfield. (1992). Application of ^{210}Pb to sedimentation studies. In: *Uranium Series Disequilibrium: Applications to Earth, Marine and Environmental Sciences* (Eds M. Ivanovich and R. S. Harman). Clarendon Press, Oxford, UK, pp. 731–738.

Bacchi, O. O. S., K. Reichardt, and G. Sparovek. (2003). Sediment spatial distribution evaluated by three methods and its relation to some soil properties. *Soil and Tillage Research*, 69, 117–126.

Bacchi, O. O. S., K. Reichardt, G. Sparovek, and S. B. L. Ranieri. (2000). Soil erosion evaluation in a small watershed in Brazil through ^{137}Cs fallout redistribution analysis and conventional methods. *Acta Geologica Hispanica*, 35(3–4), 251–259.

Bachhuber, H., K. Bunzl, W. Schimmack, and I. Gans. (1982). The migration of Cs-137 and Sr-90 in multilayered soils: Results from batch, column and fallout investigations. *Nuclear Technology*, 59, 291–301.

Belyaev, V. R., P. J. Wallbrink, V. Golosov, A. S. Murray, and A. Y. Sidorchuk. (2004). Reconstructing the development stages of a gully in the Upper Kalaus basin, Stavropol region (Southern Russia). *Earth Surface Processes and Landforms*, 29, 323–341.

Belyaev, V. R., P. J. Wallbrink, V. Golosov, A. S. Murray, and A. Y. Sidorchuk. (2005). A comparison of direct measurement, USLE and caesium-137 based methods for evaluating soil redistribution from severe sheet and ephemeral gully erosion, Stavropol region, southern European Russian. *Geomorphology*, 1, 173–193.

Bernard, C., and M. R. Lavardiere. (2000). Using ^{137}Cs as a tool for the assessment and the management of soil erosion/sedimentation risks in view of the restoration of the rainbow smelt (*Osmerus mordax*) fish population in the Boyer river basin. *Acta Geologica Hispanica*, 35(3–4), 321–327.

Bernard, J. M., and T. A. Iivari. (2000). Sediment damages and recent trends in the United Sates. *International Journal of Sediment Research*, 15, 135–148.

Blake, W. H., D. E. Walling, and Q. He. (1999). Fallout beryllium-7 as a tracer in soil erosion investigations. *Applied Radiation and Isotopes*, 51, 599–605.

Blake, W. H., D. E. Walling, and Q. He. (2002). Using cosmogenic beryllium-7 as a tracer in sediment budget investigations. *Geografiska Annaler: Series A, Physical Geography*, 82A, 89–102.

Bujan, A., O. J. Santanatoglia, C. Chagas, M. Massobrio, M. Castiglioni, M. S. Yanez, H. Ciallella, and J. Fernandez. (2000). Preliminary study on the use of the ^{137}Cs method for soil erosion investigation in the Pampean region of Argentina. *Acta Geologica Hispanica*, 35(3–4), 271–277.

Bujan, A., O. J. Santanatoglia, C. Chagas, M. Massobrio, M. Castiglioni, M. S. Yanez, H. Ciallella, and J. Fernandez. (2003). Soil erosion evaluation in a small basin through the use of the ^{137}Cs technique. *Soil and Tillage Research*, 69, 127–137.

Burch, G. J., C. J. Branes, I. D. Moore, R. D. Braling, D. H. Mackensie, and J. M. Olley. (1988). Detection and prediction of sediment sources in catchments: Use of ^{7}Be and ^{137}Cs. *Proceedings of the Symposium on Hydrology and Water Resources, Canberra, ACT*, ANU, Canberra ACT, Australia, pp. 146–151.

Campbell, B. L., R. J. Loughran, and G. L. Elliott. (1982). Caesium-137 as an indicator of gemorphic processes in a drainage basin system. *Australian Geography Studies*, 20, 49–64.

Clark, E. H. (1985). The off site cost of soil erosion. *Journal of Soil and Water Conservation*, 40, 19–22.

Crickmore, M. J., G. S. Tazioli, P. G. Appleby, and F. Oldfield. (1990). *The Use of Nuclear Techniques in Sediment Transport and Sedimentation Problems*. IHP-III Project 5.2 SC-90/WS-49, UNESCO, Paris.

de Jong, E., C. B. M. Begg, and R. G. Kachanoski. (1983). Estimates of soil erosion and deposition for some Saskatchewan soils. *Canadian Journal of Soil Science*, 63, 607–617.

Elliott, G. L., B. L. Campbell, and R. J. Loughran. (1990). Correlation of erosion measurement and soil caesium-137 content. *Journal of Applied Radiation and Isotopes*, 41, 713–717.

Elliot, W. J., G. R. Foster, and A. V. Elliot. (1991). Soil erosion: Processes, impacts and prediction. In: *Soil Management for Sustainability* (Eds R. Lal and R. J. Pierce). Soil and Water Conservation Society, Ankeny, IA, pp. 25–34.

Farenhorst, A., S. K. Papiernik, I. Saiyed, P. Messing, K. D. Stephens, J. A. Schumacher, D. A. Lobb, S. Li, M. J. Lindstrom, and T. E. Schumacher. (2008). Herbicide sorption coefficients in relation to soil properties and terrain attributes on a cultivated prairie. *Journal of Environmental Quality*, 37, 1201–1208.

Food and Agriculture United Nations Organization (FAO) (1977). *Assessing Soil Degradation.* Soils Bulletin No. 34, FAO, Rome.

Fukuyama, T., C. Takenaka, and Y. Onda. (2005). ^{137}Cs loss via soil erosion from a mountainous headwater catchment in Japan. *Science of the Total Environment*, 350, 238–247.

Fulajtar, E. (2000). Assessment of soil erosion through the use of ^{137}Cs at Jaslovske Bohunice, Western Slovakia. *Acta Geologica Hispanica*, 35(3–4), 291–300.

Fulajtar, E. (2003). Assessment of soil erosion on arable land using ^{137}Cs measurements: A case study from Jaslovske Bohunice, Slovakia. *Soil and Tillage Research*, 69, 139–152.

Garcia Agudo, E. (1998). Global distribution of ^{137}Cs inputs for soil erosion and sedimentation surveys. In: *Use of ^{137}Cs in the Study of Soil Erosion and Sedimentation.* IAEA TECDOC-1028. IAEA Publications, Vienna, pp. 117–121.

Gibbs, M. (2005). Sediment tracking method refinement. Technical Publication TP 294. Auckland Regional Council, New Zealand. URL: www.arc.govt.nz

Gibbs, M. (2006). Sediment source mapping in Mahurangi Harbour. Technical Publication TP 321. Auckland Regional Council, New Zealand. URL: www.arc.govt.nz

Golosov, V. (2002). Special considerations for areas affected by Chernobyl fallout. In: *Handbook for the Assessment of Soil Erosion and Sedimentation Using Environmental Radionuclides* (Ed. F. Zapata). Kluwer Publications, Dordrecht, The Netherlands, pp. 165–183.

Golosov, V. (2003). Application of Chernobyl-derived ^{137}Cs for the assessment of soil redistribution within a cultivated field. *Soil and Tillage Research*, 69, 85–97.

Govers, G., T. A. Quine, P. J. J. Desmet, and D. E. Walling. (1996). The relative contribution of soil tillage and overland flow erosion to soil redistribution in agricultural land. *Earth Surface Processes and Landforms*, 21, 929–946.

Guimarães, M. F., V. F. Nascimento Filho, and J. C. Ritchie. (2003). Application of cesium-137 in a study of soil erosion and deposition in southeastern Brazil. *Soil Science*, 168(1), 45–53.

Haciyakupoglu, S., T. A. Ertek, D. E. Walling, F. Ozturk, G. Karahan, A. E. Erginal., N. Celebi, L. Gureli, H. Kaya, and H. Akgun. (2006). Using caesium-137 measurements to investigate soil erosion rates in western Istanbul (NW Turkey). *Catena*, 64, 222–231.

Higgitt, D. L. (1991). Soil erosion and soil problems. *Progress in Physical Geography*, 15, 91–100.

Horowitz, A. J., and D. E. Walling (Eds). (2005). *Sediment Budgets* 1 & 2. International Association of Hydrological Sciences Publication 291 and 292, IAHS Press, Wallingford, UK.

International Atomic Energy Agency (IAEA). (1991). *The International Chernobyl Project: An Overview, Assessment of Radiological Consequences and Evaluation of Protective*

Measures. Report by an International Advisory Committee. IAEA Publications, Vienna.

International Atomic Energy Agency (IAEA). (1995). *Use of Nuclear Techniques in Studying Soil Erosion and Siltation*. IAEA-TECDOC-828. IAEA Publications, Vienna.

International Atomic Energy Agency (IAEA). (1998). *Use of* ^{137}*Cs in the Study of Soil Erosion and Sedimentation*. IAEA TECDOC-1028. IAEA Publications, Vienna.

International Atomic Energy Agency (IAEA). (2002). *Report of the IAEA/RCA Project Formulation Meeting of the Regional Technical Co-operation Project for East Asia and the Pacific on "Restoration of Soil Fertility and Sustenance of Agricultural Productivity (RAS/5/039), Part II: Measuring Soil Erosion/Sedimentation and Associated Pesticide Contamination, Beijing, China, February 2002*. IAEA Publications, Vienna.

Ionita, I., and R. M. Margineanu. (2000). Application of ^{137}Cs for measuring soil erosion/deposion rates in Romania. *Acta Geologica Hispanica*, 35(3–4), 311–319.

Kachanoski, R. G. (1993). Estimating soil loss from changes in soil caesium-137. *Canadian Journal of Soil Science*, 73, 515–526.

Lal, R. (1988). *Soil Erosion Research Methods*. Soil and Water Conservation Society, Ankeny, IA.

Lal, R. (2000). Soil management in the developing countries. *Soil Science*, 165, 57–72.

Lal, R. (2007). Farming carbon. *Soil and Tillage Research*, 96, 1–5.

Le Cloarec, M.-F., P. Bonté, J.-M. Mouchel, I. Lefevere, and S.-L. Colbert. (2007). Distribution of ^{210}Pb, ^{7}Be and ^{137}Cs in watersheds of different scales in the Seine basin: Inventories and residence times. *Science of the Total Environment*, 375, 125–139.

Li, S., D. A. Lobb, M. J. Lindstrom, and A. Farenhorst. (2007). Tillage and water erosion on different landscapes in the northern North American Great Plains evaluated using the ^{137}Cs technique and soil erosion models. *Catena*, 70, 493–505.

Li, Y., and M. J. Lindstrom. (2001). Evaluating soil quality–soil redistribution relationships on terraces and steep hillslope. *Soil Science Society of America Journal*, 65, 1500–1508.

Li, Y., M. J. Lindstrom, J. Zhang, and J. Yang. (2000). Spatial variability of soil erosion and soil quality on hillslopes in the Chinese Loess Plateau. *Acta Geologica Hispanica*, 35(3–4), 261–270.

Li, Y., J. Poesen, J. C. Yang, B. Fu, and J. H. Zhang. (2003). Evaluating gully erosion using ^{137}Cs and Pb/^{137}Cs ratio in a reservoir catchment. *Soil and Tillage Research*, 69, 107–115.

Li, Y., Q. W. Zhang, D. C. Reicosky, L. Y. Bai, M. J. Lindstrom, and L. Li. (2006). Using ^{137}Cs and ^{210}Pbex for quantifying soil organic carbon redistribution affected by intensive tillage on steep slopes. *Soil and Tillage Research*, 86, 176–184.

Li, Y., Q. W. Zhang, D. C. Reicosky, M. J. Lindstrom, L. Y. Bai, and L. Li. (2007). Changes in soil organic carbon induced by tillage and water erosion on a steep cultivated hillslope in the Chinese Loess Plateau from 1898–1954 and 1954–1998. *Journal of Geophysical Research*, 112, G01021.

Lobb, D. A., R. G. Kachanoski, and M. H. Miller. (1995). Tillage translocation and tillage erosion on shoulder slopes landscape positions measured using ^{137}Cs as a tracer. *Canadian Journal of Soil Science*, 75, 211–218.

Loughran, R. J. (1989). The measurement of soil erosion. *Progress in Physical Geography*, 13, 216–233.

Loughran, R. J., S. J. Curtis, G. L. Elliott, B. L. Campbell, K. Kiernan, and M. G. Temple-Smith. (1992). A reconnaissance survey of soil erosion in Australia. *Proceeding of the 7th ISCO Conference*, Sydney, September 1992, Vol. 2.1, pp. 52–63.

Loughran, R. J., and G. L. Elliott. (1996). Rates of soil erosion in Australia determined by the caesium-137 technique: A national reconnaissance survey. In: *Erosion and Sediment Yield: Global and Regional Perspectives* (Eds D. E. Walling and B. W. Webb). IAHS Publication No. 236, pp. 275–282.

Loughran, R. J., G. L. Elliott, B. L. Campbell, and D. J. Shelly. (1988). Estimates of soil erosion from caesium-137 measurements in a small cultivated catchment in Australia. *Journal of Applied Radiation and Isotopes*, 39, 1153–1157.

Mabit, L., C. Bernard, and M. R. Laverdière. (2007). Assessment of erosion in the Boyer River watershed (Canada) using a GIS oriented sampling strategy and ^{137}Cs measurements. *Catena*, 71(2), 242–249.

McCarty, G. W., and J. C. Ritchie. (2002). Impact of soil movement on carbon sequestration in agricultural ecosystems. *Environmental Pollution*, 116(3), 423–430.

McDowell, R. W., and A. N. Sharpley. (2003). Phosphorus uptake and release from stream sediments. *Journal of Environmental Quality*, 32, 937–948.

McDowell, R. W., A. N. Sharpley, and G. Folmar. (2001). Phosphorus export from an agricultural watershed: Linking source and transport mechanisms. *Journal of Environmental Quality*, 30, 1587–1595.

Menzel, R. G. (1960). Transport of strontium-90 in runoff. *Science*, 131, 499–500.

Menzel, R. G., and S. J. Smith. (1984). Fertility erosion. In: *Isotopes and Radiation in Agricultural Sciences* (Eds M. F. L'Annunziata and J. O. Legg). Academic Press, London, UK.

Mou, J. (1996). Recent studies of the role of soil conservation in reducing erosion and sediment yield in the loess plateau area of the Yellow River Basin. In: *Erosion and Sediment Yield: Global and Regional Perspectives* (Eds D. E. Walling and B.W. Webb). IAHS Publication No. 236, pp. 541–548.

Nguyen, M. L. (2008). *Integrated Approaches for the Assessment of Land Use Impacts on Soil Loss/ Sediment Production and Related Environmental Problems*. Report of the Consultant's Meeting Held in Vienna, Austria, 5–7 November 2007, IAEA, Vienna, Austria.

Nouira, A., E. H. Sayouti, and M. Benmansour. (2003). Use of ^{137}Cs technique for soil erosion study in the agricultural region of Casablanca in Morocco. *Journal of Environmental Radioactivity*, 68, 11–26.

OECD. (2004). Agricultural Impacts on soil erosion and soil biodiversity: Developing indicators for policy analysis. *Proceedings from an OECD Expert Meeting*, Rome, Italy, March 2003.

Onda, Y., H. Kato, Y. Tanaka, M. Tsujimura, G. Davaa, and D. Oyunbaatar. (2007). Analysis of runoff generation and soil erosion processes by using environmental radionuclides in semiarid areas of Mongolia. *Journal of Hydrology*, 333, 124–132, doi:10.1016/j.jhydrol.2006.07.030.

Onda, Y., Y. Takenaka, M. Furuta, T. Nonoda, and Y. Hamajima. (2003). Use of ^{137}Cs for estimating soil erosion processes in a forested environment in Japan. *Transactions, Japanese Geomorphological Union*, 24(1), 13–25.

Organization for Economic Co-operation and Development, OECD. (2003). Agriculture and land conservation: Developing indicators for policy analysis. *Proceedings of Japan/ OECD Expert Meeting*, Kyoto, Japan, May 2003.

Pennock, D. J. (1998). New perspectives in the soil erosion–soil quality relationship. In: *Use of ^{137}Cs in the Study of Soil Erosion and Sedimentation*. IAEA TECDOC-1028, Vienna, pp. 13–25.

Pennock, D. J. (2000). Suitability of ^{137}Cs redistribution as an indicator of soil quality. *Acta Geologica Hispanica*, 35(3–4), 213–217.

Pennock, D. J. (2003). Terrain attributes, landform segmentation and soil redistribution. *Soil and Tillage Research*, 69, 15–26.

Pennock, D. J., and A. H. Frick. (2001). The role of field studies in landscape-scale applications of process models: An example of soil redistribution and soil organic carbon modeling using CENTURY. *Soil and Tillage Research*, 58, 183–191.

Pennock, D. J., D. S. Lemmon, and E. de Jong. (1995). Caesium-137 measured erosion rates for five parent material groups in southwestern Canada. *Canadian Journal of Soil Science*, 75, 205–210.

Pennock, D. J., and F. Zapata. (1995). *Report of the FAO/IAEA Consultants Meeting on the Use of Isotopes in Studies of Soil Erosion.* IAEA Report CT-2665. IAEA Publications, Vienna.

Phan, S. H., N. T. Binh, and T. C. Tu. (2006). The correlation between loss of ^{137}Cs and soil erosion rates in Central Highlands. *Journal of Vietnam Soil Science*, 26, 92–94.

Pimentel, D., C. Harvey, P. Resosudarmo, K. Sinclair, D. Kurz, M. McNair, S. Crist, L. Shpritz, L. Fitton, R. Saffouri, and R. Blair. (1995). Environmental and economic cost of soil erosion and conservation benefit. *Science*, 267, 1117–1123.

Porto, P., D. E. Walling, G. Callegari, and F. Catona. (2006). Using fallout ^{210}Pb measurements to estimate soil erosion in three small catchments in southern Italy. *Water, Air and Soil Pollution: Focus*, 6, 657–667.

Porto, P., D. E. Walling, V. Ferro, and C. Di Stefano. (2003a). Validating erosion rate estimates provided by caesium-137 measurements for two small forested catchments in Calabria, Southern Italy. *Land Degradation and Development*, 14, 389–408.

Porto, P., D. E. Walling, V. Tamburino, and G. Callegari. (2003b). Relating Cs-137 and soil loss from cultivated land. *Catena*, 53, 303–326.

Queralt, I., F. Zapata, and E. Garcia Agudo. (Eds) (2000). Assessment of soil erosion and sedimentation through the use of the ^{137}Cs and related techniques. *Acta Geologica Hispanica* (special issue), 35(3–4), 195–367.

Rezzoug, S., H. Michel, F. Fernex, G. Barci-Funel, and V. Barci. (2006). Evaluation of ^{137}Cs fallout from the Chernobyl accident in a forest soil and ist impact on Alpine Lake sediments, Mercantour Massif, S.E. France. *Journal of Environmental Radioactivity*, 85(2–3), 369–379.

Ritchie, J. C., and G. W. McCarty. (2003). ^{137}Cs and soil carbon in a small agricultural watershed. *Soil and Tillage Research*, 69, 45–51.

Ritchie, J. C., and J. R. McHenry. (1973). Determination of fallout Cs-137 and natural gamma-ray emitting radionuclides in sediments. *International Journal of Applied Radiation and Isotopes*, 24, 575–578.

Ritchie, J. C., and J. R. McHenry. (1990). Application of radiation fallout ^{137}Cs for measuring soil erosion and sediment accumulation rates and patterns: A review. *Journal of Environmental Quality*, 19, 215–233.

Ritchie, J. C., and J. R. McHenry. (1995). Fallout Cs-137: A tool in conservation research. *Journal of Soil and Water Conservation*, 30, 283–286.

Ritchie, J. C., and C. Ritchie. (2008). Bibliography of publications of ^{137}Cs studies related to erosion and sediment deposition. URL: http://www.ars.usda.gov/Main/docs.htm?docid = 16037

Schuller, P., A. Ellies, A. Castillo, and I. Salazar. (2003). Use of the ^{137}Cs technique to estimate tillage- and water-induced soil redistribution rates on agricultural land under different use and management in central-south Chile. *Soil and Tillage Research*, 69, 45–51.

Schuller, P., A. Iroumé, D. E. Walling, H. B. Mancilla, A. Castillo, and R. E. Trumper. (2006). Use of ^{7}Be to document soil redistribution following forest harvest operations. *Journal of Environmental Quality*, 35, 1756–1763.

Schuller, P., A. Sepulveda, A. Trumper, and A. Castillo. (2000). Application of the ^{137}Cs technique to quantify soil redistribution in Paleohumults from Central-South Chile. *Acta Geologica Hispanica*, 35(3–4), 285–289.

Schuller, P., D. E. Walling, A. Sepulveda, A. Castillo, and I. Pino. (2007). Changes in soil erosion associated with the shift from conventional tillage to a no-tillage system, documented using Cs-137 measurements. *Soil and Tillage Research*, 94(1), 183–192.

Schuller, P., D. E. Walling, A. Sepulveda, R. E. Trumper, J. L. Rouanet, I. Pino, and A. Castillo. (2004). Use of Cs-137 measurements to estimate changes in soil erosion rates associated with changes in soil management practices on cultivated land. *Applied Radiation and Isotopes*, 60, 759–766.

Schumacher, J. A., T. C. Kaspar, J. C. Ritchie, T. E. Schumacher, D. L. Karlen, E. R. Venteris, G. M. McCarty, T. S. Colvin, D. B. Jaynes, M. J. Lindstrom, and T. E. Fenton. (2005). Identifying spatial patterns of erosion for use in precision conservation. *Journal of Soil and Water Conservation*, 60(6), 355–361.

Sepúlveda, A., P. Schuller, and D. E. Walling. (2007). Use of 7Be to document erosion associated with a short period of extreme rainfall. *Journal of Environmental Radioactivity*, doi:10.1016-j.jenvrad.2007.06.010.

Sogon, S., M.-J. Penven, P. Bonte, and T. Muxart. (1999). Estimation of sediment yield and soil loss using suspended sediment load and ^{137}Cs measurements. *Hydrobiologia*, 410, 251–261.

Tamura, T. (1964). Selective sorption reactions of ^{137}Cs with soil minerals. *Nuclear Safety*, 5, 262–268.

Theocharopoulos, S. P., H. Florou, P. Kritidis, D. Belis, F. Tsoloucha, M. Christou, P. Kouloumbis, and T. Nikolaou. (2000). Use of the ^{137}Cs isotopic technique in soil erosion studies in Central Greece. *Acta Geologica Hispanica*, 35(3–4), 301–309.

Turner, J. (2000). *Report of the Advisory Group Meeting on Sediment Tracing (Fingerprinting) by Nuclear Techniques and Their Application to the Planning and Design of Erosion and Sedimentation Remediation Strategies and the Assessment of Their Effectiveness With Emphasis on Dam Sustainability*. IAEA Report AG-1090. IAEA Publications, Vienna.

United Nations Environment Programme (UNEP). (2000). *Global Environment Outlook 2000*. Earthscan Publications Limited, London, UK.

Van Pelt, R. S., T. M. Zobeck, J. C. Ritchie, and T. E. Gill. (2007). Validating the use of ^{137}Cs measurements to estimate rates of soil redistribution by wind. *Catena*, 70(3), 455–464.

Venteris, E. R., G. W. McCarty, J. C. Ritchie, and T. J. Gish. (2004). Soil organic carbon and soil redistribution: Influence of site history and site variables. *Soil Science*, 169(11), 787–795.

Wallbrink, P. J., W. Blake, S. Doerr, R. Shakesby, G. Humphreys, and P. English. (2005). Using tracer based sediment budgets to assess redistribution of soil and organic material after severe bush fires. *Proceedings of the Symposium held at Foz do Iguacu, Brazil, April 2005*, IAHS Publication 292 (Eds A. J. Horowitz and D. E. Walling), IAHS Press Wallingford, UK, pp. 223–230.

Wallbrink, P. J., C. E. Martin, and C. J. Wilson. (2003). Quantifying the delivery of sediment sediment-P and fertilizer-P from forested, cultivated and pasture areas at the land use and catchment scale using fallout radionuclides and geochemistry. *Soil and Tillage Research*, 69, 53–68.

Wallbrink, P. J., and A. S. Murray. (1993). The use of fallout radionuclides as indicators of erosion processes. *Hydrological Processes*, 7, 297–304.

Wallbrink, P. J., and A. S. Murray. (1996). Distribution and variability of ^{7}Be in soils under different surface cover conditions and its potential for describing soil redistribution processes. *Water Resources Research*, 32, 467–476.

Wallbrink, P. J., A. S. Murray, and J. M. Olley. (1999). Relating suspended sediment to its original soil depth using fallout radionuclides. *Soil Science Society of America Journal*, 63, 369–378.

Wallbrink, P. J., B. P. Roddy, and J. M. Olley. (2002). A tracer budget quantifying soil redistribution on hillslopes after forest harvesting. *Catena*, 47, 179–201.

Walling, D. E. (1998). Use of ^{137}Cs and other fallout radionuclides in soil erosion investigations: Progress, problems and prospects. In: *Use of* ^{137}Cs *in the Study of Soil Erosion and Sedimentation*. IAEA TECDOC-1028, Vienna, pp. 39–62.

Walling, D. E. (2001). Recent advances in the use of environmental radionuclides in soil erosion investigations. In: *Nuclear Techniques in Integrated Plant Nutrient, Water and Soil Management*. Proceedings of FAO/IAEA International Symposium, Vienna, October 2000, IAEA C&S Papers Series 11P, pp. 279–301.

Walling, D. E. (2005). Tracing suspended sediment sources in catchments and river systems. *Science of the Total Environment*, 344, 159–184.

Walling, D. E., and S. B. Bradley. (1988). *The Use of Caesium-137 Measurements to Investigate Sediment Delivery From Cultivated Areas in Devon*, IAHS Publication 174. IAHS Press, Wallingford, CT, pp. 325–335.

Walling, D. E., A. L. Collins, P. A. Jones, G. J. L. Leeks, and G. Old. (2006). Establishing fine-grained sediment budgets for the Pang and Lambourn LOCAR study catchments. *Journal of Hydrology*, 330, 126–141.

Walling, D. E., A. L. Collins, and H. M. Sichingabula. (2003a). Using unsupported ^{210}Pb measurements to investigate soil erosion and sediment delivery in a small Zambian catchment. *Geomorphology*, 52, 193–213.

Walling, D. E., and Q. He. (1999a). Improved models for estimating soil erosion rates from cesium-137 measurements. *Journal of Environmental Quality*, 28, 611–622.

Walling, D. E., and Q. He. (1999b). Use of fallout ^{210}Pb measurements to estimate soil erosion on cultivated land. *Soil Science Society of America Journal*, 63, 1404–1412.

Walling, D. E., and Q. He. (2000). *The global distribution of bomb-derived ^{137}Cs reference inventories*. Final Report on IAEA Technical Contracts No. 10361/R0-R1. University of Exeter, Exeter, UK.

Walling, D. E., Q. He, and P. G. Appleby. (2002). Conversion models for use in soil-erosion, soil-redistribution and sedimentation investigations. In: *Handbook for the Assessment of Soil Erosion and Sedimentation Using Environmental Radionuclides* (Ed. F. Zapata). Kluwer Publications, Dordrecht, The Netherlands, pp. 111–164.

Walling, D. E., Q. He, and P. J. Whelan. (2003b). Using ^{137}Cs measurements to validate the application of the AGNPS and ANSWERS erosion and sediment yield models in two small Devon catchments. *Soil and Tillage Research*, 69, 27–44.

Walling, D. E., P. N. Owens, and G. J. L. Leeks. (1998). The role of channel and floodplain storage in the suspended sediment budget of the River Use, Yorkshire, UK. *Geomorphology*, 22, 225–242.

Walling, D. E., and T. A. Quine. (1991). The use of caesium-137 measurements to investigate soil erosion in arable fields in the UK: Potential applications and limitations. *Journal of Soil Science*, 42, 147–162.

Walling, D. E., and T. A. Quine. (1993). *Use of ^{137}Cs as a Tracer of Erosion and Sedimentation: Handbook for the Application of the ^{137}Cs Technique*. Report to the UK Overseas Development Administration. Exeter, UK.

Walling, D. E., and T. A. Quine. (1995). The use of fallout radionuclide measurements in soil erosion investigations. In: *Nuclear Techniques in Soil–Plant Studies for Sustainable Agriculture and Environmental Preservation*. Proceedings of International FAO/IAEA Symposium, Vienna, October 1994, IAEA Proceeding Series STI/PUB/947. Vienna, pp. 597–619.

Yang, M. Y., D. E. Walling, J. L. Tian, and P. L. Liu. (2006). Partitioning the contributions of sheet and rill erosion using beryllium-7 and cesium-137. *Soil Science Society of America Journal*, 70(5), 1575–1590.

Zapata, F. (2001a). *Final Report of the Co-ordinated Research Project on Assessment of Soil Erosion Through the Use of the Cs-137 and Related Techniques as a Basis for Soil Conservation, Sustainable Agricultural Production and Environmental Protection*. IAEA-311-D1-RC-629.4. IAEA, Vienna.

Zapata, F. (2001b). *Report of the FAO/IAEA Consultants Meeting on Assessment of Soil Conservation Technologies for Sustainable Agricultural Production.* IAEA Report-311-01CT01553. IAEA, Vienna.

Zapata, F. (Ed.) (2002). *Handbook for the Assessment of Soil Erosion and Sedimentation Using Environmental Radionuclides.* Kluwer Publications, Dordrecht, The Netherlands.

Zapata, F. (Ed.) (2003). Field application of the Cs-137 technique in soil erosion and sedimentation. *Soil and Tillage Research* (special issue), 69, pp. 1–153.

Zapata, F. (2007). Use of environmental radionuclides to monitor soil erosion and sedimentation in the field, landscape and catchment level before, during and after implementation of soil conservation measures. In: *Monitoring and Evaluation of Soil Conservation and Watershed Development Projects* (Eds J. de Graaff, et al.). Science Publishers, USA, pp. 301–317.

Zapata, F., and E. Garcia Agudo. (2000). Future prospects for the ^{137}Cs technique for estimating soil erosion and sedimentation rates. *Acta Geologica Hispanica*, 35(3–4), 197–205.

Zapata, F., E. Garcia Agudo, C. Hera, K. Rozanski, and K. Froehlich. (1995). Use of nuclear techniques in soil erosion and siltation studies. In: *Nuclear Techniques in Soil–Plant Studies for Sustainable Agriculture and Environmental Preservation.* Proceedings of International FAO/IAEA Symposium, Vienna, October 1994, IAEA Proceedings Series STI/PUB/947. IAEA, Vienna, pp. 631–642.

Zapata, F., E. Garcia Agudo, J. C. Ritchie, and P. G. Appebly. (2002). Introduction. In: *Handbook for the Assessment of Soil Erosion and Sedimentation Using Environmental Radionuclides* (Ed. F. Zapata). Kluwer Publications, Dordrecht, The Netherlands, pp. 1–13.

Zapata, F., and Y. Li. (2007). *Report of the Fourth and Final Research Coordination Meeting of the FAO/IAEA Coordinated Research Project "Assess the Effectiveness of Soil Conservation Techniques for Sustainable Watershed Management Using Fallout Radionuclides".* IAEA-311-D1-RC-888. IAEA, Vienna.

Zhang, X. B., Y. Q. Qi, D. E. Walling, X. He, A. B. Wen, and J. Fu. (2006). A preliminary assessment of the potential for using ^{210}Pbex measurement to estimate soil redistribution rates on cultivated slopes in the Sichuan Hilly Basin of China. *Catena*, 68, 1–9.

Zhang, X. B., T. A. Quine, D. E. Walling, and A. B. Wen. (2000). A study of soil erosion on a steep cultivated slope in the Mt. Gongga region near Luding, Sichuan, China using the ^{137}Cs technique. *Acta Geologica Hispanica*, 35(3–4), 229–237.

Zhang, X. B., and D. E. Walling. (2005). Characterizing land surface erosion from cesium-137 profiles in lake and reservoir sediments. *Journal of Environmental Quality*, 34, 514–523.

Zhang, X. B., Y. Zhang, A. B. Wen, and M. Feng. (2003). Assessment of soil erosion losses on cultivated land by using the ^{137}Cs technique in the Upper Yangtze River Basin of China. *Soil and Tillage Research*, 69, 99–105.

Zheng, J.-J., X. He, D. E. Walling, X. Zhang, D. Flanagan, and Y. Q. Qi. (2007). Assessing soil erosion rates on manually-tilled hillslopes in the Sichuan hilly basing using ^{137}Cs and ^{210}Pbex measurements. *Pedosphere*, 17(3), 273–283.

CHAPTER 8

Isotopic Tracers in Climatology

Prosenjit Ghosh[1,*], S.K. Bhattacharya[2] *and* Klaus Froehlich[3]

Contents

1. Introduction

The climate of the Earth is the product of differential latitudinal solar heating. As a result, energy is constantly transferred from the equator (relatively hot) towards the poles (relatively cold) through two transporting agencies: the atmosphere and the oceans. The atmosphere responds to

*Corresponding author. Tel.: +91-080-2293 2599 (O), +91-080-2356 1730 (R); Fax: +91-080-23600865
E-mail address: pghosh@caos.iisc.ernet.in

[1] Centre for Earth Sciences, Indian Institute of Science, Bangalore, 560 012, India
[2] Physical Research Laboratory, Navarangpura, Ahmedabad, 380 009, India
[3] Viktor-Wittner-Gasse 36/7, A-1220 Vienna, Austria

Radioactivity in the Environment, Volume 16
ISSN 1569-4860, DOI 10.1016/S1569-4860(09)01608-8

internal and external forces in a matter of days, months or a few years. The oceans, however, have a longer response time. The surface ocean can change over months to a few years, whereas the deep oceans take decades to centuries. From a physical point of view, in terms of volume, heat capacity and inertia, the deep ocean is the only viable agency for driving and sustaining long-term climate change on centennial to millennial timescales.

Another important factor in climate issues is the change in abundance of greenhouse gases in the atmosphere. The growth of greenhouse gases like CO_2, water vapour and methane in the atmosphere has a strong impact on temperature and climate. Here also, the oceans have a major role to play in determining the CO_2 content of the atmosphere. There are four regulating processes governing the CO_2 concentration in the atmosphere:

(1) Ocean temperature, which controls the amount of carbon dioxide dissolved in the ocean water;
(2) Ocean productivity, as the CO_2 is scavenged by marine organisms to form organic matter via photosynthesis;
(3) Balance between the down-welling and upwelling of ocean water.
(4) Amount of calcium carbonate produced in the oceans by living organisms.

Corals, planktonic foraminifera and nannofossils form carbonate shells and thus regulate the amount of atmospheric carbon dioxide that can remain dissolved in the oceans. Surface water contains both dissolved carbon dioxide and organic matter. When this is down-welled, the carbon gets transferred to the deep ocean. Most importantly, most deep and intermediate water formation occurs at high latitudes where the sinking surface water is extremely cold, and cold water contains more dissolved gases than warm water. In addition, oxygen is also dissolved in the water and subsequently reacts with organic matter to liberate carbon dioxide. Hence, when the water is finally up-welled, it is extremely rich in carbon dioxide.

Since the processes of oceanic heat transfer and carbon dioxide regulation occur on decadal to century scale time periods, historical records are too short to provide any information on the ocean system prior to human intervention. Hence, we must turn to marine sediments to provide information about ocean-driven climate change. Such archives often contain a continuous record on a variety of timescales, especially suitable for reconstruction of past climate prior to anthropogenic influence. Another source of information for deciphering past climate change with high resolution is provided by ice core samples. Ice accumulated in the colder region of the planet preserves a continuous record of change in precipitation, temperature or trace gas composition of the past atmosphere and serves as a useful climate proxy. Information about temperature and precipitation can be obtained from isotopic analyses of ice samples, whereas trapped gases in the ice core can be used for estimating the composition of the atmosphere during periods of ice formation. Apart from ice and sea sediments, records from continental lake sediments, annual bands in coral and tree rings, seasonal bands in varves and speleothems also act as important proxies for climate change.

During the last few decades, our ability to reconstruct past climate has improved considerably because of the development of new technology. Key ingredients of this progress include high-precision determination of trace gas concentrations and stable isotope ratios in samples of air, water, rock and soil using chromatographic and mass spectrometric techniques. Stable isotopes are useful in several important aspects of global change research. Complementary to stable isotopes, environmental radionuclides, such as ^{14}C, ^{36}Cl and uranium series radionuclides, are used to provide the timescales of the climate records. A major objective of climate studies is to understand the contemporary climate change based on studies of atmospheric air and water from a global network of ground stations as well as from firn or snow sampled in polar regions. A second goal is to reconstruct past environmental conditions from climate proxies when a compound with stable isotopes of C, O, N or H is laid down and preserved over time in reservoirs like groundwater, ice, lake and ocean sediments, corals, soils, speleothems and tree rings.

Specific objectives of studying past climate are

- to reconstruct the whole range of natural climatic variability, which includes abrupt changes (i.e. occurring on a timescale comparable to a human life span) and beyond the period covered by instrumental observations,
- to test models using past conditions and compare the results with observations in order to understand how the climate system works.

2. Role of Environmental Isotopes in Understanding Climate Changes

2.1. Stable isotopes

Atmospheric CO_2 provides a link between biological, physical and anthropogenic processes. Carbon and oxygen are exchanged among various reservoirs like the atmosphere, the oceans, the terrestrial biosphere and, on a longer timescale, the sediments and sedimentary rocks. A serious present concern is about carbon dioxide due to its large anthropogenic contribution, which includes fossil fuel combustion, deforestation, agriculture and cement production (Andres et al., 1999). Each year, approximately

120 Gt C are exchanged (in each direction) between the atmosphere and the terrestrial ecosystem, and another 90 Gt are exchanged between the atmosphere and ocean (Watson and Noble, 2002). The anthropogenic contribution is about 6 Gt yr^{-1} which goes directly to the atmosphere. Currently, both the oceans and the terrestrial system show a net uptake of carbon which offsets this contribution to some extent, but the budget is not wholly accounted for and apparently there is a 'missing carbon sink'. The oxygen and carbon isotopic compositions of individual components in these systems, in particular air CO_2, provide a powerful tool for quantifying the contribution of different components to ecosystem exchange. When this is used in conjunction with concentration or flux measurements, further insight can be gained into the sources and sinks of CO_2 in the ecosystem (Keeling, 1961; Keeling et al., 1995).

Plant photosynthesis discriminates against ^{13}C, that is plant carbon tends to have less ^{13}C than the CO_2 from which it is formed (Figure 1). This discrimination provides a tool for interpreting changes in $\delta^{13}C$ of atmospheric CO_2 which is generally done in one of three ways:

- First, it can be used to partition CO_2 fluxes between the land and the oceans and thus help to constrain the location and processes involved in creating the 'missing carbon sink' (Sundquist, 1993). This can be done because there is little carbon isotope discrimination associated with exchange of CO_2 between the ocean and atmosphere. Consequently, if the sink is in the 'land', changes in atmospheric CO_2 concentrations will be accompanied by large changes in $\delta^{13}C$, whereas if the sink is due to absorption of CO_2 by the ocean, changes in CO_2 concentrations will have little effect on $\delta^{13}C$.
- Second, discrimination by C_3 plants is influenced by environmental factors such as availability of light, water and nutrients, and, consequently, provides a means of interpreting changes in $\delta^{13}C$ of atmospheric CO_2 in terms of environmental changes, for example drought, El Nino and global warming.
- Third, since $\delta^{13}C$ of atmospheric CO_2 has changed over time due to addition of ^{13}C-depleted fossil fuel, carbon isotope ratios of respired CO_2 differ slightly from those of photosynthesis. This is termed the 'isotope disequilibrium effect' and may provide a means of determining the relative importance of respiration and photosynthesis in causing changes in atmospheric CO_2.

The majority of land plants employ the C_3 photosynthetic pathway which results in stored carbon being approximately 18‰ depleted in ^{13}C with respect to the atmosphere (Farquhar et al., 1982). Around 21% of modern carbon uptake by plants occurs via C_4 photosynthesis where ^{13}C discrimination is less (about 4‰), so that the global mean discrimination is slightly smaller than the pure C_3 value, around 15‰ (Lloyd and Farquhar,

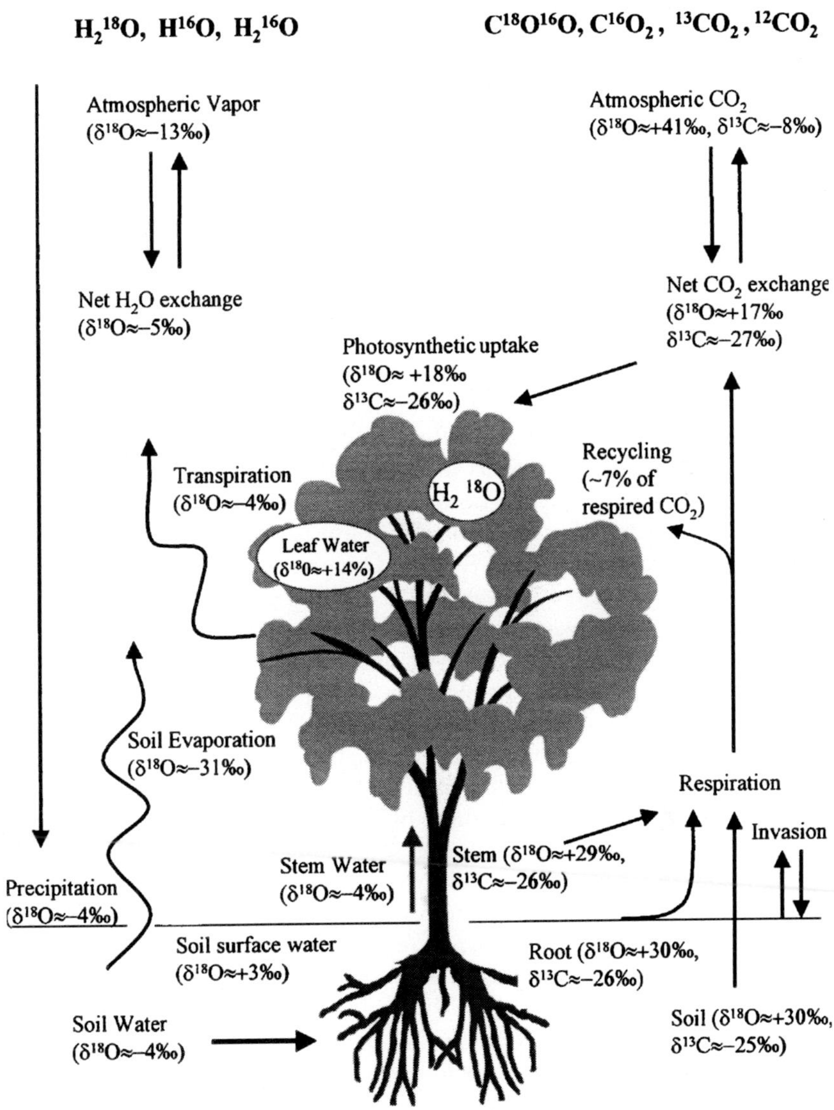

Figure 1 Isotopic composition of C, O and H pools in the plant ecosystem. The values are general averages (modified from Yakir and Sternberg, 2000, with permission from elsevier) and vary considerably with geographical location and environmental conditions. $\delta^{18}O$ and δD values are given relative to VSMOW and $\delta^{13}C$ values relative to VPDB.

1994). Release of CO_2 from fossil fuel involves combustion of coal and petroleum, which are mainly products of C_3 photosynthesis but modified slightly by fractionation processes (Tan, 1981). Since isotopic fractionations during fossil fuel combustion, biomass burning or soil respiration are negligible, both industrial and land-use activities impose similar changes on the $^{13}C/^{12}C$ ratio of the global atmosphere.

In contrast to photosynthesis, diffusion across the air–sea interface results in much less fractionation of the $^{13}C/^{12}C$ ratio (Mook et al., 1974). The dissolved CO_2 in general is assumed to be in isotopic equilibrium with dissolved inorganic carbon (DIC) which constitutes the bulk (~99%) of the mixed layer carbon. For the typical open ocean, primary production marine photosynthesis has minimal impact, globally, on atmospheric carbon isotope values (Francois et al., 1974). Thus, atmospheric $^{13}C/^{12}C$ changes can be used to partition the uptake of fossil fuel carbon between oceanic and terrestrial reservoirs (Pearman and Hyson, 1986). They can also be used in studies of natural variability in the carbon cycle (Keeling et al., 1989) and in calibrating global carbon budget models (Heimann and Maier-Reimer, 1996). Figure 1 summarises the influence of biospheric processes on stable isotope ratios of water and CO_2 in individual compartments.

Unlike carbon, the oxygen isotope ratio ($^{18}O/^{16}O$) of atmospheric CO_2 is primarily determined by exchange with leaf water, soil water and surface sea water (Francey and Tans, 1987). Oxygen isotope exchange occurs under equilibrium conditions inside the chloroplast of a leaf cell. The equilibrium exchange reaction occurs in the presence of carbonic anhydrase acting as a potent catalyst for the hydration reaction. The chloroplast water is enriched in ^{18}O relative to soil water owing to evaporation from leaves where $H_2{}^{16}O$ evaporates preferentially relative to $H_2{}^{18}O$. The enrichment of chloroplast water is sensitive to relative humidity and temperature, both of which are highly variable in different regions of the globe. Oxygen isotopes of CO_2 exchange with those of leaf water and imbibe the signature of evaporation; CO_2 is released during soil respiration and slowly diffuses to the atmosphere. Oxygen atom exchange with sea water occurs through the exchange of CO_2 molecules across the air–sea interface. The net effect of these ecosystem-specific exchange reactions can be observed in the CO_2 of regional atmospheric samples when sampling is made with high temporal resolution.

The oxygen isotopic composition of soil and leaf water can vary considerably. Soil water tends to follow the composition of precipitation, which is progressively depleted in ^{18}O relative to sea water as it moves towards high latitudes and towards the interior of a continent. In contrast to higher latitudes, there is no correlation between surface temperature and $\delta^{18}O$ values of precipitation in the tropics (Rozanski et al., 1993a, 1993b). Tropical regions are characterised by converging air masses that are forced to move vertically rather than horizontally. As a result, they cool predominantly by convection in atmospheric towers, while surface temperature gradients remain negligible. Although temperature does not correlate with $\delta^{18}O$ in the tropics, a negative correlation has been observed between the amount of rainfall and $\delta^{18}O$ values at tropical island locations (Rozanski et al., 1993a).

In a changing globe, the isotopic composition of palaeo-water and palaeo-atmosphere CO_2 can be obtained directly from ice core samples and trapped inclusions within ice cores. Indirectly, estimating the isotopic composition of past precipitation and past atmospheres can be made from analysis of proxy records like the skeletal remains of animals, lake sediments and soil minerals that have formed in equilibrium with their surrounding environment.

2.2. Radionuclides

Radionuclides occurring naturally in palaeo-climatic archives are used to provide the timescales necessary for establishing the chronology of palaeo-climatic changes. There are several timescales of climatic records. On a long timescale of ten thousand to hundreds of thousands of years, oceanic sediments, polar ice and cave deposits preserve signatures of major climate changes (glacial–interglacial cycles) driven mainly by the cyclic variations of the orbital parameters of the Earth system. An intermediate timescale of a 1,000 to more than 10,000 years includes aperiodic events of significant amplitude of climatic parameters but with shorter duration (e.g. Younger Dryas and Little Ice Age). Finally, a short timescale of 10–1,000 years is accessible by various archives such as lake sediments, reef corals, polar and continental glaciers. This time period covers Holocene climate events including major floods, periods of regional drought or volcanic eruptions, as well as man-induced environmental changes (Rozanski et al., 1993b). Combining different dating methods with radionuclides of different half-lives, the whole time period of palaeo-climatic archives can be covered.

In the following section, various radiometric dating methods used for archives such as lacustrine and marine sediments, ice cores and speleothems are reviewed. The characterisation of these radionuclides in terms of origin (production) and distribution is given in Chapter 1, the analytical techniques are described in Chapter 9 and Chapters 2, 5, 6 and 7 address the principles of radiometric dating techniques. Here, specific aspects of these radiometric methods for dating in the selected palaeo-climatic archives are reviewed.

2.2.1. ^{210}Pb and ^{137}Cs

The ^{238}U decay series radionuclide ^{210}Pb and the anthropogenic radionuclide ^{137}Cs are most suitable to date sediments up to ages of a few hundred years. While ^{210}Pb (half-life 22 years) allows continuous dating from 0 to about 200 years, ^{137}Cs (half-life 30 years) can be used as a time-marker for the period around 1963 when the atmospheric ^{137}Cs concentration reached its maximum (Erten, 1997). It is not a straightforward procedure to calculate sediment ages from ^{210}Pb profiles because the initial amount of ^{210}Pb is difficult to estimate, and after deposition, the sediment can be re-worked by biological activities. Further details about ^{210}Pb and ^{137}Cs dating are given in Chapter 7 (Section 2.1). ^{210}Pb has also been used to date coastal marine sediments, growth rates of corals and peat, and ice cores (e.g. Suzuki et al., 1996). A recent study of ^{210}Pb accumulation at the Summit of Greenland revealed that the assumption of constant ^{210}Pb, usually made in ^{210}Pb dating, may not be valid under certain conditions (Dibb, 2002).

2.2.2. ^{32}Si

Cosmogenic ^{32}Si (half-life 140 years) is an excellent candidate for provision of a timescale in the range from 100 to 1,000 years and, thus, to fill the dating gap between the chronologies using the shorter-lived ^{210}Pb and the longer-lived ^{14}C. For sediment dating, biogenic silica is extracted from the sediment. The ^{32}Si measuring technique is described by Morgenstern et al. (2001). An example of sediment dating using ^{32}Si together with other radiometric dating tools has been given by Suckow et al. (2001). Morgenstern et al. (2000) published the first ^{32}Si accelerator mass spectrometry (AMS) measurements on glacier ice and snow. Their results show that about 1 kg of ice is sufficient to cover a time period of seven half-lives, corresponding to a dating limit of 1,000 years. An age of more than 800 years was obtained for ice at the bottom of the Fox Glacier (New Zealand, 43°S).

2.2.3. ^{14}C

Radiocarbon (^{14}C) is the most frequently used radionuclide in palaeo-climate dating. Its half-life of 5,730 years allows dating in the range from less than 1,000 years to about 50,000 years, a range of major interest in palaeo-climatic studies. The material used for ^{14}C dating includes calcite of lacustrine and marine sediments, corals (e.g. Bard et al., 1998; Paterne et al., 2004) and speleothems (e.g. Pazdur et al., 1999). For sediment dating, ^{14}C is also measured in organic material such as seeds, sticks and gyttja.

The ^{14}C ages of calcite have to be corrected for the so-called reservoir or hard-water effect. The bicarbonate reservoir from which the calcite is formed consists of a mixture between dissolved atmospheric (^{14}C-containing) CO_2 and old (^{14}C-free) carbon dissolved from carbonate of rocks and sediments. The reservoir effect is represented by the deviation of the calcite-^{14}C initial content from the atmospheric ^{14}C content. Hence, the apparent ^{14}C age of calcite may be up to hundreds of years too old. For example Rea and Colman (1995) found for sediments of the Lakes Michigan and Huron a reservoir effect of about 250 years and 440 years, respectively. To quantify the extent of ^{14}C reservoir effects, Hall and Henderson (2001) dated carbonate of lake sediments both by the ^{14}C

method and the U/Th isochrone technique. For Antarctic lakes, ^{14}C reservoir ages of up to 3–4 ka have been found. Precise ^{14}C dating requires separating the detrital fraction in sediments from the bulk carbon source to avoid a shift towards higher apparent ^{14}C ages. Further problems that need careful consideration especially in ^{14}C dating of Arctic marine sediments include corrections for marine reservoir variation, vital effects between different marine species, seasonal differences in marine phytoplankton ^{14}C activity and re-suspension of organic particulates (Domack, 2003).

In general, due to the large carbon reservoir of the oceans, radiocarbon ages of oceanic samples (e.g. shells, fish, marine mammals, etc.) are up to several hundred years older than corresponding terrestrial samples. Therefore, an age correction is necessary to compare the ages of marine and terrestrial samples. The difference between the ^{14}C age of a marine sample and the ^{14}C age of an atmospheric sample is called the reservoir ^{14}C age, $R(t)$, which is a function of time (Stuiver and Braziunas, 1993). Because of the complexities of ocean circulation and mixing processes, this ^{14}C age correction varies with the location, which is expressed by ΔR. The ΔR value of a given location can be obtained from the marine reservoir correction database (http://intcal.qub.ac.uk/marine/). Typical ^{14}C reservoir ages in corals are about 300 years (corals from Mururoa and Tahiti) and 400 years (corals from Barbados and New Guinea) (Bard et al., 1998).

Coupled ^{14}C and $^{230}Th/U$ dating (see below) of corals allows reconstruction of the temporal variations of the atmospheric $^{14}C/^{12}C$ ratio (Bard et al., 1998; Paterne et al., 2004), which have to be known for precise ^{14}C dating of palaeo-climatic changes. Paterne et al. (2004) found in two of the numerous analysed corals an extremely high variability of the $^{14}C/^{12}C$ ratio at about 12 and 15 ka ago. This finding has been suggested to be an indication of rapid changes in surface ^{14}C of the region where the corals have been collected. Highly elevated ^{14}C values have also been measured in a stalagmite from a cave in Barbados. Due to simultaneous $^{230}Th/U$ dating of the stalagmite, a broad ^{14}C peak occurred between 45 and 33 ka ago superimposed by several rapid excursions, the largest of which occurred between 44.3 and 43.3 ka. From 26 to 11 ka, the atmospheric ^{14}C value decreased by about 60%, modulated by numerous minor excursions. The major features of this stalagmite record cannot be caused by solar or terrestrial magnetic field modulation alone but by substantial fluctuations in the carbon cycle (Beck et al., 2001).

The invention of AMS (Chapter 9) also paved the way for ^{14}C dating of polar ice. Andree et al. (1984) were the first to measure the ^{14}C of CO_2 entrapped in gas (air) bubbles within the ice using AMS. Modern approaches allow determination of the total gas volume, CO_2 concentration, and $^{14}C/^{12}C$ and $^{13}C/^{12}C$ ratios after extraction of the gaseous component by sublimation of the ice within a vacuum chamber. The dating accuracy reached by this technique is between about 100 and 300 years for ice ages in the range between about 10 and 20 ka (Moorman et al., 1998). High-precision measurements enable the dating of climate variations on decadal timescales (Etheridge et al., 1996). One of the important factors controlling the time resolution is the smearing of atmospheric signals by diffusion of air in the firn and by the transition from firn to ice bubbles (Levchenko et al., 1996).

2.2.4. ^{230}Th/^{234}U dating

The ^{230}Th/^{234}U dating method is very suitable for archives that represent closed systems with a known initial amount of uranium and having no external source of thorium. Archives in which this dating method is most frequently used include calcites of speleothems, lake and marine sediments and corals.

Calcite speleothems are formed by carbonate-saturated groundwater entering caves. Slow outgassing of dissolved CO_2 lowers the carbonate solubility in these drip waters and leads to deposition of crystalline calcite together with uranium which in groundwater occurs mainly in the form of uranyl(carbonate) complexes. Because of the low solubility of thorium in groundwater, its initial content in speleothems is usually very low or negligible. Similarly, the initial amount of thorium in corals is very small or negligible. Living corals absorb uranium from sea water but no (or a negligible amount of) thorium because of the very low thorium solubility in sea water. Thus, ^{230}Th dating of corals provides an alternative to the ^{14}C dating of marine carbonate. It should, however, be noted that surface corals are limited to the tropical ocean.

Assuming that at the time $t = 0$ of formation of a calcite sample, the ^{230}Th content was zero, the following expression can be derived (see Chapter 2, Section 2.3):

$$\frac{^{230}Th}{^{234}U} = \frac{^{234}U}{^{238}U}(1 - e^{-\lambda_{230}t}) + \frac{\lambda_{230}}{\lambda_{230} - \lambda_{234}}\left(1 - \frac{^{234}U}{^{238}U}\right)(1 - e^{-(\lambda_{230}-\lambda_{234})t})$$

where ^{230}Th, ^{234}U and ^{238}U signify the activity concentrations of these radionuclides in the given sample and the subscripts '230', '234' and '238' of the decay constant λ are related to the radionuclide with the corresponding atomic mass number. There is an improved version of the ^{230}Th method that also considers addition of ^{234}U to the calcite sample through decay of initial ^{238}U as well as a possible initial incorporation of ^{230}Th (Edwards et al., 1986). An excellent example of the application of this method was presented by Ludwig et al. (1992) who dated the longest speleothem climate record thus far, a calcite vein from the Devils Hole cave, Nevada. For samples with ages from about 50 to 150 ka, a measuring precision of 1 ka was reached, and the oldest sample of 560 ka was dated with a precision of 50 ka. In general, the dating precision decreases with increasing age and decreasing uranium content of the sample. For Holocene ages between 8 and 14 ka, Bard et al. (1998) have reached an analytical precision between 30 and 60 years.

2.2.5. Other radiometric dating techniques

2.2.5.1. ^{226}Ra/Ba. Recently, the potential of using ^{226}Ra/Ba ratios as an alternative chronometer for Holocene marine mollusc shells has been explored by Staubwasser et al. (2004). The ratio between ^{226}Ra and Ba has been used because of the very similar geochemical behaviour of radium and barium, which makes the ratio rather robust against partial losses of these elements during fossilisation of the molluscs. The study has demonstrated that, when corrected for surface contamination, the ^{226}Ra/Ba ratios of fossil marine mollusc shells decreases with age and may potentially be used for dating of Holocene mollusc shells up to ages of a few thousand years. It is expected that the ^{226}Ra/Ba chronometer may become a useful dating tool for the Late Holocene especially in oceanic regions where ^{14}C dating is of limited use because of insufficient knowledge of the ^{14}C reservoir effect.

2.2.6. *In situ*-produced cosmogenic radionuclides

In surface rocks exposed to the atmosphere, radionuclides such as ^{10}Be, ^{14}C, ^{26}Al and ^{36}Cl are produced by interaction of the secondary component of the cosmic radiation with rock materials. The radionuclide content of the rock increases with duration of the exposure to the cosmic radiation and decreases with depth in the rock due to absorption of the cosmic radiation, surface erosion and radioactive decay. The half-life of the radionuclide defines the dating range. Such *in situ* exposure ages provide the timing of the end of a glacial phase. Exposure ages derived from cosmogenic radionuclides in boulders from the crests of moraines represent the time elapsed, since the boulder stabilised or emerged from the sediment. Moraines trace the margins of past glaciers and thus provide the chronology of glacier fluctuations due to past temperature and precipitation changes in the given region.

A detailed description of the fundamentals of exposure age dating is given in Chapter 6 (Section 4.3) of this volume, focusing on exposure dating of both boulders and bedrock surfaces for purposes, amongst others, of constraining the timing of past ice margin fluctuations.

3. Environmental Stable Isotopes and Radionuclides in Palaeo-Climatic Archives

In the following section, we describe palaeo-climatic aspects of changes in stable isotope ratios of O, H and C in natural environments. The timing of these changes is provided by radiometric dating (Section 2.2). In particular, the following archives will be discussed:

(1) Sea sediments,
(2) Ice core samples,
(3) Phanerozoic soil carbonate and organic matter,
(4) Himalayan foothill sediments,
(5) Modern day precipitation.

3.1. Sea sediments

Much of our quantitative knowledge of climate fluctuation in the past has been derived from records preserved in ice sheets and deep-sea sediments. The main palaeo-climatic indicator has been the abundance ratio of the oxygen isotopes ($^{18}O/^{16}O$) in water and carbonates. The difference in physical properties of these two isotopes caused by the mass difference leads to temperature-dependent isotopic fractionations during phase changes and chemical reactions. This allows the isotopic ratio to be used as a tracer for studying: (1) climate and the hydrological cycle, (2) carbonate precipitation and dissolution and (3) photosynthesis and related processes. Products formed as a result of interaction between water and its surrounding and preserved over time can be studied for their isotopic signature to derive a proxy record of past climate change. For example stable isotope analysis of carbonates enables us to study the conditions of deposition and estimate the temperature of formation. This technique has been used extensively for palaeo-climatic reconstruction since the early 1950s. Analyses of stable oxygen and carbon isotopes from foraminiferal shells have played a pivotal role in palaeoceanography since the pioneering efforts of Emiliani (1955) who built on the works of Urey (1947); and Epstein et al. (1953) interpreted the isotopic record from deep-sea cores in terms of a series of Pleistocene climate/temperature cycles. Cores from various locations in the Atlantic, Pacific and Indian Oceans provided isotopic records showing similar trends. In a further study, Shackleton and Opdyke (1973) correlated the isotope variations with magnetic stratigraphy and obtained 22 recognisable isotopic stages. Analysing not only records of planktonic foraminifera, but also of deep-sea benthic foraminifera which avoid the noise imposed by short-term temperature and salinity fluctuations, Shackleton and Opdyke (1973) demonstrated that the $\delta^{18}O$ signal predominantly reflects fluctuations in global ice volume, while temperature plays a secondary role. This discovery started widespread use of $\delta^{18}O$ records in global stratigraphic correlations (Imbrie et al., 1992; Shackleton, 1977). Turning attention to $\delta^{13}C$ variations with depth in a core, Pisias and Shackleton (1984) showed their potential significance in studying water mass movement and palaeo-productivity, and postulated a connection between climatically induced changes in the terrestrial biosphere, the observed carbonate dissolution cycles and the flux of dissolved CO_2 in the oceans. Combining the time variations of $\delta^{18}O$ around the world by stacking a great number of records in consistent manner allowed them to form a global ice volume variation curve for comparison with models of astronomically forced climate fluctuations (Prell and Kutzbach, 1986; Martinson et al., 1987; Chappell and Shackleton, 1986). Consequently, oxygen isotope stratigraphy has become not only a global correlation tool, but also an established dating technique (Figure 2). In addition, the primary ice volume control on (benthic) oxygen isotope records has led to their use in the approximation of past sea-level variations after calibration with studies of other sea-level indicators (Shackleton, 1987; Bard et al., 1996; Linsley, 1996; Rohling et al., 1998; McCrea, 1950).

Comparison of different types of palaeo-climatic records from ice sheets and ice caps, peat bogs, lake and ocean sediments, loess deposits, speleothems and tree rings from different parts of the world provides a mosaic of local responses to global climate changes. For a comparison, it is essential that the timing of the different records is accurately known and that the differences in response time for different records are considered during interpretation.

Ocean sediment carbonates are mostly comprised of plankton shells known as foraminifera and ostracods. They are microscopic in size, but control the shallow ocean chemistry in a major way. The chemistry of carbonate formation in aqueous medium involves saturation with HCO_3^- and Ca^{++} ions. In the ocean, CO_2 evolves from various processes like respiration, degassing from subsurface sources and decomposition of dead organic matter. Depending on pH, carbon dioxide gets dissolved in water and forms bicarbonate ion. On the other hand, Ca^{++} is mainly supplied from weathering of the continental rocks. The microscopic animals secrete shells around their body as a protective covering using HCO_3^- and Ca^{++} ions from the surrounding water. After death, the animal shells sink down to the ocean floor and finally become part of the sediments. The shells are made up of two types of carbonate mineral, calcite and its precursor aragonite. Calcite and aragonite are enriched in ^{18}O compared to the water from which they are precipitated. When the reaction takes place in isotopic equilibrium, the $\delta^{18}O$ value of the calcite is related to the $\delta^{18}O$ of sea water and the temperature in °C by:

$$T = 16.9 - 4.2(\delta_c - \delta_w) + 0.13(\delta_c - \delta_w)^2$$

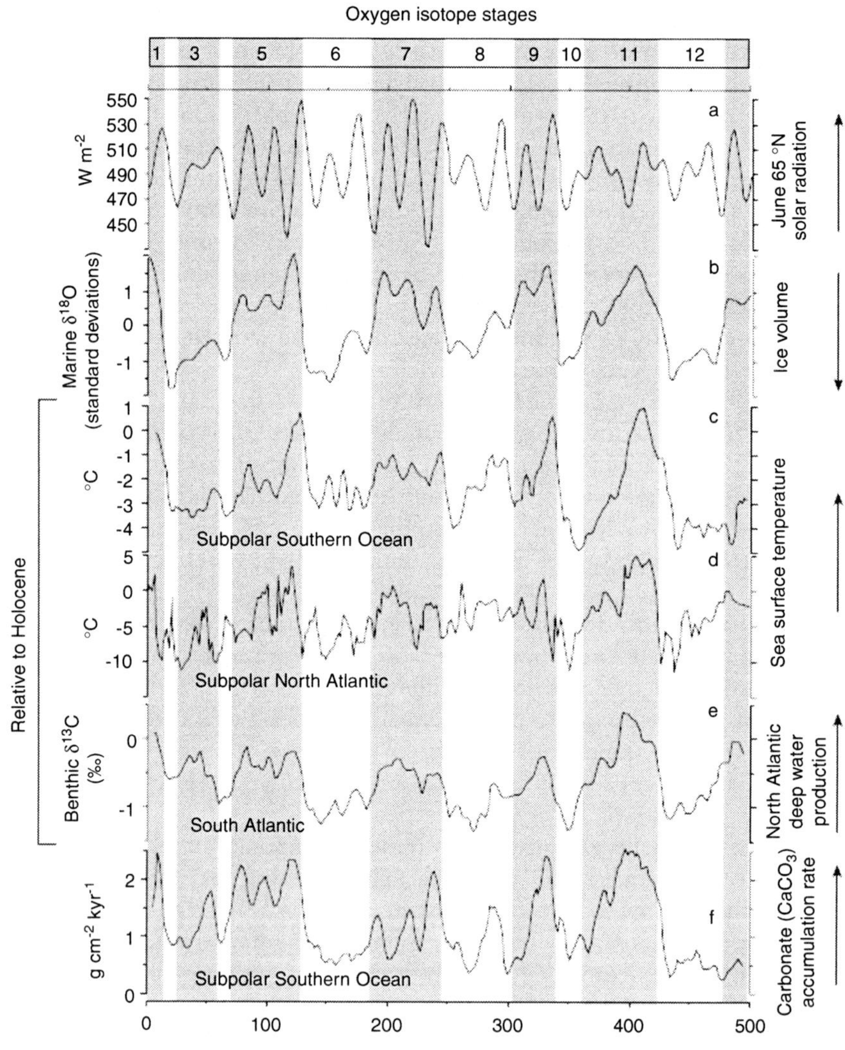

Figure 2 Oxygen isotope stages from ice core records shown in relation to variations in (a) solar insolation at 65°N in watt per square meter; (b) ice volume effect established from oxygen isotopic analyses of foraminifera from DSDP cores; (c) expected variation in sea surface temperature in °C estimated from $\delta^{18}O$ of foraminifera in DSDP core samples from the sub-polar southern ocean and from (d) the sub-polar North Atlantic Ocean. Also shown is the variation in (e) North Atlantic deep water formation and (f) the carbonate accumulation rate in the sub-polar Southern Ocean (Imbrie et al., 1993).

(Keeling et al., 1995; Sundquist, 1993) where δ_c is the $\delta^{18}O$ value of CO_2 prepared by reacting calcite with 100% H_3PO_4 at 25°C and δ_w the $\delta^{18}O$ of CO_2 in equilibrium with water at 25°C. Both δ values are on the same (vienna standard mean oceanic water [VSMOW]) scale. The oxygen isotopic palaeo-thermometer is based on the assumption that biogenic calcite and aragonite are precipitated in isotopic equilibrium with sea water. This requirement is met only by a few carbonate-secreting organisms including molluscs and foraminifera. For a 1°C rise in temperature, there is a 0.25‰ drop in $\delta^{18}O$ of the carbonate. Decreasing $\delta^{18}O$ values thus indicate increasing temperatures if the $\delta^{18}O$ of the water remains constant. The oxygen isotope ratio of sea water is intimately linked with fractionation processes within the hydrological cycle. Schematically, this cycle comprises of evaporation, atmospheric vapour transport, precipitation and subsequent return of freshwater to the ocean (directly via precipitation and indirectly via runoff/iceberg melting). Long-term storage of freshwater in aquifers and especially ice sheets modifies the sea water isotope ratios in that timescale. Formation and melting of seasonal sea ice induces strong local variability. Finally, the spatial distribution of oxygen isotopes in the world ocean depends on the processes of advection and mixing of water masses from different source regions carrying different isotopic signatures. Both these parameters, namely temperature and water composition, exhibit global and regional variations with time. The global components include changes in the sea surface temperature (SST) and the distribution of the total water between the world oceans and the continental ice sheets. The regional components depend on variations in SST and ^{18}O induced by local upwelling, mixing of water masses and the amount of fresh water discharge near coastal regions. Any change in regional climate can be estimated by subtracting the global signal from the total variation. However, in oceanic sediments, this interference can be resolved by simultaneous measurement of trace element ratios like Ca/Sr, which are solely dependent on temperature and not on other parameters like salinity of water.

The carbon isotopic composition ($\delta^{13}C$) of foraminifera is mainly determined by CO_2 forming the HCO_3^- ion. The source of CO_2 is mainly respiration and decomposition of organic matter. Both these processes produce CO_2 of similar isotopic composition. In phytoplankton, carbon isotopic composition is predominantly controlled by the net photosynthetic activity of the regional water mass. Other factors like temperature are of minor significance. $\delta^{13}C$ variations in foraminifera have global components like changes in $\delta^{13}C$ of CO_2 induced by fluctuations in the deep water formation and circulation and in the global productivity due, for instance, to changing wind strengths. The local variations in the input of terrestrial carbon and upwelling also have a significant influence on the $\delta^{13}C$ of foraminifera. Therefore, variations of $\delta^{18}O$ and $\delta^{13}C$ in marine planktonic foraminifera can be used to understand the variability of photosynthetic activity in the past. Despite the complexity of a variety of controls, natural waters tend to show a characteristic range of isotope values, which in turn are mimicked or tracked by the carbonate minerals precipitated from them.

3.1.1. A case study

The Asian monsoon over the Indian subcontinent and southeast Asia is a dominant feature of the tropical climate and impacts on the lives of nearly half of the world's population. Monsoon climate is characterised by seasonal shifts in wind pattern associated with a drastic change in the general precipitation and temperature patterns. The Asian land mass and in particular the Indian subcontinent are surrounded by the Arabian Sea on the west, the Bay of Bengal towards the east and the Indian Ocean in the south. Heat build up on the mainland and Tibetan plateau in the north results in the formation of a lower density air mass or areas of low pressure. Conversely, denser air associated with high pressure dominates ocean surfaces. Wind and ocean currents that result from air flowing from high- to low-pressure regions bring a lot of rain in the Indo-Gangetic plains during summer. During winter, this reverses. Variation in space and time of solar heating due to the Earth's tilt creates seasonal hemispheric energy imbalances having different responses from land and ocean surfaces. The hemisphere receiving the most direct radiation (during the summer months) experiences a net radiative heating (more energy is gained from the sun than is lost to space). The winter hemisphere at the same time experiences a net radiative cooling. As part of a global compensation, heat is transported from warmer to cooler areas by ocean and wind currents. Since the areas of heat surplus and deficit change throughout the year, the direction of transport must change as well (as in case of sea breezes). Figure 3a and b show low-latitude surface wind directions averaged over the summer and winter seasons, respectively. As noted earlier, regions dominated by monsoons experience pronounced seasonal wind shifts, indicative of a large land–sea temperature contrast. In the South Asian example, the rainy season, typically beginning in June, is preceded by nearly 2 months of scorching temperatures, cooled only with the commencement of the summer rains brought by the southwesterlies. January is the peak of the dry season, which is marked by a cool, dry northeasterly flow over most of the region.

In the case of the Indian subcontinent, both winter and summer monsoonal precipitations drain large quantity of fresh water into the Arabian Sea and Bay of Bengal, which finally mixes with the Indian Ocean. The north Indian Ocean is a particularly interesting region because wind strengths here reach high intensity during summer (southwest monsoon) and winter (northeast monsoon). The present-day meteorological condition over the Arabian Sea is controlled by seasonally varying insolation and

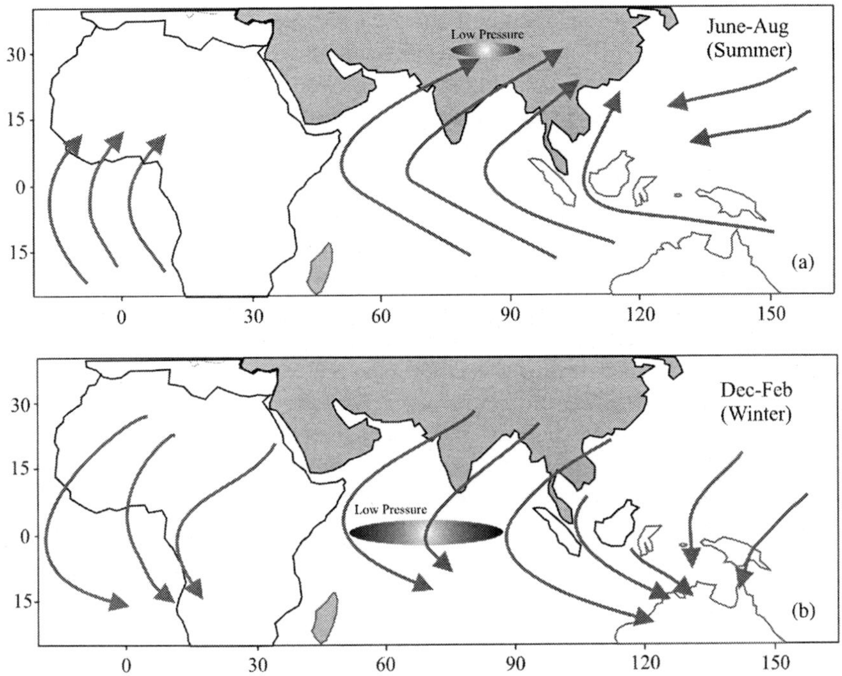

Figure 3 The wind circulation pattern in the Indian Ocean during summer (a) and winter (b) months. This process is governed by differential heating of land and water bodies adjacent to Asia during cool and warm periods.

an associated wind system, which cause both monsoon rain over the Indian region and intense upwelling in the western Arabian Sea. During summer time, fresh water discharge into the Arabian Sea through monsoon-fed rivers in south India together with evaporation in the northwestern portion controls the salinity pattern. In contrast, winter monsoon causes increased precipitation in southeast coastal India, and increased runoff from the southern rivers results in lower salinity in the southwestern Bay of Bengal. Even on an annual scale, the Bay of Bengal has a lower salinity compared to the Arabian Sea due to the larger fresh water discharge from the Ganges–Brahmaputra–Irrawaddy river systems. Interestingly, the northeast winter monsoon current over the oceans transports low salinity water from the Bay of Bengal to the eastern Arabian Sea following a route south of Sri Lanka, causing a change in salinity pattern. There are good reasons to believe that this mode of salinity distribution varied in the past. The record of past monsoonal activity in the subcontinent can be obtained from the study of oxygen isotopic composition in surface-dwelling foraminifera from the Arabian Sea.

Duplessy et al. (1981) studied the $\delta^{18}O$ variations in surface-dwelling foraminifera in 41 piston cores and 36 core top samples from the Arabian Sea, the Bay of Bengal and the Andaman Sea, and compared the values corresponding to the Holocene and the last glacial maximum (LGM) periods. They found a marked change in the salinity pattern at LGM and attributed the variation to reduced river inflow to the Bay of Bengal and increased evaporation in the northern Arabian Sea due to stronger northern winds from continent to the oceans. Furthermore, during LGM, the low salinity, low $\delta^{18}O$ area along the southwestern coast of India and cold high $\delta^{18}O$ region along the Arabian coast influenced by upwelling (both signifying southwest monsoon strengths) diminished. The two together seem to indicate that during the LGM, the SW monsoon was weaker and correspondingly the NE monsoon was relatively stronger.

To confirm the prediction of Duplessy et al. (1981) and understand the vagaries of the South Asian summer monsoon and winter monsoon during the last 25,000 years, Sarkar et al. (1990) analysed four species of foraminifera, *Globigerinoides sacculifer, Orbulina universa, Globigerinoides ruber* and *Globorotalia menardii*, isolated from two cores, SK-20-185 and SK-20-186, taken from the Arabian Sea (Figure 4). Core SK-20-185 was drilled at a location which marked the boundary of maximum influence of the ocean current from the Bay of Bengal to the Arabian Sea during winter time. In contrast, core SK-20-186 was away from the southern tip of the

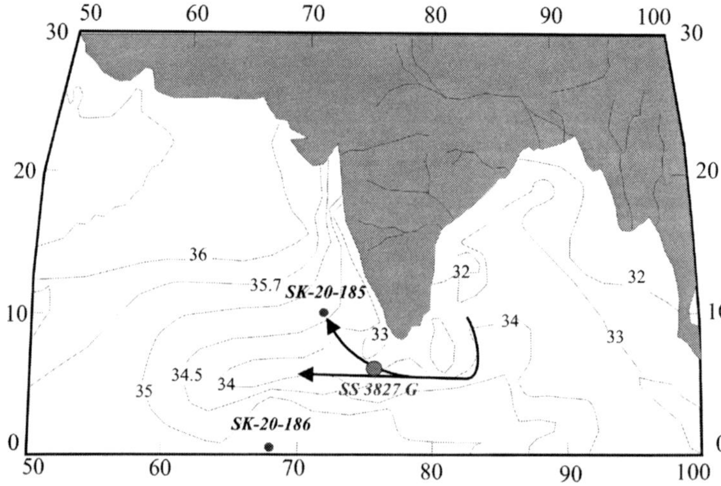

Figure 4 Schematic pattern of surface water currents in the Indian Ocean. During the monsoon period, large fresh water fluxes from rivers cause a drop in salinity in Bay of Bengal water. This low salinity water is transported to the Arabian Sea. There it mixes with the highly saline Arabian Sea water (shown by arrows). The locations of two sediment cores SK-20-185 and SK-20-186 are shown (see text).

subcontinent and represents the water mass of the equatorial Indian Ocean. A mean sedimentation rate of 2.2 cm ka^{-1} was established for core SK-20-185 from ^{14}C dating of bulk foraminifera taken from different depths. As explained, the oxygen isotopic ratio of forams is sensitive to any change in temperature and composition of sea water. Since both these parameters exhibit variations on global and regional scales, to estimate any regional change in temperature and water composition, an appropriate correction should be imposed by subtracting the influence of the global component. One can derive information about global palaeo-climatic changes from the record in the SPECMAP (http://gcmd.nasa.gov/records/GCMD_EARTH_LAND_NGDC_PALEOCL_SPECMAP.html) and subtract the temperature component. The residual component after removal of the temperature effect indicates the salinity change as a result of change in the regional circulation of rainfall/precipitation.

The isotopic record of the last 35 ka is shown in Figure 5. Sarkar et al. (1990) reported oxygen isotopes in the planktonic foraminiferal species *Gs. sacculifer, O. universa, Gs. ruber and Gr. menardii* from the core SK-20-185 from the eastern Arabian Sea (Figure 5). The core was dated by the

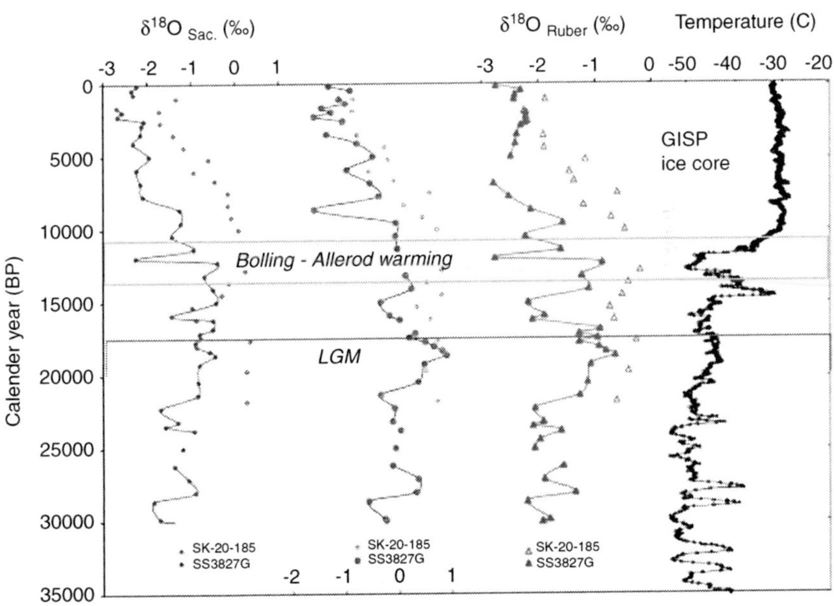

Figure 5 Temporal variations in the oxygen isotope ratios of three species of foraminifera *viz. Globigerinoides ruber, Globigerinoides sacculifer, Globorotalia menardii* separates from sediments (SK-20-185 and SS3827 G) from the Indian Ocean. The negative excursion during early deglaciations is shown by a shaded region (adapted from Sarka (1989) and Manish Tiwari (2005), with permission from Elsevier).

conventional radiocarbon method on bulk sediments. In a more recent study (Tiwari et al., 2005) on core SS3827G (close to the earlier location), age determination was made on selected planktonic foraminifera rather than bulk sediment where younger ages were reported. An important finding of the study by Tiwari et al. (2005) was the detection of an excursion of $\delta^{18}O$ towards depleted values (by ~1‰) during the 19–17 ka (early deglacial period) which adds a 2 ka offset to the Sarkar et al. (1990) age data. Results from these two studies are compared in Figure 5. This excursion lasted for 2 ka based on the recent high-resolution study. It also coincided with a drop in productivity of the ocean established independently from the $CaCO_3$ content in the sediments. Such a negative excursion can be explained if one considers that, during the early deglacial period, the SW monsoon wind was still in a weakened mode with a large enhancement of the NE monsoon. This period of strongest NE monsoon lasted for ~2,000 years starting from 19,000 calendar years ago. Also plotted in Figure 5 is an estimation of the temperatures from GISP ice core isotopic results from the northern hemisphere (http://www.ncdc.noaa.gov/paleo/icecore/greenland/gisp/gisp.html). It is evident from the nature of the curve that the NE monsoon intensity largely correlates with the variation in temperature of the polar region in the northern hemisphere. As shown in Figure 5, the timing of the sudden rise in temperature in the ice core record coincides with the negative excursion of $\delta^{18}O$ of planktonic foraminifera from the Indian Ocean – probably suggesting enhancement of the NE wind circulation.

3.2. Ice cores

The isotopic composition of oxygen and hydrogen in water of ice cores preserves climatic information relating to local temperature, precipitation rate, moisture source conditions, etc. Additionally, trapped air in ice cores provides direct information on atmospheric trace gas concentrations and indirectly allows us to estimate aerosol fluxes of marine, volcanic, terrestrial, cosmogenic and anthropogenic origins (Friedli et al., 1986; Etheridge et al., 1996; Francey et al., 1999). The ice-drilling project undertaken in the framework of a long-term collaboration between Russia, United States and France at the Russian Vostok Station in East Antarctica has provided a wealth of information (Wagenbach, 1989) for the past two glacial–interglacial cycles (i.e. for the last 400,000 years). To obtain trace gas concentration and isotopic records from ice cores, the trapped air must be released from the ice samples while carefully maintaining the quantitative composition. The CO_2 can be isolated from the released air samples for $\delta^{13}C$ analysis. $\delta^{18}O$ measured in ice cores from Greenland and Antarctic is a proxy for regional temperature at the time of the snow/ice formation. Therefore, ice cores provide a crucial input to deciphering past climate changes.

The data plotted in Figure 6a and b show $\delta^{13}C$ results obtained from trapped air in an Antarctic ice core. It represents one of the earliest results from ice cores exhibiting a general increase in CO_2 concentration with time together with depletion in $\delta^{13}C$ values. The data also made it possible

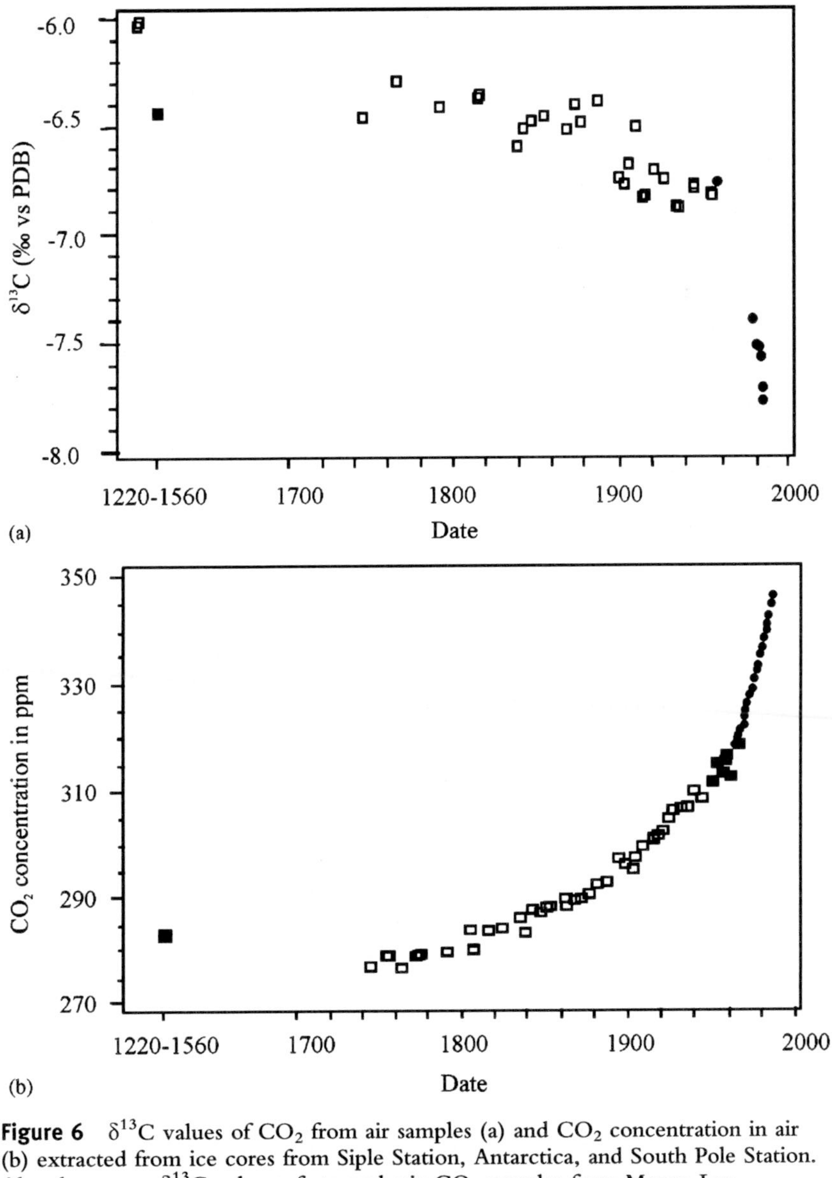

Figure 6 $\delta^{13}C$ values of CO_2 from air samples (a) and CO_2 concentration in air (b) extracted from ice cores from Siple Station, Antarctica, and South Pole Station. Also shown are $\delta^{13}C$ values of atmospheric CO_2 samples from Mauna Loa (adapted from Friedli et al. (1986), with permission from Elsevier).

to directly compare measurements of CO_2 concentrations in air at Mauna Loa (Keeling et al., 1995) with the ice core archives from Antarctica. The average $\delta^{13}C$ value of air CO_2 samples before 1,800 AD is −6.41‰ in the ice core. This observation is in excellent agreement with the extrapolated pre-industrial mean value for the South Pole inferred from direct air sampling (6.44‰). This comparison adds confidence to the use of ice cores as a tool for palaeo-climatic investigations.

One of the important factors in climate is heat flux, observable through measurement of temperature. As a result of fractionation processes, the oxygen and hydrogen isotopic content of snow is linearly related to the temperature. Hence, the temperature dependence of isotopic fractionation factors for oxygen and hydrogen can be used to reconstruct past temperature changes of the atmosphere from ice core records. As mentioned, continuous ice cores from the ice sheets of Greenland and Antarctica have provided long-term climatic information. Wagenbach (1989), Dansgaard et al. (1983) and Johnsen et al. (1997) have utilised oxygen isotope measurements in ice cores from Greenland to reconstruct local surface temperature during the last 100,000 years. Similarly, Lorius et al. (1985) and more recently Jouzel et al. (1996) have used $\delta^{18}O$ measurements from the Vostok ice core (3,310-m-thick ice deposit) to reconstruct surface temperature variations for more than the last 400,000 years.

3.2.1. A case study: Vostok ice core record

In January 1998, the Vostok project yielded the deepest ice core ever recovered, reaching a depth of 3,623 m. Drilling stopped at about 120 m above the surface of the Vostok Lake, a deep sub-glacial lake which extends below the ice sheet over a large area. Existing data indicate that the Vostok ice core record extends through four climate cycles, with ice slightly older than 400 ka at a depth of 3,310 m, thus spanning a period comparable to that covered by numerous oceanic and continental records.

Here, we discuss some of the important discoveries on climatic variability based on a series of detailed Vostok records covering this ~400 ka period. Temperature reconstruction is shown in Figure 7 for the Vostok ice core along with the observed variation in trace gas concentration. Two significant features of this record are the large amplitude of the glacial–interglacial cycles and the rapid nature of the change from glacial to interglacial temperatures. The estimation shows 8–12°C temperature change in surface air between glacial and interglacial periods. There are four glacial cycles marked by low CO_2 and CH_4 concentrations in the atmosphere associated with enriched oxygen isotope ratios in ice. The third and fourth climatic cycles are of shorter duration than the first two. Similar features are also found in the deep-sea record. Each of these glacial–interglacial cycles consists of a sawtooth sequence of warm interglacial events, followed by increasingly

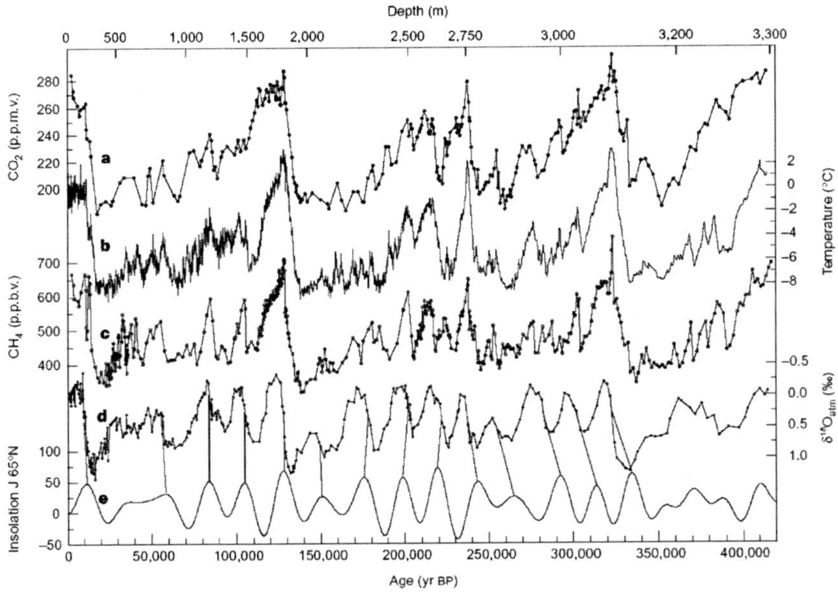

Figure 7 Variation of CO_2, CH_4 and $\delta^{18}O$ obtained from analysis of Vostok ice core samples from Eastern Antarctica. Also shown are estimates of temperature and insolation changes (adapted from Petit et al. (1999), with permission from Elsevier).

colder interstadial events culminating in a deep glacial period and finally ending with an abrupt return to the next interglacial. The data plotted in Figure 7e provide estimates of insolation changes at 65° North. The warmest time interval at 200–230 ka was slightly warmer than the Holocene (Rau et al., 1989, 1992). These observations may be important in the context of the continued contemporary CO_2 increase and possible consequences for the future of the Earth's climate. The Vostok ice core records can also be used to infer the sensitivity of global climate to future anthropogenic increases of greenhouse gas concentrations (Wagenbach, 1989).

3.3. Phanerozoic CO_2 from soil carbonate and organic matter

The long-term history of carbon dioxide concentrations in the surface oceans and the Earth's atmosphere can be delineated using some of the recent techniques involving stable isotopes. Two techniques have been reliably used in recent years to estimate the past CO_2 levels: (1) the carbon isotopic composition of alkenones in ancient ocean sediments to derive the pCO_2 level in the surface layer and (2) a CO_2 palaeo-barometer utilising $\delta^{13}C$ of pedogenic carbonates in palaeosols.

Several studies (Hinga et al., 1994; Bidigare et al., 1997) have shown that the total isotope effect associated with marine photoautotrophic carbon fixation is, in part, proportional to the concentration of available CO_2. This relationship provides a starting point for interpreting isotopic trends of marine sedimentary organic carbon (δ^{13}_{TOC} of planktonic biomass) in terms of oceanic dissolved CO_2 concentration. This, in turn, allows one to estimate ancient surface water carbon dioxide variability (Rau et al., 1991; Raymo and Horowitz, 1996 and references therein). However, using δ^{13}_{TOC} to reconstruct palaeo-CO_2 is complicated by several factors. For example the net δ^{13}_{TOC} value very often results from an integration of primary and secondary isotopic signals. In recent studies, it has been shown that the compound-specific isotopic analysis of molecular markers derived from phytoplankton can be utilised to circumvent this difficulty and provide one of the most reliable proxies for surface water CO_2 (Pagani et al., 2000). Pagani et al. (1999a, 1999b, 2000) extracted specific molecular markers (alkadienones) from organic matter preserved in Miocene ocean sediments from the North and South Atlantic and estimated the pCO_2 level from their $\delta^{13}C$ composition. The results (Figure 8) show that pCO_2 steadily increased starting from a low value of ~180 ppm at 14 Ma and stabilised during the Late Miocene (9 Ma) with concentrations between 320 and 250 ppmv. These uniformly low pCO_2 values are consistent with Middle to Late Miocene alkenone-based pCO_2 estimates and trends from other localities or other types of indirect proxies such as stomatal index and boron isotopes in foraminifera (Pagani et al., 2000).

While it is possible to estimate pCO_2 concentrations of the past from ocean sediments, there are few other pCO_2 proxies in the terrestrial ecosystem itself, which have been used for Phanerozoic pCO_2 reconstruction. Pedogenic carbonates are one such system that has found widespread application (Cerling, 1991; Ekart et al., 1999; Ghosh et al., 2001). The way palaeosol carbonates record the past CO_2 level in the atmosphere is quite interesting. These carbonates are precipitated in the root zone of plants when groundwater supersaturated with carbonate ions can release CO_2 by some process. They are common in regions receiving less than 800 mm annual rainfall. The addition of CO_2 in the groundwater during plant respiration and subsequent evaporation and transpiration of water from a plant can induce supersaturation of dissolved salts and carbonate precipitation. Palaeosol carbonates record the isotopic composition of local soil CO_2, which primarily reflects the type of vegetation in the ecosystem as well as the contribution from atmospheric CO_2 (Cerling, 1984). Thus, the soil CO_2 is a mixture of two components: plant-respired CO_2 and atmospheric CO_2. It is important to note that the $\delta^{13}C$ of atmospheric CO_2 (~−7‰) is very different from that of soil CO_2 (~−25‰). Atmospheric CO_2 penetrates inside the soil by diffusion and mixes with the soil CO_2 resulting in its isotopic change. Today, except for ecosystems with

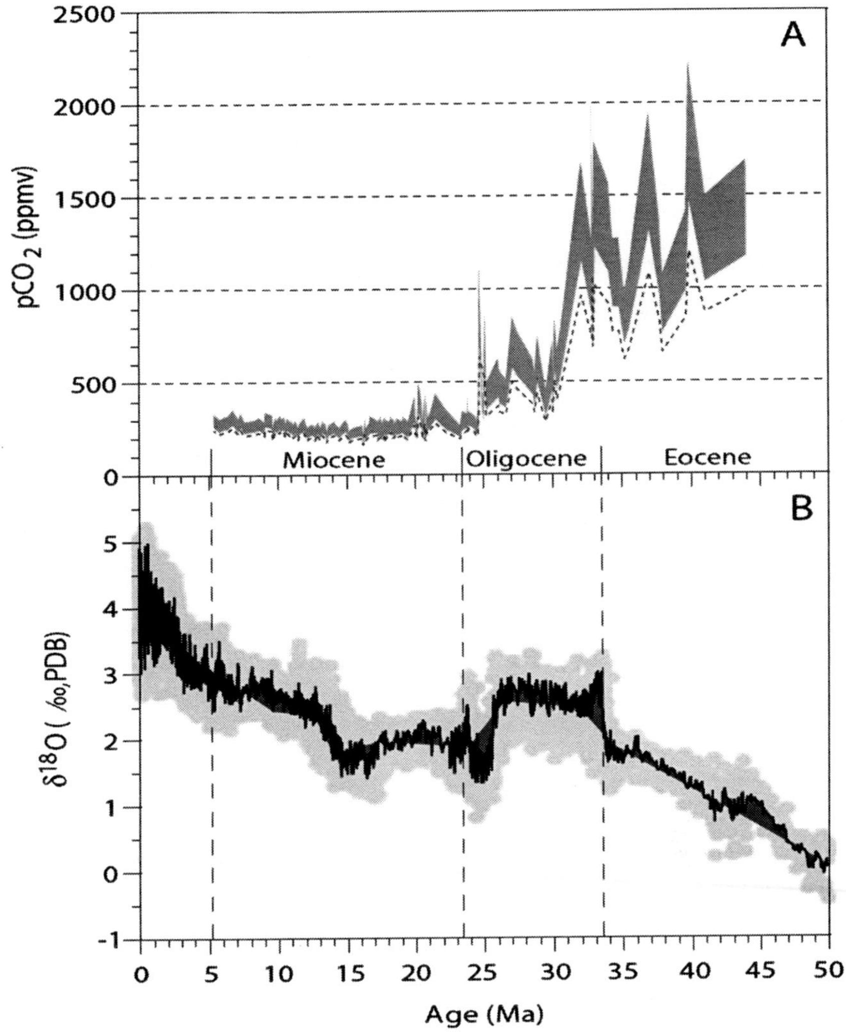

Figure 8 (A) pCO_2 estimates calculated on the basis of the carbon isotopic composition of the unsaturated alkenones (adapted from Pagani et al. (1999), with permission from Elsevier). The record shows an increase in pCO_2 concentration during the Miocene period; (B) Oxygen isotope ratio in deep-sea carbonates as a function of time (adapted from Zachos et al. (2001), with permission from Elsevier).

very low productivity such as deserts, the atmospheric contribution to the total soil CO_2 is small due to the low concentration of CO_2 in the modern atmosphere. However, if atmospheric CO_2 is high, as occurred in the distant past, it can make a significant contribution to total soil CO_2 resulting in a measurable isotopic shift (towards the positive side) in the $\delta^{13}C$ of soil carbonate precipitated in isotopic equilibrium with soil CO_2. Therefore, $\delta^{13}C$ analysis of pedogenic carbonates can provide evidence for large variations in pCO_2 in the geologic past (Pagani et al., 1999; Pagani et al., 2000; Ghosh et al., 1995; Mora et al., 1996). Cerling (1991) developed a detailed model for this purpose as described below.

3.3.1. Model for estimation of atmospheric pCO_2

Cerling's (1991) palaeo-barometer model is basically an isotopic mixing model where the soil CO_2 is made up of atmospheric CO_2 (through diffusion) and plant-respired CO_2 from vegetative sources. Since this plant-respired CO_2 flux dominates, soil CO_2 in well-aerated soils can be modelled as a standard diffusion–production equation (Baver et al., 1972; Kirkham and Powers, 1972; Cerling, and Quade, 1993):

$$\frac{\partial C_s}{\partial t} = D_s \frac{\partial^2 C_s}{\partial z^2} + \phi_s(z)$$

where C_s = soil CO_2 (molecule cm^{-3}), t = time, z = depth in the soil profile (cm) and $\phi_s(z)$ = CO_2 production rate as a function of depth (molecule s^{-1} cm^{-3}), D_s is the diffusion coefficient and ϕ_s the production function with respect to the depth z (Cerling, 1984). Solving this equation with a no-flux boundary at depth and C_s (at $z = 0$) set equal to atmospheric carbon dioxide yields:

$$C_s(z) = S(z) + C_a$$

where $S(z)$ is the CO_2 contributed by soil respiration and C_a the atmospheric CO_2. Cerling (1991) solved this equation for ^{12}C and ^{13}C and showed that the isotopic composition of soil CO_2 is controlled by a diffusion process (diffusion of CO_2 from soil to atmosphere) and CO_2 production. In a palaeosol, C_aCO_3 precipitates in equilibrium with soil CO_2 which is modified by an atmospheric contribution, thus allowing the $\delta^{13}C$ value of carbonate to be used for atmospheric pCO_2 calculation. The solution to the soil CO_2 equation of Cerling (1991) can be recast in terms of the isotopic composition of atmospheric CO_2:

$$C_a = S(z)\frac{\delta^{13}C_s - 1.0044\delta^{13}C_\phi - 4.4}{\delta^{13}C_a - \delta^{13}C_s}$$

under the assumption that $^{12}C_s/^{12}C_a \approx C_s/C_a$, where $\delta^{13}C_s$, $\delta^{13}C_\phi$ and $\delta^{13}C_a$ are the isotopic compositions of soil CO_2, soil-respired CO_2 and atmospheric CO_2, respectively. The term $S(z)$ is a function of depth but approaches a constant value below about 20–30 cm depth (Cerling and Quade, 1993). Soil CO_2 is enriched in ^{13}C relative to respired CO_2 by

4.4‰ due to the mass-dependent rate of diffusion (Baver et al., 1972) independent of CO_2 concentration.

The $\delta^{13}C$ of soil carbonate is governed by the $\delta^{13}C$ value of the soil CO_2 and fractionation during precipitation which is temperature dependent (Deines et al., 1974). The $\delta^{13}C$ values of the two sources of soil CO_2 (e.g. respiration and atmosphere) are quite different. Although the recent pre-industrial atmospheric CO_2 is −6.5‰, the CO_2 respired by C_3 type vegetation can have a value between −20‰ and −35‰, with a mean around −27‰ (Ehleringer, 1989). The degree of infiltration of the atmospheric CO_2 in the soil matrix (which is dependent on the atmospheric pCO_2) thus influences the $\delta^{13}C$ value of the soil CO_2 and consequently the $\delta^{13}C$ value of the pedogenic carbonate. Therefore, this model allows us to estimate the pCO_2 values of the ancient atmosphere from the $\delta^{13}C$ value of the pedogenic carbonates if the following parameters are known:

(a) The temperature of calcite precipitation in the soil,
(b) The $\delta^{13}C$ value of the plant-respired CO_2,
(c) The $\delta^{13}C$ value of the atmospheric CO_2,
(d) The difference between the concentration of the soil pCO_2 and the atmospheric pCO_2.

A detailed discussion on estimation of these parameters is given in Ghosh et al., (2001).

3.3.2. A case study: calculation of pCO_2 in the ancient atmosphere

Ghosh et al. (2001) determined the atmospheric CO_2 concentration using Gondwana soil carbonates from India. The motivation for the study came from the discovery of well-developed and well-preserved palaeosols in the Gondwana sediments of central India, which cover a period of significant CO_2 change in well-spaced intervals. The stable isotopic composition of pedogenic carbonates formed in these palaeosols was investigated to decipher the CO_2 concentrations and compare them with that predicted by the well-known Berner model. To understand the past pCO_2 variations quantitatively, Berner (1994) first proposed a model known as GEOCARB II based on geological factors like CO_2 outgassing through the mid-ocean ridge or volcanoes and CO_2 consumption through weathering. He improved it further in GEOCARB III. Both these models predict that in the Early Phanerozoic (550 Ma), the pCO_2 was 20 times the present atmospheric level (PAL). Subsequently, the pCO_2 level declined in the Middle and Late Palaeozoic (450–280 Ma) to reach a minimum (approximately similar to the PAL) at about 300 Ma. The period 300–200 Ma was again characterised by a rapid rise when the pCO_2 increased to five times the PAL. It was followed by a gradual decline to the PAL with a small peak in the Early Tertiary.

Using the mean $\delta^{13}C$ values of the palaeosol carbonates along with the above-mentioned values of the required parameters and the estimated $\delta^{13}C$ of the atmospheric CO_2 based on soil organic matter in the model of Cerling (1991), we have calculated pCO_2 values for the five time-windows mentioned above (Ghosh et al., 2001). The values are (in ppmV) 540–890 for Motur, 910–1,510 for Denwa, 880–1,460 for Tiki, 1,675–2,775 for Bagra and 1,110–1,850 for Lameta periods. The uncertainties in average $\delta^{13}C$ (OM) are not included in the quoted uncertainty of the pCO_2. However, variations in $\delta^{13}C$ (OM) (0.5‰ for Motur and Denwa, and 1.2‰ for Tiki and Bagra Formation) would change the calculated mean pCO_2 by 20% for Motur and Denwa, and 40% for Tiki and Bagra Formations, respectively. It is to be noted that the soil samples studied here belong to the same geochemical milieu as they all formed either in the same basin or in a neighbouring basin (Tiki) at different times. This implies that the relative CO_2 variation deduced here is expected to be well constrained.

Using the mean $\delta^{13}C$ values of the palaeosol carbonates, the following ranges of pCO_2 values (in ppmV) were obtained for the different geologic periods: 540–890 for 260 Ma, 910–1,510 for 240 Ma, 880–1,460 for 230 Ma, 1,675–2,775 for 160 Ma and 1,110–1,850 for 65 Ma before present. These values represent the first independent estimates of atmospheric CO_2 level during 260–65 Ma based on soils formed in a Gondwana continent located in the southern hemisphere. This is considered an important period in the evolution model (GEOCARB III), as it predicts an increase in the CO_2 level after the Early Permian (310–285 Ma) low and ascribes it to an enhanced rate of degassing. This prediction is verified by our results but the detailed nature of the variation is found to be slightly different. Figure 9 shows a plot of CO_2 concentration in ppmV against age (in Ma). It is seen that there is excellent agreement of derived concentrations with Berner's prediction between 275 and 160 Ma (corresponding to ages of Motur, Denwa, Tiki and Bagra Formations) but the abundance of carbon dioxide for 80 Ma (Lameta) is 1,480 ppm in contrast to Berner's estimate of about 1,000 ppmV (GEOCARB III). The disagreement persists even at the lower limit of our estimate. In this context, it is to be noted that all the available evidence indicates a very high CO_2 concentration in the Middle Cretaceous to Late Cretaceous atmosphere. Emiliani (1995) points out that the Middle Cretaceous was characterised by widespread anoxia, huge marine ingression indicating active mid-ocean ridges and warm oceans. All these factors indicate a large increase in the CO_2 flux to the atmosphere and consequent exacerbation of the greenhouse effect and vigorous expansion of biotic productivity. Based on these considerations, an upward revision of the model CO_2 input flux at the Late Cretaceous can therefore be considered, which will reduce the disagreement.

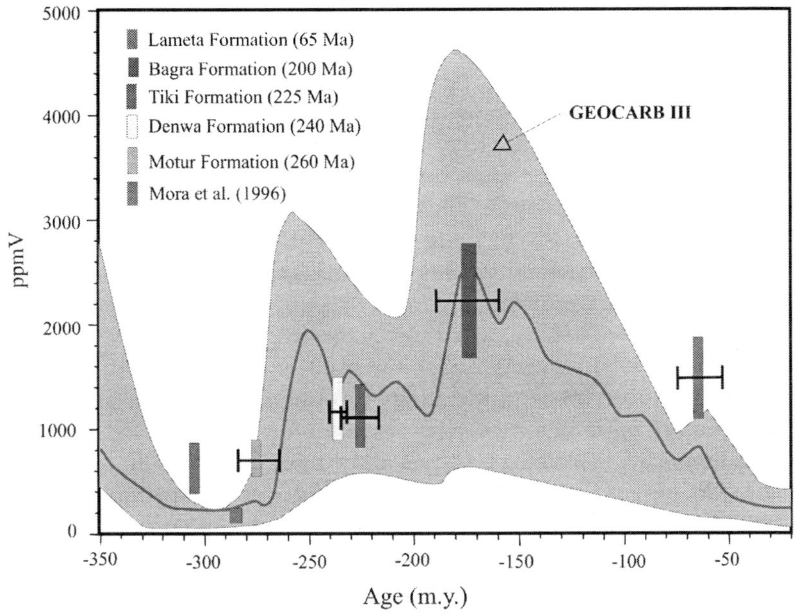

Figure 9 Estimated concentration of atmospheric CO_2 as a function of time. The points are based on isotopic analyses of pedogenic carbonates. The dashed curve along with the envelope (error range) is obtained from GEOCARB III (adapted from Ghosh (2000), with permission from Elsevier).

3.4. Himalayan foothill sediments: Story of past monsoon and Asian climate

The fore-deep created at about 20 Ma ago during the last phase of the Himalayan orogeny provided an accretionary basin for the well-known Siwalik sediments (Johnson et al., 1985). These sediments, dominantly fluvial in nature, are characterised by the presence of intercalating palaeosols, suggesting phases of non-deposition in the river floodplain (Johnson, 1977; Tandon and Narayan, 1981). These palaeosols contain a variety of components such as soil carbonate, organic matter and clay minerals, which are sensitive tools for reconstructing past climate and vegetation history. Based on carbon and oxygen isotope ratios of soil carbonates from Pakistan and Nepal Siwaliks, it was shown that a major shift in vegetation (C_3 to C_4 transition) occurred around 7 Ma ago, and this transition was probably associated with hydrological changes induced by onset of the monsoon (Quade et al., 1989, 1995). Subsequent studies from other parts of the world, especially from low-latitude areas, provided evidence that vegetational change did not occur at the same time in the different places; in lower latitudes, change was earlier (~4 Ma) compared to the higher latitude areas (Cerling et al., 1997; Latorre et al., 1997, Ding and Yang, 2000). It was suggested that at temperatures typical of low latitudes, lowering of atmospheric CO_2 concentrations acted as the trigger for sudden expansion of C_4 plants (Cerling et al., 1997). In this context, it may be mentioned that increased weathering rates during the past 40 Ma, especially in the tectonically active Himalayan–Tibetan region, might have been the cause of lowering of CO_2 (Raymo and Ruddiman, 1992). Recent studies on alkenones from Pacific sediments showed that the atmospheric CO_2 concentration increased from 15 to 9 Ma ago and stabilised at the pre-industrial value (~280 ppm) at around 9 Ma (Pagani et al., 1999). This led Pagani et al. (1999) to suggest that C_4 plant expansion might have been driven by enhanced low-latitude aridity or changes in seasonal precipitation pattern since, in most regions, C_4 grass expansion is seen to be associated with changes in rainfall. Sanyal et al. (2004) studied extensive Siwalik exposures in India to document the timing and nature of C_4 grass expansion in Indian Siwaliks and its possible relation to changes in rainfall. Three Siwalik sections, two from Kangra sub-basin (Ranital and Kotla) and one from Subathu sub-basin (Haripur Khol) in Himachal Pradesh, India, were studied. These two areas are located in a region sensitive to changes in monsoon pattern, since the seasonal rainfall is very high there with closely spaced isohyets. Isotopic studies from such an area can help in deciphering past changes in monsoon and in assessing their relation to the expansion of C_4 grass. The ages of the sedimentary succession in Subathu and Kangra sub-basins are constrained by magnetostratigraphy (Sangode et al., 1996, 2003). The age of the sampled section in the Kangra sub-basin ranges from ~11 to ~6 Ma and in Subathu sub-basin from ~6 to 0.5 Ma. There is very little temporal overlapping between the Kangra and Subathu sections and, therefore, the continuity of the sections in a composite stratigraphy may not be strictly correct. In both the areas, soil carbonate nodules are collected from the palaeosol horizons and their ages are assigned from the ages of the corresponding palaeosol beds.

The isotopic data from the three sections (Haripur, Ranital and Kotla) can be combined to make composite plots representing the regional time variation of carbon and oxygen isotope ratios (Figure 10a–c). The composite plot for $\delta^{13}C$ shows a major change in values at about 6 Ma, the values becoming more positive by about 8‰. This transition is believed to represent a shift in vegetation pattern from an exclusive C_3 regime to a mixed C_3–C_4 regime. In the composite plot, the oxygen isotope ratio (indicated by a solid line in Figure 10b) shows three evolutionary phases. At around 10.5 Ma, the $\delta^{18}O$ is characterised by a highly depleted value of about −10‰; it tends towards positive values with decrease in age, reaching −6.6‰ at around 6.5 Ma. Subsequently, the isotope ratio is characterised by a sharp depletion going up to −9‰ at 5.5 Ma and then a second phase of

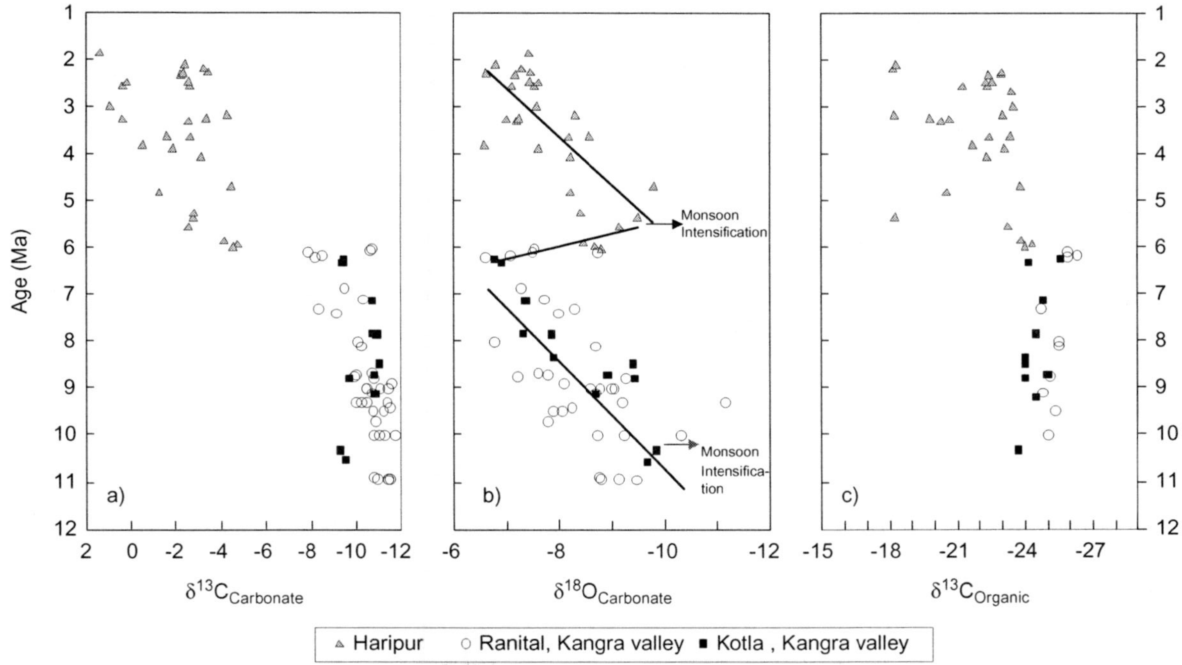

Figure 10 Composite plot of isotope ratios against age of soil carbonate nodules from three sections in Siwalik sediments (Ranital, Kotla and Haripur Khol): (a) carbon, (b) oxygen and (c) organic matter carbon isotopic ratios from the same nodules. The carbon isotope ratio of soil carbonate and associated organic matter indicates that vegetation was entirely of C_3 type from 11 to 6 Ma. The post 6-Ma period is marked by the appearance and expansion of C_4 grass. The oxygen isotope composition shows stepwise variation with a probable peak at 10.5 Ma and a clear peak at 5.5 Ma. These variations are attributed to changes in monsoon rainfall. The solid lines are drawn to guide the eye (adapted from Sanyal, 2004, with permission from Elsevier).

enrichment reaching −6.5‰ at around 2 Ma. These indicate two phases of depletion occurring around 10.5 and 5.5 Ma punctuated by a period of maximum enrichment at around 6.5 Ma. Sanyal et al. (2004) proposed that the above-mentioned two phases of depletion occurring around 10.5 and 5.5 Ma reflect an increase in the intensity of summer monsoon rainfall. The summer monsoon in north India is characterised by heavy rains during June, July, August and September, and is caused by a convective vortex of west–northwest-moving depressions originating from the Bay of Bengal operating on moist oceanic air moving in the same direction. In contrast, during winter, dry continental air blows from the northeast resulting in little rain except occasional precipitation in the Kangra region due to western disturbances bringing moisture from the Mediterranean region. The winter rains are usually more enriched relative to summer rains, as observed in the seasonal variation data for Delhi rainfall (IAEA, 2003). Furthermore, the $\delta^{18}O$ of rainwater decreases with an increase in the amount of precipitation (IAEA, 2003). At low latitudes, the average monthly rainfall and the mean monthly $\delta^{18}O$ are usually negatively correlated, an increase by 1,000 mm of precipitation being associated with a decrease in $\delta^{18}O$ by 1.5‰ (Yurtsever and Gat, 1981). Therefore, it can be argued that the two phases of depletion in the oxygen isotopic composition of soil carbonate (i.e. during 10.5 and 5.5 Ma) were the result of contemporary intensification of the summer monsoon, since an intensified monsoonal wind system would generate more intense and frequent depressions (storms) and result in more rains. It is to be noted that Quade et al. (1989) also obtained a depleted $\delta^{18}O$ value at about 10.5 Ma, and in their composite profile, the maximum depletion occurs at about 9 Ma. Subsequently, the value increases and the scatter becomes large. An increasing trend up to 6 Ma was obtained with an attendant swing after 5.5 Ma. Our study from Indian Siwalik shows that hydrological change is not always accompanied by vegetational change. The $\delta^{18}O$ of soil carbonate shows that the monsoon was intensified at around ~10.5 Ma, but the $\delta^{13}C$ of soil carbonate did not change from that of the C_3 type. After 10.5 Ma, the $\delta^{18}O$ value started showing enrichment and at around 6 Ma, it reached a high value indicating a weak phase of the monsoon, but the contemporary $\delta^{13}C$ of soil carbonate continued to be of C_3 type without any change. The $\delta^{13}C$ data indicate that during the entire period from 10.5 to 6 Ma, vegetation was exclusively of C_3 type.

After 6 Ma, the $\delta^{18}O$ suddenly shows depletion corresponding to a second phase of intensification of the monsoon. Coincidentally, the $\delta^{13}C$ of the soil carbonate also shows enrichment, indicating the appearance of C_4 plants (Figure 10). An intense monsoon implies not only high rainfall but also strong seasonality and this latter property could have ushered in the C_4 plants. After this depletion, the $\delta^{18}O$ again started a swing towards enriched values, which continued up to 1.8 Ma. This could point towards a second phase of weakening of the monsoon. However, during this whole period, the vegetation was a mixture of C_3 and C_4 type. Unlike the $\delta^{13}C$ of the time range 10.5–6 Ma, which was characterised by a low spread, the $\delta^{13}C$ for 6–1.8 Ma shows a wide range of values. This large spread in $\delta^{13}C$ probably indicates changes in growing season conditions induced by climatic oscillations (Thomas et al, 2002; Sangode et al., 2001).

From the above discussion, it is clear that during the period 10.5–6 Ma the monsoon was intensified but C_4 plants appeared at around 6 Ma, which indicates that a change in seasonality associated with a strong monsoon is probably a controlling factor for the appearance of C_4 plants, but this factor must have operated in tandem with the CO_2 decline discussed above. It is likely that the initial lowering of CO_2 created an environment for C_4 grass that was facilitated by a change in precipitation pattern and caused the expansion of C_4 plants in the Indian Siwalik.

4. Conclusions

It is important to note that, in interpretation of palaeo-climatic data, it is essential that we have a good knowledge of the ages of the sedimentary archive that we are using. In that context, the dating techniques using the variety of radioactive isotopes discussed in Section 2.2 play a crucial role. In the present chapter, we have shown how information on past climate changes can be obtained from geological records by using stable isotope variations, in particular, the oxygen and carbon isotope ratios. There are several other isotopes which have been used recently with the advent of modern high-precision mass spectrometric methods. But oxygen and carbon isotopes continue to find new applications in many areas of earth sciences due to their ubiquitous occurrence and ease of determination.

ACKNOWLEDGEMENTS

We sincerely acknowledge the painstaking work of Dr. A. Sarkar, Dr. P. Sanyal and Dr. Manish K. Tewari who worked in the Stable Isotope Laboratory of PRL during their doctoral thesis to generate the data reviewed here. Mr. R. A. Jani helped in maintaining the Isotope Laboratory.

REFERENCES

Andree, M., E. Moor, J. Beer, H. Oeschger, B. Stauffer, G. Bonani, H. J. Hofmann, E. Morenzoni, M. Nessi, M. Suter, and W. Wölfli. (1984). ^{14}C dating of polar ice. *Nuclear Instruments and Methods B*, 5(2), 385–388.

Andres, R. J., D. J. Fielding, G. Marland, T. A. Boden, N. Kumar, and A. T. Kearney. (1999). Carbon dioxide emissions from fossil-fuel use, 1751–1950. *Tellus*, B51, 759.

Bard, E., M. Arnold, B. Hamelin, N. Tisnerat-Laborde, and G. Cabioch. (1998). Radiocarbon calibration by means of mass spectrometric $^{230}Th/^{234}U$ and ^{14}C ages of corals: An updated database including samples from Barbados, Mururoa and Tahiti. *Radiocarbon*, 40(3), 1085–1092.

Bard, E., B. Hamelin, and M. Arnold. (1996). Deglacial sea-level record from Tahiti corals and the timing of global meltwater discharge. *Nature*, 382(6588), 241–244.

Baver, L. D., W. H. Gardner, and W. R. Gardner. (1972). *Soil Physics*. Wiley, New York, 498pp.

Beck, J. W., D. A. Richards, R. L. Edwards, B. W. Silverman, P. L. Smart, D. J. Donahue, S. Hererra-Osterheld, G. S. Burr, L. Calsoyas, A. J. T. Jull, and D. Biddulph. (2001). Extremely large variations of atmospheric ^{14}C concentration during the last glacial period. *Science*, 292(5526), 2453–2458, doi:10.1126/science.1056649.

Berner, R. A. (1994). GEOCARBII: A revised model of atmospheric CO_2 over Phanerozoic time. *American Journal of Science*, 294, 56–91.

Bidigare, R. R., A. Fluegge, and K. H. Freeman. (1997). Consistent fractionation of C-13 in nature and in the laboratory: Growth-rate effects in some haptophyte algae. *Global Biogeochemical Cycles*, 11, 279–292.

Chappell, J., and N. J. Shackleton. (1986). Oxygen isotope and sea level. *Nature*, 324(6093), 137–140.

Cerling, T. E. (1984). The stable isotopic composition of modern soil carbonate and its relationship to climate. *Earth and Planetary Science Letters*, 71, 229–240.

Cerling, T. E. (1991). Carbondioxide in the atmosphere evidence from Cenozoic and Mesozoic paleosols. *American Journal of Science*, 291, 377–400.

Cerling, T. E., M. J. Harris, J. B. MacFadden, G. M. Leakey, J. Quade, V. Eisenmann, and R. J. Ehleringer. (1997). Global vegetational change through the Miocene/Pliocene boundary. *Nature*, 389, 153–158.

Cerling, T. E., and J. Quade. (1993). Stable carbon and oxygen isotopes in soil carbonates. In: *Climate Change in Continental Isotopic Records* (Eds P. K. Swart, K. C. Lohmann, J. A. McKenzie, and S. Savin). American Geophysical Union, Washington, DC, pp. 217–231.

Dansgaard, G. S., J. W. C. White, and S. J. Johnsen. (1983). The abrupt termination of the younger dryas climate event. *Nature*, 339, 532.

Deines, P., D. Langmuir, and R. S. Harmon. (1974). Stable carbon isotope ratios and existence of a gas phase in the evolution of carbonate ground waters. *Geochimica et Cosmochimica Acta*, 38, 1147–1164.

Dibb, J. E. (2002). The accumulation of ^{210}Pb at Summit, Greenland since 1855. *Tellus B*, 44(1), 72–79.

Ding, Z. L., and S. L. Yang. (2000). C_3/C_4 vegetation evolution over the last 7.0 Ma in the Chinese Loess Plateau: Evidence from pedogenic carbonate $\delta^{13}C$. *Palaeogeography, Palaeoclimatology, Palaeoecology*, 160, 291–299.

Domack, E. W. (2003). Review of radiocarbon dating in Antarctic marine sediments. XVI INQUA Congress, Reno, NV, USA, Session 86, Abstract 86-2.

Duplessy, J. C., G. Delibrias, J. L. Turon, et al. (1981). Oxygen and carbon isotopic composition and biogeographic distribution of planktonic foraminifera in the Indian Ocean. *Palaeogeography, Palaeoclimatology, Palaeoecology*, 33(1–3), 9–46.

Edwards, R. L., J. H. Chen, and G. J. Wasserburg. (1986). ^{238}U-^{234}U-^{230}Th-^{232}Th systematics and the precise measurement of time over the past 500,000 years. *Earth and Planetary Science Letters*, 81, 175–192.

Ehleringer, J. R. (1989). Carbon isotope ratios and physiological processes in arid land plants. In: *Applications of Stable Isotopic Ratios to Ecological Research* (Eds P. W. Rundel, J. R. Ehleringer, and K. A. Nagy). Springer, New York.

Ekart, D. D., T. E. Cerling, I. P. Montanez, and N. J. Tabor. (1999). A 400 million year carbon isotope record of pedogenic carbonate: Implications for paleoatmospheric carbon dioxide. *American Journal of Science*, 299, 805–827.

Emiliani, C. (1955). Pleistocene temperatures. *Journal of Geology*, 63, 538.

Emiliani, C. (1995). *Planet Earth Cosmology, Geology and the Evolution of Life and Environment.* Cambridge University Press, Great Britain, 718pp.

Epstein, S., R. Buchsbaum, H. A. Lowenstam, and H. C. Urey. (1953). Revised carbonate water isotopic temperature scale. *Geological Society of American Bulletin*, 64, 1315.

Erten, H. N. (1997). Radiochronology of lake sediments. *Pure and Applied Chemistry*, 69(1), 71–76.

Etheridge, D. M., L. P. Steele, and R. L. Langenfelds. (1996). Natural and anthropogenic changes in atmospheric CO2 over the last 1000 years from air in Antarctic ice and firn. *Journal of Geophysical Research – Atmosphere*, 101(D2), 4115–4128.

Farquhar, G. D., M. H. O'Leary, and J. A. Berry. (1982). On the relationship between carbon isotope discrimination and the inter-cellular carbondioxide concentration in leaves. *Australian Journal Plant Physiology*, 9, 121.

Francey, R. J., M. R. Manning, and C. E. Allison. (1999). A history of delta C-13 in atmospheric CH4 from the Cape Grim air archive and antarctic firn air. *Journal of Geophysical Research – Atmosphere*, 104(D19), 23631–23643.

Francey, R. J., and P. P. Tans. (1987). Latitudinal variation in oxygen-18 of atmospheric CO_2. *Nature*, 327, 495.

Francois, R., M. Altabet, R. Goericke, D. McCorkie, C. Brunet, and A. Poisson. (1974). Changes in the $\delta^{13}C$ of surface water particulate organic matter across the subtropical convergence in the SW Indian Ocean. *Global Biogeochemical Cycles*, 7, 627.

Friedli, H., H. Lotscher, and H. Oeschger. (1986). Ice core record of the C13/C12 ratio of atmospheric CO2 in the past 2 centuries. *Nature*, 324, 237–238.

Ghosh, P. (2000). Isotopic and elemental studies of Gondwana carbonates and their implications. Unpublished Ph.D. Thesis of (Physical Research Laboratory) Gujarat University, Ahmedabad.

Ghosh, P., S. K. Bhattacharya, and R. A. Jani. (1995). Paleoclimate and palaeovegetation in central India during the upper cretaceous based on stable isotope composition of the palaeosol carbonates. *Palaeogeography, Palaeoclimatology, Palaeoecology*, 114, 285–296.

Ghosh, P., S. K. Ghosh, and S. K. Bhattacharya. (2001). CO_2 levels in the Late Palaeozoic and Mesozoic atmosphere from soil carbonate and organic matter, Satpura basin, Central India. *Palaeogeography, Palaeoclimatology, Palaeoecology*, 170, 219–236.

Hall, B. L., and G. M. Henderson. (2001). Use of uranium-thorium dating to determine past ^{14}C reservoir effects in lakes: Examples from Antarctica. *Earth and Planetary Science Letters*, 193, 565–577.

Heimann, M., and E. Maier-Reimer. (1996). On the relations between the oceanic uptake of CO2 and its carbon isotopes. *Global Biogeochemical Cycles*, 10, 89.

Hinga, K. R., M. A. Arthur, and M. E. Q. Pilson. (1994). Carbon -isotope fractionation by marine phytoplankton in culture – the effects of CO2 concentration, pH, temperature and species. *Global Biogeochemical Cycles*, 8, 91.

IAEA. (2003). http://isohis.iaea.org

Imbrie, J., E. A. Boyle, S. C. Clemens, et al. (1992). On the structure and origin of major glaciation cycles 1. Linear Responses to Milankovitch Forcing. *Paleocenography*, 7, 701.

Imbrie, J., A. Berger, E. A. Boyle, et al. (1993). On the structure and origin of major glaciation cycles. *Palaeoceanography*, 8, 699–735.

Johnson, G. D. (1977). Paleopedology of *Ramapithecus*-bearing sediments, North India. *Geologische Rundschau*, 66, 192–216.

Johnson, M. N., J. Stix, L. Tauxe, P. F. Cerveny, and R. A. K. Tahirkheli. (1985). Paleomagnetic chronology, fluvial process and tectonic implications of the Siwalik deposits near Chinji Village, Pakistan. *The Journal of Geology*, 93, 27–40.

Johnsen, S. J., H. B. Clausen, W. Dansgaard, et al. (1997). The delta O-18 record along the Greenland Ice Core Project deep ice core and the problem of possible Eemian climatic instability. *Journal of Geophysical Research, Oceans*, 102(C12), 26397.

Jouzel, J., C. Waelbroeck, B. Malaize, et al. (1996). Climatic interpretation of the recently extended Vostok ice records. *Climate Dynamics*, 12, 513.

Keeling, C. D. (1961). The concentration and isotopic abundances of carbon dioxide in rural and marine air. *Geochimica et Cosmochimica Acta*, 24, 277.

Keeling, C. D., R. B. Bacastow, A. F. Carter, S. C. Piper, T. P. Whorf, M. Heimann, W. G. Mook, and H. Roeloffzen. (1989). Aspects of climate variability in the Pacific and the Western Americas. In: *Geophysical Monogr. Ser 55* (Ed. D. H. Peterson). AGU, Washington, DC.

Keeling, C. D., T. P. Whorf, M. Wahlen, and J. van der Plicht. (1995). Interannual extremes in the rate of rise of atmospheric carbon dioxide since 1980. *Nature*, 375, 666.

Kirkham, D., and W. L. Powers. (1972). *Advanced Soil Physics*. Wiley-Interscience, New York.

Latorre, C., J. Quade, and C. W. McIntosh. (1997). The expansion of C_4 grasses and global change in the late Miocene: Stable isotope evidence from the Americas. *Earth and Planetary Science Letters*, 146, 83–96.

Levchenko, V. A., R. J. Francey, D. M. Etheridge, C. Tuniz, J. Head, V. I. Morgan, E. Lawson, and G. Jacobson. (1996). The "bomb spike" determines the age spread and age of CO_2 in Law Dome firn and ice. *Geophysical Research Letters*, 23(23), 3345–3348.

Linsley, B. K. (1996). Oxygen-isotope record of sea level and climate variations in the Sulu Sea over the past 150,000 years. *Nature*, 380, 234–237.

Lloyd, J., and G. D. Farquhar. (1994). C-13 discrimination during CO_2 assimilation by terrestrial biosphere. *Oecologia*, 99, 201.

Lorius, C., J. Jouzel, C. Ritz, et al. (1985). A 150,000 year climatic record from Antarctic ice. *Nature*, 316, 591.

Ludwig, K. R., K. R. Simmons, B. J. Szabo, I. J. Winograd, J. M. Landwehr, A. C. Riggs, and R. J. Hoffman. (1992). Mass-spectrometric ^{230}Th-^{234}U-^{238}U dating of the Devils Hole calcite vein. *Science*, 258, 284–287.

Martinson, D. G., N. G. Pisias, and J. D. Hays. (1987). Age dating and the orbital theory of the ice ages –development of a high resolution 0 to 300,000 year chrono stratigraphy. *Quaternary Research*, 27(1), 1–29.

McCrea, J. M. (1950). On the isotopic chemistry of carbonates and a paleotemperature scale. *Journal of Chemical Physics*, 18, 849.

Mook, W. G., J. C. Bommerson, and W. H. Staverman. (1974). Carbon isotope fractionation between dissolved bicarbonate and gaseous carbon dioxide. *Earth and Planetary Science Letters*, 22, 169.

Mora, C. I., S. G. Driese, and L. A. Colarusso. (1996). Middle to late Paleozoic atmospheric CO_2 levels from soil carbonate and organic matter. *Science*, 271, 1105–1107.

Moorman, B. J., F. A. Michel, and A. Wilson. (1998). ^{14}C dating of trapped gases in massive ground ice, Western Canadian Arctic. *Permafrost and Periglacial Processes*, 7(3), 257–266.

Morgenstern, U., L. K. Fifield, and A. Zondervan. (2000). New frontiers in glacier ice dating: measurement of natural ^{32}Si by AMS. *Nuclear Instruments and Methods in Physics Research, B*, 172, 605–609.

Morgenstern, U., M. A. Geyh, H.-R. Kudrass, R. G. Ditchburn, and I. J. Graham. (2001). ^{32}Si dating of marine sediments: Application to the shelf of Bangladesh. *Radiocarbon*, 2B, 909–916.

Pagani, M., M. A. Arthur, and K. H. Freeman. (1999a). Miocene evolution of atmospheric carbon dioxide. *Palaeoceanography*, 14, 273.

Pagani, M., K. H. Freeman, and M. A. Arthur. (1999b). Late Miocene Atmospheric CO2 Concentrations and the Expansion of C4 Grasses. *Science*, 285, 876.

Pagani, M., K. H. Freeman, and M. A. Arthur. (2000). Isotope analyses of molecular and total organic carbon from Miocene sediments. *Geochimica et Cosmochimica Acta*, 64, 37–49.

Paterne, M., L. K. Ayliffe, M. Arnold, G. Cabioch, N. Tisnérat-Laborde, Chr. Hatté, E. Douville, and E. Bard. (2004). Paired ^{14}C and $^{230}Th/U$ dating of surface corals from the Marquesas and Vanuatu (sub-equatorial Pacific) in the 3000 to 15,000 cal yr interval. *Radiocarbon*, 46(2), 551–566.

Pazdur, A., T. Goslar, M. Pawlyt, H. Hercman, and M. Gradzinski. (1999). Variations of isotopic composition of carbon in the Karst environment from Southern Poland, present and past. *Radiocarbon*, 41(1), 81–97.

Pearman, G. I., and P. Hyson. (1986). Global transport and inter-reservoir exchange of carbon dioxide with particular reference to stable isotopic distributions. *Journal of Atmospheric Chemistry*, 4, 81.

Petit, J. R., J. Jouzel, D. Raynaud, et al. (1999). Climate and atmospheric history of the past 420,000 years from the Vostok ice core, Antarctica. *Nature*, 399, 429.

Pisias, N. G., and N. J. Shackleton. (1984). Modeling the global climate response to orbital forcing and atmospheric carbon dioxide changes. *Nature*, 310, 757–759.

Prell, W. L., and J. E. Kutzbach. (1986). Sensitivity of the Indian monsoon to forcing parameters and implications for its evolution. *Nature*, 323, 526–528.

Quade, J., M. L. J. Cater, P. T. Ojha, J. Adam, and M. T. Harrison. (1995). Late Miocene environmental change in Nepal and the northern Indian subcontinent: Stable isotopic evidence from paleosols. *GSA Bulletin*, 107(12), 1381–1397.

Quade, J., T. E. Cerling, and J. R. Bowman. (1989). Development of Asian monsoon revealed by marked ecological shift during the latest Miocene in the northern Pakistan. *Nature*, 342, 163–166.

Rau, G. H., T. Takahashi, and D. J. Des Marais. (1989). Latitudinal variations in plankton delta C-13 implications for CO2 and productivity in past oceans. *Nature*, 341, 516–518.

Rau, G. H., P. N. Froelich, T. Takahashi, and D. J. Des Marais. (1991). Does Sedimentary Organic $\delta^{13}C$ Record Variations in Quaternary Ocean [CO2(aq)]? *Palaeoceanography*, 6, 335–347.

Rau, G. H., T. Takahashi, and D. J. Des Marais. (1992). The relationship between delta C-13 of organic matter and CO2 (aq) in Ocean surface water – data from a JGOFS site in the northeast Atlantic ocean and a model. *Geochimica et Cosmochimica Acta*, 56, 1413–1419.

Raymo, M. E., and M. Horowitz. (1996). Organic carbon paleo-pCO(2) and marine-ice core correlations and chronology. *Geophysical Research Letters*, 23, 367–370.

Raymo, M. E., and W. F. Ruddiman. (1992). Tectonic forcing of late Cenozoic climate. *Nature*, 359, 117–122.

Rea, D. K., and S. M. Colman. (1995). Radiocarbon ages of pre-bomb clams and the hard water effect in lakes Michigan and Huron. *Journal of Paleolimnology*, 14(1), 89–91.

Rohling, E. J., M. Fenton, and F. J. Jorissen. (1998). Magnitudes of sea-level lowstands of the past 500,000 years. *Nature*, 394, 162–165.

Rozanski, K., A. L. Araguas, and R. Gonfiantini. (1993a). Isotopic patterns in modern global precipitation. In: *Climatic Change in Continental Isotopic Records* (Eds P. K. Swart, K. C. Lohmann, J. McKenzie, and S. Savin), Geophysical Monograph Series (IUGG), pp. 1–36.

Rozanski, K., S. W. Fowler, and E. M. Scott. (1993b). Global ocean studies, the greenhouse effect, and climate change: Investigating interconnections. *IAEA Bulletin*, 2, 25–32.

Sangode, S. J., J. Bloemendal, R. Kumar, and S. K. Ghosh. (2001). Plio-Pleistocene pedogenic change in the Siwalik Palaeosols: A rock magnetic approach. *Current Science*, 81(4), 387–392.

Sangode, S. J., R. Kumar, and S. K. Ghosh. (1996). Magnetic polarity stratigraphy of the Siwalik sequence Haripur area (H.P.), NW Himalaya. *Journal of Geological Society of India*, 47, 683–704.

Sangode, S. J., R. Kumar, and S. K. Ghosh. (2003). Magnetic polarity stratigraphy of the Late Miocene Siwalik group sediments from Kangra re-entrant, H.P., India. *Himalayan Geology*, 24(1), 47–61.

Sanyal, P. (2004). Isotope geochemistry of Siwalik sediments and signature of past climate change. Unpublished Ph.D. Thesis (Physical Research Laboratory), Gujarat University, Ahmedabad.

Sanyal, P., S. K. Bhattacharya, R. Kumar, S. K. Ghosh, and S. J. Sangode. (2004). Mio-Pliocene monsoonal record from Himalayan foreland basin (Indian Siwalik) and its relation to vegetational change. *Palaeogeography, Palaeoclimatology, Palaeoecology*, 205, 23–41.

Sarkar, A. (1989). Oxygen and carbon isotopes in Indian Ocean sediments and their palaeoclimatic implications. Unpublished Ph.D. Thesis (Physical Research Laboratory), Gujarat University, Ahmedabad.

Sarkar, A., R. Ramesh, and S. K. Bhattacharya. (1990). Oxygen isotope evidence for a stronger winter monsoon current during the last glaciation. *Nature*, 343, 549–551.

Shackleton, N. J., and N. D. Opdyke. (1973). Oxygen isotope and palaeomagnetic stratigraphy of Equatorial Pacific core V28-238: Oxygen isotope temperatures and ice volumes on a 105 year and 106 year scale. *Quaternary Research*, 3, 39.

Shackleton, N. J. (1977). Oxygen isotope stratigraphic record of late Pleistocene. *Philosophical Transactions of the Royal Soc. of London, Series B, Biological Sciences*, 280, 169.

Shackleton, N. J. (1987). Oxygen isotopes, Ice volume and sea level. *Quaternary Science Reviews*, 6(3–4), 183–190.

Staubwasser, M., G. M. Henderson, P. A. Berkman, and B. L. Hall. (2004). Ba, Ra, Th, and U in marine mollusc shells and the potential of ^{226}Ra/Ba dating of Holocene marine carbonate shells. *Geochimica et Cosmochimica Acta*, 68(1), 89–100.

Stuiver, M., and T. F. Braziunas. (1993). Modeling atmospheric ^{14}C influences and ^{14}C ages of marine samples to 10,000 BC. *Radiocarbon*, 35(1), 137–189.

Suckow, A., U. Morgenstern, and H.-R. Kudrass. (2001). Absolute dating of recent sediments in the cyclone-influenced shelf area off Bangladesh: Comparison of gamma spectrometric (^{137}Cs, ^{210}Pb, ^{228}Ra), radiocarbon, and ^{32}Si ages. *Radiocarbon*, 43(2B), 917–927.

Sundquist, E. T. (1993). The global carbon dioxide budget. *Science*, 19, 934.

Suzuki, T., A. Maki, K. Ohta, K. Kamiyama, Y. Fujii, and O. Watanabe. (1996). Determination of lead-210 in an ice core from an Arctic Glacier by an alpha-ray measurement. *Analytical Sciences*, 12, 923–926.

Tan, P. P. (1981). Carbon cycle modelling. In: *SCOPE 16* (Ed. B. Bolin). Wiley, Chichester.

Tandon, S. K., and D. Narayan. (1981). Calcrete conglomerate, casehardened conglomerate and cornstone – A comparative account of pedogenic and non-pedogenic carbonates from the continental Siwalik Group, Punjab, India. *Sedimentology*, 28, 353–367.

Thomas, J. V., B. Prakash, and R. Mahindra. (2002). Lithofacies and paleosol analysis of the Middle and Upper Siwalik Groups (Plio-Pleistocene), Haripur-Kolar Section, Himachal Pradesh, India. *Sedimetary Geology*, 150, 343–366.

Tiwari, M. (2005). Quaternary geochronology and palaeoclimatology of the Indian Ocean. Unpublished Ph.D. Thesis (Physical Research Laboratory), Gujarat University, Ahmedabad.

Tiwari, M., R. Ramesh, B. L. K. Somayajulu, A. J. T. Jull, and G. S. Burr. (2005). Early deglacial (19–17 ka) strengthening of the northeast monsoon. *Geophysical Research Letters*, 32, L19712.

Urey, H. C. (1947). The Thermodynamic Properties of Isotopic Substance. *Journal of Chemical Society*, 562.

Wagenbach, D. (1989). *Environmental Records in Alpine Glaciers and Ice Sheets*. Wiley, Dahlem Konferenzen, pp. 69–83.

Watson, R. T., and I. R. Noble. (2002). *Challenges of a Changing Earth*. Springer Verlag, (IGBP Series).

Yakir, D., and L. Sternberg. (2000). The use of stable isotopes to study ecosystem gas exchange. *Oecologia*, 123, 297.

Yurtsever, Y., and J. R. Gat. (1981). Atmospheric waters. In: *Stable Isotope Hydrology: Deuterium and Oxygen-18 in the Water Cycle* (Eds J. R. Gat and R. Gonfiantini). IAEA, Vienna, pp. 103–142.

Zachos, J., M. Pagani, L. Sloan, E. Thomas, and K. Billups. (2001). Climate response to orbital forcing across the oligocene-miocene boundary. *Science*, 292, 686.

CHAPTER 9

Analysis of Radionuclides

Axel Suckow[1,2,*]

Contents

1. Introduction

This chapter deals with the methods by which radionuclides are detected and quantified. It is a little bit in contrast to the rest of this book,

*Corresponding author. Tel: +49 (0)511 643-2527; Fax: +49 (0)511 643-3665
E-mail address: Axel.Suckow@liag-hannover.de

[1] Leibniz-Institut für Angewandte Geophysik (LIAG), Stilleweg 2, 30655 Hannover, Germany
[2] Isotope Hydrology Laboratory, Department of Nuclear Sciences and Applications, International Atomic Energy Agency (IAEA), Wagramerstrasse 5, A-1440 Vienna, Austria

Radioactivity in the Environment, Volume 16
ISSN 1569-4860, DOI 10.1016/S1569-4860(09)01609-X

which focuses mainly on how to use and interpret the radionuclide data obtained from environmental samples. Why should it be of interest for the reader how the data are produced? Can't the scientist just take the samples, send them to a laboratory of her or his choice, and be lucky with the numbers that come back? Why bother about the methods that make numbers out of samples?

Indeed this kind of loveless treatment of one of the most crucial steps in science – the measurement itself – is an increasing trend in our modern world. And probably with more and more specialized laboratories and standardized methods, the number of persons who have a detailed insight into the measurement process and who really understand this crucial step will further decrease. But besides sentimental feelings about a development not very favorable for the "machine slaves," there are some serious reasons why it is useful for those who apply the data to know some details about how the data are produced.

The most important reason is probably that the most severe errors during any environmental investigation occur during the sampling itself and during sample storage and transport. And this way of taking, handling, and storage is directly related to the measurement technique applied. A water sample taken through a metal tube will hardly be suitable to investigate the concentration of this metal in the environment. A sample for a gas tracer has to be treated differently than one for a dissolved mineral. A sample for a well soluble gas like CO_2 in total dissolved inorganic carbon (TDIC) can be easily precipitated as $BaCO_3$ or $SrCO_3$ in the field. But nevertheless laboratories need the carbonate originating from between several milliliters and more than a hundred liters depending on which carbon isotope (^{13}C or ^{14}C) is to be measured and how the laboratory does the measurement. A radioactive noble gas like radon can perhaps be taken by fast filling and tightly screwing of a glass bottle, whereas for sampling of a volatile less soluble noble gas like helium even a glass ampoule flame sealed in the field under ultra pure nitrogen atmosphere will not be sufficient – if the wrong kind of glass is used.

In the event that the samples are taken correctly and reach the home base in favorable shape, it is still useful for those who apply the data to know how the data are produced. ^{14}C to be measured in a laboratory using proportional counters will normally take at least 4 weeks until the sample can be measured and needs normally gram amounts of carbon. In contrast the more expensive AMS technique enables the scientist in principle to work with the numbers after less than a week and will do the job with milligram amounts of carbon. But the scientist will have to pay a higher price, which is justified due to the higher measurement effort and the more expensive equipment necessary.

But imagine even if this step is also taken: the samples are taken correctly, reached the lab in good shape; a spreadsheet of numbers is delivered back to the customer. Still it is very useful for the submitter to know some details about the measurement process. A tritium value of 2 TU (tritium units) from a 1-L groundwater sample has to be interpreted very differently if it originates from a laboratory measuring with liquid scintillation counting without enrichment than when the same value from the same sample originates from a laboratory using the ^{3}He ingrowth technique: in the former case, this value means "No modern water was detected" in the latter case this means "there definitely is a contribution of modern water, being in the northern hemisphere approximately equal to an admixture of 20% recent rain."

This chapter is contained in this book to help you understand why sampling techniques and data interpretation will be very different if different measurement techniques are applied.

2. Radiometry

There are different ways to measure radioactive nuclides in environmental samples. The oldest and still most common methods use the radioactive decay itself. They use techniques of detecting the ionizing particles released during radioactive decay. This "activity," the number of disintegrations per unit time, is evaluated quantitatively in terms of how many atoms disintegrate per gram or cubic centimeter of sample and time, which is called "specific activity." Normally, but not necessarily, these methods are specialized to the kind of radioactivity (alpha, beta, gamma radiation), and they use certain properties of the radiation, like the kind of radiation or particle energy. This field of laboratory techniques is known as radiometry.

2.1. Basics of radiometry

From the introductory chapters of this book (as well as from the high-school physics courses), the reader knows already that three common types of particles can originate in radioactive decay. Figure 1 again illustrates them with the example of the natural decay chain of uranium, namely of ^{238}U. In this picture, the arrows pointing down to the left indicate the emission of alpha particles (helium cores), the arrows to the upper left indicate the emission of beta particles (electrons), and the wave-like arrows to the upper left indicate that electromagnetic waves in the form of gamma particles are emitted. From this picture, the reader can learn some important things: For instance that the product of a radioactive decay can be radioactive again like ^{222}Rn, which is the daughter of ^{226}Ra. It can also be stable like ^{206}Pb originating from ^{210}Po. The reader can learn that radioactive decay always

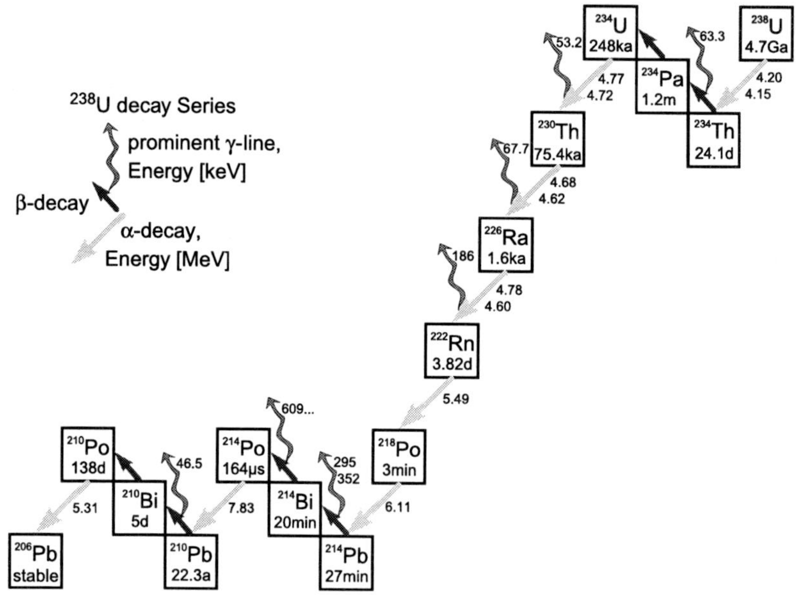

Figure 1 The natural radioactive decay series of ^{238}U.

changes the chemical element: This is due to the fact that a charged particle leaves the atomic nucleus during radioactive decay. Since the number of charges in the nucleus equals the number of electrons in an atom, and since the chemical properties are determined by the number of electrons, chemical behavior can change drastically after radioactive decay. For instance, the heavy metal radium (^{226}Ra) converts into the noble gas radon (^{222}Rn). Or the reader can learn from Figure 1 that a certain nuclide emits either alpha or beta particles, but not both. Therefore, if we have a beta-detecting device at hand, we might be unable to detect a nuclide like ^{210}Po. Or that both alpha and beta decay sometimes are associated with emission of a gamma particle – but there is no nuclide emitting only gamma particles alone. And obviously there also exist nuclides like ^{218}Po or ^{238}U where the radioactive decay seems not to be associated with the emission of a gamma particle. Maybe we still know from school physics that the ranges of alpha, beta, and gamma radiation are very different. As we will see when explaining the detectors in detail, this is one of the crucial points for the measurements and the reason for much of the effort needed to measure radioactivity.

Meanwhile Hollywood and movies like "007 James Bond: Dr. No" taught everybody how radioactivity is detected: There exist devices looking like tubes wired to a loudspeaker and when they come near to a source of radioactivity they start to make a noise and crackle like grandma's old radio set! No, that is NOT the reason why it is called radioactivity! But the basic electronic principle behind nearly all detectors of alpha, beta, or gamma radiation is fairly simple and can be explained easily with this device of the Dr. No movie: A large resistor is connected to a power supply and the circuit is closed with something that is able to detect pulses of voltage or current. Figure 2 illustrates the principle and should be simple enough that those readers not very interested in electronics (like me) can also understand it. In the case of the James Bond movie, the power supply is a battery, a tube filled with a special gas is the resistor, and the loudspeaker is the detecting device. But how does this arrangement work? How the pulses come into being, if there is nothing other than a constant voltage connected to a resistor which is so large that no current can flow? The secret is that radioactive particles have enough energy to produce ions. If they hit atoms like the ones in the gas-filled tube, they are able to separate

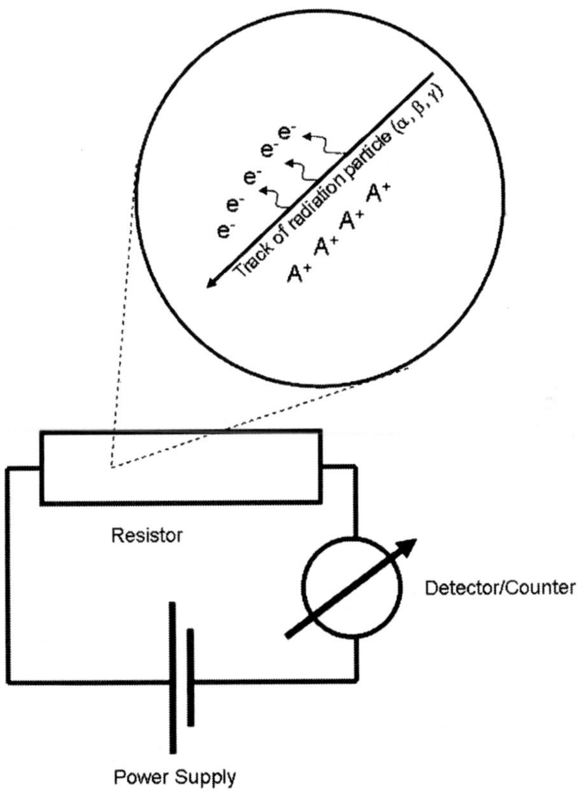

Figure 2 The electronic principle of a detector for radioactive particles. A resistor, large enough that no current can flow, is connected to a power supply and a detector. A radioactive particle produces a cloud of negative and positive charges in the resistor and subsequently a short flow of electric current in the circuit, which is detected.

electrons from atoms, thus producing pairs of charged particles – the negatively charged electrons and the positively charged atoms with a missing electron, the ions. With this pair of charges the resistor is no longer a real resistor, but something in which current flows – and every such event is heard as one crack in the speaker.

Those with knowledge of electronics will have noticed that the picture was pretty much simplified; the power supply normally is one for high voltage, the signal from the resistor has to be amplified, the pulses might be shaped and all these things. Yes, you are right, and these details can be handed over to the wizards in electronics but the basic principle behind them stays the same. But there is one important refinement applied in many of the devices to be discussed and that goes a little bit further than the James Bond crackles; there are some special kinds of these resistors in Figure 2 that not only produce a simple indifferent "crack" but that produce different pulses depending on the energy of the ionizing particle. These detectors are known as "proportional detectors" or "proportional counters" because the charge released within the resistor is proportional to the energy of the particle in the radiation. This gives the measurement device not only the ability to detect *that* there was a radioactive decay but also the possibility to see *what kind* of decay occurred. To be more exact, they measure the energy the particle had. If we look back to Figure 1, we will notice that this refinement makes it possible to distinguish the gamma radiation of ^{210}Pb with an energy of 46.5 keV from the one produced by ^{214}Bi having more than ten times this energy, namely 609 keV. We will see an example of how this looks later when discussing the detection of alpha and gamma radiation.

Radioactivity is thought of normally as something existing only in very specialized environments like a nuclear power plant, a test site for nuclear bombs or perhaps in a hospital where the doctors apply such substances against cancer or to gain deeper insight into our body. But the fact that radionuclides are applied in environmental studies with noncontaminated samples tells us that our whole surroundings are more or less radioactive. Yes, radioactivity is even a common phenomenon in our own body. This has some implications for the measurement. If our surroundings are radioactive to a certain degree, if even the measurement device itself contains a certain amount of radioactivity, and if we keep in mind that the ranges of some kinds of radiation like gamma rays can be in the order of several meters, how can we then distinguish between the radioactivity in the sample and that from the device and its surroundings? The answer is called passive and active shielding. And the technique of measuring radioactivity lower than the normal ambient radioactivity is called low-level radiometry.

The first thing to be done in low-level radiometry is of course to construct the counting device itself with special materials containing very low amounts of radioactivity. This problem in itself gave and gives enough work for several master and diploma theses, to find construction material such as copper, tin, glass, Teflon, etc. with a low content of radioactive nuclides.

The second action is to place the counting device into a passive shielding. Normally this involves at least 10 cm of special lead. The lead itself contains only a low amount of radioactive nuclides if it is molten from ore originating from special mines or if it is recycled from ballast of sunken ships. But high-energy background radiation absorbed within the lead can induce secondary radiation of minor energy. Therefore inside the lead chamber, another shielding of some millimeter of copper is often used, to shield the device from this secondary radiation. Copper itself can produce X-rays that can be shielded with a third layer inside the copper, for example, made of Plexiglas. Also boron-loaded paraffin is used to stop secondary neutrons induced from radiation reacting in the primary shielding. This kind of shielding with materials becoming lighter to the inside is known as "gradient-Z shielding," because Z is the letter that chemists use for the number of protons in an atom and because this number Z decreases from outside to inside of the shielding. The left side of Figure 3 shows a possible arrangement of shielding materials and counters.

In the case of low-level gamma spectrometry, one counter in a gradient-Z shield might already give good results. In case of beta emitting nuclides to be detected, with low ambient concentration, long half-lives

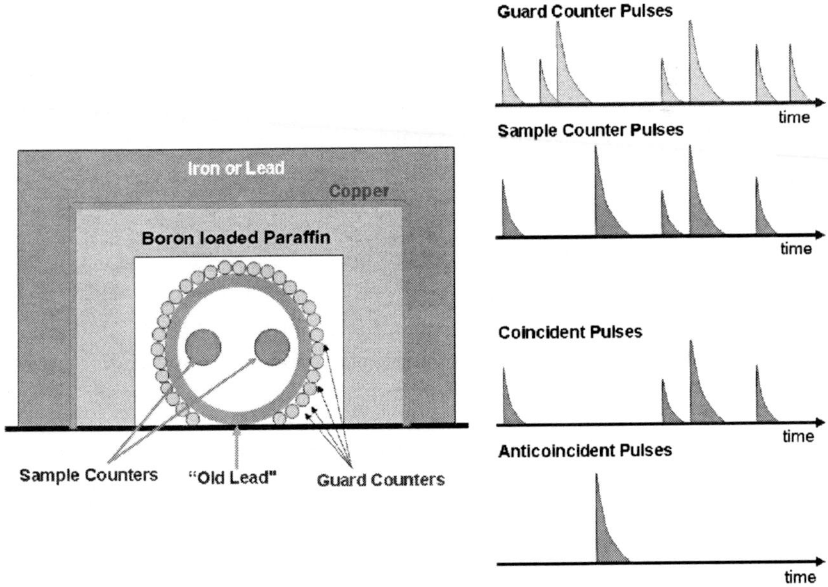

Figure 3 Principle of passive shielding and active shielding with anticoincidence guard counters.

and badly defined particle energy, it is necessary to apply additional construction to diminish the background of cosmic radiation, which easily penetrates the lead shielding. For instance, the Austrian National Low Level Tritium laboratory in the Arsenal in Vienna is situated in the basement and is constructed with roughly 1 m wall thickness of special low-level concrete. The low-level laboratory in the University of Bern is situated in a side tunnel of the railway tunnel under nearly 100 m of granite. The counting devices for the Gallex experiment investigating solar neutrinos are situated in a highway tunnel near Gran Sasso in Italy. As a rule of thumb, every meter of earth or rock above the counting device weakens the cosmic part of background radiation by 10%.

Besides this passive shielding, very often an active shielding against ambient radiation is applied. For this, the counter for the sample is surrounded by counters detecting ambient radiation. This principle is illustrated on the right side of Figure 3; if a high-energy particle from cosmic radiation penetrates the passive shielding, then it should produce simultaneous counts in the outside guard counter as well as in the inner counter measuring the sample. These pulses are obtained at the same time since the electronic devices are generally not able to resolve counts following each other faster than something like a nanosecond. A special electronic system selects only pulses that are obtained in the inner counter alone without a corresponding pulse in the outer counter. Since these counts are those not coinciding with outside counts, these are called "anticoincident" counts and this kind of active shielding is called "anticoincidence shielding." To give an idea about the decrease in background following from this trick, for a 1 L radiocarbon gas proportional counter in the Hannover geochronology laboratory typically 1,000 counts per min are detected, of which more than 990 are coincident and discarded, and less than ten anticoincident counts are really attributed to the sample.

2.2. Alpha radiation

2.2.1. Basics

Alpha particles consist of two protons and two neutrons. From the chemical point of view, the alpha particle is the nucleus of a helium atom, without electrons. Alpha particles of a certain radionuclide exit the nucleus with a very well-defined energy which is in the order of several megaelectronvolts. So proportional counting methods for alpha radiometry can identify the energy since the counts for this energy form a well-defined peak within the spectrum and have the great advantage that they can identify the nuclide from which the particle is emitted. The alpha energy is high relative to beta particles, which generally have an energy of only several tens to hundreds of kiloelectronvolts. Although the energy as well as the mass of alpha particles is higher than for beta particles, their range in matter is generally lower, the reason being that the alpha particles carry two charges and that the energy loss in matter is proportional to the square of the particle charge. So alpha particles of 1 MeV have only 1/4 of the range of betas of this energy. Due to this general low range of the radiation, it is normally not possible to measure the alpha activity of a sample "as it is"; absorption of the radiation within the sample itself diminishes the energy of the particle to be detected and makes spectrometry impossible. Another important fact for the measurement is that most relevant alpha-emitting natural radionuclides are metals, which can be deposited as a thin layer on disks. Alpha spectrometers are, therefore, optimized to measure the particles emitted from such metal disks. So the measurement problem can be defined as to "extract the element in question by a chemical process from the sample, and bring it to an environment where the energy of the alpha particle can be detected without prior energy loss."

2.2.2. Chemical pretreatment

Since the chemical pretreatment is of crucial importance for the radiometric measurement, the combination of chemical pretreatment and radioactivity measurement was given its own name – "Radiochemistry". For nearly all environmental samples to be measured by alpha spectrometry, the sample pretreatment starts with dissolution of the sample material. Chemists have developed a whole bunch of methods to transform solid samples into aqueous liquids, mostly involving acids. They range from the simple dissolution of carbonates with phosphoric or hydrochloric acid up to the treatment with hydrofluoric acid and heat, which even dissolves silicates. It is also a common technique to expose the sample to subsequent steps of different combinations of solvents that each dissolves a certain part of the sample and to analyze each of these soluble fractions separately. This pretreatment of course is dependent on the kind of sample (carbonate, soil, peat, marine sediment, ...) and on the scientific question to be answered with the analysis (age-dating, determination of mobile phases, alteration of exposed surfaces, ...). Therefore, a very close cooperation between submitter ("problem owner") and laboratory is required.

After transforming the sample into a liquid, the elements to be measured have to be separated. Today this is mostly done using ion exchanger resins, where sometimes several precipitation and dissolution steps are intercalated. As examples, the purification and separation process for U, Th, and Po analysis are shown in Figure 4 which is a combination of descriptions in Flynn (1968), Hennig et al. (1980), and Heijnis and van der Plicht (1992).

After the element in question is separated chemically, it is electroplated onto the surface of a small disc. Here special care is necessary to obtain a homogeneous layer. Also the layer must not be too thick to avoid self-absorption of the emitted alpha particles in the layer itself. In certain

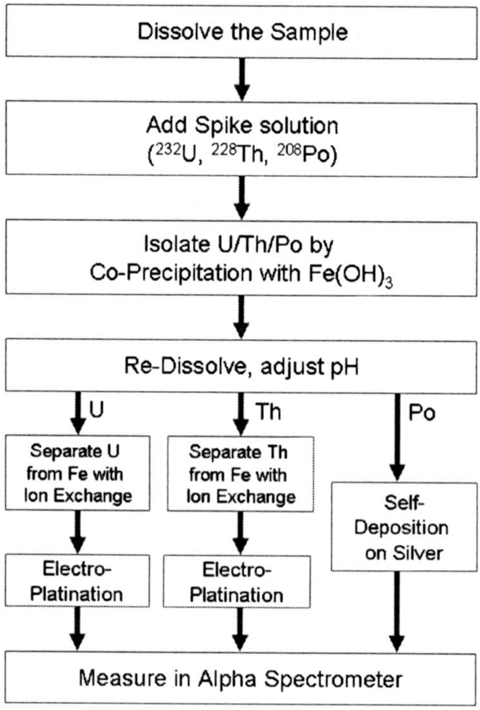

Figure 4 Chemical steps for alpha-spectrometric analysis of U, Th, and Po.

cases, electroplating is not necessary, for example, polonium for the analysis of ^{210}Po (or ^{210}Pb) can be spontaneously deposited on silver discs.

Since all this chemical pretreatment is normally not as reproducible as desired for precise and quantitative analysis of radioactivity in natural samples, it is usual to add a yield tracer ("spike") in the very first step of the preparation process. This is a solution containing well-defined concentrations of the radioactive element in question but with a different isotopic composition, for example, for the analysis of ^{234}U, ^{238}U, and ^{235}U, spike solutions containing mainly ^{232}U can be used; for the analysis of thorium isotopes, ^{228}Th can be used. It can be assumed that the detector efficiency is identical for the detection of alpha particles from the added spike isotope as for the isotope under investigation. Therefore, the use of a spike as yield tracer reduces the absolute determination of activities to the relative comparison of peak areas in the alpha spectra, which is possible with much higher precision.

2.2.3. Measurement system

For some decades now, solid-state silicon semiconductor detectors have been and still are state of the art for the detection of alpha particles. These usually are either ion-implanted silicon or surface barrier silicon detectors. Details of the advantages and disadvantages of the detector type can easily be requested from the supplier. Measurement of the plated disks containing the sample is carried out in a vacuum chamber. Here the sample is exposed directly to the detector, being separated only by a few millimeters from the disc surface. Neither active nor passive shielding is necessary. This is because of several factors; first the detectors are very small, so their detection efficiency for ambient radiation is very low, especially in the high-energy range of the alpha particles where electrons, protons, muons, and gamma radiation of this energy mostly cross the detector without interaction; secondly the energy of the alpha particles in question is very well defined, so any counted event with a different energy can be distinguished in the alpha spectrum. Evaluation of activity is done using the peak area of the spike nuclide and the known amount of spike added (in Bq or dpm) and comparing these numbers to the peak area of the nuclide in question. From these and the sample mass prior to dissolution, the absolute specific activity of the sample in Bq/kg or dpm/g can be calculated.

It is evident that this measurement configuration can detect only a maximum of 50% of the decay events for the principle reasons that every alpha particle emitted in the direction away from the detector cannot be counted. Also the detection efficiency varies slightly with the distance of the disc from the detector. And finally the silicon detectors tend to age due to the radiation damage in the sensitive layer. The use of a yield tracer, therefore, accounts not only for variations within the chemical preparation but also for sample-to-sample variations during the detection process. Additional pitfalls are that the detector surface can be contaminated, for instance if the sample disc comes in contact with it when the samples are changed. Therefore, frequent but time-consuming background measurements of the system without a sample are necessary for quality assurance. Also blank sample preparations are usual where the whole chemical procedure is undertaken without a sample, to recognize if any of the chemical substances used contains small amounts of the radionuclides in question.

2.2.4. Applications

Alpha spectrometry is a standard technique in nuclear chemistry and therefore has a wide range of applications. Among the most important is the study of timescales of natural processes. An example is the investigation of sedimentation rates by disequilibrium methods; here the fact is used that sedimentation processes often cause radioactive disequilibrium. This can be due to a missing nuclide in the decay chain, maybe because the nuclide in question is removed by natural processes prior to the sedimentation. One of the most prominent nuclide pairs in use here is $^{230}Th/^{234}U$, which uses the radioactive deficiency of ^{230}Th in freshly precipitated carbonate materials;

the particle-reactive thorium is quickly removed from the soil or ocean water but the uranium stays in solution and is incorporated into the carbonate precipitated. This can be used to date carbonate sediments like speleothems (Hennig et al., 1980) and corals, but also to date peat (Heijnis and van der Plicht, 1992). Another often used method is based on having a surplus of a radionuclide relative to the mother nuclide. Relevant methods use excess ^{230}Th, excess ^{231}Pa, or excess ^{210}Pb (via determination of the granddaughter nuclide ^{210}Po). But natural transport processes are also investigated using nuclides of the U/Th decay chain and alpha spectrometry as the measurement technique. Examples are scavenging processes in the ocean and rock/water interactions between groundwater and rocks. The standard textbook prior to this series, which also deals with many theoretical details of these methods, was by Ivanovich and Harmon (1992).

In recent studies especially those using the radionuclides with long half-lives, alpha spectrometry tends to be replaced by mass-spectrometric methods like thermal ionization mass spectrometry (TIMS) or multi-collector inductive coupled plasma mass spectrometry (MC-ICP-MS). Both methods need much smaller sample size since they can count the atoms directly and do not have to wait until the atom decays (see Section 4 for details). Where for TIMS, it is still necessary to perform many complicated chemical pretreatment steps similar to Figure 4, MC-ICP-MS can often handle samples after simple dissolution.

2.3. Beta radiation

2.3.1. Basics

One of the important difficulties with beta radiation is that, due to the nuclear physics of this decay process, there is no well-defined emission energy of the electron released during the decay. There is a defined maximum energy and a broad distribution of electron energies beneath. Spectrometry during the measurement, therefore, is not as useful as for alpha and gamma radiations since there are no sharp peaks. Instead, broad windows are defined, and the events in these windows are registered. It is nevertheless an advantage if the counts are collected and stored as an energy spectrum in a multichannel buffer, because only this enables later variations of the window parameters during evaluation. Especially for low-level beta radiometry, it can also be advantageous to record the charges released by the disintegration as a function of time, since this makes it possible to evaluate the rise time or shape of the pulse. There are selection methods for each peak after rise time or pulse shape analysis that can considerably decrease the background of the method (Pandola et al., 2004).

A second important difference is that it does not make much sense to talk about the "range" of beta radiation, since the emitted electrons are reflected by interactions with matter and perform a random walk.

Therefore for beta radiometry, the sample in most cases is not outside a detector but inside, such that the electron energy is released to some kind of detector material.

2.3.2. Gas proportional counters

For gas proportional counters, the sample is converted to a gas and this gas is filled into a detector. The detector consists of a cylindrical metallic tube containing a thin wire in the center (see Figure 5, left). A high voltage is applied between the wire and the tube wall. The electron resulting from any disintegration creates a cloud of electron–ion pairs in the gas, which drift to the wall and wire, respectively. This charge pulse is amplified, normally sorted by height and recorded.

2.3.2.1. Counter construction and counting gases. The high voltage between wire and tube has to have a certain range depending on the construction (tube and wire diameter, kind and pressure of gas); if it is too small the electrical field in the counter is not sufficient to accelerate the electrons to the wire before they are caught by gas molecules. However, if the high voltage is too high, the electric field will create electron–ion pairs and pulses with residual ions from the last disintegration (the ions travel much slower than the electrons). If the high voltage is even stronger, the field will create ions without any radiation, as in a fluorescent lamp. In between there is a range of high voltage values where each disintegration creates exactly one count, called the "plateau region." In the plateau region, the amount of charge in the pulse is proportional to the energy of the detected particle – which is where the name of the detector comes from.

Many parameters influence the performance of the gas counter. Qualitatively, it is evident that the minimum and maximum voltage will be smaller for a thinner wire; since the electrical field is radially symmetric it is strongest at the wire and here it is stronger the thinner the wire is. Any proportional gas counter has a certain "dead time," which is mainly the time the electrons need to travel to the wire and the ions need to travel sufficiently away from the wire – finally to the wall of the tube. This means that the dead time becomes smaller if the diameter of the wire is smaller (higher strength of electrical field). It also becomes longer if the gas pressure is higher. The dead time is typically in the range of several tens of microseconds and within this time no other disintegration can be measured. This means, however, that the maximum measurable count rates are in the order of 10,000/s.

Each counter has a certain "dead" volume, where disintegrations will not produce a count. This for instance is near the walls, because the particle can hit the wall before an electron cloud is created, and also near the ends, because here the electrical field does not have the necessary cylindrical

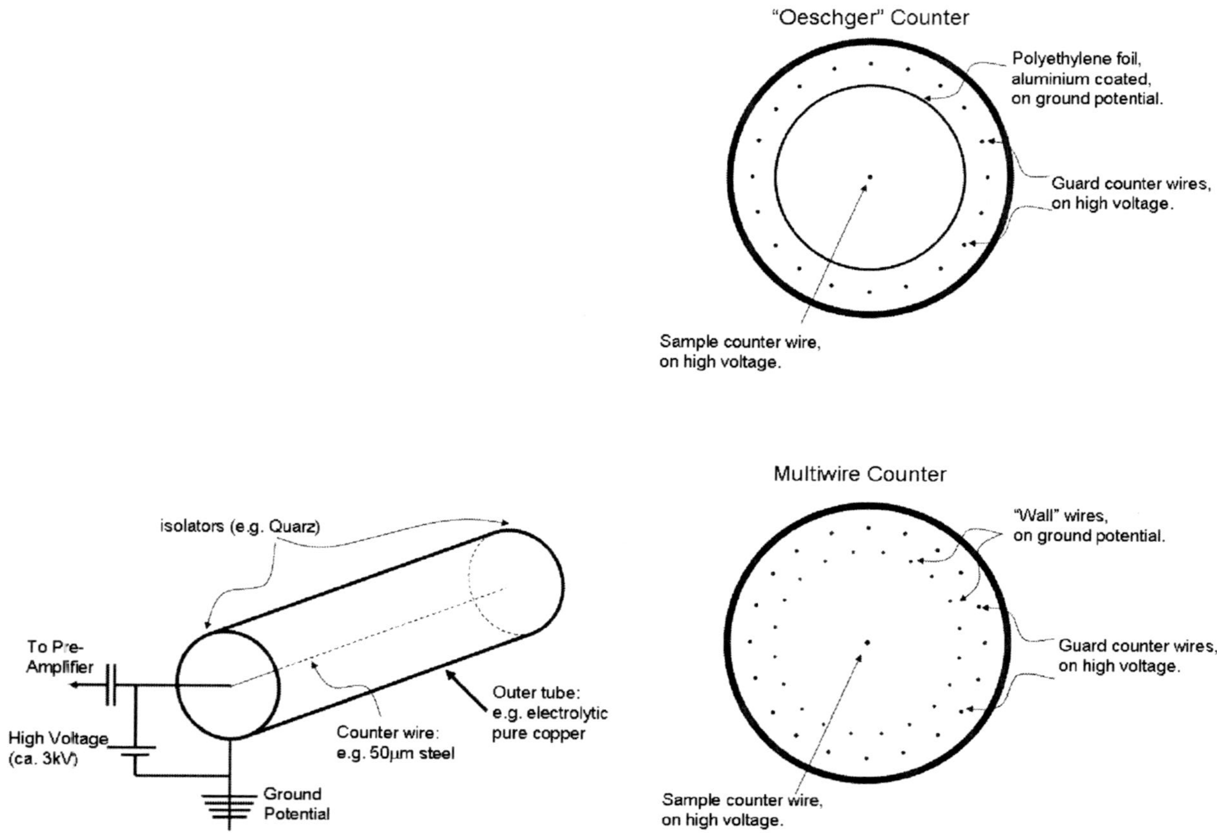

Figure 5 Construction principles of gas proportional counters.

symmetry. Besides, the total volume, or better the number of atoms inside the counter, influences sensitivity – the more atoms, the more disintegrations per time occur, leading to a better signal-to-noise ratio. These facts lead to large counters filled with high gas pressures, and which have the shape of long cylinders. But the attempt to optimize in this direction also leads to large sample sizes, which might not be available. It is also counteracted by the fact that background counts in part enter through the surface, which means that small counters have advantages for low-level applications. Sometimes also the guard counter has a defined size, shape or volume, which places constraints on the maximum size of the sample counter.

The counting gases that are used in gas proportional counters depend on the sample material and into what gas the nuclide in question can readily be converted. In any case, it has to have certain properties to be suitable as counting gas; for a given counter the onset voltage of the plateau has to be small enough, the plateau has to be sufficiently flat, long and stable. These properties are also influenced by the kind and purity of the counting gas; the drift times of electrons in gases differ, mainly due to differences in the cross-sections of electron–molecule interactions. Short drift times result in short rise-times of the pulses and any impurities in the counting gases have less probability to catch the electrons and to disturb the counting process. Argon and methane have very small cross-sections of electron-interaction ("Ramsauer-effect") and therefore a mixture of 10% methane in argon, known as "P-10" is a quite ideal counting gas. But also acetylene (C_2H_2), ethane (C_2H_6), CO_2, or mixtures of a sample gas and P-10 are used. For measurement of tritium in water, hydrogen is prepared by reduction with magnesium. Either this hydrogen is used directly together with P-10 or it is converted to methane or ethane using a catalyst. For measurement of radiocarbon, the sample is first converted to CO_2. If this is used as counting gas directly, it has to be purified, since the interaction of CO_2 with electrons is strong. In some laboratories, the CO_2 is subsequently converted to acetylene using water and an intermediate step forming lithium carbide. Other laboratories convert the CO_2 to methane using a ruthenium catalyst or even the acetylene to ethane using a platinum or palladium catalyst. Of course care has to be taken that the other chemical adducts forming the counting gas are free of radioactivity, for example, the CO_2 which is used to form methane to measure a tritium sample has to be free of radiocarbon, and if radiocarbon is to be measured, the water to form acetylene from carbide has to be free of tritium.

Since beta radiation has no defined energy, it allows no good energy discrimination between sample and background counts. Therefore, passive and active shielding of background radiation is very important here. Passive shielding is obtained by placing the laboratory into the basement of the building or even deep underground, and by lead surrounding the detector. Active shielding is realized by positioning guard counters around the sample counter as shown in Figure 3. Of course this construction needs a certain effort and therefore often many sample counters are positioned within the same guard counter and lead housing. One of these constructions was realized as the "Heidelberg Organ" (Schoch et al., 1980) with nine sample detectors of 4 L volume within a housing of quadric-shaped guard counters. In principle, it is possible to use the surrounding sample counters as guard counters for any other sample counters as well, excluding every count that happens to be coincident with one in any other counter, but in practice this gave no real advantage. The background of the system is between 3 and 3.6 cpm.

Later several applications made the need for smaller counters evident. One of these systems used a well-type NaI detector (see the section on gamma spectrometry below) as guard counter (Jelen and Geyh, 1986). This allowed for four to six miniature radiocarbon counters of 15 cc volume and restricted in diameter and length to be placed in the guard counter. Since NaI is a solid material, it has a higher efficiency to detect penetrating particles than gas counters, which together with the small size reduces the background of the sample counters to 0.02 cpm. Other groups used clear plastic materials as guard counter, since some of these emit light when penetrated by radioactivity.

One of the problems of the system of different devices as sample counter and guard counter is that the wall material in between can be one of the sources of background, since it may contain small amounts of radioactivity. Of course a very careful selection of materials is necessary in low-level radiometry. Besides that, the principle of "Oeschger" type or "multiwire" counters avoid the walls between guard and sample counter since they combine both in one device. The right side of Figure 5 illustrates the principle: within one large diameter tube, a metal-coated plastic foil creates a wall between the inner (sample) and outer (guard) counter. Instead of the foil, a system of wires is also sometimes used. Both guard and sample counter are filled with sample gas. The advantage of the system is a lower background due to the missing wall and shielding material between the two counters. The drawback here is the higher effort during construction and maintenance and that the guard counter is also filled with sample gas. This results in the need for more sample material.

2.3.2.2. Applications. Gas proportional counters had their widest application in earth sciences for the measurement of radiocarbon and tritium. For a long time, they were the main workhorses in radiocarbon dating laboratories, some of them processing around a thousand samples per year. Gas proportional counters were the first to measure radiocarbon with accuracy in the range of one per thousand for a modern sample, which was needed for studies on ocean circulation. Still gas proportional counters are the equipment of choice if environmental tritium is to be measured by radiometry without prior isotope enrichment (see below), since they reach detection limits well below 1 TU. For noble gas radionuclides, gas proportional counters are still the method of choice to measure ^{85}Kr, ^{39}Ar, and ^{37}Ar. Nevertheless, it is seldom that new laboratories are installed with gas proportional counters as detecting devices today. In radiocarbon, they are more and more replaced by liquid scintillation counting (see next paragraph) since the measurement device is commercially available, and by accelerator mass spectrometry (AMS) since the necessary sample size is much smaller. For tritium also liquid scintillation counting is more important today due to the simpler sample preparation, and in future the ingrowth method will gain importance (see below). For basic science needing special ultra low-level detectors, proportional counters are still the method of choice; during the solar neutrino studies of the Gallex experiment, miniature counters were specially developed. Situated in an underground laboratory with 1 km rock cover in a special tunnel of the Gran Sasso, these counters have an active volume of approximately 1 cc and, after energy-discrimination, rise-time, and pulse-shape analysis using neuronal networks, they deliver a background of better than 0.06 counts per day (d'Angelo, 1989; Pandola et al., 2004; Belli et al., 2005), a number not yet reached by any other technique.

2.4. Alpha and beta radiation: liquid scintillation counting

Besides the semiconductor detectors for alpha spectrometry and besides the gas proportional counters for beta radiometry, liquid scintillation counting ("LSC") is an old and established method to detect radioactivity. As the name implies, the sample is measured in liquid form here. This is easy, e.g. for water to be measured for tritium. Other samples have to be either absorbed or dissolved in a liquid (e.g., dissolved CO_2) or converted into a liquid, for example, radiocarbon samples are often converted into benzene. Therefore, again chemical preparation of the sample plays an important role although it will not be discussed here.

2.4.1. Basic principles and detector construction

As the name implies, LSC is a method that detects not the charge created by the radioactive particle but that detects light emission. The radioactive nuclide is contained in a liquid in which other components create light from the radioactive decay. The radionuclide can be dissolved in the liquid ("solvent") or be part of the liquid molecule. The substances which create photons of light from the decay energy are called "scintillators" (sometimes also "fluors").

One of the advantages of this approach is that the nuclide to be measured this way is introduced in a medium, which acts as a small detector with 4π solid angle. This is similar to the sample being the counting gas of a gas proportional counter, but with the advantage that the density of the detecting medium is higher and thus the range of the radioactive particle in the medium is lower. For alpha and beta particles, it therefore can be assumed that the total decay energy is transferred to the liquid, which for high-energy particles is not always the case in gas proportional counters. After being stopped in the liquid, the radioactive particle has transferred its energy to the solvent molecules. These in turn transfer the energy to scintillator molecules which emit light. This whole process takes place within some 10 ns, which means that LSC counters can quantify much higher activities than gas proportional counters. The drawback of the technique is the efficiency of energy conversion to light and that the detector is external to the medium; the light emitted is detected by photomultiplier tubes (PMTs) to which the sample vial is exposed (Figure 6). The photocathode in the PMT in turn creates electrons from the light that are then multiplied in an avalanche from dynode to dynode, finally creating an electric pulse.

Low-level detector systems contain at least two PMTs. To decrease background counts, the event is registered only if both record light in coincidence. Depending on the kind of fluor, the emitted light has wavelengths between 300 and 500 nm, corresponding to energies of 2.5–3 eV. So disintegrations like that of tritium having maximum electron energy of 18.1 keV in principle could create several thousand quanta of light. One of the problems of the technique is that energy losses-called "quenching"-occur during energy transport from the emitted particle to the fluor ("chemical quenching") and in the light path from the fluor to the PMT ("color quenching").

Since the range of charged particles is much smaller in liquids than in gases and since the radionuclides are surrounded by the liquid, it can be assumed that the particle energy is completely transferred to the liquid molecules. But the mechanism of energy transfer can differ between the interactions; it can be heat, excitation of electrons, ionization, phonon waves, etc. Not all of these interactions with the solvent are able to transfer

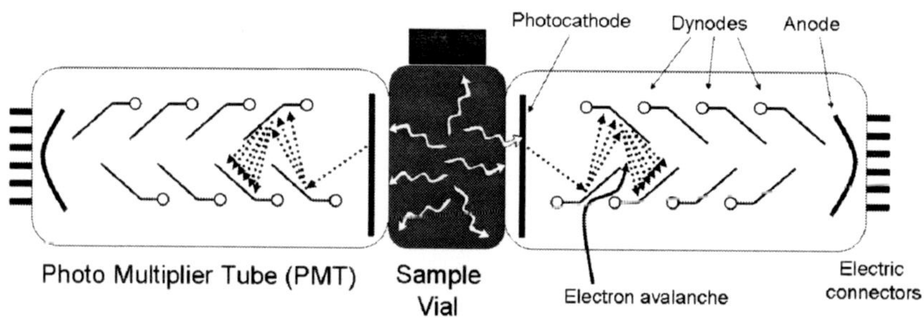

Figure 6 Sample vial and photomultiplier arrangement in liquid scintillation counting.

energy to the fluor, which causes the chemical quenching. In fact the very details of the energy transfer to the fluor molecules are not fully understood. But much empirical knowledge exists on the best combinations of sample liquid, solvent molecules, and scintillator type. The fluors are normally organic molecules like naphthalene and many others and the solvent also is organic like toluene or benzene. Probably one ideal combination is for the case of radiocarbon measurement where the sample for LSC measurement is converted into benzene, so solvent and sample are identical and only scintillator has to be added. Also for other liquid sample materials like water, cocktails of solvent and fluor were developed that are less poisonous than benzene and that are biodegradable.

In ideal cases like radiocarbon or radium, detection efficiency can be higher than 95%, which is comparable to gas proportional counters. In other cases like tritium, where the sample is water which shows strong quenching and where the decay energy is lower, detection efficiencies are around 30%. So only a third of the tritium disintegrations are detected, and gas proportional counters have superior efficiencies. Nevertheless, LSC is also the preferred method today for low-level tritium measurement. The main reason is that sample preparation is normally very easy; the water can be used as it is, or for low-level applications it has only to be distilled. Sample-to-sample memory effects are negligible since the measurement vials are cheap single-use disposable products. For the detection of environmental tritium, isotopic enrichment procedures are necessary as discussed below. Typical sizes for the liquid scintillation vials are 20 ml, but for small radiocarbon samples 10 ml vials are also common. To enable measurement of environmental tritium without prior enrichment, large volume counters were also constructed, using vials up to 150 ml. For continuous monitoring of nuclear facilities, mobile systems were developed or constructions with flow-through vials that enable continuous measurement.

2.4.2. Applications

Liquid scintillation counters are the standard workhorses of radiation protection. Nuclear facilities routinely control their effluents using LSC. Also in life sciences where, for example, biochemical or protein reactions are investigated using radioactive marked materials, LSC is the method of choice to measure this artificial radioactivity. LSC in many cases was used in environmental studies with active labeling, for example, for transport processes in the water cycle using tritiated water (Zimmermann et al., 1966; Athavale et al., 1980). Due to restrictions for the protection of the environment, this kind of study has become very rare.

In environmental studies, the majority of environmental tritium samples are still measured using LSC after prior isotopic enrichment. Many conventional dating laboratories use the technique for radiocarbon. Also for radiation protection studies investigating the natural content of nuclides from the uranium and thorium decay series like ^{226}Ra and ^{222}Rn in water, or in the investigation of acid mine drainage tailings, liquid scintillation counting is used (Galán Lopez et al., 2004; Prichard and Gesell, 1992; Salonen and Hukkanen, 1993; Zhou et al., 2001).

2.5. Gamma radiation

As stated before, the ranges of the three different kinds of radiation are very different. Whereas alpha particles are typically absorbed within a sheet of paper or some centimeters of air, beta particles can have a range up to meters in air, depending on their energy. Gamma particles are electromagnetic waves with a shorter wavelength than the more commonly known roentgen radiation (X-rays). Depending on their energy, they can penetrate kilometers of air or several decimeters of matter. As we will see in the following sections, this difference in range between alpha, beta, and gamma radiation explains most of the differences in the measurement techniques applied.

2.5.1. Basics

Whereas alpha and beta radiations are charged particles emitted from the atomic nucleus, gamma radiation consists of electromagnetic waves in quanta of high energy, which corresponds to very short wavelengths. The reason for this short wavelength can be explained by taking the size of the emitting region into account; electromagnetic radiation comes into being when charges are moved. The long wavelength of radio waves corresponds to a movement of charges along the length of a radio antenna, the wavelength of microwaves has a well-known size that fits into the decimeter size of a microwave oven giving a resonator. For common light, the emitting region is the movement of electrons in the shells of atoms and molecules, and for gamma radiation the emitting region is the atomic nucleus, thus producing really short wavelengths. The way that gamma waves come into being is due to a reorganization of the charged particles in the atomic nucleus when one or more of these charges gets lost due to the emission of an electron (beta radiation), two protons and two neutrons (alpha radiation) or due to the spontaneous fission of a nucleus. So to be precise, the gamma radiation is not emitted from the mother nucleus, but really from the daughter nucleus during the reorganization of its core particles (the protons and neutrons left in the core). The fact that this reorganization is not always necessary or that sometimes a different degree of reorganization is necessary explains another property of gamma radiation; the disintegration does not always emit a gamma particle. If a radionuclide is alpha radioactive, for each decay exactly one alpha particle is emitted. But even if a certain radionuclide emits gamma radiation, this is not necessarily the case in all nuclear disintegrations. So while all kinds of radioactivity can be associated with gamma emission (since always a charged particle leaves the nucleus), not each disintegration will emit a gamma (because it might be emitted such that no reorganization of the core is necessary). Expressed in scientific terms, the emission probability is not necessarily one. And it can be that, during decay, several gamma quanta can be emitted, since the nucleus sometimes reorganizes itself in several steps. In this case, the electromagnetic radiation emitted is called a gamma cascade. Apart from these characteristics of gamma radiation that are not very comfortable from the point of view of measurement techniques, the comfortable part in gamma spectrometry is that the energies are well defined: nuclear energy states are discrete, and therefore gamma energies are discrete, since they come into being as a transition from one state to the other. For the gamma-measurement techniques, it is therefore highly desirable to obtain the gamma energy as precisely as possible. Only spectrometry gives qualitative and quantitative information about the nucleus that emitted the radiation.

The range of gamma radiation strongly depends on the gamma energy. If for instance the lead in the lead shielding contains some ^{210}Pb (see Figure 1) then the emitted low-energy gamma radiation of 46.5 keV will be completely absorbed within the several millimeters of copper shielding inside the lead. In contrast, the 1,460 keV line emitted by ^{40}K will be measurable in most low-level detectors, since potassium is a common element in nearly every brick or concrete material of buildings and since this gamma radiation penetrates the lead and copper shielding with a certain probability. The fact that gamma radiation is able to penetrate matter is also advantageous for the measurement, as a result, no or only minor pretreatment of the sample is necessary. Whereas the sample has to be transformed into a gas to be measured in gas proportional counters or into a liquid in liquid scintillation spectrometry, and whereas it has to be plated on small disks in alpha spectrometry and measured in high vacuum, things are much easier using gamma spectrometry; usually one only has to expose the sample to a detector in a defined geometry. Often the sample is only dried prior to measurement, since water does not emit gamma radiation and the determination of specific activity, therefore, should be related only to the dry mass of sample. The most typical geometries to analyze samples in the laboratory are shown in Figure 7. In the most common situation for environmental samples, an amount of some tens to hundreds of gram of sample is situated in a beaker and positioned on top of the detector cap (Figure 7A). Instead of a beaker or a petri dish filled with sediment or water, the sample for example could also be a filter paper. Of course, in this case, only those gamma particles can be detected that are emitted in the direction of the detector, which here means $<50\%$ of the total. If a large amount of sample is available (a pound or more), the sample might surround the detector in the so-called Marinelli geometry (Figure 7B).

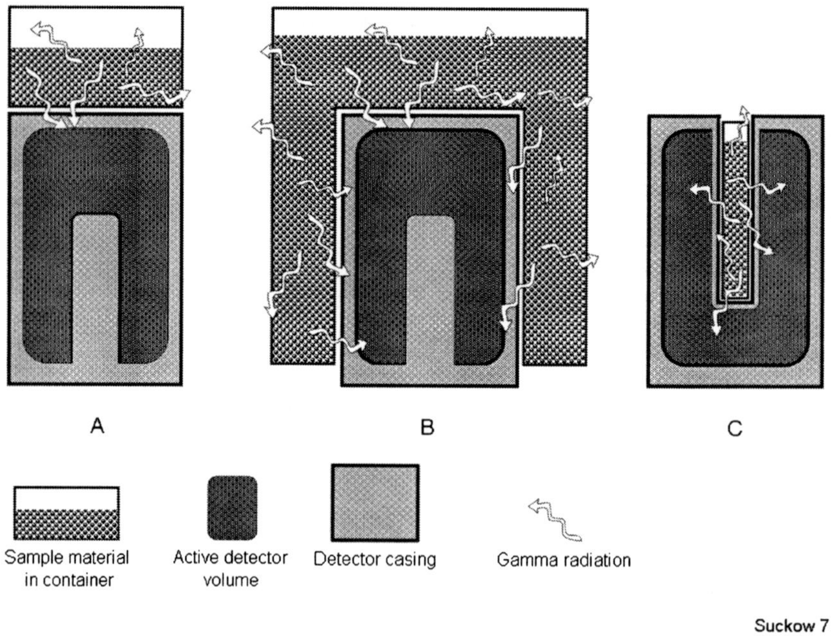

Figure 7 Some laboratory detector-sample geometries for gamma spectrometry.

In this case, an even smaller relative fraction of emitted gammas is detected, which is more than counterbalanced by the much higher sample size. If only a small amount of sample is available, the sample may also be positioned in a detector containing a hole, called well-type detector geometry (Figure 7C). In this geometry detection efficiencies can be 80% of the total emitted gammas.

Besides these laboratory geometries, it is also possible to map areas of several square kilometers by mounting a large gamma detector in a car (Geyh and Lorch, 1963), or into a plane or a helicopter (Rybach et al., 2001) and in some cases even behind a boat (Wijngaarden et al., 2002).

2.5.2. Detector types

Detectors for gamma spectrometry should have large masses; the interaction of gamma radiation with matter is stronger the more mass is penetrated by the radiation. Gamma detectors are made of solid and heavy materials, with the basic principle being that the heavier the active part of the detector, the better can high energy radiation be detected. Today the important detector materials used are thallium-activated sodium iodide (NaI), bismuth germanate (BGO), silicon, and high-purity germanium crystals (HPGe).

2.5.2.1. NaI and BGO. Sodium iodide crystals are amongst the oldest detectors, but they are still the best for many purposes. They do not convert the radiation directly into a measurable electric pulse but create an optically visible scintillation when the radiation interacts with the crystal. Therefore, a PMT has to be attached to the detecting crystal, and the crystal has to be clear enough that the light resulting from the scintillation effect can reach the photomultiplier. A disadvantage of this detection principle is that the energy resolution is quite bad. It is for instance not possible to detect the single gamma lines of the U/Th decay chain (Figure 1) with a NaI detector. Instead several lines of the gamma spectrum contribute to a broad energy peak. The advantage of NaI as detector material is that very large detectors can be built – the crystal for the airborne survey published by Rybach et al. (2001) has a volume of 16 L and accordingly a mass of nearly 60 kg – and that the detection efficiency can be very high. This makes NaI crystals very effective at measuring the total emitted radiation of a sample or for use during mapping experiments (IAEA, 2003). But they can also be used for laboratory analytical purposes such as for the determination of total uranium and thorium contents of a sample (Höhndorf 1973).

Bismuth germanate crystals (BGO, sometimes also BiGe, chemical formula $Bi_4Ge_3O_{12}$) are based on the same principle as sodium iodide. Since they have nearly twice the specific gravity of sodium iodide they have a higher stopping power and detection efficiency at the same volume. But this advantage comes at the price of an even lower energy resolution than for NaI. For small volume requirements, BGO crystals can be used for mapping purposes even in ultralight airplanes. BGO crystals are also used as anticoincidence detectors in liquid scintillation counters. And both NaI and BGO detectors are used in well-hole logging to detect gamma radiation. This can be to study natural radiation or to allow *in situ* chemical analysis by neutron activation. In the latter case, the logging tool has to contain a neutron source.

2.5.2.2. HPGe (high-purity germanium). Analytical laboratory systems for gamma analysis today normally use high-purity germanium detectors. The detector element in this case is a high purity single germanium crystal. The tremendous technical advances in semiconductor manufacture during recent decades have also made it possible to produce large single germanium crystals. Chemically very pure germanium is further purified using the zone-refining technique and high purity single crystals are then Czochralski grown directly out of the crucible containing germanium melt. Typical manufactured detector crystal sizes can be several kilograms today. The crystal is mounted in a cryostat maintaining the necessary operational temperature at around 90 K (liquid nitrogen). Electronically, the crystal acts as a diode with a voltage of approximately 3 kV against the barrier direction, and the charges created by the radiation photon are collected.

The germanium crystal becomes a diode due to lithium diffusion into the surface of the crystal, forming an n^+ contact surface. This surface is typically 600 μm thick and is a layer in which gamma radiation cannot be detected. The second, p^+ contact is formed by implanting boron ions, and this contact forms a layer with a thickness of only some tens of a micron. Since half a millimeter of germanium is able to considerably absorb low-energy gamma radiation, different detector types are distinguished; for a "P-type" detector the n^+ contact surface is exposed to the gamma radiation from the sample and the p^+ contact surface forms the primary electrical contact. For an "N-type" detector the thinner p^+ surface is exposed to the radiation and therefore this detector can register gamma photons of lower energies (below 200 keV). This is important for environmental studies since only laboratories equipped with these "N-type" detectors can measure the low-energy gamma emissions of, for example, ^{210}Pb (46.5 kev) or ^{234}Th (63.3 keV). Besides these constructions, HPGe detectors can also be "wrapped around the sample" such that the sample is positioned in a small hole in the detector crystal and cryostat, where the detector is called a "well type" (Figure 7C). Since this well for a germanium detector has only a diameter of a centimeter or less and a depth of some centimeters, this configuration can be used for small amounts of sample material only, as occurs for instance if sediments from small diameter corers are to be age-dated.

The disadvantages of germanium detectors are a higher price, smaller detector mass, and the necessity for liquid nitrogen cooling. Due to the smaller mass, germanium detectors are more transparent to high-energy radiation above 1 MeV than NaI or BGO, resulting in lower detector efficiency at these energies. But these disadvantages are more than counterbalanced by the fact that state-of-the-art HPGe detectors have an energy resolution of <1 keV for low energies like ^{210}Pb and ^{234}Th and around 2 keV for high energy radiation like the 1,461 keV radiation of ^{40}K. This means that, in contrast to NaI and BGO detectors, it is possible to directly identify the nuclide that emits the radiation. So selective radiometric analysis for several nuclides like ^{210}Pb, ^{137}Cs, ^{214}Pb, ^{234}Th, ^{228}Ac, ^{212}Pb is possible on the same sample within one measurement. This is a great advantage in sediment dating or bioturbation studies.

From the analytical point of view, environmental samples show some peculiarities that restrict the analyses to specialized laboratories; radiation from natural gamma samples can be so weak that active anticoincidence shielding may be necessary (Appleby et al., 1986) or that special ultra-low-background laboratories may be necessary (Neumaier et al., 2000). Low-energy gamma analysis like that for the 46.5 keV of ^{210}Pb also needs special attention to the self-absorption within the sample that weakens this radiation, dependent on the sample material. And in the case of well detectors and the analysis of nuclides emitting many gamma photons during one disintegration like ^{214}Bi, special caution is necessary for the efficiency calibration since these photons sum up in the detector (Debertin and Schötzig, 1979).

2.5.3. Applications

Probably the largest number of gamma analyses per year falls into the huge field of radiation protection. Here environmental and technical samples but also whole animals or humans are passed through detectors investigating the amount and energy dependence of their gamma emissions with comparable insensitive techniques. These techniques control the emission of radiation in the vicinity of nuclear facilities and investigate the impact of accidents like that in Chernobyl in 1986. For instance, using portable gamma spectrometers, it was easily possible to derive several interesting facts about the accident, like nuclide composition, kind of reactor fuel and reactor-burning state, from the fallout deposited hundreds of kilometers away. Since gamma spectrometry does not necessarily need any sample preparation, it can be used for large-scale mapping from planes or helicopters (Rybach et al., 2001). In part this is used for radiation protection, as in the vicinity of nuclear power plants. But also useful sedimentological information can be drawn from these maps and the spectra. Gamma-logging is also a standard technique for investigating geological boreholes as well as simple groundwater wells and specialized textbooks exist on this topic (Rider, 1996). Wijngaarden et al. (2002) also showed that a submerged BGO detector towed by a boat over the sediment can be used for mapping. In both the latter applications, the direct determination of radionuclides is not the target of the investigation but the sand–silt–clay content of the sediment, which is reflected in the thorium content of the sediment. Here the fact is used that nuclides of the uranium and thorium series are better absorbed on fine-grained particles. Also in other environmental studies gamma spectrometry gives valuable insight. For instance, using the bomb fallout of ^{137}Cs, it is possible to quantify soil erosion by just comparing the ^{137}Cs inventory with the area-specific fallout values (Walling and Quine, 1991). For sediment dating using ^{210}Pb, gamma spectrometry has the large advantage of giving information not only about the total ^{210}Pb activity (as alpha counting via ^{210}Po would do) but also about the supported ^{210}Pb by measuring ^{214}Pb on the same sample with the same measurement. Besides that, it was possible to use the bomb fallout of ^{137}Cs and sometimes very short-lived nuclides like ^{228}Ra as independent age indicators in sediment dating and sediment balance studies, all from the same measurement (Dukat and Kuehl, 1995; Michels et al., 1998, 2000; Suckow et al., 2001a). In special cases, the technique allows the use of bomb fallout of ^{241}Am and sometimes the Chernobyl fallout of ^{134}Cs as an independent age indicator (Suckow and Gäbler, 1997). For studies of reworking and biological activities in deep-sea sediments gamma spectrometry also proved to be a highly efficient technique (Suckow et al., 2001b).

3. Isotopic Enrichment

Although radiometric methods have very sensitive atomic abundance detection limits as compared to other methods like mass spectrometry, sometimes even this high sensitivity is not sufficient to solve environmental problems. For example, tritium is a very useful tool in hydrogeology since it can be used as a marker of modern precipitation in groundwater. Tritium content in modern rain in the northern hemisphere is about 10 TU. One TU corresponds to a specific activity of 0.12 Bq/kg or an atomic ratio of 1 tritium atom per 10^{18} hydrogen atoms. Direct state-of-the-art radiometric methods to measure tritium in water typically have a detection limit of 5 TU (LSC) or 1 TU (proportional counters) which even with long counting times cannot be lowered much more. This means that, with direct radiometric methods, it would be possible to detect an admixture of down to 10% of modern water in groundwater. In the southern hemisphere, where no testing of thermonuclear bombs took place and where much fewer nuclear facilities exist and where the larger ratio of ocean to land surface dilutes the anthropogenic tritium signal, the tritium content of rain is lower by nearly a factor of five. This means that the tritium method in hydrology would not be applicable at all in the southern hemisphere and not very meaningful in the northern hemisphere with a detection limit of 5 or 1 TU. The only solution is to enrich the isotope in question prior to radiometric measurement by a factor of ten or more. This isotopic enrichment always comes at the price of a much larger sample size.

The physical principles used for isotope enrichment on an analytical laboratory scale are the same as those used in the nuclear industry to obtain isotope separations or to produce nuclear fuel, for example, for uranium enrichment. However, the aim is very different; the nuclear industry tries to enrich large amounts of substances – typically several kilograms or tons – and the exact isotopic enrichment factor per step is less important, since several steps are performed sequentially and since the isotopic composition of educts and product is routinely measured. In contrast, the analytical laboratory enriches small amounts and if possible in only one step, and here it is essential to exactly quantify the enrichment factor of this step because it directly influences the analytical result. Only two enrichment methods have been demonstrated to meet these criteria on a routine analytical basis.

3.1. Electrolytic enrichment

The most frequent method of isotopic enrichment of tritium is the electrolysis of water. Direct electric current causes the dissociation of the water molecule into hydrogen at the cathode and oxygen at the anode. Since the tritium atom is three times heavier than the hydrogen atom and since all phase transitions tend to enrich the heavier isotope in the less mobile phase, the electrolytic process strongly enriches the tritium abundance in the liquid phase. The hydrogen reactions on cathode materials involve a strong material dependence and this material dependence is not yet fully understood in terms of the details of the physical interaction. But many laboratories maintain facilities for electrolytic tritium enrichment and a lot of empirical knowledge exists, although very few publications are known. The most common system consists of the "IAEA Batch cells," being concentric cylinders of stainless steel as anode material and mild steel as cathode material. Between 250 and 1,000 ml of water are electrolyzed in these cells down to the 15–20 ml needed for radiometric analysis. The water mass enrichment factor of 15–50 in these systems corresponds to an isotopic enrichment of roughly 90% of the mass enrichment. A sample having 10 TU prior to enrichment, electrolyzed with 90% isotopic efficiency from 300 ml down to 15 ml, has at least 180 TU after enrichment and thus is well measurable since it is now a factor of 36–180 above the prior-mentioned detection limit.

Besides the cell configuration with steel, systems using gold and platinum electrodes also exist. These tend to have lower isotopic enrichment efficiencies, but are sometimes more reproducible than the steel systems. For larger enrichment factors, some systems use more water and refill the electrolytic cell continuously during electrolysis. Due to the high currents needed, all systems have to be cooled; 1 A dissociates 2.97 g water per hour, so in order to achieve the enrichment in a reasonable time, currents of up to 10 A are used. And since a highly explosive gas mixture is produced during the process, all enrichment systems need special precautions concerning ventilation and removal of the explosive gas mixture. Nevertheless in some laboratories using this technique, explosions have indeed occurred, causing more or less severe damage. The electrolysis introduces additional sources of uncertainty due to weight determinations, constancy of the electrolytic enrichment factor for the cells and so on, and therefore electrolytic enrichment only gives higher accuracy when the activity to be determined is smaller than a system-dependent threshold value. Details of the data evaluation and uncertainty determination can be found in Gröning and Rozanski (2003).

3.2. Thermal diffusion enrichment

Another technique to enrich the isotopic ratio of a sample uses the thermal diffusion effect. Gas diffusion is mainly dependent on the speed with which gas molecules are moving and is therefore strongly mass and temperature dependent. Therefore, within a closed gas volume having an external temperature gradient, heavy gas molecules enrich in the cold region of this volume. The enrichment is the better, the stronger the temperature difference is. Laboratory systems to isotopically enrich gas samples combine this effect with thermal convection, a principle first described by Clusius and Dickel (1938). These systems contain a scorching wire at the center of a vertical glass tube, which is water-cooled at the wall. This means that, with typical wire diameters below 1 mm and inner diameters of the glass tube of <1 cm, the temperature gradient can be $500°C\,cm^{-1}$. Detailed experimental studies revealed also in this case that theory is not able to account for all the effects concerning the isotope enrichment, and that the enrichment factors have to be determined empirically.

Such a system for the enrichment of hydrogen prior to radiometric measurement of tritium in proportional counters has been realized and operated for several decades in the Department of Environmental Physics at the University of Heidelberg. It consisted of four enrichment tubes in series and could be operated also in dynamic mode where it was able to retain more than 99% of the tritium in a sample (Gonsior, 1961), which means that the isotopic enrichment directly corresponds to the mass enrichment. Thus the uncertainty additionally introduced by thermal diffusion is smaller than by electrolytic enrichment. With a combination of electrolytic enrichment using gold and platinum electrodes, thermal diffusion enrichment and proportional counters, it was possible to reach a detection limit below 0.05 TU. This detection limit was mainly caused by contamination of samples during storage and, as a minor effect, during handling (Weiss et al., 1976), which is confirmed by today's conventional laboratories reaching similar detection limits with electrolytic enrichment in large-volume batch cells (Morgenstern and Taylor, 2009).

A very similar system for the enrichment of ^{13}C and ^{14}C in CO_2 was established at the University of Groningen and is described in Kitagawa and van der Plicht (1997). With a tube length of 3 m and a pressure of 1,500 mBar, which corresponds to a sample size of 0.5 g carbon, this system allowed an enrichment of a factor of three for ^{13}C and of six for ^{14}C. This allows extension of the radiocarbon time scale by 15,000 years. Of course here also the disadvantage of the method is a larger sample size by a factor of 500, and the main problem of the application is to find uncontaminated samples of this age and size.

4. Mass Spectrometry

If we enlarge the focus a little bit from pure radiometry, meaning the "detection and analysis of radioactive decay," toward "quantification of radioactive nuclides in a certain sample material," then analysis of radionuclides in the environment has also to consider mass-spectrometric techniques. Radiometry is the most sensitive method if the activity in the sample is high and the half-life of the nuclide in question is short. This normally means that the isotopic abundance (the number of radioactive atoms in question within "normal" atoms or within the decay chain) is low. If the half-life is very long, the total activity, which is the number of disintegrations in a reasonable sample size, can be very low. In these cases, radiometry can be less sensitive than sorting the atoms by mass and counting them. This is the way mass spectrometry tries to solve this measurement problem. For those interested in the very broad range of mass spectrometry, this chapter will surely be disappointing. Many textbooks exist on mass spectrometry and only those parts of the technique that are important for the measurement of radionuclides will be described briefly.

4.1. Basics

The basic principle of mass spectrometry is known to everybody who ever tried to control different types of vehicles on a bending road; cruising with the same speed, the heavier vehicle will be the first drawn out of the curve. Mass spectrometry does the same as Figure 8 illustrates; it produces charged particles from the sample in the *ion source*, accelerates them to the same energy, and sends them to the *analyzer* where normally a magnetic field forces them around a corner. And the *detector* will find the heavy ones on the outer side of the curve, the light ones on the inner side.

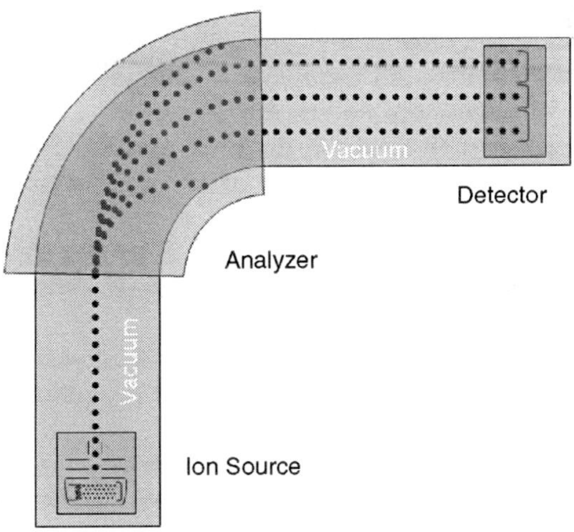

Figure 8 Principle of mass spectrometry: the *ion source* converts the sample to ions and accelerates these, in the *analyzer* these are differentiated by mass and registered in the *detector*.

There are many different kinds of mass spectrometers, some of them so small that you can hold them in one hand or send them to mars, some bulky and heavy that you need to build a house around it. But for nearly all of them the three parts indicated above can be distinguished: (1) the Ion source where the charged particles are produced from the sample and where all are accelerated to the same speed; (2) the analyzer that contains one or several curves for the mass selection; and (3) the detector system where the mass-selected charged particles are registered. All these three parts can vary considerably from type to type and that makes mass spectrometry a science of its own with its own textbooks. The ion source can create the ions by electron impact, by applying high temperatures, using plasma torches, other ions or by laser beams. The analyzer can use variable or static magnetic field or electric fields or simply the flight time or even a combination of all these to separate the masses. And there are various kinds of single or multiple detectors to register the ions. But one construction detail is common to all mass spectrometers: the process has to take place in vacuum. Mass spectrometers are always high vacuum measurement devices.

The conditions under which mass spectrometry is advantageous as compared to radiometry becomes evident if one keeps some facts in mind:

1. the method which is more sensitive will be the one producing more counts per time, where counts means detected radioactive disintegrations for radiometry or ions hitting the detector for mass spectrometry;
2. the methods each have to relate the counts from a sample to the counts of background (signal-to-noise ratio) and only if the signal is higher than the background is a reliable measurement possible;
3. the signal strength for both techniques is approximately proportional to the number of atoms admitted to the device;
4. the possible number of atoms in the measurement device is limited, in mass spectrometry ultimately by the constraint of maintaining a vacuum, in radiometry by the counter geometry;
5. the sample amount that can be admitted to the device is a factor of a thousand or more smaller for mass spectrometry compared to radiometry;
6. the sensitivity of mass spectrometry (counts per admitted atom) is two to three orders of magnitude smaller compared to radiometry (counts per disintegration);
7. the measurement time for radiometry is limited to a maximum of several days, while for mass spectrometry more than an hour will seldom be used;
8. radiometry needs a significant number of atoms to disintegrate within the measurement time.

These facts result in the following conditions under which mass spectrometry is advantageous:

- if the half-life is very long compared to typical measurement times, meaning that only a small fraction of the radioactive atoms will disintegrate within the measurement time
- if the available amount of sample is very small.

Typical radionuclides where mass spectrometry is advantageous are ^{238}U, ^{234}U, ^{232}Th, ^{230}Th, ^{129}I, ^{36}Cl, ^{14}C, and ^{10}Be.

If we reverse the arguments, it follows that in most cases where plenty of sample material is available and the half-life is short, radiometry will be the method of choice. Typical nuclides where radiometry is advantageous are ^{226}Ra, ^{222}Rn, ^{214}Bi, ^{214}Pb, ^{210}Pb, ^{137}Cs, ^{7}Be, ^{3}H, although for the last, tritium, a kind of "mixed" method will be introduced later (ingrowth technique).

There are, of course, also nuclides where both techniques have very large difficulties and amongst these are ^{81}Kr, ^{39}Ar, ^{37}Ar. The special case of ^{81}Kr will be discussed in the very last paragraph "Future Developments" of this chapter, since it can serve as a model for possible future technical development.

Just to keep in mind: radiometry of course can only measure radioactive nuclides whereas for mass spectrometry the overwhelming number of applications involve the measurement of stable isotopes or chemical compounds. Therefore, mass spectrometry is surely the more frequently applied technique – but this chapter deals with radionuclides only.

4.2. TIMS

One of the oldest and most often applied mass-spectrometric technique to measure radioactive or radiogenic nuclides is "Thermal Ionization Mass Spectrometry," in short TIMS. In the most abundant type of ion source for mass spectrometers, the "Nier" source, ions are produced from gaseous samples by the impact of electrons. In contrast, during thermal ionization, the sample is placed directly on a small filament by evaporation of solution – typically only some microliters. During measurement this filament is electrically heated and ions are produced only by the high temperature applied – which explains the name. For mass spectrometers with a Nier source, the filament belongs to the "inner" part of the machine and is seldom replaced. For some machines, the replacement of the filament is a difficult task that can cause some headache to the operator. In contrast for TIMS, the filament has to be replaced for every sample. But, even with modern high vacuum equipment, it would be most uncomfortable really to open the ion source for every sample. Therefore, several of these filaments are mounted on one "wheel" with normally one or two dozen positions.

The sample size can be very small using TIMS; down to 1 ng of the element to be analyzed is possible, which means that sample sizes for instance for uranium can be much smaller than using alpha spectrometry. Since the sample is very small, the risk of contamination is very high. So sample chemical preparation for TIMS normally needs very pure and therefore expensive substances and has to be done in a clean lab with glove boxes or flow boxes. This makes sample throughput for TIMS small as compared to other mass-spectrometric techniques. Since there is only a small amount of sample that can be placed on the filament, it is often not possible to switch very often between masses: the sample evaporates too fast. Newer generations of TIMS machines, therefore, have several collectors to measure all necessary masses of one element at the same time. Of course this construction increases the price of the measurement system.

Despite the much higher price of equipment like the MS and clean lab and the necessary greater experience as compared to alpha spectrometry, TIMS has developed to a routine technique for several geochronological methods using mother/daughter nuclides with long half-lives like $^{230}Th/^{234}U$, $^{231}Pa/^{235}U$, $^{87}Rb/^{87}Sr$, and has advantages for $^{40}K/^{40}Ca$ (Marshall and DePaolo, 1982).

4.3. MC-ICP-MS

In analytical chemistry, the mass-spectrometric technique has become one of the main "workhorses." Since the amount of samples during routine chemical analysis is very high, chemists have developed very efficient preparation techniques to "feed" the sample into the mass spectrometer. One of the most common methods today first liquefies the sample – usually using strong acids – and then uses a plasma torch whereby this liquid is sprayed into hot plasma and ionized by temperature. From this plasma torch, the technique has its name "Inductively Coupled Plasma Mass Spectrometry," in short ICP-MS. The advantage for the laboratory as compared to TIMS is evident: both techniques have to dissolve the sample, but where TIMS needs the time-consuming element of separation in the clean lab before the measurement, ICP-MS can start measuring basically with this solution.

One of the technical problems with this approach was to keep the plasma torch stable and to stabilize the extraction conditions for ions from the plasma into the high vacuum of the mass analyzer. Some generations of ingenious engineers and scientists later, it is nowadays possible to couple a sophisticated and sensitive high-resolution mass separator with several collectors to measure more than one mass at a time behind the plasma ion source. This construction consequently is called multicollector-ICP-MS or in short MC-ICP-MS. Besides the "basic" analytical chemistry that deals with any element and its stable isotopes, MC-ICP-MS develops toward a direction where it will be the main workhorse also for geochronologists using, for example, the $^{230}Th/^{234}U$ pair for dating.

4.4. AMS

"Accelerator Mass Spectrometry," in short AMS, became an important measurement technique especially for radiocarbon (^{14}C), but also for other nuclides important in geosciences like ^{10}Be, ^{32}Si, ^{36}Cl, ^{41}Ca, ^{129}J. Seen from the general principle, an accelerator mass spectrometer consists of the coupling of several mass spectrometers with a high-energy acceleration step and a stripper in between (see Figure 9). Of course the question immediately arising here is: "What can the coupling of several mass spectrometers and high particle energy achieve that a normal mass spectrometer cannot"? AMS allows the measurement of isotope ratios with higher sensitivity than normal mass spectrometry on comparably small samples. The capability to measure very rare isotopes near an abundant neighboring mass is called "abundance sensitivity" and is the main strength of AMS machines. For instance, it is no problem for a "normal" mass spectrometer to determine the isotopic ratio of ^{13}C to normal ^{12}C, which is on the order of 1% for natural carbon. If you want to use the $^{230}Th/^{234}U$ pair for dating and have to determine the $^{230}Th/^{232}Th$ ratio, then isotopic ratios of 10^{-6} or 10^{-8} have to be measured, which already needs a more sophisticated mass spectrometer like a TIMS or a MC-ICP-MS, if possible with an additional element that filters the ions not only by mass but also by energy. If you want to measure isotopic ratios of 10^{-12} which is the typical ratio of ^{14}C to ^{12}C in a modern sample or even go beyond as necessary for some of the above-mentioned radionuclides, you need an AMS. AMS can measure abundance sensitivities down to 10^{-18}!

Why are these high isotope ratios a problem for the measurement? For these high ratios both the elemental separation during sample preparation and the mass/charge separation during the measurement in a normal mass spectrometer are simply not good enough. For instance, if you want to measure ^{14}C on mass 14 and are not completely able to separate nitrogen from the sample, then you will have a signal from ordinary nitrogen (^{14}N) as well. Keep in mind that the $^{12}C/^{14}C$ ratio of 10^{-12} means that one per mill of one ppm of nitrogen will produce a thousand times larger signal than ^{14}C! Even if you can separate nitrogen from carbon completely, every high vacuum system has a hydrogen background (since hydrogen is able to diffuse through steel) and in the ion source this will form ^{13}CH or $^{12}CH_2$, both of the same mass as ^{14}C. AMS offers several tricks to do the job, and those tricks discussed here are sketched in Figure 9. The first trick is that the AMS ion source creates negatively charged ions whereas most other

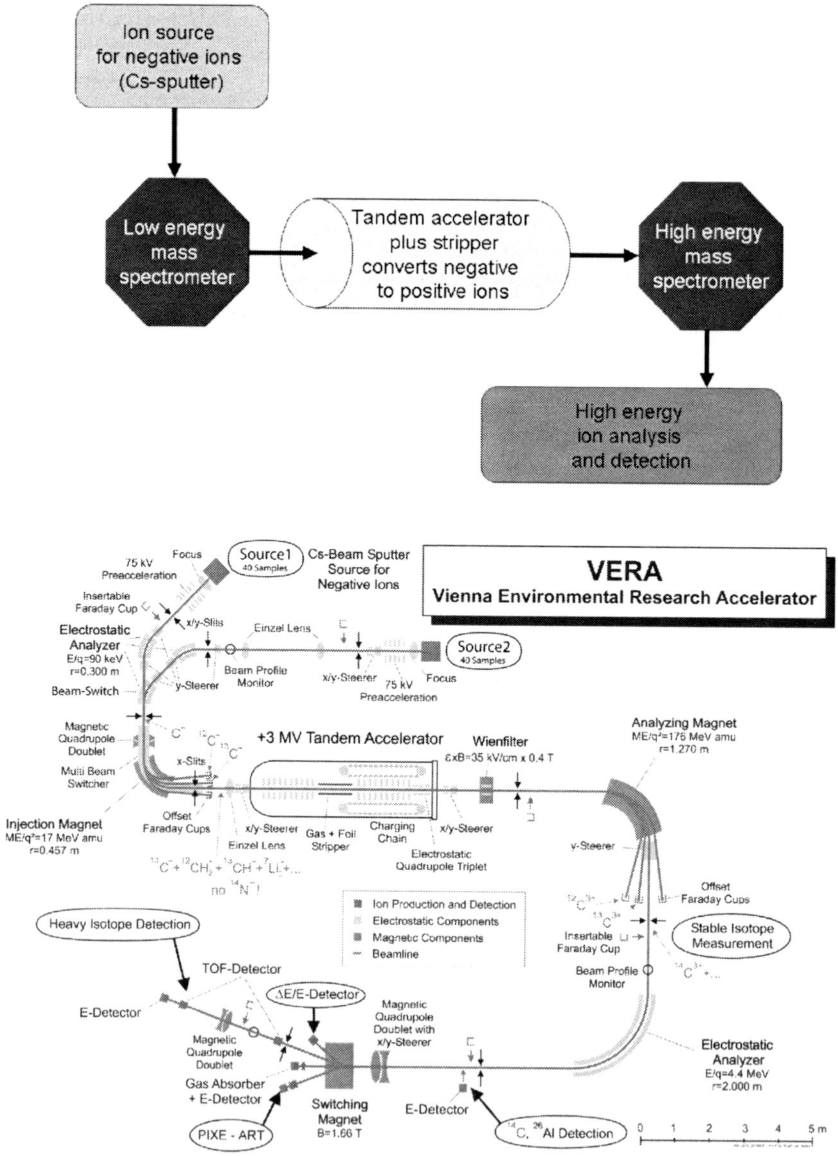

Figure 9 General scheme for accelerator mass spectrometry (AMS) and design details of the Vienna Environmental Research Accelerator (VERA: design by courtesy of Walter Kutschera, Vienna, from http://isotopenforschung.univie.ac.at/).

ion sources create positive ions. Taking radiocarbon as the example, this is a first discrimination step because ^{14}N does not form negative ions. A normal mass spectrometer then selects the desired mass (here 12, 13, or 14) and feeds it to the accelerator. Here the ions are accelerated by voltages more than a thousand times higher than the 2–10 kV in a normal MS. The second trick is here in that the ions pass a stripper (gas or a thin carbon foil), which removes electrons and converts the ions from single negatively charged to (multiple) positively charged. This step breaks up any molecular bond, which in our example is a second discrimination against ^{13}CH or $^{12}CH_2$. As a result of this acceleration, we have positively charged ions (no longer molecular) of high energy and these are again selected for mass/charge ratio in a high-energy mass spectrometer. After this selection of one charge/mass combination (e.g., $^{13}C^{4+}$ or $^{14}C^{3+}$ ...), the whole experience of high-energy nuclear physics in analyzing particles can be applied using high-energy filters and detectors for further discrimination.

There are plenty of examples in the scientific world that one person alone can learn on how to construct or to buy and run a simple radiometric device or a conventional mass spectrometer. For an AMS, this is no longer possible; you need a team of technicians and scientists to run, tune, and maintain the machine. So important for the sample submitter to know is also that most commercially working AMS laboratories are "measurement factories." You have to contact them – if possible before you take the samples. At least visit their homepage to ask not only about prices and their speed of sample processing but mainly for their requirements in terms of sample size, sample pretreatment and conservation, etc.

4.5. Ingrowth technique

A final mass-spectrometric method that has to be described here uses the mass spectrometer as detection device, although in principle it is a radiometric method. This method determines tritium, but not by detecting the tritium decay directly nor by counting the tritium atoms in a mass spectrometer. This method uses the fact that the daughter nuclide of tritium is the stable noble gas helium-3 (^{3}He), and it measures not the mother nuclide but the daughter nuclide produced by the radioactive decay during storage. It was introduced by Clarke et al. (1976), and the basic procedure is summarized in the following steps:

1. degas a water sample completely,
2. store this degassed water sample in a vacuum-tight container such that no gas can enter from outside, and
3. measure the ^{3}He accumulated in the sample after a storage time long enough that a well-measurable amount of ^{3}He has accumulated.

As can be seen from this procedure, the method actually measures the radioactive decay; only the tritium atoms that disintegrated during storage can be quantified. But the decay signal is not detected directly, but the daughter atoms generated by parent decay are summed up during the storage time. For practical applications, this has two advantages: first the sample size can be made larger than for the direct radiometric method, since there is no necessity that the sample fits into any detector. Typical sample sizes are between 40 g and 1 L of water whereas the largest LSC devices or proportional counters can only handle 10–100 g of water or an equivalent amount of counting gas. Second, the counting time of the ingrowth technique corresponds to the storage time of the degassed water sample. This can be enlarged up to months or even a year. In contrast, it is not useful to enlarge the counting time of a radiometric device to more than a week, since the measurement system cannot be kept stable for such a long time. Summed up, this means that the detection limit for the ingrowth technique is in the order of some 10^{-3} TU, which is a factor of ten better than the best state-of-the-art laboratories achieve with electrolytic enrichment and radiometry. In cases where the sample size is limited, as for instance for water distilled from sediment or plant material, the ingrowth technique is the only method to reach a detection limit below one TU.

The price for these advantages is threefold: first the measurement device itself (the mass spectrometer) is less sensitive relative to a radiometric detector; second, the device is much more expensive and complicated; and third, the time needed to process a sample is much longer due to the storage time of the degassed sample. The reason for the smaller sensitivity is that an ion source in a mass spectrometer does not produce ions as efficiently as a proportional counter or LSC detects disintegrations. Typically only one of 1,000 atoms in the mass-spectrometric system is detected, and this number can be increased by a factor of about ten only, using counting times of half an hour in a static noble gas mass spectrometer. A sample of 1 kg water with an activity of 1 TU produces 0.119 atoms ^3He per second or nearly two million atoms within a storage time of half a year. This means that the detection limit of 5 mTU achieved with this method using 500 g water (Bayer et al., 1989) corresponds to the detection of less than 5,000 ^3He atoms in more than 10^{25} molecules of water! It is difficult to be even more sensitive in the measurement of noble gas isotopes and it really needs a lot of effort and will also be discussed in the following and final section.

5. Future Perspectives

This chapter has introduced many of the basic concepts involved in the measurement of radionuclides. Some of these concepts are as old as knowledge about radioactivity itself, like the decay counting using scintillating materials, already used by Ernest Rutherford more than 100 years ago. Others like routine ^{14}C atom counting using AMS are younger than 30 years.

A perspective on the future is always very difficult since it cannot account for any future invention and has to somehow judge the existing laboratory approaches, which are not yet routinely used. This implies that the timescale for a prognosis should really not be longer than at most two decades. Any such estimate on future development is hampered also by the necessarily personal and incomplete perspective any scientific author has; the personal preferences for some techniques, originating from one's own biography, will shift the prognosis toward those well understood and looking promising to the author, whereas another author will find other techniques more promising. Moreover, one always has to face the possibility that some totally new and unseen technique will emerge to render all earlier prognoses unrealistic.

From the knowledge of the past an indication toward the future might nevertheless be derived; evidently decay counting has shown an evolution from counters designed individually in every lab and involving complicated preparations toward industrially manufactured semi- or fully-automated counters, often using semiconductor techniques and standardized preparation techniques. In the early days of these applications, tritium and ^{14}C decay counting by gas proportional counters was the method of choice. To build and maintain gas proportional counters needed technically experienced persons in this field, in most cases physicists. Later the LSC technique became more important when new laboratories were installed. For tritium, this has some evident advantages since it means one preparation step less – the conversion of water to a counting gas is not necessary. For radiocarbon the opposite is true, since samples are converted first to CO_2, which in some cases already can be used as counting gas. Other laboratories use acetylene as counting gas and need a further preparation step. But for LSC counting of ^{14}C one more step is always necessary to convert acetylene to benzene. The reason why more newly installed laboratories nevertheless use LSC, therefore, is not the ease or throughput of sample preparation. Also the price of the measurement device seems not to be important, since commercial low-level LSC counters are more expensive than dedicated proportional counters. Also the quality of results seems not to be a reason (Geyh, 1972). More important seems to be that LSC counters are available on a commercial basis including commercial service and maintenance, and that they are more versatile; the same instrument can easily measure tritium, ^{14}C, ^{226}Ra, ^{222}Rn, with just some necessary modifications in sample preparation. In contrast, a gas proportional counter has to be rebuilt if another counting gas is applied; at least the thickness of the anode wire has to be changed.

If one looks at the number of radiocarbon and tritium measurements today (2009) then it is evident that mass-spectrometric techniques have latterly gained more importance: although the number of radiometric and AMS radiocarbon laboratories worldwide is not very different, probably more than 80% of the radiocarbon samples are processed in AMS laboratories today. A laboratory that measures more than some hundred samples per year by radiometry may be considered large; many do less. A typical ^{14}C-dedicated AMS laboratory can process 2,000 samples per year or more. Again the reason is not the price per sample, which is often a factor of two higher in AMS laboratories, nor the investment costs, which are more than a factor of ten higher for AMS laboratories, nor the sample preparation, which needs comparable effort per sample in radiometric and AMS laboratories. More important seems to be sample size – AMS allows processing of samples containing 1 mg carbon and less – and again versatility because an AMS laboratory can measure nuclides that are otherwise not measurable at all (e.g., ^{36}Cl, ...). In addition, there appears to be a kind of feedback effect that supports large laboratories on a marketing basis since "everybody seems to get the samples measured there." A similar development can be seen in the case of tritium, where the laboratories using the ingrowth method via ^{3}He are growing in importance and show higher sample throughput with higher investment at comparable price and effort per sample.

In the field of mass spectrometry, recent years have seen the dawn of laser spectrometric techniques. These have already started to replace some of the applications involving measurement of stable isotopes (Kerstel et al., 1999). It seems not improbable that this might happen also for radionuclides, and as an example the history of the measurement of ^{81}Kr will be discussed briefly in the following.

For several reasons ^{81}Kr is a very ideal tracer for dating very old groundwater; it is a noble gas and therefore undergoes no chemical reactions in the aquifer during groundwater flow. Its half-life is 229,000 years that makes it ideal for studies concerning nuclear waste disposal or palaeogroundwater with slow movement. The isotope ratio $^{81}Kr/Kr$ in natural atmospheric krypton is 5.3e−13, corresponding to an activity of 0.08 dpm/L Kr. ^{81}Kr is produced naturally by spallation in the atmosphere, and neutron-capture of stable krypton. The global ^{81}Kr inventory is about 100 Mol of which 98 reside in the atmosphere. This means that ^{81}Kr seems not to be influenced by palaeoclimatic variations that are for instance important for radiocarbon. Also no anthropogenic or underground production seem to exist as is the case for ^{36}Cl, another important dating tracer of old groundwater (Lehmann et al., 2003). Despite these favorable properties, up to now <50 groundwater samples have been measured for ^{81}Kr since the analytical effort is very considerable.

The first published values of ^{81}Kr in the atmosphere by Loosli and Oeschger (1969) were measured in a very sophisticated low-level underground laboratory using small gas proportional counters filled with pure atmospheric krypton, prepared before the first nuclear bomb tests. This was soon no longer possible since ^{85}Kr, a short-lived anthropogenic radioactive isotope released during reprocessing of nuclear fuel, masked the activities of the long-lived krypton isotope. Also the application of radiometry of krypton for groundwater samples is not feasible; in 1 L of water the typical amount of dissolved ^{81}Kr is 1,300 atoms and together with the long half-life this results in one ^{81}Kr disintegration per 300 years (Lehmann et al., 1985). The sample preparation, therefore, requires field degassing of 10 m^3 of groundwater (!) or more, resulting in around 800 L of gas (this of course depends on the source in question). During laboratory preparation, the heavy noble gas fraction of these gases is purified-around 8 l of argon including all other noble gases. The krypton fraction (typically 1 cc) then has to be separated from the argon. Since mass spectrometry is not able to measure isotope ratios of 5e-13 directly, the ^{81}Kr isotope has to be enriched in the krypton sample by several orders of magnitude. This was achieved during the first measurements by introducing a small amount of the krypton fraction into a quadrupole mass-spectrometer system and impacting the ion beam of mass-81 into a target of aluminum coated kapton. This step was repeated by evaporating the krypton-bearing part of the target to produce a new krypton "educt" sample, obtaining a total enrichment factor of 4,000,000 (Lehmann et al., 1985). In a later version, a commercial isotope enrichment system with a Wien filter was used, but the principle of filling a high vacuum system with the krypton and implanting the mass in question into a target was kept similar (Lehmann et al., 1987).

Despite these isotopic enrichments, the "ordinary" mass spectrometry was not sensitive enough to detect ^{81}Kr. The reason was that the ionization probability in an electron impact source is too small. Therefore, the first mass-spectrometric measurements used a resonance ionization source (RIS-MS). Here the Kr in the MS was first frozen to a small metal spot, then released into the gas phase by a laser beam and 5–10 μs later nearly completely ionized by another laser beam tuned to the specific ionization energy of krypton. Although this method was declared a routine analysis (Thonnard et al., 1987), the total number of water samples processed and measured in this way was probably <30.

With the advance of AMS, of course there were soon attempts also to measure ^{81}Kr with the AMS technique. As discussed above, the abundance sensitivity of 1e-13 to 1e-15 needed for groundwater dating should be achievable using AMS. But most AMS machines are tandem accelerators that have to start with negative ions produced by a sputter ion source with a solid target. Noble gases like krypton do not build negative ions and a gaseous sample cannot easily be sputtered. Therefore during these measurements a synchrotron, AMS was used with a specially developed ion source for krypton (Collon et al., 2000), which made it possible for the first time to study several samples in the same aquifer. Nevertheless, the sensitivity of the AMS system was much too small for routine measurements; several million atoms ^{81}Kr resulted in 50–60 counts only. One of the reasons for the difficulties AMS has with these measurements is that mass spectrometry is sample destructive; ^{81}Kr atoms that are ionized and measured are implanted into the detector and lost for any subsequent measurement.

The next step in improving the technique was therefore a measurement that is in principle capable of measuring the same atom more than once. This was realized as atomic trap trace analysis (ATTA). Instead of the ions in a mass spectrometer, the method uses neutral atoms excited into a metastable level. A beam of these atoms is collimated by lasers after the outlet of the excitation volume and cooled with a laser beam against the atom flight direction (Zeeman slowing). These cooled atoms are then trapped in a magneto-optical trap (MOT), where they can be measured by resonance fluorescence. Concerning abundance sensitivity, this method is superior to mass spectrometry since nearly every part of the measurement can be made isotope selective; the resonance spectroscopy to detect the trapped atom can be made isotope sensitive. The MOT traps only atoms with a defined weight. And the Zeeman slowing, which needs the absorption of many photons by the cooled atom is also tuned to a certain isotope, so atoms of a different mass are not slowed. Although all krypton atoms are excited into the metastable state, it is possible to selectively push out (de-excite) the atoms of contaminating isotopes (Chen et al., 1999). Finally, the detected atoms are not necessarily lost for further measurement, since the atoms leaving the trap can be fed back to the source by a differential pumping system (Du et al., 2003). Therefore, this technique does not need difficult isotopic enrichment. An interesting feature of this development is that the ATTA technique also enables the measurement of the anthropogenic short-lived isotope ^{85}Kr; in case of ^{81}Kr the prepared sample is first measured for ^{85}Kr by low-level gas proportional counters. This is needed to detect any contamination or admixture of recent water to the sample. If proved to be initially free of ^{85}Kr, a spike of ^{85}Kr is then added to the gas sample to reduce the ATTA measurement to an isotope ratio measurement between ^{85}Kr and ^{81}Kr. In the case of a young sample, the long-lived ^{81}Kr can be treated as the stable reference for the determination of ^{85}Kr. The technique made it possible to determine the age of groundwater flowing to the great oases of the Sahara in Egypt (Sturchio et al., 2004). In principle the method also could reduce the necessary sample size, but field extraction will probably still be necessary.

Coming back to the prognosis of future developments, it needs not much power of prophecy to state that the contribution of gas proportional counters to the measurement of environmental radionuclides will further decrease. New installed labs use LSC, MS, or AMS techniques and the number of scientists able to build and tune gas proportional counters decreases, as does the number of labs that use them. In certain fields, the radiometric measurement itself will have decreasing importance as compared to mass spectrometry. Future measurement systems for nuclides with long half-lives will focus more on direct mass-spectrometric counting, used for radiocarbon and other cosmogenic nuclides measured with AMS, or on the ingrowth measurement as for tritium measured via ^{3}He. The past decades have shown tremendous advances in the development of tunable diode lasers. This made it possible for laser spectrometric methods to show their advantages in some special fields that were formerly dominated by mass spectrometry (Kerstel et al., 1999; Trigt et al., 2001, 2002). Small, field-deployable, shoebox-sized devices with a high sample throughput are under development for stable isotope measurements of water or water vapor in environmental studies (Eric Kerstel and Harro Meijer, personal communication). ATTA, a laser-based technique with fantastic abundance sensitivity, proved to be the superior technique for ^{81}Kr as compared to RIS-MS and synchrotron-AMS. ^{39}Ar is a very useful groundwater tracer that for the time being can only be measured by low-level gas proportional counting in one laboratory worldwide (Loosli and Oeschger, 1980; Loosli, 1983). Probably the near-future might show the application of ATTA to the measurement of ^{39}Ar, since several groups are addressing this problem already (Roland Purtschert and Werner Aeschbach-Hertig, personal communication). Maybe future systems measuring tritium via the ingrowth of ^{3}He will not measure with a static noble gas mass spectrometer, but using an ATTA system and enable detection limits below 1 mTU.

If this chapter has stimulated your interest to learn more details about the various techniques in radionuclide analysis, you may wish to take recourse to comprehensive textbooks on this subject such as L'Annunziata (2003), Gross (2004), and Tuniz (1998).[1] Above all, the book by Povinec (2008) is recommended which has recently been published within the series "Radioactivity in the Environment" to which this book also belongs. It covers a wide range of relevant topics including sampling techniques for environmental studies, radiometric methods, and a variety of modern technologies that have only briefly been discussed in this chapter in Section 4.

REFERENCES

Appleby, P. G., P. J. Nolan, D. W. Gifford, M. J. Godfrey, F. Oldfield, N. J. Anderson, and R. W. Battarbee. (1986). ^{210}Pb dating by low background gamma counting. *Hydrobiology*, 143, 21–27.

Athavale, R. N., C. S. Murti, and R. Chand. (1980). Estimation of recharge to the phreatic aquifers of the lower Maner Basin, India by using the tritium injection method. *Journal of Hydrology*, 45, 185–202.

[1] In 2009, an updated and broader version of this book will be published: Tuniz, C., Kutschera, W., and Fink, D. *Accelerator Mass Spectrometry*. CRC Press, Boca Raton, 2nd Ed. ISBN: 0001420068434.

Bayer, R., P. Schlosser, G. Bönisch, H. Rupp, F. Zaucker, and G. Zimmek. (1989). Performance and blank components of a mass spectrometric system for routine measurement of helium isotopes and tritium by the ^{3}He ingrowth method. *Sitzungsberichte der Heidelberger Akademie der Wissenschaften, Mathematisch-naturwissenschaftliche Klasse*, 5, 241–279.

Belli, P., C. Cattadori, R. Cerulli, S. d'Angelo, N. Ferrari, and L. Pandola. (2005). Response of low-noise miniaturized proportional counters in the keV region. *Nuclear Instruments and Methods in Physics Research Section A: Accelerators, Spectrometers, Detectors and Associated Equipment*, 541(3), 566–573.

Chen, C. Y., Y. M. Li, K. Bailey, T. P. O'Connor, L. Young, and Z. T. Lu. (1999). Ultrasensitive isotope analysis with a magneto-optical trap. *Science*, 286, 1139–1141.

Clarke, W. B., W. J. Jenkins, and Z. Top. (1976). Determination of tritium by mass spectrometric measurement of ^{3}He. *International Journal of Applied Radiation and Isotopes*, 27, 515–522.

Clusius, K., and G. Dickel. (1938). Neues verfahren zur gasentmischung und isotopentrennung. *Naturwissenschaften*, 26, 546.

Collon, P., W. Kutschera, H. H. Loosli, B. Lehmann, R. Purtschert, A. Love, L. Sampson, D. Anthony, B. Davids, D. Cole, D. Morrissey, R. Pardo, M. Paul, B. Sherrill, and M. Steiner. (2000). ^{81}Kr in the Great Artesian Basin, Australia: A new method for dating very old groundwater. *Earth and Planetary Science Letters*, 182, 103–113.

d'Angelo, S. (1989). Performance test of low-noise proportional counters for the GALLEX experiment. *Nuclear Instruments and Methods in Physics Research Section A: Accelerators, Spectrometers, Detectors and Associated Equipment*, 279(1–2), 148–151.

Debertin, K., and U. Schötzig. (1979). Coincidence summing corrections in Ge(Li)-spectrometry at low source-to-detector distances. *Nuclear Instruments and Methods*, 158, 471–477.

Du, X., R. Purtschert, K. Bailey, B. E. Lehmann, Z. Lorenzo, Z. T. Lu, P. Mueller, T. P. O'Connor, N. C. Sturchio, and L. Young. (2003). A new method of measuring ^{81}Kr and ^{85}Kr abundances in environmental samples. *Geophysical Research Letters*, 30(20), 2068, doi:10.1029/2003GL018293.

Dukat, D. A., and S. A. Kuehl. (1995). Non-steady-state ^{210}Pb flux and the use of $^{228}Ra/^{226}Ra$ as a geochronometer on the Amazon continental shelf. *Marine Geology*, 125, 329–350.

Flynn, W. W. (1968). The determination of low levels of polonium-210 in environmental materials. *Analytica Chimica Acta*, 43, 221–227.

Galán Lopez, M., A. Martin Sanchez, and V. Gómez Escobar. (2004). Application of ultra-low level liquid scintillation to the determination of ^{222}Rn in groundwater. *Journal of Radioanalytical and Nuclear Chemistry*, 261(3), 631–636.

Geyh, M. A. (1972). A Comparison: Proportional Counter and Liquid Scintillation Spectrometer for Radiocarbon Dating. *Proceedings of the VIIIth International Conference on Radiocarbon Dating, B*, Wellington, New Zealand, pp. 81–93.

Geyh, M. A., and S. Lorch. (1963). Determining the contribution of gamma rays to the natural environmental radiation at ground level. In: *The Natural Radiation Environment* (Eds J. A. S. Adams and W. M. Lowder). Rice University, Houston, TX, pp. 979–988.

Gonsior, B. (1961). Eine thermodiffusionsanlage für tritium zur verarbeitung geringer tritiumkonzentrationen. *Zeitschrift für angewandte Physik*, 13(12), 545–548.

Gröning, M., and K. Rozanski. (2003). Uncertainty assessment of environmental tritium measurements in water. *Accreditation and Quality Assurance*, 8, 359–366.

Gross, J. H. (2004). *Mass Spectrometry – A Textbook*. Springer, Heidelberg, Germany.

Hennig, G.-J., U. Bangert, and W. Herr. (1980). Dating of Speleothem by Disequilibria in the U-Decay Series. In: *Progress in Scientific Dating Methods* (Ed R. Burleigh). British Museum Occasional Papers No. 21, pp. 73–84.

Heijnis, H., and J. van der Plicht. (1992). Uranium/thorium dating of Late Pleistocene peat deposits in N. W. Europe, uranium/thorium systematics and open system behaviour of peat layers. *Chemical Geology*, 94, 161–171.

Höhndorf, A. (1973). Auswertung von gamma-spektren zur quantitativen analyse von uran, thorium und kalium in Gesteinen. *Geologisches Jahrbuch E*, 1, 81–92.

IAEA. (2003). Guidelines for radioelement mapping using gamma ray spectrometry data. IAEA-TECDOC 1363. International Atomic Energy Agency, Vienna (Available at http://www-pub.iaea.org/MTCD/publications/PDF/te_1363_web.pdf).

Ivanovich, M., and R. S. Harmon. (1992). *Uranium Series Disequilibrium. Applications to Earth, Marine and Environmental Sciences.* Clarendon Press, Oxford.

Jelen, K., and M. A. Geyh. (1986). A low-cost miniature counter system for radiocarbon dating. *Radiocarbon*, 28(2A), 578–585.

Kerstel, E. R. Th., R. Trigt, N. van Dam, J. Reuss, and H. A. J. Meijer. (1999). Simultaneous detection of the $^{2}H/^{1}H$, $^{17}O/^{16}O$, and $^{18}O/^{16}O$ isotope abundance ratios in water by the means of laser spectrometry. *Analytical Chemistry*, 71, 5297–5303.

Kitagawa, H., and J. van der Plicht. (1997). Enrichment of sub-milligram size carbon samples. *Nuclear Instruments and Methods in Physics Research*, B123, 218–220.

L'Annunziata, (Ed.) (2003). *Handbook of Radioactivity Analysis*. 2nd Ed., Academic Press San Diego, San Diego, CA.

Lehmann, B. E., A. Love, R. Purtschert, P. Collon, H. H. Loosli, W. Kutschera, U. Beyerle, W. Aeschbach-Hertig, R. Kipfer, S. K. Frape, A. Herczeg, J. Moran, I. N. Tolstikhin, and M. Gröning. (2003). A comparison of groundwater dating with ^{81}Kr, ^{36}Cl and ^{4}He in four wells of the Great Artesian Basin, Australia. *Earth and Planetary Science Letters*, 211, 237–250.

Lehmann, B. E., H. Oeschger, H. H. Loosli, G. S. Hurst, S. L. Allman, C. H. Chen, S. D. Kramer, M. G. Payne, R. C. Phillips, R. D. Willis, and N. Thonnard. (1985). Counting ^{81}Kr atoms for analysis of groundwater. *Journal of Geophysical Research*, 90(B13), 11547–11551.

Lehmann, B. E., D. F. Rauber, N. Thonnard, and R. D. Willis. (1987). An isotope separator for small noble gas samples. *Nuclear Instruments and Methods in Physics Research*, B28, 571–574.

Loosli, H. H. (1983). A dating method with ^{39}Ar. *Earth and Planetary Science Letters*, 63, 51–62.

Loosli, H. H., and H. Oeschger. (1969). ^{37}Ar and ^{81}Kr in the atmosphere. *Earth and Planetary Science Letters*, 7, 67–71.

Loosli, H. H., and H. Oeschger. (1980). Use of ^{39}Ar and ^{14}C for groundwater dating. *Radiocarbon*, 22, 863–870.

Marshall, B. D., and D. J. DePaolo. (1982). Precise age determinations and petrogenetic studies using the K-Ca method. *Geochimica et Cosmochimica Acta*, 46(12), 2537–2545.

Michels, K., A. Suckow, M. Breitzke, H. R. Kudrass, and B. Kottke. (2000). Sediment transport in the shelf canyon "Swatch of No Ground" (Bay of Bengal). *Deep-Sea Research*, 50(5), 1003–1022.

Michels, K. H., H. R. Kudrass, Ch. Hübscher, A. Suckow, and M. Wiedicke. (1998). The submarine delta of the Ganges-Brahmaputra: Cyclone-dominated sedimentation patterns. *Marine Geology*, 149, 133–154.

Morgenstern, U., and C. B. Taylor. (2009). Ultra low-level tritium measurement using electrolytic enrichment and LSC. *Isotopes in Environmental and Health Studies*, 45(2), 96–117.

Neumaier, S., D. Arnold, J. Böhm, and E. Funck. (2000). The PTB underground laboratory for dosimetry and spectrometry. *Applied Radiation and Isotopes*, 53(1–2), 173–178.

Pandola, L., C. Cattadori, and N. Ferrari. (2004). Neural network pulse shape analysis for proportional counters events. *Nuclear Instruments and Methods in Physics Research Section A: Accelerators, Spectrometers, Detectors and Associated Equipment*, 522(3), 521–528.

Povinec, P. (Ed) (2008). Analysis of environmental radionuclides. In Special Volume of "*Radioactivity in the Environment*, Vol. 11" (Ed. M. S. Baxter), Elsevier.

Prichard, H. M., and T. F. Gesell. (1992). The measurement of ^{222}Rn in drinking water by low-level scintillation counting. *Health Physics*, 33, 577–581.

Rider, M. (1996). *The Geological Interpretation of Well Logs*. Gulf Publishing Co., Houston, TX, 2nd Ed., 256 p.

Rybach, L., B. Bucher, and G. Schwart. (2001). Airborne surveys of Swiss nuclear facility sites. *Journal of Environmental Radioactivity*, 53, 291–300.

Salonen, L., and H. Hukkanen. (1993). Advantages of low-background liquid scintillation alpha-spectrometry and pulse shape analysis in measuring ^{222}Rn, uranium and ^{226}Ra in groundwater samples. *Journal of Radiochemical and Nuclear Chemistry*, 226(1–2), 361–372.

Schoch, H., M. Bruns, K.-O. Münnich, and M. Münnich. (1980). A multi-counter system for high precision carbon-14 measurements. *Radiocarbon*, 22, 442–447.

Sturchio, N. C., X. Du, R. Purtschert, B. E. Lehmann, M. Sultan, L. J. Patterson, Z. T. Lu, P. Müller, T. Bigler, K. Bailey, T. P. OÇonnor, L. Young, R. Lorenzo, R. Becker, Z. El. Alfy, B. El. Kaliouby, Y. Dawood, and A. M. A. Abdallah. (2004). One million year old groundwater in the Sahara revealed by krypton-81 and chlorine-36. *Geophysical Research Letters*, 31(5), L05503.

Suckow, A., and H.-E. Gäbler. (1997). Radiometric dating and heavy metal content of a recent sediment core from Lake Trenntsee in Northeastern Germany. *Isotopenpraxis (Isotopes in Environmental and Health Science)*, 33(4), 367–376.

Suckow, A., U. Morgenstern, and H. R. Kudrass. (2001a). Absolute dating of recent sediments in the cyclone-dominated shelf area off Bangladesh: comparison of Gamma-Ray spectrometric (^{137}Cs, ^{210}Pb, ^{228}Ra), Radiocarbon and ^{32}Si ages. *Radiocarbon*, 43(2), 917–927.

Suckow, A., U. Treppke, M. Wiedicke, and M. Weber. (2001b). Bioturbation coefficients of deep-sea sediments from the Peru Basin Determined by Gamma Spectrometry of ^{210}Pbexc. *Deep-Sea Research II*, 48, 3569–3592.

Thonnard, N., R. D. Willis, M. C. Wright, W. A. Davis, and B. E. Lehmann. (1987). Resonance ionization spectroscopy and the detection of ^{81}Kr. *Nuclear Instruments and Methods in Physics Research*, B29, 398–406.

Trigt, R., E. R. Th. van Kerstel, G. H. Visser, and H. A. J. Meijer. (2001). Stable isotope ratio measurements of highly enriched water samples by means of laser spectrometry. *Analytical Chemistry*, 73(11), 2445–2452.

Trigt, R., H. A. J. van Meijer, A. E. Sveinbjörnsdóttir, S. J. Johnsen, and E. R. T. Kerstel. (2002). Measuring stable isotopes of hydrogen and oxygen in ice by means of laser spectrometry: The Bølling transition in the Dye-3 (south Greenland) ice core. *Annals of Glaciology*, 35, 125–130.

Tuniz, C. (1998). *Accelerator mass spectrometry: Ultrasensitive analysis for global science*. CRC Press, Boca raton, FL, ISBN 0-8493-4538-3.

Walling, D. E., and T. A. Quine. (1991). Use of ^{137}Cs measurements to investigate soil erosion on arable fields in the UK: Potential applications and limitations. *Journal of Soil Science*, 42, 147–165.

Weiss, W., W. Roether, and G. Bader. (1976). Determination of blanks in low-level tritium measurement. *International Journal of Applied Radiation and Isotopes*, 27, 217–225.

Wijngaarden, M., L. B. van Venema, and R. J. De Meijer. (2002). Radiometric sand-mud characterisation in the Rhine-Meuse estuary-Part B. In situ mapping. *Geomorphology*, 43(1), 103–116.

Zhou, W., T. Iida, and X. Yang. (2001). Occurrence of ^{222}Rn, ^{226}Ra, ^{228}Ra and U in groundwater in Fujian Province, China. *Journal of Environmental Radioactivity*, 53, 111–120.

Zimmermann, U., K.-O. Münnich, W. Roether, W. Kreutz, K. Schubach, and O. Siegel. (1966). Tracers determine movement of soil moisture and evapotranspiration. *Science*, 152, 346–347.

Author Index

Subject Index